Marine mussels: their ecology and physiology

THE INTERNATIONAL BIOLOGICAL PROGRAMME

The International Biological Programme was established by the International Council of Scientific Unions in 1964 as a counterpart of the International Geophysical Year. The subject of IBP was defined as 'The Biological Basis of Productivity and Human Welfare', and the reason for its establishment was recognition that the rapidly increasing human population called for a better understanding of the environment as a basis for the rational management of natural resources. This could be achieved only on the basis of scientific knowledge, which in many fields of biology and in many parts of the world was felt to be inadequate. At the same time it was recognised that human activities were creating rapid and comprehensive changes in the environment. Thus, in terms of human welfare, the reason for IBP lay in its promotion of basic knowledge relevant to the needs of man.

IBP provided the first occasion on which biologists throughout the world were challenged to work together for a common cause. It involved an integrated and concerted examination of a wide range of problems. The Programme was co-ordinated through a series of seven sections representing the major subject areas of research. Four of these sections were concerned with the study of biological productivity on land, in fresh water, and in the seas, together with the processes of photosynthesis and nitrogen fixation. Three sections were concerned with adaptability of human populations, conservation of ecosystems and the use of biological resources.

After a decade of work, the Programme terminated in June 1974 and this series of volumes brings together, in the form of syntheses, the results of national and international activities.

INTERNATIONAL BIOLOGICAL PROGRAMME 10

Marine mussels: their ecology and physiology

EDITED BY

B. L. Bayne

Principal Scientific Officer
NERC Institute for Marine Environmental Research
Plymouth, UK

CAMBRIDGE UNIVERSITY PRESS

CAMBRIDGE
LONDON·NEW YORK·MELBOURNE

Published by the Syndics of the Cambridge University Press
The Pitt Building, Trumpington Street, Cambridge CB2 1RP
Bentley House, 200 Euston Road, London NW1 2DB
32 East 57th Street, New York, NY 10022, USA
296 Beaconsfield Parade, Middle Park, Melbourne 3206, Australia

First published 1976

Library of Congress Cataloguing in Publication Data
Main entry under title:
Marine mussels, their ecology and physiology.
   (International Biological Programme; 10)
   Bibliography: p. 411
   Includes index.
   1. Unionidae – Ecology. 2. Unionidae – Physiology. I. Bayne,
   Brian Leicester. II. Series.
QL430.7.U6M37   594'.11   75-25426
ISBN 0 521 21058 5

Printed in Great Britain at the
University Printing House, Cambridge
(Euan Phillips, University Printer)

Contents

Contents

Table des matières

Table des matières

Содержание

Содержание

Содержание

Contenido

Contenido

Contributors

Bayne, B. L.   NERC Institute for Marine Environmental Research, Citadel Road, Plymouth, Devon PL1 3DH, UK

Gabbott, P. A.   NERC Unit of Marine Invertebrate Biology, Marine Science Laboratories, Menai Bridge, Gwynedd, UK

Koehn, R. K.   Department of Ecology and Evolution, State University of New York, Stony Brook, New York 11790, USA

Levinton, J.   Department of Ecology and Evolution, State University of New York, Stony Brook, New York 11790, USA

Mason, J.   DAFS Marine Laboratory, PO Box 101, Victoria Road, Aberdeen AB9 8DB, UK

Roberts, D.   Zoology Department, Queen's University, Belfast, Northern Ireland

Seed, R.   Zoology Department, Queen's University, Belfast, Northern Ireland

Thompson, R. J.   Marine Sciences Research Laboratory, Memorial University of Newfoundland, St John's, Newfoundland, Canada

Widdows, J.   NERC Institute for Marine Environmental Research, Citadel Road, Plymouth, Devon PL1 3DH, UK

Yonge, C. M.   13, Cumin Place, Edinburgh EH9 2JX, UK

Preface

In April 1968, at a meeting in Varna Bulgaria, a committee of the International Biological Programme selected the edible mussel, *Mytilus edulis*, and related species, to constitute one of the 'themes' of the Marine Productivity, or PM, section. This selection was in response to the large number of research projects involving *Mytilus* which were submitted to the PM section by the various National Committees of IBP. More fundamentally, the selection of marine mussels to constitute a theme of IBP/PM was due to the importance of this group of animals to many aspects of the ecology of coastal waters, notably to the productivity of the shallow-water benthos and to aquaculture. Five years after the meeting in Varna there was general agreement that the aims of the final, or 'synthesis', stage of the IBP would best be met for this theme by a review of our present knowledge of marine mussels. Throughout the active period of the 'marine mussels' theme of IBP/PM, the most frequent requests made of the Convenor were those seeking a critical introduction to various aspects of research on these animals. This present volume, then, is a response to this need of a general review of the ecology and physiology of mussels.

The term 'mussels' has no firm taxonomic status. It has been used, on the one hand, to refer to all bivalve molluscs that are not 'oysters'; at the other extreme, it has been used to include only the different species of the genus *Mytilus*. In this volume we have, in general, restricted the use of the term 'mussels' to species of the three genera *Mytilus*, *Modiolus* and *Perna*. However, different authors have laid their own emphasis according to the demands of their subject-matter. The shell form of the typical mussel is shared by other bivalve species, most probably due, as C. M. Yonge discusses in Chapter 1, to the selective value of the posterior enlargement of the mantle cavity. It is pertinent, therefore, that the evolution of this shell form within the Bivalvia as a whole, and some of its implications for the function of the individual, should be considered in the first chapter. In other chapters, also, it has been necessary to consider other members of the Bivalvia than the typical mussels, when not to do so would have resulted in a superficial discussion of the subject. In all cases, however, our emphasis is on the mytilids of the three genera mentioned, for these are the 'typical mussels' of common usage, and it is these animals that have constituted the relevant IBP/PM theme over the past five years.

Mussels have been the subject of a considerable amount of research effort, reflecting both their ecological and economic importance. Their geographical distribution is world-wide and they are the dominant organisms in many littoral and shallow sublittoral ecosystems, including rocky and sediment shores on open coasts and in estuaries and marsh. Mussels have therefore aroused interest not only in the factors controlling their own population ecology, but also in their role in the structure and function of the communities of which they are a characteristic and important component. These aspects of the ecology of mussels are discussed by Seed in Chapter 2, and information on the biology of the larval stages is reviewed by Bayne in Chapter 4. Because of their ecological dominance, their inshore and estuarine distribution and their sessile mode of life, mussels are a popular source of material for pollution research. Some of the recent literature on mussels and pollution is reviewed by Roberts in Chapter 3. In recent years there has been a significant advance in understanding the genetics of populations of marine invertebrates, and much of this research has been done with mussels, notably the common 'blue' mussel, *Mytilus edulis*. As a result of these studies, which are reviewed by Levinton & Koehn in Chapter 9, research into the ecology of mussels is now entering a new phase, where the dynamics of population changes can be linked, on the one hand with community processes, and on the other with the mechanisms of population genetics.

Mussels have also been popular research material for experimental studies. They are resistant to a wide variety of environmental conditions and they show a marked capacity to adapt to extremes of environmental change. Studies on the physiology of mussels have been reviewed in Chapters 5 and 6 by Bayne, Thompson & Widdows. In the first of these chapters the processes of feeding, digestion and respiration are treated; in Chapter 6 the processes of circulation, excretion, ionic and osmotic control, and the systems of chemical and nervous co-ordination are discussed. In Chapter 7 an attempt is made to bring together some of this physiological understanding in order to provide more insight into the adaptive processes of mussels.

Many of the recent studies of the biochemistry of mussels have a physiological orientation, in the sense of being concerned with metabolic regulation at the level of the 'whole organism'. Indeed, as Gabbott points out in Chapter 8, we are fortunate that the recent literature on mussels provides a perspective 'which includes metabolic regulation at all levels of organisation'. Investigations of the seasonal changes in metabolism, and of anaerobiosis, are two examples where biochemical and physiological studies operate together to provide increased understanding of the biology of mussels. This close liaison between biochemistry and physiology, as with genetics and population ecology, provides some hope for considerable advances in knowledge in the near future.

Finally, in Chapter 10, Mason considers the aquaculture of mussels. The harvesting of mussels as food extends a long time into history, but recent developments in resource management and in the different forms of bottom and raft cultivation show promise that the considerable potential of mussels as a source of protein may soon be realised in many countries.

We have attempted in this volume to give a wide coverage to these aspects of the biology of mussels. Although the main subject of our discussions, viz. marine mussels, suggests a narrow field of scientific effort, in fact, by posing problems of general relevance we hope to appeal to a wide readership. The authors have attempted to review the literature 'in prospect rather than in retrospect' and to present, not only a detailed discussion of the published information on mussels, but also considered suggestions of some of the potentially productive areas of future research.

I should like to acknowledge the support of the IBP Central Committee. Cambridge University Press, and in particular Howard Moore, have shown considerable patience in the face of a retreating deadline. Miss R. Truscott came to our help at a critical moment and has coped with the bulk of the typing. Miss P. Smith helped with the bibliography. I should especially like to thank my wife and family for tolerating so many lost weekends.

B. L. BAYNE

April 1975

1. The 'mussel' form and habit

C. M. YONGE

Species of the Mytilacea are to be found, often in dense populations, in all seas, attached by byssal threads to hard or gravelly surfaces. The superfamily or, as he prefers, the family, Mytilidae has been thoroughly surveyed by Soot-Ryen (1955). Broadly speaking it may be divided into ' Modiolus ' and ' Mytilus ' groups, the former with umbones a little short of the anterior end, the latter with a more pointed shell and the umbones at the extremity. Generally speaking, mytilids are triangular in outline (Fig. 1.1). This is due to the heteromyarian condition, with posterior regions of the body and of the enclosing mantle and shell enlarged, with corresponding reduction of the anterior regions. The posterior adductor is greatly enlarged and the anterior adductor reduced, even in a few instances lost.

The typical ' mussel' (species of *Mytilus*, *Modiolus* or *Perna*) is attached in great numbers together by threads of tanned protein (Tamarin & Keller, 1972; Tamarin, Lewis & Askey, 1974) secreted by an associated series of glands in the foot. These glands form the byssal apparatus, which is of supreme importance in all mytilids. Certain species, such as the tropical *Septifer bilocularis*, and all boring mytilids, are solitary. In the typical mussels, threads are planted by the foot in such a manner that the animals are able to withstand water movements in any direction, although with a general alignment parallel to the major directions of water flow. In certain cases, as in *Musculus*, byssal threads are secreted in such numbers as to form a ' nest' (Merrill & Turner, 1967). But byssal attachment has also provided the means of boring into rock. This has probably originated from a ' nestling' habit but has involved change from a modiolid to a bullet shape, with the anterior adductor secondarily enlarged (Fig. 1.1*d*), a change in form due, it would appear, to the effect of a tangential component in shell growth (Yonge, 1955). In these rock borers, byssal threads are disposed in anterior and posterior series (Fig. 1.2*b*) with the byssal (pedal) muscles arranged in anterior and posterior groups (Fig. 1.1*d*) alternately pulling the shell forward and drawing it back from the head of the boring. Species of *Botula* (*Adula*) (Fig. 1.2*a*) bore by these purely mechanical means into non-calcareous rocks. Species of *Lithophaga* – the widely present and often economically important ' date mussels' – only occur in calcareous rocks where chemical penetration occurs, assisted by mechanical erosion (Yonge, 1955).

The remarkable *Fungiacava eilatensis* should be mentioned (Fig. 1.3). In this mytilid the shell, here dorso-ventrally flattened, is reduced to minimal thickness (almost exclusively periostracum) and enclosed within a pallial

1

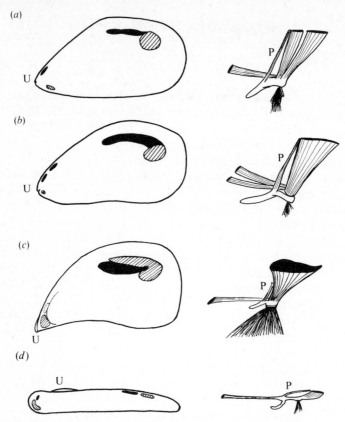

Fig. 1.1. Form of the shell (left), with disposition of the byssal and small pedal (P) retractor muscles in selected mytilids. (*a*) *Mytilus edulis*; (*b*) *Semimytilus algosus*; (*c*) *Septifer bilocularis*; (*d*) *Lithophaga plumula.* Adductor scars hatched, retractor scars black; U, umbo. (From Yonge & Campbell, 1968.)

envelope formed from the extended middle folds of the mantle margins. The adductors are greatly reduced, the gills are enlarged and the palps vestigial. The foot is relatively large with possible sensory functions. By exclusively chemical means it penetrates the skeletons of fungid corals (Goreau, Goreau & Yonge, 1972) living ventral-side uppermost in intimate symbiotic association with the coral (Goreau, Goreau, Yonge & Neumann, 1970).

The outstanding success of the Mytilacea can be attributed to posterior enlargement and consequent development of the heteromyarian condition. As in all heteromyarian bivalves, this is a direct consequence of byssal attachment. It is not, however, an inevitable consequence of this. Ark shells, such as *Arca noachina*, where the attaching byssus extends along the greater length of the flattened ventral surface, are isomyarian and retain

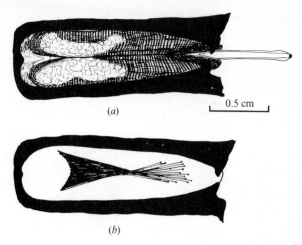

Fig. 1.2. Boring mytilid, *Botula fulcata*. (*a*) Boring (in non-calcareous rock) opened to expose dorsal surface of borer, anterior end of shell withdrawn from head of boring, siphons extended; *(b)* animal removed to display attachment of byssal threads in large anterior and posterior groups. (From Yonge, 1955.)

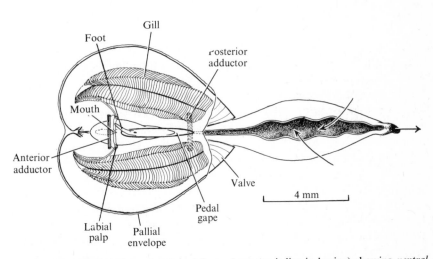

Fig. 1.3. *Fungiacava eilatensis*, viewed from above (as it lies in boring) showing *ventral* surface with siphons that extend into the coelenteron of the fungid coral. Shell valves almost completely enclosed within pallial envelope, enlarged food-collecting gills but vestigial selecting labial palps; both adductors reduced and dubiously functional, foot with byssal opening at base projecting through pedal gape. (From Goreau, Goreau, Yonge & Neumann, 1970.)

3

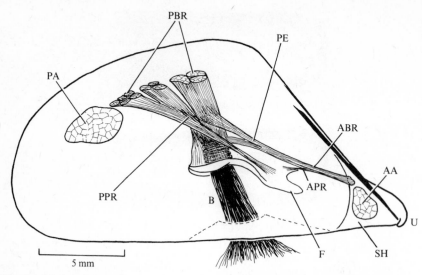

Fig. 1.4. *Dreissena polymorpha*, musculature viewed from right (i.e. within left valve); foot greatly contracted, anterior shell shelf (SH) and characteristically 'infolded' ligament also shown. Pedal constituents of retractors consist of a discrete posterior muscle (PPR) and a small section (APR) of the anterior retractor (ABR). AA, anterior adductor; B, byssus; F, foot; PA, posterior adductor; PBR, posterior byssal retractors; PE, pedal elevator; U, umbo. (From Yonge & Campbell, 1968.)

adductors of very similar size. But the posterior opening is so restricted that an additional inhalant current is drawn in anteriorly (Yonge, 1953*a*). In evolutionary history, after initial byssal attachment, there were obvious possibilities of either anterior or posterior enlargement, the shell becoming inequilateral. Anterior enlargement could have no survival value, however, because of the consequent reduction of the posterior regions where water both enters and leaves the mantle cavity. But by enlarging these openings and raising them higher above the substrate, posterior enlargement would have just the opposite effect; such changes would have selective value.

The extent of this advantage is indicated by the number of times that the heteromyarian condition has independently been assumed throughout the Bivalvia and *always* in association with byssal attachment. Even the 'typical' shell form of mussels is far from being confined to the Mytilacea; examples are to be found in at least two other superfamilies. The best known are the freshwater mussels (Dreissenacea). The well-known *Dreissena polymorpha* (Fig. 1.4) has a striking superficial resemblance to the mytilids (Yonge & Campbell, 1968; Morton, 1969*a*). The eulamellibranch ctenidium, the greater degree of mantle fusion and the structure of the opisthodetic ligament, with a unique forward extension of fused

4

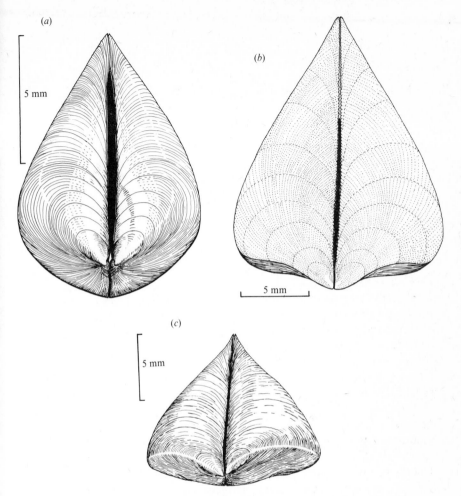

Fig. 1.5. Anterior aspects of three heteromyarians. (*a*) *Mytilus edulis*; (*b*) *Septifer bilocularis*; (*c*) *Dreissena polymorpha*. Note very flattened under-surfaces of (*b*) and (*c*), both with anterior shell shelves. (From Yonge & Campbell, 1968.)

periostracum between the originally formed periostracum and the under-lying primary ligament all cut it off totally from the Mytilacea. *D. polymorpha* is even more flattened ventrally, to the extent that shelves are formed within the umbonal regions for the attachment of the anterior adductors and pedal (byssal) retractors. Similar shelves are present in the mytilid *Septifer*, which is also more flattened than *Mytilus* (Fig. 1.5*a* and *b*) but here they function only for the attachment of the adductor muscle (Yonge & Campbell, 1968).

The Dreissenacea appear to be relatively recent migrants from the sea

5

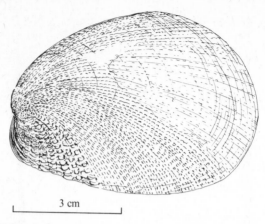

Fig. 1.6. *Cardita* (*Bequina*) *semiorbiculata*, shell viewed from left. (From Yonge, 1969.)

because they retain the pelagic veliger larva. *D. polymorpha* is reported as moving across Europe from the Black and Caspian Seas late in the eighteenth century, reaching London docks in 1824 (Morton, 1969*b*). It is extremely tolerant of silt and is a typical inhabitant of the very still waters of canals and waterworks; the extreme heteromyarian form seems likely to have been evolved under very different conditions of exposure before the species migrated into fresh waters. It is interesting that the only group of freshwater bivalves highly adapted for life in rushing waters, namely species of the family Etheriidae (Unionacea), are mainly cemented by one or other valve, with assumption of the monomyarian condition. The South American *Bartlettia stefanensis*, however, where the anterior end is reduced but capable of local extension, is certainly heteromyarian (Yonge, 1962*a*).

The distribution and habits of *D. polymorpha* are simulated by the mytilid *Limnoperna fortunei* which occurs in the rivers of south-east Asia and has recently established itself in reservoirs at Hong Kong where it is presenting problems similar to those presented by *D. polymorpha* in Europe (Morton, 1973). As in so many instances throughout the Bivalvia, convergence in form is accompanied by convergence in habit.

One of the largest 'mussels' is the very laterally (but not ventrally) compressed *Cardita* (*Beguina*) *semiorbiculata* (Fig. 1.6), a species of the Carditacea which occurs in the western Pacific, very firmly attached in clefts by a profusion of byssal threads (seldom in numbers together) within colonies of solid corals, usually species of *Porites* (Yonge, 1969). This animal has a peculiar 'modiolid' form with the umbones tending to overgrow the anterio-ventral regions where a relatively large anterior adductor is retained. The adult foot is much reduced, with the large

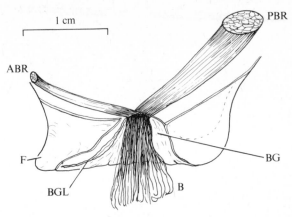

Fig. 1.7. *Cardita (Bequina) semiorbiculata,* dissection to display from the left side and the disposition of anterior and posterior retractors, exclusively concerned with the byssus. ABR, anterior byssal retractor; B, byssus; BG, byssal groove; BGL, byssal gland; F, foot; PBR, posterior byssal retractor. (From Yonge, 1969.)

posterior and smaller anterior retractors exclusively concerned with the byssus (Fig. 1.7). In related but less rigidly byssally attached species, such as *C. variegata,* the retractors are composed of both pedal and byssal components. Here again, the ligament has the typical and altogether distinctive form it assumes in the Carditacea with the characteristic subdivided cardinal crest (Yonge, 1969).

Members of two other superfamilies, *Sphenia binghami* (Myacea) and the north Pacific *Entodesma saxicola* (Lyonsidae: Anomalodesmata), both byssally attached, have much reduced and ventrally situated anterior adductors, although without the triangular outline of other hetero-myarians. The hinge line remains parallel to the ventral surface although the posterior regions are enlarged (Fig. 1.8). The former species lives byssally attached, usually within empty valves of other bivalve species, at moderate depths (Yonge, 1951); the latter lives in rock pools where it is extensively and securely attached in sheltered crevices (Yonge, 1952). Both bivalves in their different ways are sheltered against strong water movements which would favour further change in external form. As a result of byssus attachment they have become heteromyarian, but they are *not* mussels.

Although in the evolution of the Pteriacea (including the pearl oysters) and the Pectinacea (scallops) – but not of the Plicatulacea (Yonge, 1975) – the heteromyarian condition must have been a stepping stone towards monomyarianism and assumption of a horizontal posture, it was also, as seen in these five independently evolved examples, a highly successful end in itself.

The frequent appearance throughout the Bivalvia of byssal attachment

(*a*)

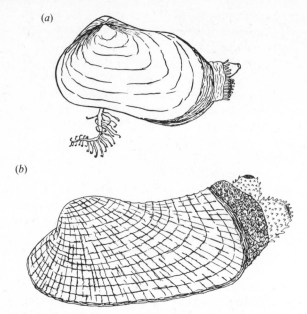

(*b*)

Fig. 1.8. Heteromyarians without a triangular shell. (*a*) *Sphenia binghami*(Myacea), byssally attached at some depth; (*b*) *Entodesma saxicola* (Lyonsidae), attached intertidally in shelter (byssus not shown). (From Yonge, 1951, 1952. Published in 1952 by the University of California Press; reprinted by permission of the University of California Press.)

(not necessarily involving heteromyarianism) indicates some common origin. So far as can be determined, the byssal apparatus is a constant post-larval structure concerned with secreting threads that temporarily attach the animal to the substrate during the vitally important period of metamorphosis. In the majority of bivalves the byssus apparatus is then lost, e.g. in Tellinacea, Solenacea, Adesmacea and other superfamilies. But it may be retained, neotenously, into adult life and this has occurred, independently, in a variety of other superfamilies (Yonge, 1962*b*). It was in this way that the initially infaunal bivalves, now equipped with the means of collecting and digesting plant plankton, were able to return to the hard substrates. This return to the epifaunistic life of their ancestors occurred very early; according to Newell (1942), *Modiolus* originated in the Devonian.

With this change of habit went major modification in the foot and the pedal retractors. The foot ceased to be an organ for penetrating through soft substrates, with its retractors exclusively employed in pulling the animal forward after temporary fixation by the sand-anchor of the dilated distal end (Trueman, 1968). Certainly in modern mytilids the foot retains the capacity for movement, especially in early life when *M. edulis* is highly mobile (Verwey, 1954). But it is now an organ adapted for movement on a

8

hard substrate, stretching forward, attaching by the tip and then pulling the body onward. Temporary byssal attachment is made whenever it comes to rest, the thread being planted by way of a groove on the under (posterior) surface of the elongate foot. When moving upward (e.g. on the side of an aquarium tank) threads are attached and then abandoned at every stage in the process. The function of the retractors is now primarily one of pulling on the byssal threads and resisting the strains imposed by mechanical forces to which the animals are exposed on the surface. These muscles must be able to resist stretching, as well as being able to contract. As shown in Fig. 1.1 (*a*) to (*d*), the actual 'pedal' content (*P*) of these muscles is extremely small. As already noted, in *C. semiorbiculata* (Fig. 1.7) the *entire* pedal musculature is attached to the byssus.

It is because not only the shell shape but also the musculature of these heteromyarians is highly adapted to the life they lead that it has been usual to think of all 'mussels' as evolving on a hard substrate. This supposition implies an initial change in the settling behaviour of the veliger so that it comes to rest and metamorphoses upon a hard instead of a soft substrate. This view has, however, been very interestingly challenged by Stanley (1970, 1972) who claims that an 'endobyssate' stage occurred prior to the present 'epibyssate' condition. He bases his view on examination of two living species, *Modiolus modiolus*, which lives from one-half to two-thirds buried in gravelly mud, and *M. demissus* which inhabits tidal marshes, assuming a posture about 30 degrees from the vertical. Attachment of the former is to coarse particles, of the latter to plant roots and coarse debris. Judging from the shell form, Stanley concludes that in the Ordovician the greatest number of bivalves were 'endobyssate', although later the proportion of such species decreased.

Stanley (1970, 1972) further points out that a variety of normally burrowing bivalves do retain the byssus to maintain some measure of stability within the substrate. This is certainly true of *Mya arenaria* in early life, and throughout life in the related *Corbula gibba* (Yonge, 1946) and in *Lyonsia* (Ansell, 1967). But all these animals are isomyarian and there is no apparent reason why such minimal attachment, with the animal vertically disposed, should favour change to the heteromyarian condition. Only if the animal settled superficially and became exposed to surface movements would the factors favouring such a condition come into play, at any rate in this writer's opinion.

Much is rightly made by Stanley of conditions in the Pinnidae (*Pinna* and *Atrina*). These are the largest heteromyarians with very great posterior extension, exclusively of the mantle and shell; they are adapted for life vertically embedded in soft substrates (Yonge, 1953*b*), i.e. they are endobyssate. They are enormously abundant in certain, usually sublittoral, habitats. To compensate for complete immobility and to offset the repeated

9

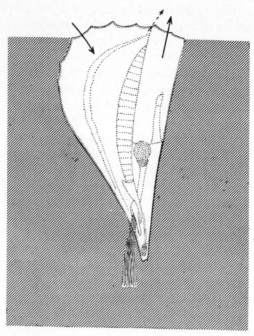

Fig. 1.9. *Pinna* sp., example of the largest heteromyarians and also infaunal, shown *in situ* revealing the great extent of mantle and shell posterior to (i.e. above) the large posterior adductor. Also note gill and waste canal (ejection of pseudofaeces indicated by broken arrow) in inhalant chamber and pallial organ in exhalant chamber. (From Yonge, 1953*b*.)

damage to which the uppermost regions of the shell are exposed, they have unusual powers of regeneration. Attachment is by long byssal threads to small stones or rubble deep in the substrate. It has been this author's contention that such animals acquired the heteromyarian condition when attached epifaunistically, only later taking advantage of the resultant body form to attach themselves to stones deeper and deeper within a soft substrate, i.e. returning to the original infaunal mode of life. The posterior regions (Fig. 1.9) became increasingly extended (reaching somewhat the same end-point as a siphonate bivalve) with prominent waste canals in the inhalant chamber for immediate upward discharge of pseudofaeces and, in the exhalant chamber, with a 'pallial organ' (Yonge, 1953*b*) probably concerned with clearing the mantle cavity when crushed shell fragments are forced into it.

The contrary view is that such animals evolved after byssal attachment within the substrate, i.e. that the present endobyssate condition was *not* preceded by any epifaunistic stage. Stanley points to the almost vertical posture of the two species of *Modiolus*. Certainly such a condition could well be intermediate between the epifaunistic condition of, say, a modern

Mytilus and the deeply buried vertical posture of the Pinnidae. But it is difficult to visualise the first stages of this change as occurring *in situ*, although the later adaptations within the mantle cavity must have done so.

Stanley also refers to the rounded ventral surface of endobyssate species which he regards as likely to have preceded the flattened ventral surface so well displayed in *Septifer* and *Dreissena* (Fig. 1.5*b* and *c*) both of which, as we have seen, possess shelves (Fig. 1.4) for attachment of the anterior adductors. There seems no doubt that the flattened under-surface is (or perhaps was, in the case of *Dreissena*) an adaptation to epifaunal life. Nevertheless, the first such animals could have been rounded ventrally had they established themselves, as appears most probable, in crevices. In other words, they may have become byssally attached 'nestlers', only gradually moving into greater exposure and acquiring the mytilid form with flattened ventral surface. It does, to the writer, appear most probable that the modiolid form – with the associated changes in adductors, foot, byssal apparatus and retractors – was evolved during a period of epifaunistic exposure rather than during endobyssate existence. Apart from moving into greater exposure, 'nestling' modiolids could either retreat within a soft substrate to occupy habitats like those of *Modiolus modiolus* or *M. demissus*, or else bore into a hard substrate like *Lithophaga*, *Botula* or *Fungiacava*. Similar changes occur in other superfamilies, e.g. in the Veneracea, from the nestling *Petricola carditoides* to the rock-boring *P. pholadiformis* (Yonge, 1958) and in the Saxicavacea, from nestling to rock-boring specimens of *Hiatella* (Hunter, 1949).

Modiolids do occur under conditions where they could not possibly have evolved, namely in the complete stillness of soft mangrove muds. With Dr Peter Bacon of the University of the West Indies, Trinidad, the author was recently able to view *Mytella guyanensis* in the Caroni swamp where (quoting from subsequent correspondence): 'The animal occurs in holes in the mud with the 'siphonal' end of the shell about 1 cm below the surface at low tide. When covered by water the mussel raises itself level with the top of the elongated, slit-like opening in the mud.' The animals secrete a profusion of very long and fine threads with which they attach themselves to rootlets of the mangroves, also producing an enclosing 'cocoon'. The animal also occurs superficially secured to rhizophores, in this case with the hind end pointing down and with a restricted byssal attachment and no enclosing cocoon. Such an animal must surely have acquired the modiolid form before assuming its present mode of life.

It is worth quoting an instance of similar change of habitat in the monomyarian (Pteriacean) *Malleus* and involving *M. regulus* with the two 'hammer oysters', *M. malleus* and *M. albus* (Yonge, 1968). All have characteristically elongate shells, the first with only a short hinge line, the animal living upright while byssally attached to a hard substrate. In the other two species the hinge line is greatly extended, the 'hammer head' being as long as the shell 'handle', both reaching 25 cm. These species live at least four-fifths buried, the former in a coarse substrate, still byssally attached to stones and rubble within it, but with the long basal extensions assisting stability. This leads to the condition in *M. albus*, which inhabits finer substrates of muddy sand and which has lost byssal attachment, including the byssal gape, to become maintained exclusively by way of the 'hammer head'. There can be no question here of the endobyssate *M. malleus* representing a stage in the evolution of the epifaunistic *M. regulus*.

Stanley's contentions are most interesting. All that is intended here is to draw attention to the conditions under which, in this writer's opinion, the triangular heteromyarian mussel with its accompanying modifications of the entire pedal complex would most probably have been selected. This does, on balance, appear more likely to have taken place during an epifaunistic than during an endobyssate mode of life, but further discussion is clearly necessary.

2. Ecology

R. SEED

Systematics

The genus *Mytilus* Linné 1758 belongs to the family Mytilidae which dates back to the Jurassic and perhaps even Devonian times (Soot-Ryen, 1969). Linnaeus gave the group an exceptionally broad base and included many dissimilar groups in addition to true mussels. Originally containing twenty-two species, its members have now been assigned to several different genera, only three being retained in *Mytilus* as presently restricted. Lamy (1936) quotes over 170 specific names previously used for *Mytilus*, though many of these have been re-allocated, some even to different families. The family is generally recognised by its shell form and sculpture, hinge structure and muscle scars, and can be divided into those having terminal umbones, the mytiliform species, and those with subterminal umbones, the modioliform species. Some forms, however, are intermediate between *Mytilus* and *Modiolus* and in development some species are modioliform before becoming mytiliform. Apart from *Mytilus* and *Modiolus* there are several other genera within the family *Mytilidae* and many species placed in one genus by some authors have been placed in quite different genera by others, suggesting that outer shell form is of limited systematic value. Indeed, shell shape is extremely variable and depends not only on environmental conditions but also on the age of the animal (Seed, 1968). Soot-Ryen (1955) stresses the importance of using as many characters as possible of both shell and soft parts when arranging species into supraspecific groups. Much of the confusion that exists in the systematics of the Mytilidae has possibly arisen by forcing species of apparently different origins into large groups based essentially on superficial similarities brought about through convergent evolution.

The genus *Mytilus* is of relatively recent origin with apparently no records older than the Pliocene. Soot-Ryen (1955) recognises three or possibly four species: *M. edulis* Linné together with its geographical subspecies; *M. californianus* Conrad; *M. crassitesta* Lischke and perhaps *M. giganteus* Nordmann. He lists the six following subspecies: *M. edulis chilensis* Hupé; *M. edulis platensis* Orbigny (= *M. canaliculus* Dall); *M. desolationis* Lamy (= *M. kerguelensis* Fletcher); *M. galloprovincialis* Lamarck; *M. edulis diegensis* Coe and *M. edulis planulatus* Lamarck.

Lamy (1936) subdivides the genus into *Eumytilus* and *Aulocomya* with their type species *M. edulis* and *M. megallanicus* Chemn respectively, and *Chloromya* for species like *M. perna* Linné (= *M. megallanicus* Röding),

13

M. viridis (= *M. smaragdinus* Chemn?) and *M. canaliculus* Martyn. Dodge (1952) also regards *Chloromya* as a subgenus of *Mytilus*. Soot-Ryen (1955), however, gives *Aulocomya* and *Chloromya* generic status, and suggests that *Chloromya* Mörch should be changed to *Perna* Retzius although *Perna* has apparently been used by subsequent authors with a quite different meaning, even as a family name. The genus *Perna* is characterised by the anterior position of the pedal retractor muscle, the absence of any anterior adductor muscle and the often green colour of the shell (Lubet, 1973).

Mytilus pellucidus Penn is usually regarded as a variety of *M. edulis*, but some authors (e.g. Dodge, 1952) consider it to be a good species. *M. edulis aeoteanus* Powell, the blue mussel found in New Zealand, is ranked as a subspecies of *M. edulis* (Fleming, 1959). *M. ungulatus* has been the source of considerable confusion in the literature and various authors (e.g. Bucquoy, Dautzenberg & Dolfus, 1887–98; Dodge, 1952) suggest that the name be abandoned.

Apart from *Mytilus*, *Perna* and *Aulocomya*, three other genera seem worthy of mention. These are *Crenomytilus* (e.g. *M. grayanus* Dunker = *M. dunkeri* Reeve), *Choromytilus* (e.g. *M. chorus* Molina and *M. meridionalis* Krauss) and *Semimytilus*, the last possibly being confined to the west coast of South America. *Mytilaster* (e.g. *M. minimus* Poli), a name sometimes used for a group of small mussels, and *Hormomya* (e.g. *M. variabilis* Krauss) are now generally regarded as subgenera of *Brachidontes* (e.g. Lamy, 1936; Klappenbach, 1965). *M. glomeratus* Gould on the other hand is the name sometimes applied to distorted *M. edulis* resulting from overcrowded conditions.

Soot-Ryen (1955) points out that the subspecies of *M. edulis* mentioned earlier could be ecological forms or genetically determined, and he maintains that at present it is impossible to circumscribe a group of specimens from one locality so well that it can be recognised in a large collection from many other localities. He suggests that widely separated populations would have acquired some characters, even though minute, which could separate them from other populations. However, he advises the use of special names for some of these geographical subspecies even where no morphological characters for their separaton can be indicated.

The systematic position of *M. galloprovincialis* has been the subject of considerable discussion in previous literature. Whilst some authorities give it specific status others regard it as a variety of the larger *edulis* complex (see Lubet, 1973, for review). Recently Seed (1971) has shown that apart from morphological and anatomical differences between the two mussels there are other physiological and biochemical differences; whether one considers *M. galloprovincialis* to be a species or a variety possibly depends to some extent on the geographical region from which it is collected. Towards the limits of its distribution, sufficient grounds exist for considering it to be a true species both on morphological and other grounds. In other regions, however, hybridisation obscures any morphological or anatomical differences that may otherwise exist between these two mussels. An interesting discussion of the phylogeny of *M. edulis* and *M. galloprovincialis* is given by Barsotti & Meluzzi (1968) who believe *galloprovincialis* to be a rather recent derivative of *M. edulis*. The warm climate which developed in the Mediterranean favoured the differentiation, and the reduced contact between the Atlantic and Mediterranean hastened this process. At present the two coexist on the Atlantic and English Channel coasts but it is frequently difficult to distinguish them, possibly due to the presence of hybrid forms. This process of differentiation between these mussels is probably still in progress.

Much of the confusion existing in the systematics of the Mytilidae has arisen because of the emphasis laid on superficial similarities in shell characters which are most subject to local and geographical variation. Although detailed morphological and anatomical studies will always be important in taxonomic studies, the use of other techniques such as electrophoresis (e.g. Seed, 1971; Koehn & Mitton, 1972) cytological studies (e.g. Lubet, 1959; Menzel, 1968; Ahmed & Sparks, 1967, 1970) and immunology (e.g. Fisher, 1969) may subsequently throw more light on the true affinities within this group.

Geographical distribution

Although the operation of certain local factors such as extreme salinity, or availability of suitable substratum, may limit the local abundance of a species, it is primarily sea temperature which controls the overall distribution of marine organisms. Temperature can affect the survival of larvae and adults, the normal reproductive behaviour and the process of metamorphosis of the species (see Thorson, 1946). Most species have a wide range of temperature within which they can survive, and a narrower range over which reproduction is possible. Those with relatively broad breeding-temperature ranges will generally be cosmopolitan in their distribution, whereas those with more limited ranges are usually more restricted.

The genus *Mytilus* provides an ideal group for studying geographical distribution since mussels are sessile and the majority occur intertidally where they can be observed readily. The *M. edulis* species-complex is circumpolar in its distribution in boreal and temperate waters of both northern and southern hemispheres (Soot-Ryen, 1955). *M. edulis* itself is a widely distributed and variable species in the northern hemisphere. It occurs in arctic waters extending south to California (Dodge, 1952) and Japan (Miyazaki, 1938a) on Pacific coasts and to North Carolina (McDougal, 1943; Dodge, 1952) on the Atlantic coast. It is present in

Greenland (Madsen, 1940), Novaya Zemlya (Madsen, 1940; Barnes, 1957), the Canadian arctic (Ellis, 1955; Lubinsky, 1958) and the White Sea (Palichenko, 1948; Nikolskii, 1966). However, in many of these most northerly locations it is found only in the sublittoral. It also occurs in European waters as far south as the Mediterranean and North Africa (Berner, 1935; Molinier & Picard, 1957). It is apparently absent only from the high arctic waters of Siberia, Franz Josefland and Spitzbergen. The subspecies *M. edulis chilensis* and *M. edulis platensis* are found respectively on the west and east coasts of South America. *M. edulis planulatus* occurs in Australia and New Zealand and *M. desolationis* in the Kerguelen Islands. *M. edulis diegensis* and *M. galloprovincialis* are taller forms than *M. edulis*, with smaller anterior adductor muscles; the former occurs especially in sheltered bays in California (Coe, 1946), the latter ranges from the Mediterranean, Adriatic and Black Seas northwards as far as northern France and south-west Britain (Fischer, 1929; Hepper, 1957; Seed, 1974).

In contrast to the cosmopolitan distribution of *M. edulis*, *M. californianus* is confined to the Pacific coast of North America principally on exposed shores from the Aleutian isles to Mexico. *M. crassitesta* is recorded in Japan (Soot-Ryen, 1955). Less appears to be known of other mytilids such as *Semimytilus*, *Crenomytilus*, *Choromytilus* and *Aulocomya* from the southern hemisphere where their commercial exploitation has largely been ignored until recent years. *Perna* (= *Mytilus*) *perna* is recorded from Brazil and Venezuela (Davies, 1970; but see also Lubet, 1973, and Molinier & Picard, 1957). *P. canaliculus* and *M. edulis aeoteanus* are recorded from New Zealand (Pike, 1971) and *M. smaragdinus* from India and south-east Asia (Davies, 1970; Obusan & Urbano, 1968; Tan, 1971).

The factor limiting the distribution of *M. edulis* appears to be temperature, the southern limit coinciding more or less with the maximum surface isotherm of 27 °C (Hutchins, 1927; Stubbings, 1954). This limit agrees with what is known of the maximum temperature for adult survival (e.g. Bruce, 1926; Ritchie, 1927; Read & Cumming, 1967; see Chapter 5). Although surviving throughout the year at Cape Hatteras in the United States, south of this point it is apparently maintained only by larvae that are swept around the point by north-easterly gales (Wells & Gray, 1960). Its distribution on other coasts is less satisfactorily explained, since records for a number of varieties or species (?) are somewhat confused. The northern limits of *M. edulis* are also less clearly defined. However, it is plentiful in Hebron Fjord in Labrador, where mussels remain frozen in ground ice at −20 °C for six to eight months of the year (Kanwisher, 1955, 1966; Williams, 1970). Stubbings (1954) has suggested that the northern boundary is set by the breeding temperature, approximating a mean monthly value of 10 °C. Barnes (1957) points out that the northern limit of the species closely parallels the limits of the maximum retreat of pack ice, whilst Dunbar (1947) maintains that the distribution of *M. edulis* suggests that much of the coast of arctic Canada is essentially subarctic, the waters of the Atlantic and Pacific extending their influence further than was previously thought.

Local distribution

Mytilus is found especially in littoral and shallow sublittoral waters though occasionally it has been recorded from deeper water. Occurring in both open water and brackish estuaries, particularly where there is significant water movement, it will live on a variety of substrata, such as rock, stones, shingle, dead shells and even compacted mud or sand, i.e. anywhere that provides a secure anchorage. Mussels are found attached to ships, pier pilings and harbour walls.

Although *Mytilus*-dominated communities are widespread on exposed shores (Lewis, 1964; Stephenson & Stephenson, 1972), sometimes even replacing barnacles from all but the upper part of the midlittoral zone, there can be no certain prediction that any given situation will support mussels in quantity. Nor, if they are present, will their local distribution and zonation follow similar patterns everywhere. They flourish in severely exposed situations, preferring gently sloping, slow-draining platforms to steep rock faces. Yet, as Lewis points out, variations between apparently similar sites suggest that physical factors alone are not the sole influence. Typically *Mytilus* dies out on more sheltered, fucoid-dominated open coasts only to reappear in extremely sheltered sea loughs and estuaries. Here individuals are fewer but larger than in more exposed situations. Whether settlement and survival is somehow inhibited amongst the long-fronded fucoids is uncertain, but mussels are known to flourish beneath a canopy of *Ascophyllum* in many sea loughs. Some typical distribution patterns are described by Lewis (1964) who suggests that the erratic local distributions of *Mytilus* possibly reflect stages in irregular cycles of settlement, competition, predation, denudation and resettlement which vary in phase from one site to another.

Mossop (1921), Kitching, Sloane & Ebling (1959) and more recently Seed (1969b) and Paine (1974) also comment upon the rather erratic local distribution of *Mytilus* and attribute considerable importance to predators in this respect, particularly in determining the downward extension of the species into the sublittoral. Its upper limit of distribution in the littoral zone is set, ultimately, by the operation of physical factors such as exposure to air and desiccation, especially on the young stages. Since protection against such desiccation is not equal in all directions – viz. the presence of humid crevices and damp pits in the rock surface – and since drifting adult

17

mussels, which may have become detached during storms, may establish themselves at levels above which younger individuals may perish, these upper limits are generally of a more irregular nature. It has been shown, however (Seed, 1969b), that many of the mussels encountered in the high shore can be of considerable age, such populations being relatively stable when compared with the more rapid turnover of animals in the lower shore. Fischer (1929) also refers to the considerable fluctuations in the abundance of mussels over successive years on the English Channel coast, and attributes this to severe predation. Paine (1974), however, has emphasised the relative stability of the upper and lower distribution limits of *M. californianus* in North America.

Whilst the extent of the vertical distribution of mussels, like many other littoral organisms, is greatest on shores encountering wave action, Baird & Drinnan (1956) stress the importance of energy requirements for basal metabolism, when exposed to air, as a factor limiting the vertical penetration of *Mytilus*.

Bagge & Salo (1967) and Young (1941) indicate that salinity is an important factor in determining the distribution of *M. edulis* and *M. californianus* respectively, whilst Kuenen (1942) points out the importance of current speed, availability of suitable substrata and food supply. Kuenen attributes less importance to desiccation, temperature and salinity and suggests that larval dispersal and distribution of enemies alone cannot explain the local distribution of mussels on the mud flats at Den Helder, Holland.

Harger (1972b) studied the competitive interactions between *M. edulis* and *M. californianus* on the Californian coast and suggested that increase in the population size of *M. edulis*, through gaining access to less favourable habitats such as wave-beaten shores ('transition zones'), can become an evolutionary asset providing the opportunity for the appearance of new genetic material with the consequent ability to colonise new locations. New genetic recombinations will be rigorously tested by interaction with the marginal physical conditions and severe competition within the transition zone. Mussels in such transition zones may, therefore, have more genetic variability, valuable to the survival of the species. During adverse conditions (storms or severe predation), however, the species may be driven back to its more favoured confines, such as harbours or estuaries. Both habitats are therefore considered to be vitally important in the success of the species. The existence of a considerable amount of genetic diversity in mussel populations, which provides the basis for a plasticity of response to environmental variability, has been demonstrated by various workers and is discussed in greater detail in Chapter 9.

Reproduction

Long-term studies of reproduction are important in ecological investigations since they provide important data relating to distribution and population structure and also enable accurate predictions to be made concerning recruitment to the population. Although the subject of considerable research, few of the older investigations of the reproductive cycle in *Mytilus* go further than establishing the times of spawning or settlement, and many of their conclusions are conflicting and lack precision. Particular attention is, however, drawn to the works of Battle (1932), Chipperfield (1953), Savage (1956), Lubet & Le Gall (1967), Seed (1969*a*) and Le Gall (1970) for *M. edulis*; Berner (1935), Bouxin (1956), Lubet (1957), Renzoni (1961, 1962), Guiseppe (1964), Bourcart & Lubet (1965), Uysal (1970), Valli (1971) and Seed (1975) for *M. galloprovincialis*; Stohler (1930), Whedon (1936), Coe & Fox (1942) and Young (1946) for *M. californianus*; Sawaya (1965), Lunetta (1969) and Umiji (1969) for *Perna perna*; Wisely (1964) and Wilson & Hodgkin (1967) for *M. planulatus* and Penchaszadeh (1971) for *M. platensis*.

Although several workers have analysed seasonal changes in the biochemical composition of tissues (e.g. Fraga, 1956; Alvarez, 1965; Moreno, Moreno & Malaspina, 1971; Dare, 1973*a*; see Chapter 8) surprisingly few attempt to relate such changes to the state of the gonad. In view of this, the work of Daniel (1921, 1922), Chipperfield (1953), and more recently Bourcart, Lubet & Ranc (1964), Bourcart & Lubet (1965) and Gabbott (1975) are therefore particularly significant. Whilst reproductive cycles should ultimately be analysed in the light of both the ecology and physiology of the animal, this present account is concerned entirely with ecological aspects of reproduction.

Methods of assessing the reproductive cycle

The term reproductive cycle is here defined as the entire cycle of events from activation of the gonad, through gametogenesis to spawning (release of the gametes) and subsequent recession of the gonad. It can be divided into a reproductive period, which starts with the initiation of gametogenesis and culminates with the emission of gametes, and a vegetative or resting period. The reproductive period may be characterised by a series of gametogenic cycles, each followed by a spawning, which may occupy only a relatively short part of the entire cycle.

Several methods of assessing the course of the reproductive cycle in marine invertebrates have been used. These may involve direct observations of spawning in natural or laboratory populations or the appearance of the gonad throughout the year. Alternatively, inferences can be made from the appearance of larvae in the plankton or recruitment of spat (planti-grades) to the population. A major objection to these indirect methods is the lack of assurance that larvae or spat are spawned in the same locality; they may be carried considerable distances by currents from localities having quite different hydrographic conditions. Studies of larval abundance and settlement can, nevertheless, serve as valuable checks on data obtained by more direct methods. Although observations of spawning in field populations provide the most reliable evidence of the culmination of natural reproduction, such data are generally difficult or impossible to obtain. Observations of spawning under laboratory conditions on the other hand, whether 'naturally' or experimentally induced, are of limited value and may lead to dubious conclusions concerning spawning in the natural habitat, where interactions of numerous variables may modify the reproductive cycle. Probably the most reliable information is that obtained from microscopic preparations of the gonad. However, since the animals are sacrificed, continuous observations cannot be made on the same individuals. Studies of natural reproductive cycles are of greatest ecological significance when carried out over a number of years, since the onset and duration of the annual cycle can vary considerably from year to year.

Anatomy and histology of the gonad

Apart from the occasional hermaphrodite, sexes in *Mytilus* are separate with no external signs of dimorphism. Mussels can mature in their first year, but the size at which this occurs depends on the rate of growth. The reproductive system consists of numerous ducts which ramify throughout most of the body, each terminating in a genital follicle. Paired gonoducts open onto papillae situated anterior to the posterior adductor muscle in the angle between the mesosoma and inner gill lamellae. Gonoducts lead into five major canals with convoluted walls forming longitudinal ciliated ridges. These lead in turn into a series of minor canals which have part of their walls of ciliated columnar epithelium.

Oogonia and spermatogonia are budded off from the follicular germinal epithelium. The early oocyte is connected to the epithelium by a broad stalk but this gradually becomes more slender and finally ruptures to leave the mature ovum free within the follicle. Spermatogonia give rise in turn to spermatocytes, spermatids and spermatozoa, the latter covering towards the centre of the follicles in the form of dense lamellae.

Several schemes for classifying gonad condition in *Mytilus* have been used. Johnstone (1898) outlined a method which was subsequently developed by Chipperfield (1953), Lubet (1957) and Lunetta (1969). This system provides a simple yet precise method for distinguishing the onset and duration of the reproductive cycle. Other schemes have been used by Wilson & Hodgkin (1967) and Seed (1969*a*). Since details are to be found elsewhere in the literature, only a brief description of the last scheme will be given here.

From histological preparations, four main stages in the reproductive cycle can be recognised; developing, ripe, spawning and spent. Developing and spawning stages can be further subdivided, resulting in a total of ten stages into which any individual can be assigned. For each sample a mean gonad index is determined by multiplying the number in each stage by the numerical ranking of the stage and dividing the sum of these products by the total of individuals in the sample. The index can vary from zero, if the entire population is spent or resting, to five when fully ripe. An increase in the index generally denotes development; a decrease signifies that spawning is in progress. It should be noted that since the mean gonad index is not an interval measure of gonad development, a skew distribution of rankings may occur. In such cases the median or modal values may provide better estimates of the central tendency of the sample than does the mean index used here.

Brief description of gonad:

(*a*) *The resting or spent gonad*:
  Stage 0.   Inactive or neuter. This stage includes virgin animals as well as those which have completed spawning.

(*b*) *The developing gonad*:
  Stage 1.   Gametogenesis begins, though no ripe gametes are yet visible.
  Stage 2.   Ripe gametes first appear. The gonad is now about one-third of its final size.
  Stage 3.   There is a general increase in the mass of the gonad to about half the fully ripe condition. In area, each follicle contains approximately equal proportions of ripe and developing gametes.
  Stage 4.   The gonad is two-thirds or more its final size. Gametogenesis is still progressing but follicles contain mainly ripe gametes.

(*c*) *The ripe gonad*:
  Stage 5.   Fully ripe. Early stages of gametogenesis are greatly reduced. Ova are compacted into polygonal configurations whilst the

21

male gonad is distended with morphologically ripe sperm. The time between morphological and physiological ripeness may vary by up to several months under different conditions.

(*d*) *The spawning gonad*:

Stage 4. Active emission commences, as evidenced from the general reduction in sperm density and rounding off of the ova as pressure within the follicles is reduced.

Stage 3. The gonad is approximately half empty.

Stage 2. Further general reduction in the area occupied by the gonad is evident. The follicles are about one-third full of ripe gametes.

Stage 1. Only residual gametes remain, some of which may be undergoing cytolysis by amoebocytes.

Criteria that exist for separating these arbitrary stages in what is a continuous process are unquestionably rather subjective, and intermediate stages inevitably occur. Only by following the entire annual cycle does it become possible to recognise each stage with any confidence. Frequently, periods of partial emission may be followed by further periods of gametogenesis and it is usually the separation of spawning and redeveloping stages where most subjectivity is encountered. A further drawback is that whilst a gonad may appear to be in a spawning condition, there is no guarantee that active emission is in progress.

The annual cycle

In describing the sequence of events in the reproductive cycle, particular reference is made to *M. edulis* from several exposed (Filey Brigg) and sheltered (Filey Bay) habitats on the North Yorkshire coast of England (Seed, 1975). Fig. 2.1 shows the mean gonad index and total percentages in the spawning, developing and resting conditions for mussels from five different habitats between 1964 and 1969, from which it will be noted that the cycle is subject to considerable variation both annually and from habitat to habitat. Redevelopment of the resting gonad commences during October and November, gametogenesis occurring over the winter until by the following early spring the gonad is morphologically ripe. During the spring months a period of partial spawning occurs. This is followed by rapid gametogenesis until by early summer the gonads are once again fully ripe. Whereas in mussels growing in favourable feeding conditions (e.g. the low shore) the index frequently re-attains a value comparable with that earlier in the year (Fig. 2.1*c* and *e*), in those mussels situated in the poorer conditions of the high littoral, full re-ripening is often accomplished only by a minority of the population (Fig. 2.1*a* and *d*) and consequently the index may never reach its earlier spring value. In certain cases, further,

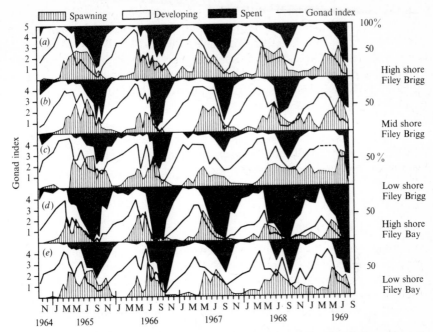

Fig. 2.1. The gonad index and the percentage distribution of spawning, developing and spent individuals of *Mytilus edulis* in five populations over five years. (From Seed, 1975.)

less intense, periods of spawning may occur even later in the summer. Successive periods of ripening and spawning were especially evident during 1965–6 when more frequent samples were taken. During July and August complete evacuation of gametes occurs. By late August or September the index in all populations reaches its lowest value as the majority of the population enter the resting condition. Throughout late August, September and October the connective tissue becomes thickened with varying amounts of glycogen and fat. That repeated spawning occurs in these populations was confirmed, in females at least, by actual egg counts, periods of gametogenesis being indicated by the relative increase in numbers of early oocytes.

A significant feature of the spawning period in these mussels is its duration. Whilst more than 50% of the Filey population rarely spawn simultaneously, the period over which 25% were spawning extends over four to six months. Records of extended spawning by *Mytilus* are not uncommon, but many of these are based upon systematic plankton observations or periods of spatfall and these methods do not always accurately reflect the natural spawning periods in the locality. Males generally seemed to be more advanced in gametogenesis than females at any particular time, though this is possibly a consequence of the arbitrary

23

nature of the classification rather than of any chronological differences in rates of development. It is not improbable, however, that production of sperm occurs faster than that of ova with their relatively large reserves of yolk. Apart from a tendency for certain high-shore populations to spawn somewhat before those in the lower shore (see also Whedon, 1936; Young, 1946) no marked or consistent differences due to habitat were recorded. Lubet (1957) was unable to find any difference in spawning time between permanently submerged mussels and those periodically exposed to air. Spawning in small and large mussels appeared to be simultaneous (see also Chipperfield, 1953) although Jensen & Spärck (1934) and Thorson (1946) maintain that young mussels matured and spawned later than old mussels in Danish waters.

Geographical variations

Previous records of reproduction for several mytilids from various parts of their geographical range are summarised in Tables 2.1 to 2.3. Since not all workers have used the same criteria for assessing reproduction and since records have not been made over the same period, comparisons should be carried out with caution. Nevertheless, these records do suggest that mussels from the warmer more southerly waters of the northern hemisphere generally spawn earlier than those further north.

A complex of physical variables is thought by Giese (1959) to influence, if not control, the sequence and timing of reproduction in marine invertebrates. He points out that temperature is widely recognised as an

Table 2.1. *Spawning periods of mytilids* (M. edulis, *UK*)

Locality	Authority	Spawning period: comments
England		
General	White (1937)	Spring, with second, possibly more important, spawning later in summer
	Raymont (1963)	Especially spring – July/Aug.; but almost any time according to conditions
	Baird (1966)	Over at least 8 months of the year in England/Wales; Mar.–Oct.
Plymouth	Matthews (1913)	Early spring; Jan.–Mar. in 1911
	White (1937)	Jan.–Mar.; spent mussels common May and Aug.
	Lebour (1938)	Apr.–June; some ripe at almost any time; veligers dominant late spring – early summer
	Chipperfield (1953)	May–early June
Padstow	Seed (1971)	Mar.–Aug. but especially May–June
Brixham	Chipperfield (1953)	Late Apr.–early June

Table 2.1. (*cont.*)

River Dart	Chipperfield (1953)	May
Southampton Water	Raymont & Carrie (1964)	Mar./Apr.–Aug./Sept.; extended spawning possibly due to local increase in water temperature by power station
	CEGB (1965)	Starts May with intermittent spawning over following 18 weeks
Brighton	Campbell (1969)	Apr.–May; secondary phase in Oct.
The Wash	Savage (1956)	Apr.
Brancaster	Chipperfield (1953)	May–early June
N. Yorkshire	Seed (1969a, 1975)	Mar.–Nov. but especially Apr.–May and July–Aug.
Northumberland	Lebour (1906)	May–Aug./Sept.
	Mitchell (cited in Lebour, 1906)	Two spawnings, one in spring another about Aug.
	King (cited in Lebour, 1906)	July–Sept.; earlier in warmer years
N. E. coast	McIntosh (1891)	Mar.–May/June
Lancashire	Johnstone (1898)	July–early Aug.; second spawning earlier in year with possibly a slow emission from Apr. onwards; variable
	Daniel (1921)	Apr.–May; as early as Mar. in warm years
Liverpool	Chipperfield (1953)	May–early June
Morecambe Bay	Herdman & Scott (1895)	May
	Dare (1973b)	Apr.–May but some spawning till Sept.–Oct.
Isle of Man	Scott (1901)	Early May–mid July
	Bruce (1926)	May
	Chipperfield (1953)	May
	Roberts (1972)	Mid June–early July. Also Oct.
Wales		
Conway	Cole (cited in Graham, 1956)	Major settlements in Mar.–Apr.
	Chipperfield (1953)	May
	Savage (1956)	Especially Apr., but also May, June, Aug.
	Baird (1966)	Apr.–May
Menai Bridge	Bayne (1964b)	Apr.–June
Cardigan Bay	Wright (1917)	Spawning begins late Apr. or May
Scotland		
St Andrews	McIntosh (1885)	Mature Apr.; spawn late May–June; larvae abundant July–Aug.
	Wilson (1886)	Mature Apr.–May; nearly empty July
Bay of Nigg	Williamson (1907)	Apr.–June
Millport	Elmhirst (1923)	Apr.–June
	Pyefinch (1950)	Heavy settlements June–July suggest May–June spawnings
Loch Sween	Mason (1969)	Spring
Isle of Skye	Chipperfield (1953)	May–early June
Ireland		
Cromaine	Crowley (1970)	Spring–summer
River Boyne	Meaney (1970a)	Late spring–early summer
Carlingford and Belfast Loughs	Wilson & Seed (1975)	Especially Apr.–June, but repeated spawnings may continue throughout summer until Oct.–Nov.

Table 2.2. *Spawning periods of mytilids* (M. edulis, *other than UK*)

Locality	Authority	Spawning period: comments
USSR		
E. Murman	Kuznetzov & Mateeva (1948)	May–Oct; larvae abundant July–Aug.
	Mateeva (1948)	May–Oct.
White Sea	Palichenko (1948)	June–Aug.
	Savilov (1953)	Mid July–early Sept.
Greenland	Madsen (1940)	Oct.
Norway		
Oslofjord	Bøhle (1965)	May–June; also autumn; larvae abundant June–July
	Bøhle (1971)	Starts early May; second spawning mid June
Trondheimsfjord	Jensen & Sakshaug (1970)	Spring–autumn; extended spawning
S. Norway	Runnström (1929)	Mar.–June
Finland	Heinonen (1962)	Maximum spawning late May–late June
Denmark		
Limfjord	Spärck (1920)	May–June
The Sound	Jørgensen (1946)	Veligers appear end May–early June and are abundant till Sept–Oct.; also in Dec.–Jan. and even spring
Isefjord	Jørgensen (1946)	Veligers appeared end May; spawning rapidly completed; by mid July veligers no longer found
General	Vorobiev (cited by Palichenko 1948)	May–Sept.
Germany		
Helgoland	Kändler (1926)	Larvae common Apr.–Dec., especially Apr.–June
Cuxhaven	Kühl (1972)	Starts May–June; larvae common May–Sept.
Kiel	Boje (1965)	Starts June
Holland		
General	Berner (1935)	Mar.–Apr.; not later than early May
Zeeland	Lambert (1935)	Mainly June; starts Mar.; smaller emissions May and July
Zuidersee	Havinga (1929)	Summer
General	Havinga (1964)	Most of the year but especially spring–early summer
North Sea	Werner (1939)	Spring–autumn
France		
General	Field (1909)	Feb.–Sept.
Normandy	Le Gall (1970)	End Dec.–early Jan. and Mar.–Apr., even as late as June
Luc-sur-Mer	Lubet & Le Gall (1967)	Sometimes as early as late Jan.–early Feb.; mainly Mar.–Apr.; partial emissions May–June
Brittany	Kriaris (1967)	Larvae abundant during winter, less abundant in summer. Esp. late Oct. and mid May

Table 2.2. (*cont.*)

Arcachon	Lubet (1957)	Feeble spawning Dec. and early Jan.; Main spawning late Mar.–June, especially Apr.–May
Baie d'Aiguillon	Herdman (1893)	Early spring
	Berner (1935)	Feb.–Mar.; not later than early Apr.
	Lambert (1950)	Early spring
N. Atlantic	Stubbings (1954)	Feb.–Sept., especially Apr.–June
N.W. Spain	Andreu (1963)	*edulis* (?) Apr.–May and Nov.–Jan.
	Andreu (1968*c*)	Extended spawning; starts Feb. and ends early summer; secondary spawning in autumn
Mediterranean	Fox (1924)	May–June and Jan.–Feb.
Black Sea	Vorobiev (cited by Palichenko, 1948)	*edulis* (?) Feb.–Oct.
N. America		
W. Canada	Stafford (1912)	Veligers abundant June–late autumn
New Brunswick	Battle (1932)	Mid June–mid Sept., especially Aug.
Long Island	Engle & Loosanoff (1944)	May; settle early June–end Aug.
Atlantic Coast	Field (1922)	Apr.–Sept.
Woods Hole	Field (1909)	Early Feb.–end Aug.
	White (1937)	June–Sept.
W. USA	Ahmed & Sparks (1970)	Late Apr./early May–late Aug.
Pacific coast	Field (1922)	Mainly cold months
	Sommer (cited by White, 1937)	Mainly winter
California	Moore & Reish (1969)	Mature ♂ throughout the year but especially Oct.–Feb.; mature ♀ Nov.–May but especially Nov.–Feb.
Japan	Sagiura (1959)	Mature Nov.–Apr.; recently spent mussels May–July
	Hirai (1963)	Dec.–May

important exogenous factor and the concept of a causal relationship between this factor, reproduction and geographical distribution, has become established as a zoogeographical principle. Orton (1920) and Kinne (1963) also stress the importance of temperature in regulating reproduction, whilst Wilson & Hodgkin (1967) and Lubet & Le Gall (1967) both conclude that reproduction in *Mytilus* varies with latitude. In a recent examination of *Mytilus* from European waters, Seed (1975) presents further evidence which points to temperature being a principal factor in controlling the broader aspects of the annual cycle. Bayne (1975*a*) found a linear relationship between the rate of gametogenesis and 'day-degrees', which are a function of temperature and time; the duration of the spawning stage, however, was more variable, and probably related to the nutritional condition of the individual and to its fecundity.

Whilst southern species usually reproduce later in the year and have a

Table 2.3. *Spawning period of mytilids* (*other than* M. edulis)

Locality	Authority	Spawning period: comments
M. galloprovincialis		
S.W. England	Seed (1971)	Especially July–Aug.
Britanny	Bouxin (1956)	Extended spawning starts Mar. (sometimes Feb.) ends July, sometimes Aug.
Arcachon	Lubet (1957)	Starts Sept. arrested by low temp.; restarts late Mar. and continues till early July
Provence	Berner (1935)	Extended spawning Sept.–Oct., Dec.–Jan.
Toulon	Bourcart & Lubet (1965)	Autumn, winter and early spring, ends May–June
Naples	Lo Bianco (1899)	Mature Mar.–Apr.
	Renzoni (1961)	Mature and spawn Feb.–Apr.
	Renzoni (1963)	End winter–early spring; second spawning in autumn
Trieste	Favretto (1968)	Mature end winter–early spring
	Valli (1971)	Mid Oct.–mid Dec.; mid Jan.–early Feb.; Mar.
Messina	Guiseppe (1964)	Nov.–May; inactive till following Sept.
Syracuse	Renzoni (1962)	Ripen and spawn Apr.–May; again ripe Oct.–Nov.
Yugoslavia	Lubet (1961)	Spawning extended but especially Mar.–Apr. and autumn
Black Sea	Kiseleva (1966a)	Mass spawning spring/autumn; larvae always abundant
	Ivanov (1971)	Late May–Dec.
Adriatic	Hrs-Brenko (1971)	Especially late Jan.; several less intense spawnings later
M. californianus		
California	Stohler (1930)	Two periods; max. in July and Nov.–Dec.
	Coe (1932)	Larval abundance suggests June–Sept.
	Whedon (1936)	Oct.–Nov.; also Jan.–Feb. and May–June; more or less continuous
	Young (1942)	Starts early Sept. and at max. in mid winter; at min. from May–Aug.; some limited summer spawning
	Fox & Coe (1943)	More or less any season but especially spring and autumn
	Young (1946)	Oct.–Mar.; less intense Mar.–Sept.; some spawning more or less throughout the year
M. viridis		
India	Paul (1942)	More or less continuous but mainly Mar.–Nov.; especially Aug.–Sept.
M. edulis diegensis		
California	Coe (1946)	All seasons but especially Mar.–June and early winter
M. edulis planulatus		
Sydney	Wisely (1964)	June–mid Aug.

Table 2.3. (*cont.*)

W. Australia	Wilson & Hodgkin (1967)	Ripe Apr.–July; some spawning Apr.–May, main spawning July; secondary spawning Sept. with other minor periods
Perna perna (= *M. perna*)		
Venezuela	Carvajal (1969)	Three peaks of larval abundance correspond to three peaks of spawning; Dec.–Jan., Mar., June–July
Brazil	Lunetta (1969)	More or less continuous but especially Apr.–June and Sept.
M. aeoteanus		
New Zealand	Pike (1971)	Especially spring
M. platensis		
Argentina	Moreno *et al.* (1971)	Especially early spring
Perna canaliculus		
New Zealand	Pike (1971)	Spawning rare before Jan.; minor spawning over summer with another heavy spawning in autumn
M. crassitesta		
Japan	Miyazaki (1935)	Variable, but especially late Dec.–early Apr.
M. smaragdinus		
Philippines (?)	Obusan & Urbano (1968)	Throughout year but especially May and Nov.
Crenomytilus grayanus		
Japan (?)	Suburo & Sakamoto (1951)	Starts July; max. late July and Aug.
Peter the Great Gulf	Sadykhova (1970*b*)	Extended but with two peaks

progressively restricted season further northwards, northern species exhibit the reverse trend – spawning earlier and with a more extended season further south. This is now a fairly widely acknowledged principle and *Mytilus* is apparently no exception (see Madsen, 1940; Palichenko, 1948). In south-west England *M. galloprovincialis* spawns several weeks later than *M. edulis* and only with the onset of maximum sea temperature (Seed, 1971). Lubet (1957) records differences in the reproductive cycles of these mussels at Arcachon and suggests that ecological factors alone could not be responsible.

Factors controlling the reproductive cycle

Various stimuli have been suggested as being important in controlling the reproductive cycles of marine lamellibranchs. These factors can be grouped as exogenous or endogenous.

Exogenous factors

Of all the factors which may influence reproduction, sea temperature has received most attention. Whilst extreme temperatures may inhibit spawning, these seem to be less limiting in warmer climates than in temperate waters (Paul, 1942; Young, 1945; Allen, 1955; Heinonen, 1962; Moore & Reish, 1969). The effects of severe winters on spawning and their possible adaptive significance are discussed by Savage (1956).

It is widely suggested that spawning in lamellibranchs may occur only over a critical temperature range which is constant for each species; some authorities (e.g. Nelson, 1928a) even propose that systematic grouping is possible on this basis. Species are more likely, however, to have several critical spawning temperatures depending on their physiological condition or geographical distribution, and adaptation to spawn at different temperatures, especially at the limits of distribution, could be an important factor in speciation.

Although some relationship between spawning and sea temperature in *Mytilus* seems evident; just how this is mediated is still unclear. Chipperfield (1953) stresses that rate of temperature change may be the important factor in determining both the intensity and duration of spawning. Results are often contradictory: whilst increasing and/or decreasing temperatures are reported to stimulate spawning, other workers have been unable to show any response to temperature change. Bayne (1965) used a combination of elevated temperatures and chemical stimulation to condition mussels to spawn in the laboratory, and later (1975a) described a relationship between rate of gametogenesis and rate of change of temperature.

Mechanical stimuli such as scraping or chipping the shell, pulling the byssus and pricking or cutting the adductor muscle are also reported to stimulate spawning in *Mytilus* (Field, 1922; Young, 1942, 1945, 1946; Bouxin, 1956; Wilson & Hodgkin, 1967; Hrs-Brenko & Calabrese, 1969; Ahmed & Sparks, 1970; Loosanoff & Davis, 1963). Strong wave action caused *M. californianus* to spawn (Young, 1942) and this could be an important factor in the 'brown mussel' during monsoons (Jones, 1950). Evidence for lunar periodicity in *Mytilus* is affirmed (Battle, 1932; Lubet, 1957) and denied (Fox, 1924; Whedon, 1936; Bouxin, 1956; Lubet & Le Gall, 1967). A review of much of the earlier literature is given by Korringa (1947). The actual mechanism involved in lunar cycles is unclear; light, temperature and pressure have all been implicated. Chemicals (Iwata, 1951a, b; Sagara, 1958; Baird, 1966) and electrical stimulation (Iwata, 1950; Sagiura, 1962) will also induce spawning though neither could be considered natural stimuli. The presence of gamones in the surrounding water may lead to epidemic spawning, thereby enhancing the chance of fertilisation. Algal extracts may be effective in some bivalves (Miyazaki,

1938*b*), suggesting that spawning might be synchronised to coincide with maximum food availability for the developing larvae. Pollution (Breese, Millemann & Dimick, 1963; Davies, 1972; Roberts, 1972) and salinity changes (Berner, 1935; Paul, 1942; Wisely, 1964; Baird, 1966; Kühl, 1972; Wilson & Seed, 1975) may also affect spawning, the latter possibly being an important factor in estuaries and tropical waters.

Endogenous factors

Lubet (1955*a*, *b*, 1956, 1957, 1965, 1966) suggests that spawning in *Mytilus* is controlled by a combination of internal and external factors. Disappearance of neurosecretion from cerebral ganglia, possibly provoked by elevated temperatures or salinity changes, appears to be essential if the mussel is to become receptive to external stimuli, the nature of which may not be important provided that peripheral receptors are stimulated. Neurosecretion is maximal in ripe animals, in contrast to the empty, vacuolated appearance of the neurosecretory cells in spent individuals. By removing the apparent source of inhibition, ablation of the ganglia results in precocious spawning. Neurosecretory cells are also described in *Mytilus* by Gabe (1955, 1965), whilst more recently Sawaya (1965) and Umiji (1969) have postulated a relationship between neurosecretory and reproductive cycles in *M. (Perna) perna*.

Conflicting opinions exist, therefore, as to the precise factors stimulating spawning in *Mytilus*. Some confusion has arisen in the search for a single factor, since in a complex process like reproduction, interactions of factors are to be expected. Whilst attention has generally focused on spawning, the factors controlling the entire reproductive cycle have received little attention. Clarke (1965) draws attention to the lack of priority given to endocrine co-ordination of reproduction, and discusses how external factors may be involved in co-ordinating the annual cycle in polychaetes. His conclusion that spawning, whether well or poorly co-ordinated, must depend ultimately on environmental factors, is supported by Chipperfield's (1953) observations that the period of maximal occurrence of ripe individuals, even during the same year, varies from one locality to another.

Recruitment

The literature concerning larval behaviour and settlement of mussels is rather sparse, though attention is drawn to the publications of Maas Geesteranus (1942), Verwey (1952), Savage (1956), De Blok & Geelen (1958), and Bayne (1964*a*, *b*, 1965). Regular monitoring of the abundance of planktonic larvae and the subsequent settlement of plantigrades not only serves to check spawning data but may also enable periods of maximum recruitment to the adult mussel population to be predicted.

Marine mussels

Whilst the literature generally reveals that *Mytilus* larvae are seasonally abundant, e.g. in spring and early summer in British waters, records of larvae over a considerable period of the year are not uncommon (Kändler, 1926; Raymont & Carrie, 1964; Edwards, 1968; Hrs-Brenko, 1971; Mason, 1971; Kühl, 1972). Such extended periods of larval abundance are in accordance with the lengthy spawning periods to which attention has already been drawn. The duration of larval life appears to vary with temperature and food supply, and Bayne (1965) has shown that *Mytilus* can also delay settlement until suitable substrata are encountered. Although three to four weeks seems to be the normal duration of planktonic existence, up to ten weeks can elapse between fertilisation and settlement (see Chapter 4).

Primary and secondary settlement

Much of the earlier work on settlement fails to distinguish between the primary settlement of early plantigrades on filamentous substrata and the secondary settlement of later plantigrades on established mussel beds. Associations between recently settled mussels and filamentous substrata have been known for some time (McIntosh, 1885; Wilson, 1886; Johnstone, 1898) and have subsequently been recorded by numerous workers (Delsman, 1910; Wieser, 1952; Chipperfield, 1953; Verwey, 1952, 1954). The full significance of these observations was not appreciated until De Blok & Geelen (1958) showed that early plantigrades were directed not onto existing mussel beds but rather onto filamentous substrata such as hydroids and various algae from which they subsequently disappeared. Experiments with various substrata indicated a preference by early plantigrades for filamentous surfaces, but this preference subsequently changed. Suitability of the substratum appears to be related to its general morphology rather than to any chemical attraction. De Blok & Geelen further suggest that since the chance of plantigrades encountering a suitable substratum after their detachment may be quite small, a large part of the population is probably continually on the move. Maas Geesteranus (1942), however, showed that young mussels settled on all types of substrata, providing these were firm and had either a rough or discontinuous surface. He also showed that plantigrades attached and detached themselves many times before finally arriving at the established bed. He argued that mussel beds attract further recruits by virtue of their surface texture, and the byssus threads seemed to be particularly important. During larval transport the foot is generally fully extended in order to seize any suitable substratum that may be encountered. In addition to restating much of the earlier work, Verwey (1952) also stressed the importance of currents in larval transport. Paine (1971, 1974) has shown that both *Perna*

32

canaliculus and *Mytilus californianus* show a preference for filamentous substrata when recruiting into rocky shore populations; patches of suitable algal substrata can act as focal points for recruitment of larvae and plantigrades.

Bayne (1964*b*) conclusively demonstrated that mussels pass successively from the plankton to sites of temporary attachment on filamentous algae and from these, via a secondary pelagic phase, to sites of more permanent attachment on adult beds. Few mussels larger than 1.5 mm occurred on the algae, whilst at their time of settlement on the mussel beds the majority measured between 0.9 and 1.5 mm in shell length. It is suggested that this primary phase of attachment is a natural prelude to final settlement and not a wasteful settlement on unsuitable substrata. Thorson (1957) discusses the difficulties that newly settled bivalves have in competition with adults and suggests that a primary attachment away from adult stocks, during which time they could grow, might have a significant survival value. Hancock (1973) discussed the relationship between adult stocks and level of recruitment in lamellibranchs. For mussels, he concluded that the presence of adults seems not to be of great significance to final settlement, although large numbers of settling plantigrades can adversely affect individuals already established in the population. Bøhle (1971) was unable to demonstrate a secondary migratory phase preceding settlement in Norwegian waters; the plantigrades appeared to sink to the bottom directly from the algae. This particular pattern of behaviour may be due to the somewhat unusual conditions of the fjords. Dare (1975) found that in Morecambe Bay (West Lancashire coast, England) heavy winter settlements of plantigrades resulted from the overwintering, in deep water, of plantigrades derived from autumn spawning of the adults, together with a fortuitous combination of hydrographic and weather factors which brought these late plantigrades inshore and, at the same time, made available large expanses of favourable substrata for final settlement.

The relatively marked seasonal abundance of early plantigrades on filamentous substrata shown by Bayne (1964*b*) was not observed by Seed (1969*a*) working on exposed rocky shores where high densities occurred more or less throughout the year. These and other algal substrata such as *Corallina* and *Gigartina* could possibly provide extensive 'reservoirs' of young mussels, many of which may migrate onto adult beds at any time during the year, so accounting for the sporadic outbursts of settlement that are observed in many localities throughout the year. Dare's (1975) study of mussels in Morecambe Bay is an illustration of this. Generally, plantigrades remain on these algae until they measure between 1 and 2 mm. The time taken to attain this size depends on local conditions and individual growth rates. Settlement late in the year, when conditions for growth are unfavourable, often results in many plantigrades remaining on their

33

Table 2.4. *Settlement periods of M. edulis*

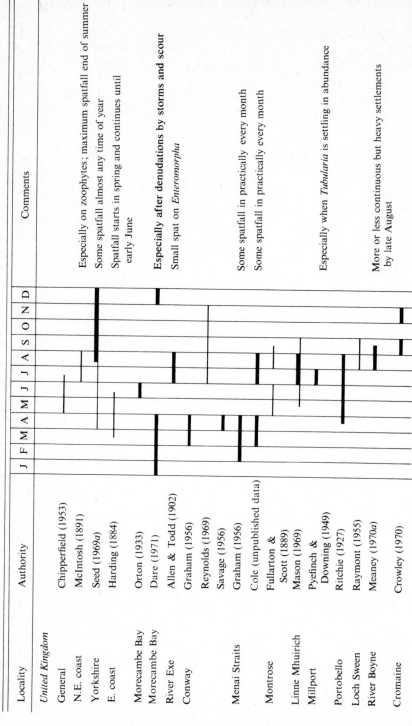

Locality	Authority	J F M A M J J A S O N D	Comments
United Kingdom			
General	Chipperfield (1953)		Especially on zoophytes; maximum spatfall end of summer
N.E. coast	McIntosh (1891)		Some spatfall almost any time of year
Yorkshire	Seed (1969a)		
E. coast	Harding (1884)		Spatfall starts in spring and continues until early June
Morecambe Bay	Orton (1933)		
Morecambe Bay	Dare (1971)		**Especially after denudations by storms and scour**
River Exe	Allen & Todd (1902)		Small spat on *Enteromorpha*
Conway	Graham (1956)		
	Reynolds (1969)		
	Savage (1956)		
Menai Straits	Graham (1956)		Some spatfall in practically every month
	Cole (unpublished data)		Some spatfall in practically every month
Montrose	Fullarton & Scott (1889)		
Linne Mhuirich	Mason (1969)		
Millport	Pyefinch & Downing (1949)		Especially when *Tubularia* is settling in abundance
Portobello	Ritchie (1927)		
Loch Sween	Raymont (1955)		
River Boyne	Meaney (1970a)		More or less continuous but heavy settlements by late August
Cromaine	Crowley (1970)		

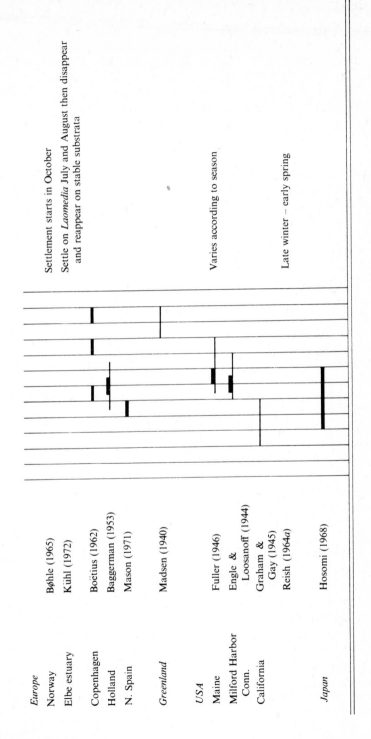

temporary substrata over winter, leaving only with the resumption of more favourable conditions the following spring. Table 2.4 summarises the timing of settlement of *M. edulis* in different localities.

Recruitment of plantigrades on to established mussel beds

Valuable data concerning seasonal spatfall can be acquired by using experimental surfaces set out periodically through the year. Such studies indicate that smooth surfaces are unattractive to spat and maximum settlement occurs on roughened or fibrous substrata (Davies, 1974). Thiesen (1972) suggests that the smooth shells of young mussels could prevent them from being smothered by plantigrades by presenting a poor surface for settlement. Heavy spatfalls on older mussels, where the periostracum is worn away and thus the shell rougher, can have disastrous consequences, as witnessed on the Conway beds during the spring of 1940 (Savage, 1956).

Although population studies can also provide useful qualitative data concerning recruitment if sampling is continued regularly throughout the year, this method does not give a true quantitative picture of settlement, since no account is normally taken of mortality or growth out of the size categories sampled. Fig. 2.2 shows that during winter, when growth is slow and mortality (from predation) is reduced, the population of plantigrades (< 3 mm) remains relatively stable. In spring and summer, heavy mortality and rapid growth reduces the numbers of plantigrades, though periodically this is offset by recruitment. These data therefore represent a balance between recruitment of plantigrades and their subsequent loss through mortality and growth. Bouxin (1956) and Reynolds (1969) also comment on the problems involved in determining the duration of settlement from population studies, due to difficulties in separating small, slow-growing members of the population from newly settled plantigrades. A marked increase in the number of small mussels is indicative of a period of settlement, but if large numbers are present on successive observations there is no way of telling whether these are the same individuals or whether substantial mortality in the intervening period has been offset by continued recruitment.

In a commercial fishery it is useful to be able to predict when maximum recruitment can be expected, in order to exploit the resource most effectively. However, the time and duration of settlement not only varies annually, but also from one region to another according to local conditions. Whilst it may be possible to anticipate periods of maximum settlement from a knowledge of the reproductive cycle and periods of larval abundance, recruitments can, nevertheless, occur at almost any time of the year and these are not always predictable from such information. A

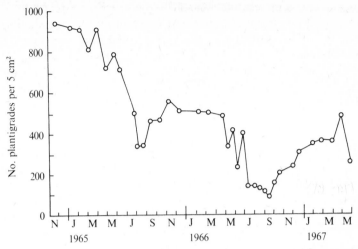

Fig. 2.2. The number of plantigrades of *Mytilus edulis* per 5 cm² in one population in North Yorkshire over three years. (From Seed, 1969*a*.)

possible explanation for such sporadic spatfall has already been discussed (p. 33). Dare (1969, 1971, 1975) suggests that the timing and intensity of settlement intertidally may be controlled by the natural cycle of physical and biological events on the mussel beds themselves rather than the availability of plantigrades in the sea. Dare found that the major spatfalls in Morecambe Bay occurred from December to April after the grounds had been denuded by winter storms. When the spring plantigrades are ready to settle there is little or no room, and few survive the competition from the established young adults. Plantigrades later in the year are therefore forced to settle sublittorally. Dare also stresses that larval transport and settlement are influenced by water movements; heavy spatfalls in Lancashire and Cumberland (England) are directly related to the wide catchment area from which larvae are received. It has previously been suggested (Harding, 1884; Savage, 1956) that unusually heavy spatfalls may follow particularly severe winter conditions. However, Reynolds (1969) concludes that at present insufficient information is available on the variations in intensity and extent of mussel settlements from year to year to draw any firm conclusions as to the actual significance of weather conditions. This is clearly an area where more critical research is required.

Field observations generally indicate a preference by *Mytilus* for flat shores which either drain slowly or which receive constant wave splash. Fewer mussels occur on rapidly draining vertical faces, though dense clusters on pier supports and harbour walls indicate that such habitats are not excluded to them. Mussels will settle on practically any stable substratum especially when the surface is roughened or scarred. Moisture-retaining cracks and pits are especially favourable and often permit mussels to extend locally further into the littoral than usual. Ross & Goodman (1974) suggest that *M. edulis* is excluded from higher regions of the shore because of the failure of plantigrades to survive there.

Mytilus is gregarious, and dense settlements are often found around the edges of, and in between individual mussels in, existing populations. Such behaviour probably has adaptive value since *Mytilus* is predominantly a littoral or shallow-sublittoral species and will therefore be subjected to mechanical forces of water movement, especially on exposed shores. The reduced surface area which is exposed to such forces by animals within dense clusters, together with the mutual support of neighbouring individuals, makes groups of mussels better adapted to withstanding wave action than isolated individuals (Harger, 1970a, 1972b; Paine, 1974). Plantigrades, once settled, provide loci for further recruitment and so the colony gradually extends. Young mussels reach established beds by their ability to attach and detach themselves from unsuitable substrata until eventually encountering a surface, usually in the crevices and amongst byssus threads provided by their own species, that satisfies their thigmotactic requirements. Once established, mussel beds increase in dimensions both through gregarious settlement and growth of individual mussels. Paine (1971) remarks that the competitive dominance of *Perna canaliculus* in New Zealand (and, by implication, that of other mytilids) is due in large part to their ability to move extensively over the substratum even after initial settlement.

Growth

There are two principal methods of assessing growth. Firstly, the size of the whole organism can be related to age. When this represents the cumulative increase with respect to time it is termed absolute growth, whilst the percentage increase per unit of time is termed relative growth. Secondly, allometric growth, the rate of growth of one parameter related to that of another, can be measured.

Since growth is defined as increase in body size, weight or volume might be the most appropriate parameters for its measurement. However, the shell is such a prominent feature of molluscan anatomy that growth is generally measured in terms of shell length. This parameter has the additional advantage of being easily measured without having to detach the animal from the substratum. In comparative investigations, however, extreme caution must be exercised in using single linear parameters to measure growth. For example, although mussels may be increasing in biomass at approximately equal rates, their growth in terms of other parameters such as length, width or height may vary considerably according to age and local environmental conditions.

Methods of measuring growth

Several methods have been used to assess growth in bivalves, but three have received particular attention. These are:

(1) The use of mean or modal size-frequency distributions.
(2) Measurements of marked mussels.
(3) The use of annual disturbance rings on the shell.

Other methods, including X-ray techniques and rates of incorporation of radioactive isotopes or fluorescent materials into the shell, are discussed by Wilbur & Owen (1964).

Size-frequency distributions for some bivalves are polymodal, each mode representing an individual year class. Changes in position of these modes with time often permit population growth to be assessed. This method is usually possible, however, only for species with a relatively restricted period of recruitment, and for which individual growth rates within each year class are fairly uniform. In *Mytilus*, where spatfall is sometimes prolonged and individual growth rates variable, this method has limited application. Nevertheless, it has been used successfully by several workers (Raymont, 1955; Boëtius, 1962; Baird, 1966; Reynolds, 1969; Thiesen, 1968, 1973). With this technique Seed (1969*b*) followed a cohort of plantigrades, which settled on previously bare rock, over a period of two years. Whilst the year classes could be identified over this period, by the time of the third major spatfall separation became progressively more difficult due to overlap in size of fast-growing individuals from one year class, with slower growing members of the previous year group. Dare (1969) points out that growth curves constructed from population data of this kind may have been modified not only by natural mortality factors, some of which may be selective in their action, but also by continual recruitment. The slow growth of the majority of a population – perhaps due to large numbers of mussels ensnared in the byssus threads of larger mussels – frequently produces populations containing an abundance of very small mussels (Richards, 1928; Whedon, 1936; Madsen, 1940; Raymont, 1955; Bouxin, 1956; Sadykhova, 1967; Reynolds, 1969).

Various types of experimental cages have been used to analyse seasonal growth in mussels of different initial size and in different local conditions (Coulthard, 1929; Coe & Fox, 1942; Coe, 1945; Bøhle, 1965; Baird, 1966; Sadykhova, 1967; Seed, 1969*b*). Harger (1970*a*) found that whereas caging did not appear to hinder growth in sublittoral populations, intertidally growth was affected according to the amount of open surface of the cage, which could modify the effects of wave action.

Whilst disturbances in growth due to seasonal rhythms of the environment have been widely used in age determinations, in *Mytilus* the presence of annual shell rings has been both affirmed (Mossop, 1922; Mateeva, 1948;

Stubbings, 1954; Lubinsky, 1958; Seed, 1969b; Le Gall, 1970) and denied (Richards, 1928; Coe & Fox, 1942; Raymont, 1955; Craig & Hallam, 1963; Reynolds, 1969; Harger, 1970a). Harger (1970a) suggests that check rings may confer additional strength to the shell, and a relationship might therefore be expected between ring frequency and exposure to wave action. Generalisations concerning the reliability of such rings in age determination are thus somewhat tenuous and depend largely on seasonal environmental conditions. In relatively uniform conditions, well-defined rings may be absent, whereas in fluctuating conditions several rings may be produced each year. Annual rings are most likely to be encountered in areas having a prolonged period of suspended shell growth. Such periods may be associated with extremes of temperature, prolonged stormy weather or even with the reproductive cycle. It is important, therefore, that the reliability of growth rings should be tested for each locality investigated. Yet, even in localities where rings are shown to be reliable indicators of age, they may not occur in all individuals and their value may be further limited by shell abrasion. Earlier rings may then have to be inferred from a series of shells of different ages within the same locality. Furthermore, growth in shell length in old mussels may be so slow that rings on the posterior margin are produced so close together that accurate age determination becomes impossible. Details of the fine structure of bivalve shells are given by Wilbur (1964) and Kobayashi (1969). Shuster (1956) suggests that growth lines are produced during periods of slow growth when carbon dioxide levels in mantle tissues are higher than normal.

From the above discussion it is evident that although several methods can be used to determine growth, each has its own intrinsic problems. Probably the most reliable estimates are those obtained using a combination of methods.

One of the features of mussel growth is its variation in rate, not only between localities but also within similar size and age groups in the same population. Even mussels grown under apparently identical conditions can exhibit widely different rates. Variation is such that whilst in ideal conditions *M. edulis* may grow very rapidly, sometimes exceeding 60–70 mm within twelve to eighteen months (Reish, 1964a; Edwards, 1968; Bøhle & Wiborg, 1967; G. Davies, 1969; Mason, 1968, 1969, 1971) under less favourable conditions growth may be exceedingly slow, some individuals measuring only 20–30 mm after perhaps fifteen to twenty years (Seed, 1969b). Very rapid growth rates have also been demonstrated in other mytilids (e.g. Coe & Fox, 1942; Coe, 1945; Millard, 1952; Pike, 1971).

Quantitative expressions of growth

Each habitat seems to impose, by its environmental conditions (physical and biotic), an upper size limit to the local mussels, and whilst fast-growing individuals approach this limit relatively quickly, in areas of slow growth this limit may be approached only by much older individuals. The maximum potential size (L_∞) of individuals within a population may be estimated by plotting size at t years against size at $t+1$ years, i.e. the Ford–Walford plot (see Beverton & Holt, 1957; Hancock, 1965b). L_∞ is given where the line of best fit with a slope of $[\exp(-k)]$ intercepts the 45° line (where $L_t = L_t + 1$). Although regressions by 'least-squares' are generally used to fit such data, their use may be spurious since the two variables are not independent of each other and some alternative technique (Bartlett, 1949) might therefore be more appropriate. The parameters of the Ford–Walford plot are basic to many growth equations, one of which, the von Bertalanffy, has been extensively used in fisheries biology:

$$L_t = L_\infty [1 - \exp(-kt)],$$

where $t =$ time and k is a constant. Thiesen (1973) suggests, however, that the von Bertalanffy equation may only be valid for *Mytilus* above one-third of their maximum size, and for smaller mussels the Gompertz equation, using the logarithm of length rather than length proper, may provide a better fit. These equations do not, however, reflect seasonal variations in growth caused by changes in temperature. Ursin (1963) showed that the product of temperature, measured from a specified zero, and time (i.e. day-degrees) could be incorporated into the von Bertalanffy equation provided that L_∞ was independent of temperature. As Thiesen (1968) points out, the influence of temperature on the L_∞ of mussels is unknown, but since arctic mussels grow slowly and generally attain a large size (Jensen, 1912; Vibe, 1951; Lubinsky, 1958) it is likely to be greater at lower temperatures. Nevertheless, Thiesen applied the modified von Bertalanffy equation, incorporating day-degrees, and found that it afforded an extremely good picture of mussel growth in the Danish Wadden Sea. It should be remembered, however, that equations such as those mentioned above are based on determinate growth whereas growth in many bivalves is frequently indeterminate, at least over their normal life span, and may not, therefore, cease at any fixed adult size. Knight (1968) criticises the use of growth equations whilst O'Connor (1975), using eight different mathematical techniques to fit the same basic growth data, obtained eight different estimates of growth. Despite these criticisms, curve fitting by means of these equations is a perfectly acceptable technique providing (a) that data are relatively complete, i.e. there is some evidence for the existence of a real asymptote, and (b) that it is appreciated that some degree of

Table 2.5. *Some values for* L_∞ *and* k *(constants in the von Bertalanffy growth equation) for* Mytilus edulis

Locality	L_∞	k	Authority	Comments
Danish Wadden Sea	77.60	0.5611	Thiesen (1968)	Mussels at low-water spring tides
Conway, North Wales	72.71	0.3426	Thiesen (1968) based on data in Savage (1956)	'Bank mussels'
Greenland	74.44	0.3927	Thiesen (1973)	'Channel mussels'
	77.5–	0.022–		Mussels from 12 very varied sites
	283.9	0.162		
Menai Straits, North Wales	67.20	1.1378	G. Davies & P. J. Dare (personal communication)	Mussels suspended on ropes from rafts
Morecambe Bay, England	62.45	0.8103	Dare (1975)	Mussels near low-water mark of spring tides

uncertainty is always associated with such estimates. The von Bertalanffy equation, without the incorporation of day-degrees, has been fitted to length versus age data for *M. edulis* by Thiesen (1968, 1973), Dare (1975) and G. Davies & P. J. Dare (personal communication). The constants of the resulting expressions of growth are compared in Table 2.5, which illustrates the low rates of growth (due to overcrowding) observed by Savage (1956), the low growth rates in Greenland, and the relatively large size attained by mussels in the Arctic.

Factors influencing growth

Seasonal and annual cycles

Various factors affect growth in length and these are generally such, in temperate waters at least, that growth is rapid during the spring and summer and slight or absent during the colder winter months (Palichenko, 1948; Boëtius, 1962; Bøhle, 1965; Meaney, 1970*a*, *b*; Mason, 1969; Dare, 1969). Seasonal variations in growth are illustrated for several localities in Fig. 2.3. Whether winter cessation in linear growth is related to the gametogenic cycle is at present uncertain. Reduced availability of food in the winter could be significant, but low temperature is unlikely to be the sole controlling factor, since *Mytilus* occurs under extreme arctic conditions where the temperature is at or below 0 °C for much of the year (Thiesen, 1973). In addition to seasonal differences, marked annual variations in growth rates occur according to local conditions.

Temperature

Temperature has been widely acknowledged as an important factor in controlling growth rate. Coulthard (1929) found that growth of *Mytilus* occurred between 3 and 25 °C, with an optimum between 10 and 20 °C. *M. californianus* has an optimum between 15 and 19 °C (Coe & Fox, 1942, 1944) with growth declining sharply above 20 °C. Similarly, declining growth rates at high temperatures have also been recorded by Coe (1945) for *M. edulis diegensis* and Richards (1946) for *M. edulis*.

Dependence of growth on temperature was more clearly demonstrated by Boëtius (1962) when shell length was plotted against age in day-degrees. Length then appeared to be linearly related to age in day-degrees ($D°$) at least for small mussels from Copenhagen ($L = 0.007\,D° - 2.59$, where $L = $ length in mm). Such a linear relationship between shell length and $D°$ seems to be characteristic when growth data are available for low values of $D°$ (Ursin, 1963). G. Davies (1969) similarly found a day-degree transformation useful when considering growth, but he showed this to be sigmoidal for Welsh mussels. Some linear relationships did exist for portions

43

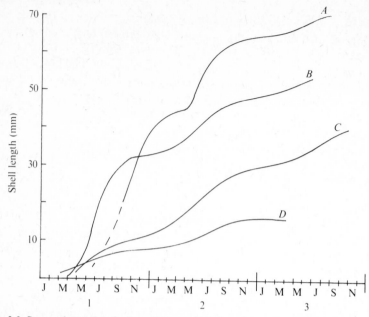

Fig. 2.3. Seasonal growth in four populations of *Mytilus edulis. A.* Linne Mhuirich, suspended culture 1966–8 (Mason, 1969); *B*, Morecambe Bay (Heysham) 1968–70 (Dare, 1973*b*); *C*, Conway, 1940–2 (Savage, 1956); *D*, Filey Brigg, 1965–7 (Seed, 1969*b*). Populations *B, C* and *D* were at low water of Spring tide.

of the data, suggesting that a doubling of the number of day-degrees should result in a corresponding increase in size. However, comparisons between Welsh and Spanish mussels suggested that factors other than temperature (possibly food supply) must be involved, a conclusion also drawn by Wilson (1971) who found that relative growth was greater during periods of increasing temperature than when temperature was decreasing through the same range. Thiesen (1973) compared some of his own and published values of size of *M. edulis* related to D°. In the Danish Wadden Sea mussels required approximately 6000 D° (*c.* 1.8 years) to reach a length of 50 mm. In the Menai Straits (G. Davies, 1969) 4000–7000 D° (mean about 1.2 years) were required to reach the same length, whilst in Disko, Greenland (Thiesen, 1973) 50 mm length was reached between 6300 and 9000 D° (7–10 years). When related to D°, therefore, the growth rates from these three areas were comparable. G. Davies & P. J. Dare (personal communication) recorded 50 mm length after 4500 D° in mussels suspended from rafts in the Menai Straits.

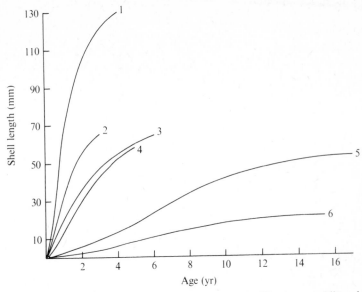

Fig. 2.4. Growth rates in six populations of *Mytilus*. 1, *M. californianus*; sublittoral (Coe & Fox, 1942). 2, *M. edulis*; River Boyne, sublittoral (Meaney, 1970*a*). 3, *M. edulis*; Chichester harbour (Stubbings, 1954). 4, *M. edulis*; Conway, sublittoral (Savage, 1956). 5, *M. edulis*; Filey Bay, high littoral (Seed, 1969*b*). 6, *M. edulis*; Robin Hoods Bay, high littoral (Seed, 1969*b*).

Age and size

Typical growth curves for *Mytilus* from various localities are illustrated in Fig. 2.4, from which it will be seen that increasing age is accompanied by a decline in growth rate. This may occur to such an extent that in very old individuals growth in length may virtually cease. This might be explained in terms of reduced relative metabolic activity in older mussels or of the greater increase in mass relative to shell length. In order to maintain even constant growth in length, larger mussels would require either longer or more efficient feeding periods. Senility itself, however, is not always the primary cause of reduced growth, since transplantation of old, non-growing mussels to more favourable situations can often result in renewed growth (Mossop, 1922; Seed, 1968).

Light

Huntsman (1921), Coulthard (1929) and Seed (1969*b*) all found a detrimental effect of light on growth suggesting that light may injure the exposed growing edge of the mantle. Shells of mussels grown in the dark are generally thinner and less densely pigmented than those grown in light.

Marine mussels

Berner (1935) suggested that shade may inhibit growth whilst Andreu (1960) found that moderate light may occasionally be beneficial. Boëtius (1962) and Dodd (1969) could find no evidence that light affects growth, although the latter did suggest that growth in light is more variable. Thus, whilst the evidence suggests that sunlight may influence growth in *Mytilus*, the results are somewhat contradictory and the direct effect of solar radiation in natural populations is difficult to assess in view of its coincident indirect effects via temperature and food supply.

Population structure

Growth of recently settled spat in populations containing two year classes was shown to be only 40% of that of spat settling on bare rock (Seed, 1969*b*). Even greater reductions in growth rate are found in populations of mixed ages, where the majority of small mussels amongst the byssus threads of larger individuals are at a disadvantage in the competition for food. Harger (1970*b*) showed that *M. californianus* confined within a clump grew very slowly, and Wilson & Hodgkin (1967) further demonstrated the detrimental effects of high densities. The process of 'thinning out' is widely practised in commercial fisheries. Using Savage's (1956) data, G. Davies (1969) found that volumetric growth of crowded mussels was only a fraction of that in scattered mussels of similar initial size and tidal level. Ironically, *M. edulis* was shown to grow much larger on exposed shores when found in association with *M. californianus* than when grown alone (Harger, 1972*a*).

Food supply and level on the shore

That mussels in the upper littoral grow more slowly than those from the low shore or sublittoral reflects the reduced amount of time available for feeding. This factor probably also determines the upper extension of mussels on any shore, since a point will be attained when the energy demands for metabolism during exposure will exceed the calories made available during the feeding period. The level at which this becomes critical will, however, vary from shore to shore depending on local conditions such as wave splash and the water-retaining properties of the shore itself. Baird (1966) examined the growth rates at different shore levels and by extrapolation found that the point of zero growth coincided with the tidal level corresponding to 56% aerial exposure. This was also found to be the natural limit of the beds on the shore investigated. Coulthard (1929) similarly found that mussels encountering 50% exposure showed little or no significant growth. Tidal level also has a marked influence on the ratio between shell and tissue weight in *M. edulis* (Baird & Drinnan, 1956; Seed, 1973), shell weight generally increasing and tissue weight decreasing for

46

any given length with increasing aerial exposure. This is thought to be due to the energy demands of metabolism during the period of exposure reducing meat weight faster than shell weight.

Food supply is probably the most important single factor in determining growth rate. If food is scarce, then growth is retarded regardless of all other conditions. The literature concerning nutritional aspects of bivalve growth is, however, surprisingly scarce.

Whilst changes in mussel growth often parallel increases in dinoflagellate populations, it has been suggested (e.g. Fox & Coe, 1943; Coe, 1945) that since these are not assimilated they contribute little to the effective ration. Such correlations may simply indicate that both organisms thrive under similar environmental conditions. Boje (1965) assumes that large spiny dinoflagellates are not utilised, rapid growth being determined by the availability of nanoplankton and 'recent detritus'. He found that mussels living under otherwise very similar conditions in the Kieler Fjord and the Nord-Ostee Canal showed very different growth rates according to the amounts of food present. Campbell (1969) and Jensen & Sakshaug (1970) showed that increase in the carotenoid content of mussels coincided with phytoplankton blooms. Since *Mytilus* cannot synthesise carotenoids *de novo* this strongly suggests that they have a mechanism for assimilating them from their food. The nature of the food and the rate of growth of *M. californianus* at different ages and under different environmental conditions over four successive years is reported by Coe & Fox (1942, 1944) and Fox & Coe (1943), who conclude that organic detritus supplies a large part of the mussels' diet, a conclusion also reached by Blegvad (1914) and Fraga & Vives (1960). Stubbings (1954) points out that the proportions of plankton and fine detritus taken by mussels vary according to their supply. There is also some evidence that bacteria may be utilised as food (Zobell & Feltham, 1938), while from more recent kinetic studies (Péquignat, 1973) it appears that *Mytilus* is capable of assimilating dissolved organic solutes present in the environment, though what proportion of their net calorific intake this represents must, for the time being, remain speculative.

Other factors

In addition to the factors discussed above, there are probably many others which influence growth. Brackish estuaries are known to be suitable for mussel growth but this is probably a function of improved feeding conditions rather than lower salinities, which are reported to have a detrimental effect on growth (e.g. Paul, 1942; Lubinsky, 1958; Bagge & Salo, 1967; Thiesen, 1968; Bøhle, 1972). *Mytilus* can, however, survive considerably reduced salinities and this frequently provides substantial protection against less tolerant predators. Mussels growing at 4–5 ‰ have a low growth rate and a maximum length of only about 40 mm (Remane &

Marine mussels

Schlieper, 1971). Bayne (1965) found that optimum salinity for larval growth varied with temperature. Storms and wave action can also be detrimental (Coe & Fox, 1944; Coe, 1945) possibly by interfering with feeding. Harger (1970*a*) showed that whereas growth in *M. edulis* was significantly affected by waves, this was not the case in *M. californianus*, a species better suited to exposed situations. Genetic differences in growth rates are recorded for certain bivalves (Chanley, 1955; Walne, 1958) and a similar explanation may account for growth variation amongst individual mussels living under apparently identical conditions. Pollutants may also be detrimental to growth (Alyakrinskaya, 1967) although mussels are known to flourish near sewage outfalls (Nair, 1962; Bøhle, 1965) possibly through an increase in the potential food supply.

Allometric growth

Since the concept of allometric growth was first introduced it has been extensively applied to many bivalves, including *Mytilus* (Richards, 1928; Coe & Fox, 1942; Fox & Coe, 1943; Coe, 1946; Genovese, 1965; Hancock, 1965*a*; Seed, 1968, 1973; Lavallard, Balas & Schlenz, 1969; Sadykhova, 1970*c*). Fig. 2.5 compares the growth of various parameters relative to shell length and shows that whilst in this instance a more or less isometric relationship exists between length and width, the rate of increase in shell height falls off in the larger size categories. Volume, shell and tissue weights on the other hand, exhibit characteristically exponential relationships. When a logarithmic scale is used, many of these parameters assume a more or less linear function over much of their length. Disproportionate growth in *Mytilus* is reflected in a gradual change in body proportions with increasing size and age, older mussels generally having relatively heavy, elongate shells where width frequently exceeds shell height. Various factors such as growth rate, population density and degree of exposure to wave action also influence differential growth, producing a considerable array of shell forms. The possible adaptive advantages that certain forms may confer, particularly in relation to adverse environmental conditions, are discussed by Lubinsky (1958).

Rao (1953) demonstrated a linear relationship between shell weight and tissue weight in *M. californianus*. Further, he found that shell weight increased, relative to tissue weight, in different populations with decreasing average annual temperature. Rao's (1953) results (also Boje, 1965) illustrate the wide range of variation to be expected within a given species. In addition to their importance in comparative morphometrics, allometric studies also emphasise the care that must be exercised in using individual parameters to measure and compare growth in different mussel populations.

48

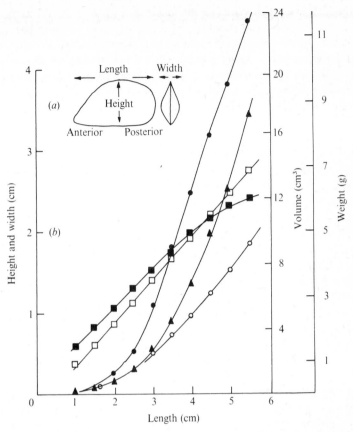

Fig. 2.5. (*a*) Terminology used to describe the mussel shell. (*b*) The relationships between various growth parameters as functions of shell length for a high-littoral population of *Mytilus edulis* at Filey Bay, Yorkshire. ●, shell weight; ○, tissue weight ($\times 10$); ▲, volume; ■, height; □, width. (From Seed, 1973.)

Mortality

The two most critical problems in any investigation of mortality are the construction of survivorship curves and the determination of the major causes of mortality. Relevant population statistics summarised in the form of life tables can be obtained from various sources of ecological data:

(i) From the age at death of a random population sample. This method is not generally applicable to mussels where individuals cannot always be aged accurately and dead individuals are quickly removed from the population.

(ii) From the shrinkage of year classes in standard population samples taken at successive time intervals. Also not applicable to mussels, since polymodal distributions are usually lost due to the overlap of year classes. However, the use of probability paper (Harding, 1949; Cassie, 1954; Harris, 1968) can sometimes be helpful in this respect.

(iii) From directly observing the survival of cohorts of approximately similar age throughout their existence. This technique is particularly useful for sessile organisms like mussels.

Seed (1969b) followed the survival of marked cohorts of mussels at three levels on an exposed rocky shore over a period of one to two years. Mortality in the low shore was severe due mainly to predation, and mussels in this region rarely survived beyond their second or third year. The relative absence of predators at higher shore levels led to enhanced survival and mussels over fifteen to twenty years of age were not uncommon. Nikolskii (1966) obtained some measure of mortality of White Sea mussels from comparisons of size-frequency data of dead and living mussels. Thiesen (1968) estimated mortality of *M. edulis* in a fishery in the Danish Wadden Sea, by observing the amounts and size of mussels laid and refished on the different mussel banks. He concluded that mortality was size-dependent, and calculated the following values for annual mortality rate, M:

Length (mm)	M
25	0.6757
30	0.5631
40	0.4223
50	0.3379

Thiesen (1968) cautioned about extrapolating beyond these lengths for predicting M. Dare (1975) measured the mortality of *M. edulis* at Morecambe Bay (England) at monthly intervals, and derived estimates of mean annual mortality (M) of between 0.937 and 0.977. In this area few mussels survive into their third year.

Both Thiesen (1973) and Dare (1975) comment that the migration of late plantigrades, either away from filamentous primary settlement sites, or as part of a general movement and redistribution on the mussel bed, can suggest a spurious 'mortality'. This may show up in a regular sampling programme as an apparent size-dependent mortality, though it must be considered as emigration.

Whereas survivorship curves are relatively easily constructed for sessile species like mussels, quantifying the actual causes of death is generally more difficult.

Potential mortality factors

Senility

It is suggested elsewhere (p. 41) that longevity may be reduced in areas of rapid growth. Palichenko (1948) cited evidence suggesting that *Mytilus* from the White Sea are longer lived (9–10 years) than those from the North (6–7 years) or Black Seas (4–5 years). Thiesen (1973) recorded *M. edulis* from Greenland as being between 18- and 24-years-old. Such observations suggest an endogenous senescence, i.e. life span fixed in terms of a developmental programme, but the rate of ageing capable of modification by environmental factors. Comfort (1957), however, maintained that in general, closely related species from temperate and tropical regions have similar life spans at the temperatures to which they are adapted. Reports of the life span in *Mytilus* vary from 4–5 to 24 years (Kuznetzov & Mateeva, 1948; Stubbings, 1954; Thiesen, 1973) whilst *Crenomytilus* may attain ages of 50 to 60 years (Sadykhova, 1970a). In the relative absence of predators in the high littoral zone, mussels of 15 to 20 years or more are not uncommon and it is perhaps only in such habitats that senility is likely to be a significant factor in mortality.

Physical factors

Whilst *Mytilus* is extremely tolerant of a wide range of environmental conditions – an important factor in explaining its cosmopolitan distribution – extremes of physical factors such as storms, salinity, excessive silt and temperature are known to contribute to its mortality. These factors are subject to continual variation, however, and it is only when they lead to mass mortality (e.g. Blegvad, 1929; Bøhle, 1965; Dare, 1973b) that their importance has been noted. Storms are known to act in a density-dependent manner on *M. californianus* (Harger & Landenberger, 1971) and since *M. edulis* is less firmly attached than the Californian mussel, the effects of storm damage on mixed populations depend on the proportions in which the two species occur (Harger, 1970c). The shearing power of waves (Harger, 1970a) and the effects of wave-driven logs (Dayton, 1971) contribute significantly to the mortality of *M. californianus* on the west coast of America.

Natural enemies

The natural enemies of mussels fall into four main categories: predators (e.g. crabs, starfish, birds); competitors for food and space (e.g. barnacles, *Crepidula*, tunicates); forms which attack the shell (e.g. *Cliona*, *Polydora*); and parasites (e.g. *Mytilicola*, *Pinnotheres*).

Predators. Of all potential mortality factors, predation plays a major role. Many species feed on mussels, and amongst the most important are gastropods, starfish, crabs and birds. The dogwhelk *Thais* (= *Nucella*) *lapillus* is a widely distributed littoral predator in northern Europe, especially abundant on exposed rocky shores, where it is known to feed extensively on mussels (Fischer-Piette, 1935; Pleissis, 1958; Kitching, Sloane & Ebling, 1959; Largen, 1967; Seed, 1969b). Mussels that have been attacked by dogwhelks can generally be identified by the presence of a small hole drilled through the shell. The distribution of *Thais* on the mussel beds is markedly seasonal – during the winter, few adult whelks are found actively feeding since at this time of the year they aggregate in cracks and pools in order to breed (Feare, 1972). Emergence from winter aggregations, on the Yorkshire coast, occurs in the springtime and over the following months several hundred individuals per square metre can be observed feeding on *Mytilus* in the middle and lower shore. Examination of drilled shells shows that the thinnest parts of the shell are most commonly attacked, especially at the umbonal end and around the region where the adductor muscles are inserted. Laboratory experiments indicate that adult whelks could each consume an average of 2.17 mussels (1–3 cm in length) per week during the summer months whilst immature individuals took an average of 1.01 mussels of this size range (Seed, 1969b). Whether such experiments reflect the normal feeding rate in natural populations is, however, uncertain.

Kitching, Sloane & Ebling (1959) studied the predation of mussels in Lough Ine (Ireland). *Thais lapillus* was the most conspicuous predator of mussels (*M. edulis*) on the open coast nearby, but was absent in the lough proper, probably due to intense predation by crabs. Paine (1971) reported that, rather surprisingly, *Neothais scalaris* seemed to have little influence on *Perna canaliculus* populations in New Zealand, although the gastropod was capable of eating both large and small mussels, and was present on the shore in high densities. He concludes that small *Perna* are more accessible to predators than larger mussels, and these small individuals are the preferred prey of the starfish *Stichaster*. Also, whereas *Stichaster* can attack numerous prey simultaneously, *Neothais* must attack mussels individually. Finally, Paine suggests that by attacking large mussels high in the intertidal zone the gastropod exposes itself to greater physical hazards than does the starfish. The interaction between *Neothais* and *Stichaster* as predators of *Perna* remains to be fully elucidated. Harger (1972a) observed that *Thais emarginata* showed a strong preference for *M. edulis* over *M. californianus* in laboratory experiments (Fig. 2.6) and Paine (1974) implicated *Thais canaliculata* as a major predator of *M. californianus* in the low intertidal. Several other gastropods, such as *Ocenebra* (Chew & Eisier, 1958), *Urosalpinx* (Human, 1971) and *Acanthina, Ceratostoma* and

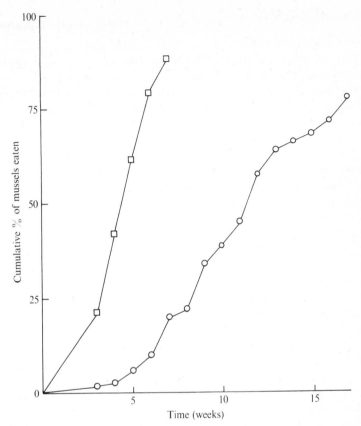

Fig. 2.6. Rate of predation by *Thais emarginata* (20 individuals) acting on a mixed-species clump of mussels (maximum size of mussels 4.0 cm shell length). ○, *M. californianus*; □, *M. edulis*. (Redrawn from Harger, 1972a.)

Jaton (Harger, 1972c) are known to eat mussels. Gastropods are not considered to be important predators of mussels in the Danish Wadden Sea, however (Thiesen, 1968), nor in Morecambe Bay, England (Dare, 1975).

Asteroid starfish are major predators of mussels in many areas. Although *Asterias rubens* is usually present on most rocky shores in northern Europe in low densities, periodically their numbers rise dramatically such that they form a blanket over much of the middle and lower shore. Such areas may then become denuded of *Mytilus* (Seed, 1969b; Dare, 1973b, 1975). Dare (1975) recorded large invasions of *Asterias* onto beds of mussels just above low-water mark in Morecambe Bay (England); starfish densities were up to 200–450 animals m^{-2}, and the swarm, at one time, covered 2.25 ha of ground, and may have cleared up to 4000 t of first-year mussels between June and September. Such swarms of starfish are clearly a major factor in controlling the distribution of *M. edulis* in the low shore and sublittorally.

As a result of some interesting experiments, Hancock (1965a) concluded that *Asterias* had difficulty in opening mussels from Denmark, which had larger adductor muscles than mussels of comparable size from British waters. However, these results might also be explained by differences in chemical attraction. Castilla (1972) has shown that *Asterias* orientates towards *Mytilus* in Y-maze experiments, especially between November and May, but less readily between June and October. He suggested that this seasonal difference may be due to a lowered chemosensitivity, or to seasonal changes in the production of attractants by the prey.

Kitching, Sloane & Ebling (1959) concluded, as a result of transplantation experiments, that the seastar *Marthasterias glacialis* was responsible for preventing the establishment of *Mytilus* sublittorally in Lough Ine (Ireland). Paine (1966, 1969, 1971, 1974) has studied the predation of *Stichaster australis* on *Perna canaliculus* in New Zealand and *Pisaster ochraceus* on *Mytilus californianus* on the west coast of America. In both cases the character of the intertidal community is dependent in part on the predatory activities of the starfish and, in particular, on their preferential consumption of the mussels (Landenberger, 1968; Paine, 1969; Feder, 1970). Removal of the starfish from the shore results in encroachment by the mussels, both vertically downwards and horizontally, into areas not previously occupied, eventually producing a virtual monoculture of mussels (see later). Predation by *Pisaster* and *Stichaster* controls the distribution of the mussels on the low shore.

Oystercatchers (*Haematopus*) feed extensively on *Mytilus* (Webster, 1941; Drinnan, 1958; Tinbergen & Kruuk, 1962; Tinbergen & Norton-Griffiths, 1964; Dare, 1966; Norton-Griffiths, 1967; Heppleston, 1971), particularly over the winter months. This frequently results in heavy mortalities on commercial mussel beds. On exposed shores, however, although mussels are taken in small numbers, oystercatchers seem to feed chiefly on limpets and dogwhelks (Feare, 1971). Sandpipers (Feare, 1966), knot (Prater, 1972), various species of duck (Belopolskii, 1961; Manikowski, 1968; Thiesen, 1968; Nilsson, 1969) and gulls (Oldham, 1930; Rooth, 1957) are also known to feed on *Mytilus*. Milne & Dunnet (1972) record around 70% of the net annual production of a mussel bed passed to bird predators (oystercatchers, eider and gulls).

Laboratory and field experiments show that crabs (*Cancer* and *Carcinus*) can take large numbers of mussels in their diet (Kitching *et al.*, 1959; Seed, 1969b; Walne & Dean, 1972; Harger, 1972c). The results suggest that size selection occurs and that the upper size limit which can be opened is directly related to the size of the crab. Kitching *et al.* (1959) and Ebling,

Kitching, Muntz & Taylor (1964) reported extensive crab predation in Lough Ine and tentatively attributed the absence of mussels sublittorally in many localities to this cause (see also Kitching & Ebling, 1967). The littoral crab population, however, varies seasonally, with an offshore migration into deeper water during winter, (Naylor, 1962). In their experiments with crabs in the Menai Straits (North Wales) Walne & Dean (1972) found that the maximum numbers of mussels were eaten between May and September. Mortality from crab predation is generally most intense in the low shore and sublittoral where crabs are more abundant and where they can feed for longer periods. Since all size ranges of crabs can crush small mussels whilst larger mussels are available only to larger, stronger crabs, a disproportionate mortality amongst smaller mussels is to be expected. Perkins (1967) showed that *Carcinus* would feed on a greater proportion of smaller mussels even though larger ones could be opened without difficulty. He suggested that a learning process may be involved, which enables crabs to feed with the minimum effort. Heavy mortality of plantigrades due to crab predation has been demonstrated by Edwards (1968), G. Davies (1969), Reynolds (1969) and Harger (1972c). Spatfall may be effective only when there are sufficient plantigrades to satisfy the needs of the predators and also provide a surplus to stock the mussel beds. Growth in *M. edulis* would therefore appear to be accompanied by a relative decrease in crab predation. The size at which this occurs is about 40–50 mm shell length, but the age of the mussels of this size will vary with local growth rates. Feeding habits of *Carcinus* were examined by Ropes (1968), who showed that feeding is influenced by abundance, size and type of food. Temperature, tides and time of day also appeared to be important. Walne & Dean (1972) also demonstrated that crabs can discriminate between the size of prey when given a choice, but the numbers eaten could be modified by experimental conditions. Prey density seemed to be important, and a competitive element between the crabs may also have been involved.

Harger (1972c) showed that both *Cancer antennarius* and *Pachygrapsus crassipes* had a preference for *M. edulis* over *M. californianus*. The maximum length of the mussels eaten by both crabs was dependent on the size of the predator. Predation rates were such that the mussels required six to eight weeks from settlement before they become large enough to escape predation by the crabs, and Harger (1972c) concluded that to survive on most rocky shores inhabited by crabs, mussels must settle at densities in excess of 10 000 per square metre. When the two species of mussel occurred together, *M. californianus* was afforded some protection from predation by the presence of *M. edulis*, but the latter species only settled in high enough densities to survive during the summer months.

Apart from the predators already mentioned, various fish (e.g. plaice, flounder) also feed on mussels, especially in flat sandy areas. The grazing activities of limpets (Connell, 1972) and sea urchins may also account for some mortality, particularly amongst young mussels on the low shore. Mammals such as seals, sea otters and walrus are also reported to take limited numbers of mussels in certain localities.

Competitors for space. Intense spatfalls of young plantigrades can constitute a major mortality factor through intraspecific competition (e.g. Savage, 1956; Thiesen, 1968), since the underlying mussels suffocate, thereby loosening the entire population from the surface of the substratum. Under such conditions, large areas are denuded of mussels, especially during stormy weather (Dare, 1975). Small mussels may become attached amongst larger individuals where they find competition too severe and die. Alternatively, the attachment of plantigrades around the bases of adult mussels may afford the former some protection from predators (Kitching & Ebling, 1967). Competition for space can be especially acute in areas of fast growth, and this occasionally leads to 'hummocking', mussels in the centre of the hummock often having no direct contact with the substratum. This in turn may lead to instability, with much of the population being easily torn away during rough seas. However, in terms of general population dynamics this should be considered as emigration rather than mortality, since some of these mussels will survive to colonise other areas. Knight-Jones & Moyse (1961) concluded that intraspecific competition may be more severe in colder than in warmer latitudes, or in difficult environments, where species are few and primary production less. On most rocky shores space is the major resource (Dayton, 1971; Connell, 1972) and competition for space amongst barnacles, algae and mussels may be intense. Under these conditions, mussels are often the competitive dominants (Paine, 1971, 1974). The allocation of space in the intertidal community is discussed later.

Parasites. The numerous parasites which *Mytilus* may harbour are not generally thought to cause substantial mortality, though infected mussels may occasionally show symptoms of disease. Only a brief account of some of the more important parasites will be attempted here; for further details see Cole (1956), Cheng (1967) and Sindermann (1970).

Numerous larval trematodes have been described from *Mytilus* (Nicol, 1906; Lebour, 1912; Jameson & Nicoll, 1913). Whilst the encysted metacercaria do little harm, the presence of rediae and sporocysts may injure the molluscan host. Sporocysts of certain forms, e.g. *Bucephalus*, can damage the gonad and may even lead to castration.

Pinnotheres (pea crabs) are commonly encountered in the mantle cavity of *Mytilus* and whilst their presence does not seem to affect growth or mortality of the host, they could be an important factor when food is in short supply. Wright (1917), however, points out that *Pinnotheres* is rarely encountered in poorly nourished mussels. Hancock (1965*a*) and Seed (1969*c*) found that tissue weights of infected mussels were significantly lower than those of non-infected mussels. The relationship is a parasitic one, the crab often causing extensive gill damage (McDermott, 1966; Seed, 1969*c*). In a related species, *Fabia*, Pearce (1966*a*) found palp damage and mantle blisters to be additional problems. Some species of pea crab are known to be host-specific, but there is considerable variation amongst the Pinnotheridae (Pearce, 1966*a*). The rate of infection is related to the size (= age?) of the host (Houghton, 1963; Seed, 1969*c*). Seed also found differences in infection rates within coexisting populations of *M. edulis* (highly infected) and *M. galloprovincialis* (poorly infected).

Of all the parasites of *Mytilus*, the 'red-worm', *Mytilicola intestinalis*, has received the greatest attention. Much of this work was stimulated after it was thought that the parasite was responsible for the heavy mortality of mussels on the Dutch beds in 1950. First described by Steuer (1902) in *M. galloprovincialis*, *Mytilicola* is a cyclopoid copepod which occurs in the gut, often as many as several dozen being found in a single mussel. It was first recorded in Britain by Ellenby (1947) but since that time it has been shown to be widespread in northern Europe (e.g. Grainger, 1951; Hockley, 1951; Korringa, 1951; Thomas, 1953; Bolster, 1954; Waugh, 1954; Hepper, 1955; Leloup, 1960; Hancock, 1969*a*). Genovese (1958) maintained that *Mytilicola* causes little or no damage in *M. galloprovincialis*. However, it is more generally accepted that the presence of the parasite may lead to loss of condition and even death, although the degree of infestation may be important (Cole & Savage, 1951; Meyer-Waarden & Mann, 1951). Andreu (1963) found an inverse relationship between flesh weight and number of parasites present. Hepper (1955) suggested that infection is not always harmful, particularly if conditions are generally favourable. Infection in more stressful situations, on the other hand, can cause serious harm. Reduced filtration rates by parasitised mussels have been reported by Caspers (1939) and Meyer-Waarden & Mann (1951). Mann (1956) drew attention to the adverse effects on gonad development. Electron microscope studies (Giusti, 1967) have shown *Mytilicola* to be a true parasite causing mechanical removal of the microvillar border of the intestinal epithelium. *Mytilicola* is estuarine, occurring especially in sandy or muddy bays (Meyer-Waarden & Mann, 1954*a*, *b*) where water movement is sluggish and salinity slightly lowered (Vilela & Monteiro, 1958). Campbell (1970) suggested that the amount of silt in the intestine of the mussel may be important in controlling the number of parasites present, and she also maintained that juvenile stages in the hepatopancreas may cause most damage. Andreu (1963) found greatest infestations in areas encountering little mixing with oceanic waters, but suggested that low salinity is probably not a decisive factor. Williams (1967, 1968) and Hrs-Brenko

(1967) found a relationship between size of host and degree of infection; the latter worker also examined some of the factors which might influence the degree of infection and spread of the parasite. Mussels higher in the littoral zone, and those raised from the bottom, are generally less infected since the infective copepodid stage crawls close to the sea bed. Williams (1969a) examined the breeding cycle of *Mytilicola*, especially in relation to temperature, and suggested that the disaster observed in the north European shellfish industry, and in which *Mytilicola* infection was implicated, could have been due to the unusually high sea temperatures.

Production

There have been surprisingly few studies on production by mussel populations. Kuenzler (1961a) studied the production of a population of *Modiolus demissus* in a salt marsh in Georgia, USA. The mean population density was estimated as 7.8 m^{-2}, equivalent to a biomass of organic matter of 11.5 g m^{-2}. There was a net annual population growth of 445 mg m^{-2} (dry body weight) and a net annual mortality loss of 1200 mg m^{-2}. Calculation of the flux of energy in this *Modiolus* population showed a total production of 16.7 kcal m^{-2} yr^{-1}, and total respiration of 39 kcal m^{-2} yr^{-1}, summing to an annual assimilation of 56 kcal m^{-2}. Rather surprisingly, more than half of the total respiration occurred while the mussels were exposed to air. Net growth efficiency for the population (growth/assimilation) was calculated as 25% and reproductive efficiency as 5%. Kuenzler (1961a) suggests that *Modiolus* is a true 'functional dominant' in the salt marsh ecosystem. In a later paper, Kuenzler (1961b) further suggests that the mussel population has an even more significant role in the function of the ecosystem, namely as an agent of the recycling of nutrients. Kuenzler calculated that filtration by the *Modiolus* population contributed to a turnover time of only 2.5 days for suspended particulate phosphorus.

Milne & Dunnet (1972) studied the productivity of an intertidal, estuarine population of *Mytilus edulis* in the Ythan estuary (Scotland). They were able to distinguish three age cohorts, viz. the spatfall of the previous year, the spatfall of two years previously and a composite group of larger mussels from three to ten years old. They measured growth rates and mortality, and calculated production as follows:

Production of survivors at each census $= N \cdot W$
Production of those dying between censuses $= N' \cdot \frac{1}{2}W$,

where $N =$ numbers surviving through the inter-census period, $N' =$ numbers removed during the inter-census period, and $W =$ growth increment. The result was a figure for gross production of 1300 kcal m^{-2} for the period April to September, and an estimated net annual production of 700 kcal m^{-2} (the difference representing body reserves, accumulated in the summer, but utilised in maintenance metabolism and gametogenesis in the winter). Virtually all of this annual production was removed by predators, approx. 70% by birds and 30% by man.

The most recent assessment of production by a mussel population is due to Dare (1975), working in Morecambe Bay, England. Dare based his calculations on monthly data for population survival, rates of growth, length:dry weight relationships, and a mean energy equivalent of 5.5 kcal per g ash-free dry wt. The data suggested that the 'carrying capacity' in one area was approx. 1.2 kg m^{-2} or 6.6×10^3 kcal m^{-2}. Production by two year classes (1968 and 1969) amounted to 2.73 and 3.75 kg m^{-2} yr^{-1} respectively, or 15.01×10^3 and 20.62×10^3 kcal m^{-2} yr^{-1}. The production:biomass ratio was therefore 2.5 to 3. Most production occurred in the first year after settlement, and production had virtually ceased after sixteen months, due to the high rates of mortality that characterise this particular population (pp. 51, 53). Production of organic matter in the shell was 32–34% of organic flesh production; although the organic component of the shell of *M. edulis* is small (probably less than 5% of the total shell weight), production is high by virtue of the large bulk of shell in a population.

Dare (1975) draws attention to the large amount of production that is made available to the decomposers and to the detrital food chains in mussel communities of this sort. The populations studied by Dare had a very considerable rate of turnover due to the, possibly rather unique, ecological conditions, viz. consistent annual settlements, rapid growth and extreme rates of mortality. Under these conditions a very high level of production was presumably utilised directly by decomposers, following mortality of the mussels from physical factors.

In addition, the large accretions of mud deposits around the mussels, composed at least in part of pseudofaeces, represent a considerable source of energy for detritovores. In contrast, Milne & Dunnet (1972) concluded that predation accounted for most of the production of the population that they studied. Presumably, where the physical (and 'biophysical', see Dare, 1975) conditions permit the maintenance of permanent, or semi-permanent mussel populations, near steady-state conditions of energy flow can be established, based on the predation food chain. Other situations can occur, however (see following section on community structure) where rapid growth of the population leads to physical instability of the mussel bed and consequent large fluctuations in biomass. Under these conditions much of the organic production of the mussel population is dissipated to decomposers and detrital food chains. This can total a very considerable amount of organic material; Dare (1975) suggests that the production potential of mussels in Morecambe Bay far exceeds published values for other natural animal populations.

Marine mussels

Community structure

Despite the importance of mussels in many littoral and shallow sublittoral ecosystems, surprisingly little information is available concerning their population and community structure. Several workers have drawn attention to interactions within these communities, but it is only recently that the dynamic nature of these interactions, and their role in controlling community structure, has begun to be understood (Paine, 1969, 1971, 1974; Dayton, 1971; Harger, 1972b; Connell, 1972; Lewis, 1972).

Earlier in this chapter, attention was drawn to the general absence of polymodal size-frequency distributions in *Mytilus*. This is due to the extended period of recruitment, coupled with variable individual growth rates. In addition, slow growth of the majority of the population – especially amongst those individuals ensnared in the byssus threads of larger mussels – frequently produces populations containing an abundance of small mussels (e.g. Richards, 1928; Madsen, 1940; Raymont, 1955; Sadykhova, 1967; Reynolds, 1969). Whedon (1936) and Bouxin (1956) also drew attention to the persistence of small (though not necessarily young?) individuals in the population throughout the year. The relative absence of small mussels reported by other workers (e.g. Mossop, 1921; Lubet, 1961; Meaney, 1970a; Nixon, Oviatt, Rogers & Taylor, 1971) is therefore surprising, but sampling techniques, rapid growth or selective mortality are possible explanations.

The annual recruitment of *Mytilus* is particularly difficult to monitor satisfactorily, not only because the timing and intensity of spatfall is often unpredictable, but also because very heavy spatfalls can sometimes blanket areas of shoreline to the virtual exclusion of all else – occasionally with drastic consequences to the mussel population itself. It has been suggested (Levinton, 1972) that the fluctuations in population size as well as patchy distribution amongst many suspension feeders could be due to their dependence on a relatively unpredictable energy source.

Mussels occur in a variety of shore habitats, from the sediment shores of protected bays and estuaries, through 'gravel' or 'pebble' shores in semi-exposed conditions, to true rocky shores which are exposed to considerable wave action. Mussels may also occur sublittorally, in the natural sediment or attached to pier pilings, buoy chains and the like. Although Shelford (1930) maintained that true succession does not occur in intertidal rocky communities (see also Pequegnat, 1963; Reish, 1964a; Connell, 1972), other workers claim to have demonstrated successional changes, with *Mytilus* representing the climactic biotic condition (Hewatt, 1935; Newcombe, 1935; Scheer, 1945; Dexter, 1947; Millard, 1952; Hosomi, 1969; Wilson & Hodgkin, 1967).

The presence of mussels on any shore drastically modifies the local

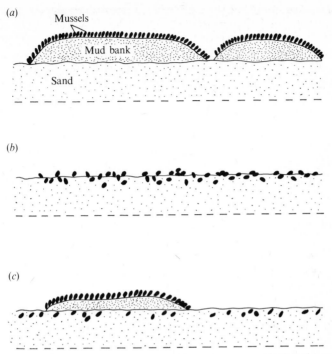

Fig. 2.7. Diagram of the effects of a storm on mussel banks. (*a*) Normal state; mussels lying as a carpet covering low banks of mud. (*b*) Just after the storm; the carpet of mussels has been disrupted and the mussels dissipated over and within the sand bottom, the mud swept away. (*c*) Some time later; the surviving mussels have gathered in carpets and new mud banks are forming, but some individuals have died in the sand. (Redrawn from Thiesen, 1968.)

environment. Water retention between individual mussels allows many species to penetrate further into the littoral zone than would be possible in the absence of mussels. The biodeposition of 'mussel mud', which is a characteristic of most mussel populations, together with the shelter afforded by the mussels themselves, encourage a species enrichment on shores where mussels are present. Furthermore, the mussel shells themselves provide 'secondary space' (Dayton, 1971) for colonisation by many species, including the algae *Ceramium* and *Callithamnion* in European waters. Some of the commoner community associates of *Mytilus* are described by Field (1916), Newcombe (1935), Shelford (1935), Zenkevitch (1963), Reish (1964*b*), Ricketts & Calvin (1965) and Davis & White (1966). Thiesen (1968) describes the destruction and reconstitution of a mussel bed on a sediment shore in the Danish Wadden Sea (Fig. 2.7). The mussels are normally found as a carpet covering a layer of mud which is partly deposited by the mussels themselves (as faeces and pseudofaeces) and partly the result of accretion (Verwey, 1952; Savage, 1956; Baird, 1966).

During storms, however, these banks may be severely damaged. The mud is swept away and the mussels dissipated over the bottom, with some buried within the sand. According to Kuenen (1942), mussels covered by more than 2 cm of sand are unable to regain the surface; these individuals perish. The mussels on the surface of the sediment, however, by crawling and gregarious behaviour, gather together in patches; biodeposition and accretion of mud begins again, and the mussel banks are reconstituted. Dare (1971, 1975) has described a more extreme case of this process of 'biophysical' mortality. In Morecambe Bay, dense settlements of mussels in the winter and early spring, and rapid growth in summer, cause large quantities of faeces, pseudofaeces and washed sand to be deposited in a layer between and beneath the mussels at a rate of around 15 cm per month. To maintain position on the mud surface, these mussels need to keep moving their position upwards, resulting in feeble attachment to one another by long, tenuous byssus threads. Many individuals fail in this competition for space, and are suffocated. Eventually, the entire mussel bed becomes unstable and is ultimately destroyed by autumn and winter gales, so making available, for later settlements, more stable pebble and shell-gravel substrata.

The instability of beds or carpets of mussels is not confined to populations on sediment or pebble shores. Dayton (1971) suggests that the main processes involved with the 'provision, procurement and subsequent utilisation' of space by mussels (*M. californianus*) on a rocky shore are (1) physical stress, (2) competition and (3) predation. Important physical factors are wave exposure, battering by drift logs and physiological stresses such as desiccation and heat (see also Seed, 1969b). In areas where food is abundant, a carpet of mussels up to 25 cm thick can develop. However, relatively few of these individuals are actually attached to the substratum; the majority of the mussels are attached to each other's shell valves and byssus threads. Dayton suggests that such a mussel bed is partially dependent on its spatial continuity for security. If a log (or a predator) causes a clearing within the carpet of mussels, wave action may tear away the remaining mussels from the substratum. These processes can lead to considerable patchiness in the distribution of mussels on the shore.

As indicated earlier, the single most limiting resource in the intertidal zone, especially for sessile organisms, is two-dimensional space. Since *Mytilus* is potentially a competitive dominant in the intertidal community (Paine, 1966) it could, in theory, monopolise this resource. In practice, however, this is prevented by physical disturbances, and by the biological processes of competition and predation.

Harger (1968, 1970a, b, c, 1972a, b, c) has made a thorough study of competition between two species of *Mytilus* (*M. edulis* and *M. californianus*) on the coast of southern California (see also Harger & Landenberger, 1971). If populations on the coast are viewed as a whole, the two species coexist. However, each species possesses a potentially exclusive refuge, *M. californianus* on extremely wave-swept shores, where its very strong byssus threads and thick shells allow it to withstand heavy wave impact, and *M. edulis* in extreme shelter, where its ability to crawl out of clumps of mussels allows it to escape burial by silt. At intermediate points between these two extremes, varying proportions of *M. californianus* and *M. edulis* are found. Harger comments that even though competition can be demonstrated between the two species, 'competitive elimination of one by the other is rare in intermediate environments'. The factors that permit this coexistence can be considered within the general concept of heterogeneity in space and time. Harger (1972*a*) summarises some of these factors as:

(1) A multiplicity of microhabitats presenting different aspects of wave action.

(2) A periodicity in the occurrence of rough weather, exposing new patches of substratum for colonisation.

(3) A species-ratio-dependent effect of such storms.

(4) Structural complexity in the secondary space made available by colonisation of the shore by mussels.

(5) Variations in predation pressure.

Environmental predictability, or constancy, is approached only at the extremes of the spectrum of wave exposure, i.e. within the spatial refuges of the two species. Where environmental predictability is low, and the advantages enjoyed by each species are continually modified, coexistence is possible. Harger (1972*a*) concludes: 'All the active relationships involved have yet to be quantified: nevertheless, the assumption of change in either, or both, wave action and predation pressure in locations outside extremes of shelter and exposure, coupled with variations in proportional representation of species in clumps, together with variation in size and age of constituent individuals, provides an infinitely variable background from which co-existence emerges.' (p. 405.)

Harger's studies are valuable, not only in the broad context of competition and the ecological niche, but also in the more specific circumstances of co-existence between different species of mussels in different geographical areas, e.g. *M. edulis* and *M. galloprovincialis* in Europe, and *M. edulis aoteanus* and *Perna canaliculus* in New Zealand (Morton & Miller, 1968). Some of the points raised by these studies were considered by Ross & Goodman (1974) in a discussion of the factors affecting the vertical intertidal distribution of *Mytilus edulis* in British Columbia, Canada. They concluded that the upper limit was related to the survival of settling plantigrades. Above mean tide level there was active competition for space between *M. edulis* and the barnacle *Balanus glandula* (see also Paine, 1969; Connell, 1972); this resulted in the mussels crawling out of clumps during the summer and being washed away by wave action during autumn and early winter. Below mean tide level, however, *M. edulis* grew more rapidly and developed firmer byssal attachments. Here the mussels eventually smothered the barnacles. This study, and others (Warren, 1936; Kitching & Ebling, 1967; Landenberger, 1968; Seed, 1969*b*; Paine, 1971, 1974) are agreed that the rapid turnover, or even complete absence, of mussels in the lower littoral and shallow sublittoral zones, may usually be attributable to intense predation. On the other hand, stability and longevity of many littoral populations, including *Mytilus*, often prove to be related in an inverse manner to recruitment intensity, growth rate and predation, increasing as one moves progressively upshore. Predators capable of switching from one source of food to another can potentially stabilise the prey species and may lead to the indefinite coexistence of competing prey species, since predation will be heaviest on whichever species is dominant at any particular time (Fischer-Piette, 1935; Landenberger, 1968; Murdoch, 1969).

Paine (1971, 1974) in particular, has made an experimental study of the role of predation in structuring the intertidal community. By means of manipulation experiments in the field, namely the exclusion of the major starfish predators from areas of the shore, Paine has shown, for both *Perna canaliculus* in New Zealand and *Mytilus californianus* in America, that 'predator removal is accompanied both by the development of a virtual monoculture [of mussels] indicating a major reduction in species richness, and by impressive changes in zonation' (1974; p. 113). This competitive superiority by the mussels is based on rapid recruitment and growth, the capacity to migrate from occupied areas into spaces made available either by physical forces or by the absence of predators, and by their ability to detach and re-attach to the substratum with the byssus. Paine suggests that the pattern of zonation on shores normally dominated by mussels may be related to the character and intensity of predation. The upper limit to the intertidal distribution of the mussels is set by little-understood 'physiological factors'. In the middle intertidal, the mussels are capable of excluding other species by successfully competing for the most important limiting resource, which is space. In the absence of the main starfish predator (*Stichaster* in New Zealand and *Pisaster* in America) the mussels dominate the space, and although other species of animals become associated with the mussels, they are not directly involved in space allocation. In the presence of the starfish, the community becomes more complex, because the starfish nullify the competitive advantage of the mussels and so serve to make space relatively more available to other species.

The populations of *M. californianus* studied by Paine (1974) in Washington, USA, have shown a remarkably constant upper and lower limit to their distribution over many years. This band of mussels can be considered to constitute a refuge from predation, existing at the upper fringe of the potentially acceptable distribution. Some individuals survive below the main band, by virtue of their large size; these mussels have escaped predation by growing too large to be eaten by *Pisaster*. Both a spatial and a temporal (in terms of growth) refuge are therefore available to the mussels. Dayton (1971) likewise records that *Mytilus* may escape predation by *Thais* by growing too large to be eaten, although when this does occur it is because the mussels initially numerically swamp the local *Thais* populations; 'escape by growth' follows. Of course, prey-predator interactions may be more complex than this. Menge (1972; see also MacArthur, 1972) maintains that the predictability of food supply both in time and space, probably determines whether a predator will be a food 'specialist' or 'generalist'. The lower is the predictability of supply of the attractive prey, the more general the predator's diet becomes. Predator pressure may therefore be important in causing irregular recruitment of the prey population, so providing an alternative means for survival in a prey species not having temporal or spatial refuge. Paine (1971), after comparing the *Perna/Stichaster* and *Mytilus/Pisaster* interactions, however, concludes that community structure is less dependent on the relative breadth of the predator's diet than on the competitive abilities of the preferred prey species. The influence of the predator is especially pronounced when the prey is competitively dominant and also provides, in its growth, a significant component of spatial heterogeneity. Another interesting, and more indirect, interaction between prey and predator is discussed by Paine (1969) and Dayton (1971). By consuming *Mytilus*, *Pisaster* procures space for the alga *Endocladia*, which is the single most important settling substratum for the mussel plantigrades. In the absence of *Pisaster*, the alga is eliminated by *Mytilus*. The predator therefore functions, to some extent, to ensure the recruitment of its major prey. Further, *Pisaster* will eat limpets (*Acmea* spp.), which in turn graze the *Endocladia*. *Pisaster*, therefore, by including in its diet a secondarily preferred prey has an indirect, positive effect on its major prey.

There is little evidence from these studies that the members of intertidal mussel communities are 'physically controlled' in the sense of Sanders (1968). Paine (1974) concludes that the predictable nature of the dynamic interaction between a major competitive dominant (*Mytilus*) and its major predator (*Pisaster*) results in a 'biologically rich and vibrant community'. This community is particularly amenable to study and experiment (Connell, 1972) and the early period of descriptive shore ecology is now progressively being replaced by a more dynamic study of the interactions that constitute the function of the intertidal communities.

3. Mussels and pollution

The class Bivalvia is of interest to ecologists studying pollution, as it comprises sedentary filter-feeding invertebrates which are likely to accumulate pollutants from the environment. This has long been appreciated for bacterial contamination (Dodgson, 1928) and pollution by heavy metals (Boyce & Herdman, 1897) and, since mussels are major fouling organisms, they have been the subject of many toxicological investigations. However, until recently there have been few studies of the effects of other potential pollutants on this group of animals.

Mussels are tolerant of a wide range of environmental contaminants (Alyakrinskaya, 1967). O'Sullivan (1971) described *Mytilus edulis* as an 'indifferent' species, that is one in which distribution is not greatly affected by pollution.

Fouling, bacterial contamination and shellfish poisoning

Because of the extensive available literature, these topics will be discussed only briefly.

Mussels, particularly *M. edulis*, are amongst the most important fouling animals in seawater cooling culverts in temperate latitudes (Anonymous, 1952; Holmes, 1970). They are also amongst the most difficult of fouling organisms to control, frequently withstanding conditions which eliminate other species (Clapp, 1950). Among the various methods used to control mussel fouling, low-level continuous chlorination is probably the most widely used. Chlorination is effective in weakening the mussels' byssal attachment, mainly by depressing pedal activity, but also by interfering with the quinone tanning process of thread formation (Holmes, 1970).

Because of their coastal and estuarine distribution, commercially exploited mussel beds are often subject to sewage pollution. The main risk associated with such pollution is the presence of bacterial and viral pathogens, particularly when the mussels are eaten raw, or only lightly cooked (Wood, 1972). Mussels may also be contaminated with bacteria of non-faecal origin; *Vibrio parahaemolyticus*, a widely distributed free-living species, is responsible for up to 20% of all cases of food poisoning in Japan. The major bacterial pathogens are the *Salmonellae*, which include the organism responsible for typhoid fever; the *Shigella* species which can cause dysentery; and certain species of *Clostridium* which can produce exotoxins pathogenic to man (Wood, 1972). Although viral pathogens are present in both raw and treated sewage, the only well-documented viral infection associated with shellfish is infective hepatitis (Mason & McLean, 1962).

Another risk associated with the consumption of mussels is 'paralytic shellfish poisoning'. This is associated with the ingestion by the mussel of certain dinoflagellates, notably *Gonyaulax veneficium*, *G. catanella*, *G. tamarensis*, *Gymnodinium breve* and *Pyrodinium pheneus*. The chemical and physical properties of paralytic shellfish poison (PSP), the toxin isolated from shellfish tissue, and saxitoxin, isolated from *G. catanella*, are almost identical (Steidinger, Burklew & Ingle, 1973). *Mytilus californianus* is killed by high concentrations of *G. breve*, whereas *Crassostrea virginica* (the American oyster) can accumulate *G. breve* toxin and suffer no mortality (Sievers, 1969; Steidinger *et al.*, 1973). PSP and saxitoxin are both water-soluble, and if the liquor is discarded when shellfish are cooked, the risk of human fatality is reduced (Steidinger *et al.*, 1973). The extensive literature on naturally occurring marine biotoxins has been reviewed by Halstead (1965).

Oils and detergents

Molluscs which have been exposed to sublethal concentrations of hydrocarbons accumulate these in their lipid pools (Stegeman & Teal, 1973). Lee, Sauerheber & Benson (1972) found that *M. edulis* assimilated greater quantities of paraffinic hydrocarbons (mineral oil and heptadecane) than aromatic compounds (toluene, tetralin, 3,4-benzopyrene and naphthalene), although autoradiography indicated that none of these compounds was metabolised. Hydrocarbons were rapidly taken up by the gill tissues and, after longer periods, high concentrations were found in the alimentary canal. Lee *et al.* (1972) suggested that the digestive gland was the site of hydrocarbon storage; they also found that on transfer to clean seawater, after previous exposure to hydrocarbon solutions, 80–90% of the hydrocarbon was eluted. In the case of mineral oil, the short-chain fractions were eluted more rapidly than others. Similarly, Stegeman & Teal (1973) found that *Crassostrea virginica* rapidly, but incompletely, discharged accumulated hydrocarbons when transferred to clean seawater. Fossato & Siviero (1974) recorded that *Mytilus galloprovincialis* accumulated aliphatic hydrocarbons from the water in the Lagoon of Venice. The different hydrocarbon contents of the mussels reflected the concentrations in the water, and the authors suggested that *M. galloprovincialis* might be utilised as a 'self-integrating index' of hydrocarbon pollution.

Crude oil does not appear to be highly toxic to mussels. Smith (1968) reported that *M. edulis* could survive in 1000 mg crude oil per litre for 24 h, but that there was no byssal attachment. Alyakrinskaya (1966) found that *M. galloprovincialis* could tolerate up to 2% oil in seawater. Zitko (1971)

Table 3.1. *A comparison of the toxicity to* Mytilus edulis *of several surfactants*

Compound tested	24 h LC_{50} (mg l^{-1})	Temperature (°C)
Slickgone	640	6.0–7.5
Petrofina	530	6.0–7.5
BP 1002	290	6.0–7.5
BP 1002+oil (ratio 2:1)	81	8.0–10.0

After Perkins (1968).

recorded between 77 and 103 μg fuel oil per gram in tissues of *M. edulis* collected from the site of a recent major oil spill. Lee *et al.* (1972) found that although mineral oil appeared to have no effect on mussels, aromatic compounds were more toxic than paraffinic hydrocarbons and suggested that the former may inhibit filtering ability. They found that tetralin, even in relatively low concentrations, caused paralysis and prevented shell closure. Zitko (1971) also emphasised that the aromatic fractions of oils are likely to have a greater effect on aquatic fauna than aliphatic compounds.

Although oil is only slightly toxic to mussels, it can affect the marketing of the animals by tainting, which may arise by oil contamination of the shells, by ingestion of oil droplets, or by the uptake of dissolved fractions of the oil (Simpson, 1968; Brunies, 1971).

Davis (1961) investigated the effects of two oils, orthodichlorobenzene and trichlorobenzene, on clam eggs and larvae. Trichlorobenzene was more toxic to eggs than orthodichlorobenzene, but even at 10 mg l^{-1} over 50% of the eggs developed normally. Neither of these two oils had an appreciable effect on survival and growth of clam larvae at concentrations up to 5 mg l^{-1}.

In contrast with crude oils, detergents and oil emulsifiers may be highly toxic to marine species, and their indiscriminate use during oil spills may have resulted in greater damage than was caused by the original pollutant. It is suggested, for example, that low levels of detergent may inhibit the natural cleansing mechanisms of mussels and so reduce their tolerance of oil (Alyakrinskaya, 1966). Crapp (1971*a, c*) demonstrated in field trials that *M. edulis* was unaffected by oiling but severely adversely affected by emulsifiers, and suffered high mortalities.

The main problem in the use of adult mussels in toxicity studies is the readiness with which the animal closes the shell when stimulated by an obnoxious substance, or by any deleterious change in the environment (Simpson, 1968; Crapp, 1971*b*). This leads to very variable assessments of

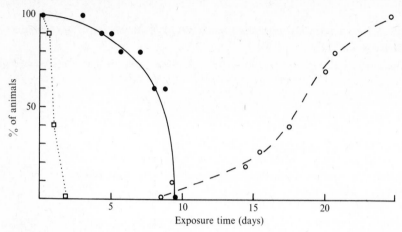

Fig. 3.1. Byssal thread formation (□), shell-closure ability (●) and mortality (○) in *Mytilus edulis* exposed to the surfactant NP10EO, 5 ppm at 6–8 °C. (From Swedmark *et al.*, 1971.)

toxicity. Foret-Montardo (1970) reviewed work on the biological effects of detergents, and studied the toxicity of thirty-seven detergent compounds to *M. galloprovincialis* taken from polluted and unpolluted waters. He suggested that at high concentrations mussels may detect the detergent and shut their valves. He also found that mussels from polluted waters were less sensitive than those from unpolluted areas. Of the major groups of compounds tested by Foret-Montardo (1970), cationic detergents were the most toxic, and anionic the least toxic, to mussels; non-ionic detergents were of intermediate toxicity. The LC_{50} values of several surfactants to *M. edulis*, as determined by Perkins (1968) are shown in Tables 3.1. Swedmark, Braaten, Emanuelsson & Granmo (1971) studied the effects of various surfactants on some marine animals, including *M. edulis*. Non-ionic surfactants were more toxic than anionic compounds; eggs and veliger larvae were more sensitive than adults. Swedmark *et al.* (1971) also examined the chronic toxicity and the sublethal effects of the surfactants. For *M. edulis*, byssal activity and ability to close the shell were useful indicators of sublethal effects (Fig. 3.1), and heart beat may be even more sensitive to surfactants than byssal activity. Swedmark *et al.* suggest that the surfactants accumulate on gill surfaces and so impair gas exchange and ionic regulation.

The toxicity of oil emulsifiers may be reduced by the evaporation of the toxic components. Perkins (1968) recorded without explanation a reduction in toxicity to mussels by BP 1002 kept in stoppered bottles. Smith (1968) found that mussels survived 24 h of exposure to 5 mg l^{-1} BP 1002, although there was a 40% reduction in byssal attachment. Exposure to 10 mg l^{-1} was lethal within 24 h. Griffith (1972) observed that a mixture of oil and Dispersol was initially very toxic to *M. edulis* at 4.6 °C, although there was a subsequent phase of partial recovery followed by a progressive mortality. At higher temperatures, toxicity was reduced. Griffith suggested that at low temperatures a volatile fraction of the Dispersol might persist.

The growth and survival of clam and oyster veligers was reduced by surfactant concentrations between 0.01 and 5.0 mg l^{-1}, depending on the compound (Hidu, 1965). Cationic surfactants were, in general, the most toxic and non-ionic the least toxic to these larvae; anionic compounds were of intermediate toxicity. Although the more recently developed LAS (linear alkylate sulphonate) compounds are at least as toxic as ABS (alkyl benzene sulphonate) compounds to oyster larvae, they are almost completely degraded by effective sewage treatment and thus considered safer to use (Calabrese & Davis, 1967).

Metals

The concentrations of several trace metals in mussels have been investigated by Vinogradov (1953), Brooks & Rumsby (1965), Segar, Collins & Riley (1971) and Graham (1972). Results of some of these analyses are shown in Table 3.2. There have also been studies on the concentration, in mussel tissues, of particular metals suspected of having metabolic significance or of presenting a potential pollution hazard (Lopes & Lorenzo, 1957; Molins, 1957; del Vecchio, Valori, Alasia & Gualdi, 1962; Hobden, 1967, 1969; Jones, Jones & Stewart, 1972; Schultz-Baldes, 1973).

Studies on the uptake of metals by mussels include those by Sautet, Ollivier & Quicke (1964) on the assimilation and elimination of arsenic; Hobden (1967) and Winter (1972) on the assimilation of iron; Pringle, Hissong, Katz & Mulawka (1968) on the uptake of cadmium, copper, iron, manganese and zinc; Majori & Petroni (1973*a*, *b*, *c*, *d*, *e*) on the accumulation and loss of cadmium, copper, lead and mercury; and Schultz-Baldes (1974) on lead. In addition, the accumulation from seawater of ^{65}Zn, ^{54}Mn, ^{59}Fe and ^{58}Co by *M. edulis* have been studied in relation to stable element levels by Pentraeth (1973). Hobden (1969) suggested that iron uptake could occur either directly into the blood across the gill or mantle epithelia, or by way of the normal feeding mechanism but, since the site of greatest accumulation varied for different metals, more than one system was probably involved. The entry of metals into the mussels will be affected by the physiological condition of the animal, by environmental factors such as temperature and salinity, and by the physicochemical form of the metal. These will all affect, in turn, the relative toxicity of the metal. Tabata (1970) found that the toxicity of heavy metals was reduced by Ca^{2+}, Mg^{2+}, CO_3^{2-} and OH^- ions, and he suggested that metallic precipitates may not be harmful to aquatic animals. However,

Table 3.2. Concentrations of metals in mussels (M. edulis) from different areas

Locality of sampling	Concentration (µg g^{-1} dry wt)													Tissue	Author
	Fe	Mn	Co	Ni	Cd	Cu	Pb	Zn	Ag	Cr	Al	V	Mo		
South coast, England	290	3.6	0.15	2.1	0.95	2.0	0.4	0.04	0.1	—	76.0	—	—	Shell	Segar et al. (1971)
South coast, England	1700	3.5	1.6	3.7	5.1	9.6	9.1	91.0	0.03	1.5	1230	—	—	Soft parts	
South coast, England	—	—	—	—	—	5–26	—	60–81	—	—	—	—	—	Soft parts	B. Brown (personal communication)
East coast, England	—	—	—	—	—	9.50	—	167–312	—	—	—	5.0	—	Soft parts	
New Zealand	1960	27.0	—	7.0	10	9.0	12	91.0	1.0	16.0	—	—	0.6	Soft parts	Brooks & Rumsby (1965)
California	—	6–28	—	—	3–7	5–11	2–8	204–341	1–1.3	2.8	—	—	—	Soft parts	Graham (1972)
California	—	9.3–45.8	—	—	2.5–5.8	5.8–8.6	9–21.2	6.9–14.2	4.4–6.3	5–7	—	—	—	Shell	Graham (1972)
California	—	5.9–7.8	—	—	2.0–4.9	9.0–303	2.2–23.4	164–310	1–5.5	1.5–7.8	—	—	—	Soft parts	Graham (1972)
California	—	8.4–14.2	—	—	2.5–9.2	8.1–18.6	9–19.4	8.6–26.5	5.0–7.9	5.7–14.2	—	—	—	Shell	Graham (1972)

a precipitate of zinc was toxic to *M. edulis* (Tabata, 1970) and marked effects in mussels exposed to ferric hydroxide flakes have been reported by Winter (1972). Schultz-Baldes (1974) found that the uptake of lead into the tissues of *M. edulis* was a linear function of the lead concentration in the medium; when the animals were subsequently placed in clean seawater the rate of loss was linearly dependent on internal concentration. The uptake of lead was most marked into the kidney. Schultz-Baldes presents a 'calibration curve' for relating lead concentrations in mussels and in seawater, for use of the mussels as indicators of lead pollution.

Heavy metals, even those essential in trace amounts, are toxic to marine organisms at relatively low concentrations. However, there is surprisingly little information on the lethal doses for mussels (Marks, 1938; Clarke, 1947; Wisely & Blick, 1967; Newell & Brown, 1971; Scott & Major, 1972). Wisely & Blick (1967) reported that for the larvae of *M. edulis planulatus* the LC_{50} values for Hg^{2+} and Cu^{2+} at 19 °C were 6.5×10^{-5} M and 3.5×10^{-4} M respectively. However, Wisely (1963) had earlier concluded that the anti-fouling action of cuprous oxide paints on mussels was due to a repellent rather than to a lethal effect.

Investigations into the sublethal effects of metals on mussels include those by Shapiro (1964), Brown & Newell (1972), Winter (1972) and Scott & Major (1972). Winter (1972) found that although iron hydroxide flakes were not acutely toxic to mussels, they caused a decline in body weight. He suggested that this could be due to the following three factors: (i) rejection of potential food together with the flakes in pseudofaeces; (ii) organic loss due to increased mucus production; (iii) further organic loss when excretory spheres were formed in the mid-gut to remove the indigestible flakes. Cupric ions were shown to cause respiratory and cardiovascular depression in *M. edulis*, although live animals and heat-killed tissue homogenates could detoxify solutions of limited copper content (Scott & Major, 1972). Scott & Major suggested that this was due to the passive binding of cupric ions with organic ligands. Shapiro (1964) also recorded a reduction in the oxygen consumption by *M. galloprovincialis* in seawater containing magnesium, copper, nickel and molybdenum ions, and he thought this might be due either to valve closure or to a direct metabolic effect. Brown & Newell (1972) overcame the problem of shell closure experimentally by severing the posterior adductor muscle of *M. edulis*, but they still observed reduced rates of oxygen consumption in solutions contaminated with copper and zinc. They concluded that a reduced energy demand was caused by a suppression of ciliary activity rather than by a direct effect on respiratory enzymes. The period of recovery of *M. galloprovincialis* on transfer to clean seawater, after previous exposure to metallic toxins, was always marked by an increase in the rate of oxygen consumption similar to that known to occur in many bivalves after

73

exposure to air (Shapiro, 1964). However, the increased oxygen consumption rate continued for a longer period after exposure to molybdenum in seawater than after exposure to air. Shapiro (1964) concluded that the suppression of respiratory activity, followed by increased oxygen demand on the amelioration of the external conditions could not be explained solely as the repayment of an oxygen debt, and he suggested that a long period of increased oxygen consumption was necessary to restore normal metabolic processes damaged by the metals.

Comparisons made by Calabrese & Nelson (1974) suggest that the sensitivity of different species of bivalve embryos to heavy metals may be quite variable. For example, embryos of *Mercenaria mercenaria* and *Crassostrea virginica* were more sensitive to mercury and zinc than embryos of *Mytilus* species (Okubo & Okubo, 1962; Calabrese, Collier, Nelson & MacInnes, 1973; Calabrese & Nelson 1974). Wisely & Blick (1967) found that the larvae of *M. edulis* exposed to 13 mg mercury per litre suffered 50% mortality within 2 h and concluded that this high resistance was due to withdrawal into the shells.

Radioactive wastes

As a result of nuclear weapon tests and the growing use of nuclear reactors, the problem of radioactive wastes is becoming increasingly important (Lucu, Jelisavcic, Lulic & Strohal, 1969; Wolfe & Schelske, 1969). The extensive literature on the radioecology of aquatic organisms has been summarised and reviewed by Polikarpov (1966). Rice & Wolfe (1971) have listed the naturally occurring and artificial radionuclides, together with their half-lives.

The uptake by animals of radionuclides that are released into the marine environment can take place directly by adsorption or indirectly by the intake of contaminated silt or food. Mussels growing on rocks or pilings are useful for detecting minute changes in certain radioisotopes in the environment; even before suitable methods for direct analysis of seawater became available, labelled compounds were detected in *M. californianus* (Young & Folsom, 1967). Relatively high levels of ^{65}Zn at the mouth of the Columbia river (USA) were attributed by these authors to the activation of stable zinc in river water circulated through nuclear reactors. ^{65}Zn appears to be concentrated in the kidneys of both *M. californianus* (Young & Folsom, 1967) and *M. edulis* (Van Weers, 1973). Within the soft parts of *M. edulis* ^{60}Co was found to be concentrated in the kidney and digestive gland (Van Weers, 1973). The digestive gland was also found to be the site of highest concentration of ^{59}Fe in *M. edulis* (Hobden, 1969) and of ^{233}Pa in *M. galloprovincialis* (Lucu *et al.*, 1969).

Variable and often high fractions of radioactive elements may also be assimilated by the shell and byssus of mussels. This has been reported for ^{233}Pa, ^{58}Co, ^{60}Co and ^{59}Fe (Lucu *et al.*, 1969; Hobden, 1969; Van Weers, 1973; Shimizu, Kajihara, Suyama & Hiyama, 1971). According to Shimuzu *et al.* (1971), experimental evidence suggests that these isotopes are bound in the structural proteins of the periostracum and byssus. Several factors have been shown to affect isotope uptake; for example salinity, but not temperature, was found to influence the rate of ^{137}Cs uptake by *M. galloprovincialis* (Lucu & Jelisavcic, 1970). The uptake of ^{106}Ru was dependent on the physicochemical nature of the isotope, chloride complexes being assimilated more rapidly by *M. galloprovincialis* than nitrosyl-nitrato complexes (Keckes, Pucar & Marazovic, 1967). Since the site of greatest accumulation varies for different isotopes, it seems likely that more than one uptake system is involved, and this may be dependent on the potential metabolic role of the element. However, since discrepancies are recorded between estimated and expected equilibration periods, it is possible that only part of the stable element is available for exchange with the radionuclide (Van Weers, 1973).

According to Polikarpov (1966), the specific feature that distinguishes the biological effect of ionising radiation from that of chemical contaminants is the absence of a threshold; that is, living systems are affected by any radiation dose, however small. However, the information relevant to marine invertebrates is scant, despite the fact that damaging and lethal effects of radiation have long been recognised. Available data for bivalves suggests that the median lethal dose at 30 days is high compared with other groups (Woodhead, 1971), being 100 krad for *C. virginica* and 110 krad for *M. mercenaria* (Price, 1962). The amount of information on the chronic effects of irradiation is negligible, although it is usually assumed that there is no safe radiation dose from the point of view of genetic damage. Chipman (1972) has reviewed some of the effects of ionising radiation on animals.

Halogenated hydrocarbons

It is convenient to group non-pesticides such as polychlorinated biphenyls (PCBs) with the halogenated pesticides, since many of their properties, and in particular their environmental persistence, are similar. Table 3.3. shows the concentration of organochlorine residues reported for *Mytilus edulis* from different sites in North America and Europe. The highest reported concentrations are for DDT in animals from California, probably reflecting the intensive agricultural use of the compound there. *M. edulis* was used by Koehman & van Genderen (1972) as an indicator organism for persistent organochlorine residues in the North Sea. They recorded highest residues near the mouth of the Rhine and suggested that these were derived from

Table 3.3. *Organochlorine residues in the soft parts of mussels (M. edulis) from different areas*

Locality and date of sampling	Concentration in mussels (μg per g fresh wt)						Author
	HEOD	DDT	DDE	Telodrin	Endrin	PCBs	
Northumberland coast, England, 1965	0.023	—	0.024	—	—	—	Robinson *et al.* (1967)
North Sea, 1966	0.01–0.1	—	0.01–0.04	0.001–0.004	0.01–0.36	—	Koehman *et al.* (1968)
Clyde, Scotland, 1966	0.1–0.3	—	—	—	—	0.2–0.8	Holdgate (1971)
West coast, Scotland 1966	0.08–0.1	—	—	—	—	0.1	Holdgate (1971)
East coast, Scotland 1966	0.06–0.16	—	—	—	—	0.01–0.3	Holdgate (1971)
East coast, England 1968	0.02–0.14	—	—	—	—	0.01–0.2	Holdgate (1971)
California, 1967–72	0.005–0.031	0.833–3.97	—	—	Detected	Detected	Butler (1973)
Maine, 1965–70	—	0.064–0.359	—	—	—	—	Butler, (1973)
New York, 1966–72	0.075–0.104	0.112–0.588	—	—	—	—	Butler (1973)
Canadian Atlantic coast, 1967	—	0.01–0.04	0.05–0.06	—	—	—	Sprague & Duffy (1971)

insecticide manufacture, although agricultural run-off from the Rhine basin was not excluded. In recent years there has been a reduction in the use of persistent organochlorines and a corresponding decline in detectable residues in the USA, after a peak in 1968 (Butler, 1973).

Differences exist in the rate of uptake of chlorinated pesticides by different species, and it has been suggested by Butler (1971) that mussels assimilate these compounds more readily than many other species. Brodtmann (1970) suggested that their tendency to flush out pesticides from their tissues in clean water renders bivalves questionable 'environmental integrators', but Butler (1971) maintained that this tendency, together with their sedentary nature and rapid assimilation of pesticides makes them superior to fish and inanimate substrates in monitoring pesticide pollution in the marine environment.

Laboratory studies of the assimilation of pesticides by bivalves include Brodtmann (1970) and Lowe, Wilson, Rick & Wilson (1971) on *C. virginica*, Butler (1971) on many species including *Mya arenaria* and *Mercenaria mercenaria*, and by Roberts (1972) on *M. edulis*. Brodtmann (1970) suggested that the gills were the primary site of entry of DDT into *C. virginica*. Butler (1971) commented that a possible cause of variations in pesticide uptake was the amount of body surface available for their sorption. Roberts (1972) summarised possible uptake mechanisms; he concluded that since the major site of concentration of Endosulfan (an organochlorine) in *Chlamys opercularis* and *M. edulis* appeared to be the gut and digestive gland, concentration of such compounds results principally from the ingestion of pesticide adsorbed on food. This would also account for the rapid elution recorded on transfer to clean seawater. Butler (1966) on the other hand, found that the gonad stored approximately twice as much DDT as the digestive tract and associated organs, and Lowe *et al.* (1971) used this evidence to suggest that a drop in residue level was due to spawning.

Initial investigations into the toxicity of pesticides to bivalves estimated lethal levels (Portmann, 1968, 1970). More recently, however, there have been a number of investigations into the chronic effects of pesticides. Oysters, for example, have been shown to respond to low levels of chlorinated insecticides by changes in shell movements and ventilation rate (Butler, 1965, 1966). Lowe *et al.* (1971) demonstrated that the mean weight of oysters that were exposed to a mixture of pesticides (DDT, Toxaphene, Parathion) was not significantly different from control oysters until after twenty-two weeks of exposure. After thirty-six weeks the difference represented about 10% of the total body weight. Butler (1971) exposed young oysters to pesticide concentrations approximately one-tenth the amount required to cause a 50% decrease in shell growth in 96-h bioassays, and he found that after six months some individuals were larger than the controls, although median growth and mortality were not statistically different.

A condition index (wet flesh weight as a percentage of total wet weight) of *M. edulis* was observed to decline more rapidly with increasing concentrations of Endosulfan (Roberts, 1972). This was possibly due to reduced feeding efficiency, or to other metabolic effects, such as those reported by Engle, Neat & Hillman (1972), who suggested that sublethal doses of chlorinated pesticides may lead to metabolic alteration indicative of an increase in glucose degradation and a suppression of gluconeogenesis in *Mercenaria mercenaria*. Byssus formation in *M. edulis* and *Chlamys opercularis* is reduced in increasing concentrations of chlorinated pesticides and PCBs (Roberts, 1973). This is probably due to reduced pedal activity, although direct interference with the synthesis or combination of different components of the byssus is possible.

Larvae which are produced from adult oysters exposed to 1.0 μg DDT per litre may contain 20–30 μg DDT per gram (Butler, 1966). Davis (1961) had earlier demonstrated 100% mortality of oyster larvae exposed to 1.0 mg DDT per litre. The effects of pesticides on embryos and larvae of *M. mercenaria* and *C. virginica* have been investigated by Davis & Hidu (1969). They found that most of the compounds tested affected embryonic development more than the survival and growth of the larvae. However, some compounds drastically reduced larval growth at concentrations which had little effect on embryonic development. Ukeles (1962) showed that some of the best algal foods for bivalve larvae were considerably less tolerant of some pesticides than the larvae themselves, so that larval growth might be indirectly affected in nature. Lindane was the least toxic, to bivalve eggs and larvae, of the insecticides tested by Davis (1961); approximately 60% of clam eggs and 43% of oyster eggs developed normally at concentrations up to 10 mg l^{-1}, which is essentially a saturated solution. Guthion and Parathion significantly reduced the growth rate of oyster larvae at 1.0 mg l^{-1}, but lower concentrations (0.025 mg l^{-1}) had a stimulatory effect on growth.

Other pollutants

In addition to those pollutants already discussed, physical wastes such as heated water and inert solids, as well as a variety of domestic and industrial wastes of ill-defined chemical composition, may constitute pollution of the environment.

Information on the effects of thermal effluents on marine and estuarine organisms has been reviewed by Kinne (1963) and Naylor (1965). Physiological and biochemical effects of thermal stress on *M. edulis* have been investigated by Bayne (1973a) and Gabbott & Bayne (1973). The genus *Mytilus* has wide powers of thermal acclimatisation (Rao, 1953; Widdows & Bayne, 1971). Temperatures between 26.7 and 28.9 °C were found to be lethal for *M. edulis* by Read & Cumming (1967). Tidal effects on thermal discharges may result in animals being subjected to periodic fluctuations in temperature rather than a continued high temperature. *M. edulis* adults survive over 1000 h exposure to 30 °C in six-hourly cycles, but can only survive 9–12 h continuous exposure to the same temperature (Pearce, 1969). The effects of such cyclical changes in temperature on reproduction are not known. However, the larvae and plantigrades of mussels were far more tolerant of increased temperature than were the adults (Pearce, 1969). Growth and survival of mussel larvae are discussed further in Chapter 4, and temperature as a lethal factor in Chapter 5.

Large amounts of suspended silt in the water adversely affect filter-feeding molluscs (Wilber, 1969). *M. edulis* died when kept in turbid waters for some weeks (Loosanoff, 1962). However, Loosanoff's experiments were carried out at 20 °C, which would have imposed an additional stress. Bivalve larvae are also affected by suspended silt; survival of oyster larvae exposed to 0.25 g silt per litre was reduced to 73% of the survival in clear water (Galtsoff, 1964).

The operation of a paper-pulp mill results in the discharge of large volumes of wastes, including sulphites, sulphates and particulate matter, which can cause excessive turbidity and oxygen depletion in the receiving waters (Wilber, 1969). Early investigations into the ecological and physiological effects of pulp-mill effluents on bivalves include those by Galtsoff, Chipman, Engle & Calderwood (1947), who found that the effluents affected shell movements in oysters, with resultant reduction in feeding efficiency. Paper pulp mill effluents were also found to induce spawning in both *M. edulis* and *M. californianus* in the laboratory (Breese, Milleman & Dimick, 1963). However, it is unlikely that the concentrations of these effluents would be high enough to trigger spawning of mussels in their natural habitats.

Conclusions

There is a lack of ready comparability of much of the data on mussels and pollution, largely due to a failure to find simple methods of describing pollutant levels and their effects. There is common use of the word 'pollution' to describe a wide range of environmental contaminants of ill-defined composition. Schafer (1963) recorded differences in the free amino acid content of muscle tissue of *M. edulis* collected from 'polluted' and 'non-polluted' waters. However, the exact nature of the pollutants was not defined, although the presence of oil was mentioned. Moreover, much of the data are fragmentary, of mainly academic interest and of little

4-2

direct value in the control of pollution (Reish, 1972). Nevertheless, information on the levels of various pollutants in the environment, and in a wide variety of marine species, including mussels, is gradually being accumulated and will be of value in establishing base-line data for monitoring programmes. In spite of their tendency to close their shell valves when experimentally subjected to a pollutant, mussels will continue to be used as 'indicator organisms' and in toxicity tests. Several indices of the toxic effects mentioned above appear promising for future studies. These include the extent of byssal attachment, growth and 'condition' measurements, larval development and various physiological indices of stress (see Chapter 7). Future research should use these indices for assessing the effects of insidious pollutants such as organohalines, various heavy metals, and the possible synergistic and antagonistic effects of mixtures of different pollutants. The guidelines suggested by Sprague (1969, 1970, 1971) for pollutant studies on fish should be followed with modification, if necessary, for use with bivalves. In addition, as Lewis (1972) has pointed out, it is important that effects at the community level be examined. The widespread distribution of the Mytilidae, and particularly of the genus *Mytilus*, together with their epifaunal communities, suggest that mussels are well-suited for large-scale studies of this kind.

4.   The biology of mussel larvae

B. L. BAYNE

Lamellibranch larvae comprise a significant component of the coastal meroplankton. Lebour (1933, 1938) discussed the importance of lamellibranch larvae both as competitors with other zooplankton for food and as themselves providing food for fish and invertebrate carnivores. Thorson (1946) recorded that about 57% of all invertebrate larvae in the Oresund were lamellibranchs (> 3000 m^{-3}) and they dominated the meroplankton in summer and in winter. Rees (1951) counted maxima of 3000 lamellibranch larvae per cubic metre in the North Sea, and Zakhvatkina (1959) found these larvae to be dominant in the zooplankton of the Bay of Sebastopol (Black Sea) in September, comprising 4–9% on average (maximum 19%) of the biomass of all plankton organisms. Mileikovsky (1968, 1970) recorded 3–200 lamellibranch larvae per cubic metre in the Norwegian Sea (surface to 50 m) and up to 1500 larvae per cubic metre in Velikaya Sound (White Sea).

The planktonic larvae of species of *Mytilus* have often been recorded as dominant, in number, over all other lamellibranch larvae. Fish & Johnson (1937) recorded a peak abundance of > 25 000 *M. edulis* larvae per cubic metre in August (1932) in the Bay of Fundy. Legaré & Maclellan (1960) later found these larvae to be a significant component of the food of herring in this area. The larvae of *M. edulis* have been observed to dominate the meroplankton by Lebour (1938; Plymouth, England); Jørgensen (1946, Øresund); Rees (1954; North Sea) and Schram (1970; Oslofjord). Schram counted up to 40 000 *M. edulis* larvae per cubic metre. Hrs-Brenko (1971, 1973) studied the occurrence of *M. galloprovincialis* larvae in the Adriatic, and reviewed other studies of these larvae in the Mediterranean.

Rees (1954) discussed the distribution of *M. edulis* larvae in the North Sea in detail. Large concentrations of larvae in the middle of the North Sea appeared to have originated on the northern coasts of Britain and dispersed some hundreds of kilometres to the south-east by the dominant current streams of mixed Atlantic and North Sea Water. In 1950 there was a prolonged supply of these larvae into the North Sea, and the absence of any large adult populations in the area not only supports Rees' argument for a coastal origin of these larvae, but also suggests that their dispersal represented a wastage to the adult stocks. In other areas (e.g. the Norwegian and Barents Seas; Mileikovsky, 1968) the distribution of larvae may be more closely related to the distribution of the adults, and less obviously associated with the major hydrographical features.

The occurrence of mussel larvae in the plankton is seasonally variable,

81

and this pattern of abundance may often be related, in a general way, to the spawning of the local adult populations (Battle, 1932: Thorson, 1946; Rees, 1954; Schram, 1970; Hrs-Brenko, 1973). Mileikovsky (1970) has discussed some of the factors that affect the timing of the occurrence of larvae in near-shore areas. He identified differences in the spawning times of the adults due to the zoogeographical origin of the species, the correlation of these spawning times with local seasonal temperature changes, and the operation of various types of lunar and diurnal spawning rhythms.

Methods

The relative paucity of data on the occurrence in plankton samples of particular species of lamellibranch is due, in part, to the difficulties of accurate identification. Also, until recently, lamellibranch larvae were difficult to maintain in the laboratory; fertilisation and the culture of the early embryonic stages were easily accomplished, but once the larvae became dependent on planktonic food, cultivation was more problematical (Stafford, 1912).

The identification of mussel larvae

The larva of *M. edulis* is one of the most frequently described of all lamellibranch larvae (Borisjak, 1909; Stafford, 1912; Kändler, 1926; Werner, 1939; Jørgensen, 1946; Rees, 1951; Loosanoff & Davis, 1963; Loosanoff, Davis & Chanley, 1966; Chanley & Andrews, 1971). The larva of *M. galloprovincialis* has been described by Zakhvatkina (1959) and Lubet (1973). Miyazaki (1935), Yoshida (1953) and Tanaka (1958) describe the larvae of *M. crassitesta*, and Yoo (1969) illustrates the larva of *M. coruscus*. Chanley (1970) includes a literature review of some larval characteristics of the Mytilidae.

The identification of lamellibranch larvae is undertaken either by 'indirect' or 'direct' methods. The indirect method is due to Werner (1939) as developed by Ockelmann (1965 and personal communication). By studying the umbonal region of the shell of young post-larval stages (that have themselves developed to a size where positive identification to species is possible) Ockelmann was able to deduce much of the history of larval shell formation, including the size of the prodissoconch I, or first larval shell, and the shape and size of prodissoconch II. This then allowed identification of the larvae to species, by comparison. Ockelmann showed a correlation to exist between the diameter of the mature oocyte and the length of prodissoconch I (greatest dimension in a line parallel to the hinge) in ten species of mytilid. The size of prodissoconch I was related, in turn, to the type of larval development (Fig. 4.1).

82

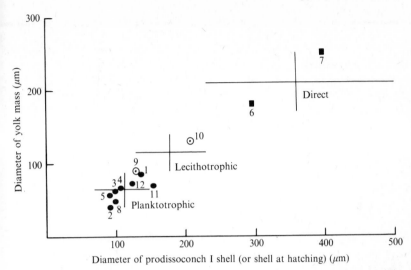

Fig. 4.1. The relationship between the diameter of the yolk mass of the ripe egg and the length of the first larval shell in twelve species of mytilid, with indications of the type of larval development. (Adapted from Ockelmann, 1965.) 1, *Modiolus modiolus*; 2, *M. adriaticus*; 3, *M. phaseolinus*; 4, *Mytilus edulis*; 5, *Musculus marmoratus*; 6, *M. discors*; 7, *M. niger*; 8, *Idasola argentea*; 9, *Dacrydium vitreum*; 10, *Crenella decussata*; 11, *Dacrydium panamensis*; 12, *Mytilus coruscus*.

The direct method of identification is due mainly to Loosanoff and his colleagues (Loosanoff & Davis, 1963; Loosanoff *et al.*, 1966; Chanley & Andrews, 1971). This depends on the rearing of larvae from known parents in the laboratory, linked with a comparison of sizes and dimensions between the 'known' larvae and larvae captured in the plankton. As Rees (1951) observed, this is 'the surest method of determining the species of a larva'.

The means of identification are based entirely on the shell. They are: the shape of the shell, the position of the ligament and the morphology of the larval hinge apparatus (Rees, 1951). Loosanoff *et al.* (1966) provide information on the shape of twenty species of larvae (including *M. edulis* and *Modiolus demissus*) from North America. Although the dimensions of the shell are useful in identification, the maximum size of the larval shell is very variable and is not a reliable aid in identification to species (Jørgensen, 1946; Sullivan, 1948; Bayne, 1965).

Werner (1939), Rees (1951) and Loosanoff *et al.* (1966) have defined the terms useful in describing lamellibranch larvae. The *length* of a larva is its greatest dimension on a line parallel to the hinge. The *height* is the greatest distance from the tip of the umbo to the ventral margin; where such a line deviates from a right-angled intersection with the hinge line, this should be recorded. Loosanoff *et al.* (1966) term this dimension the *width* of the shell.

83

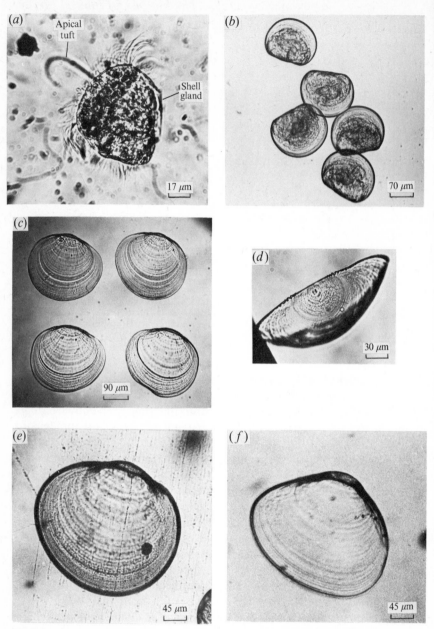

Fig. 4.2. (a) The veliger larva of *Mytilus edulis*. (b) Veliconcha larvae of *Mytilus edulis*, 12 days after fertilisation, at 16 °C. (c) Shell valves of full-grown larvae of *Mytilus edulis*. (d) Dorsal view of a shell valve of *Mytilus edulis* veliconcha larva, showing the larval hinge. (Courtesy of K. Ockelmann.) (e) The left valve of a veliconcha larva of *Modiolus modiolus*. (Courtesy of D. B. Quayle.) (f) The left valve of a veliconcha larva of *Modiolaria marmorata*. (Courtesy of D. B. Quayle.)

Convexity is a measure of the maximum distance from one valve to the other. The *shape* of the larva is assessed in terms of the prominence of the umbos, the outline of the shell, its convexity and the ratio between length and height. In mytilid larvae, the anterior shell margin is narrow, or 'bluntly pointed'; the posterior margin is broader, more rounded and often extended ventrally; the umbos are often not prominent (Fig. 4.2).

The larva develops into a *trochophore*, and thence to a *veliger*, before secretion of a shell (Fig. 4.2*a*). The first larval shell, the *prodissoconch I*, is secreted by the shell-gland of the veliger; the larva is now at the 'straight-hinge' or D-shaped' stage. Secretion of the second larval shell, or *prodissoconch II*, begins immediately. This shell is secreted cyclically by the mantle, and exhibits growth lines (Millar, 1968). Once secretion of prodissoconch II has started, the larva is called a *veliconcha*. At this stage locomotion is by means of a velum only, but as the larva approaches metamorphosis a pedal organ develops, and when this becomes functional the larva is called a *pediveliger* (Carriker, 1961). After metamorphosis, the secretion of the adult shell, the *dissoconch*, begins; the young post-larva is called a *plantigrade*. Some of these larval stages are illustrated in Field (1922).

The position of the larval ligament can be useful in identification (Rees, 1951). In the Mytilidae it is located posteriorly on the hinge line. The prodissoconch I has a weakly developed hinge structure. In the veliconcha, however, a distinct hinge is often present, and provides an important aid to identification. The straight part of the hinge is the provinculum. In mytilids this is thickened and bears several small rectangular teeth. For the Mytilidae, Rees (1951) used the term provinculum to include the thickened part of the shell that extends beyond the straight portion of the hinge. These lateral regions of the provinculum also bear teeth in mytilid larvae (Fig. 4.3; also Fig. 4.2*d*). The characteristic shape of mytilid larvae is illustrated in Fig. 4.2 with photomicrographs of the shells of the larvae of *M. edulis*, *Modiolus modiolus* and *Modiolaria marmorata*. Zakhvatkina (1959) records that the larvae of *M. galloprovincialis* are 'identical in structure to the larvae of . . . *M. edulis*'. Yoo (1969) has photographs of different larval stages of *M. coruscus*.

When preserving lamellibranch larvae for identification, care must be taken to avoid acid conditions which corrode the shell. Rees (1951) used formalin buffered with borax, and Carriker (1950) developed a special preservative. Chanley & Andrews (1971) used a seawater solution of 10% sugar (sucrose), 1% formalin and 0.05% sodium bicarbonate. Rees (1951) suggested the use of a sodium hypochlorite solution for separating the shell valves from the soft tissues for observation of the hinge. Chanley & van Engel (1969) have described a means for the three-dimensional representation of the dimensions of lavae.

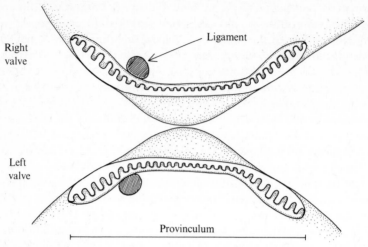

Fig. 4.3. Drawing of the inside view of the larval hinge of *Mytilus edulis* showing the provinculum and the position of the ligament. (Based on Rees, 1951.)

The rearing of mussel larvae

Any attempt to rear mussel larvae, either as an aid to identification or in more general studies, must include the acceleration of gametogenesis in the adult, inducing the adults to release ripe gametes (spawning), and the cultivation of suitable organisms for food, as well as the maintenance and feeding of the larvae and their protection from disease. Space does not allow a full discussion of these methods here; instead, reference will be made to the more useful recent studies.

Modern methods for culturing lamellibranch larvae were developed by Loosanoff & Davis (1963) in North America and by Bruce, Knight & Parke (1939) and Walne (1956, 1964, 1966) in Great Britain. Bayne (1965), Hrs-Brenko & Calabrese (1969), Yoo (1969) and Hrs-Brenko (1973) have applied these techniques in the study of *Mytilus* larvae. Lough & Gonor (1971) reared the larvae of *Adula californiensis* (Mytilidae) in the laboratory, and the rearing of other mytilid larvae is described by Chanley (1970) and Culliney (1971).

In attempts to secure a reliable supply of adults in spawning condition, techniques have been developed for accelerating gametogenesis by gradually raising the ambient temperature whilst supplying adequate food (Loosanoff & Davis, 1950, 1951). Bayne (1965) obtained adult *M. edulis* in a condition in which they were responsive to a spawning stimulus by raising the water temperature from 7 to 13 °C over 25–30 days; Hrs-Brenko (1973) was equally successful by raising the temperature from 1 to 18 °C within 13 days, whilst feeding the mussels with *Phaeodactylum*. The required period for conditioning an adult stock in this way is dependent on temperature and on the stage of gametogenesis reached by the population.

Methods of inducing *Mytilus* to spawn have been described by Young (1945), Iwata (1950, 1951*a*, *b*), Lubet (1959), Sugiura (1962), Breese *et al.* (1963), Loosanoff & Davis (1963), Bayne (1965) and Hrs-Brenko (1973). These methods include the administration of a mild electric shock, injection into the mantle tissue of 0.5 M potassium chloride or ammonium chloride, mechanical irritation of the posterior adductor muscle, and exposure either to a rapid change of temperature or to the genital products of other individuals. The basis for the effectiveness of these different stimuli is not well understood. Iwata (1952) found that an electrical stimulus of 30 volts, administered for only 20s, induced maturation and spawning after a latent period of 30–40 min. If the oocytes were removed from the ovary and then stimulated in the same way, maturation did not occur. Treatment with potassium and ammonium ions led to similar observations. Iwata concluded that these stimuli act through the follicle cells of the ovary, possibly initiating the secretion of a substance that causes the oocytes to begin the meiotic division and the stem that connects the oocyte to the follicle wall to break down. Spawning would then follow automatically. Electrical stimulation and treatment with cations may also cause 'cytoplasmic maturation' of the sperm. In either case, the commonly observed latent period between stimulus and spawning suggests an indirect action on the spawning process.

Articles by Loosanoff & Davis and by Walne, already referred to, serve as a sufficient guide to the basic techniques of larval culturing. However, these techniques are continually being modified and improved, especially in the context of oyster culture. Recent discussions include attempts to produce a dried algal food (Hidu & Ukeles, 1964) or artificial foods (Chanley & Normadin, 1966); the treatment of seawater to remove harmful dissolved substances (Millar & Scott, 1968) or the preparation of artificial seawater (Courtright, Breese & Krueger, 1971); the description of optimal conditions of aeration (Helm & Spencer, 1972); and attempts to culture larvae in bacteria-free conditions (Millar & Scott, 1967*a*). Walne (1970*a*) discusses some of the present problems in larval culture. It is to be expected that further understanding of larval biology will lead to the improvement of cultivation techniques.

Morphology of the gametes and larvae

The gametes

Spermatogenesis in *Mytilus edulis* has been described by Field (1922) and Lubet (1959), and in *M. galloprovincialis* by Lubet (1959) and Lucas (1971). Franzen (1955) described the appearance of the spermatids and sperm of

87

M. edulis. The ultrastructure of the sperm was studied by Niijima & Dan (1965; *M. edulis*) and Bourcart, Lavallard & Lubet (1965; *Perna perna*), and these studies reviewed by Idelman (1967). Longo & Dornfeld (1967) described the ultrastructure of spermatid differentiation in *M. edulis*, including the formation of the acrosome. Some interest has centred on the acrosome and 'acrosome reaction' in *Mytilus* sperm (Niijima, 1963), in which the acrosome disintegrates at fertilisation to form a slender acrosomal process, whilst simultaneously releasing a substance capable of lysing the vitelline membrane of the egg. The triggering and the mechanism of this reaction have been discussed by Dan, Kakizawa, Kushida & Fujita (1972), and Hauscha (1963) purified an egg-membrane lysin from *Mytilus* sperm.

Reverberi (1971) reviewed some of the literature on the structure of the egg of *Mytilus*, and Humphreys (1962) gave a detailed description. The spawned egg of *M. edulis* is 68–70 μm in diameter with a vitelline coat of between 0.5 and 1.0 μm thick. The coat is separated from the cytoplasmic surface of the egg by a perivitelline space about 0.2 μm wide. The plasma membrane of the egg forms a border of microvilli 0.7–1.0 μm long, which pass through the vitelline coat into an outer 'jelly layer' (7–10 μm thick) where their tips give rise to numerous fibrils (Dan, 1962). A large germinal vesicle occupies much of the egg; its membrane is double, with many pores allowing communication with the cytoplasm. The cytoplasm of the unfertilised egg contains numerous cortical granules in a layer extending from the base of the microvilli for a distance of about 1.5 μm. Humphreys (1967) ascribed to these granules the function of secreting a material which helps to maintain the thickness of the vitelline coat during early embryogenesis. The 'endoplasmic region' of the egg (Humphreys, 1962) consists of mitochondria, lipid droplets, yolk granules and cytoplasmic vesicles between 20 and 400 μm in diameter. The small vesicles are near the periphery of the egg, the larger vesicles deeper in the cytoplasm; Humphreys (1962) suggests that further study is necessary to clarify their function. Worley (1944) has described the formation of yolk in the oocytes of *M. californianus*. According to Longo & Anderson (1969*a*) heterogeneous aggregations of yolk bodies, lipid droplets and mitochondria in the mature egg constitute a form of ooplasmic segregation. Annulate lamellae, found in close association with the nuclear membrane, and which are also of uncertain function, have been described in *M. edulis* oocytes during vitellogenesis by Durfoot (1973). Ahmed & Sparks (1970) examined chromosomes from the eggs of *M. edulis* and *M. californianus* from the USA, and recorded a common diploid chromosome number of twenty-eight. They also observed autosomal polymorphisms in both species (see Chapter 9).

Fertilisation

In *Mytilus edulis* the two main reproductive ducts from each gonad unite into a common duct which opens on the genital papilla (Field, 1922; White, 1937). In *M. californianus* the ducts open separately (Young, 1945). In both species, eggs and sperm are discharged directly from the genital ducts to the exterior, and fertilisation occurs in the water. Spawning is not accompanied by unusual shell movements, and ciliary activity in the genital ducts appears to be the main propulsive force for ejecting the gametes. Heart rate increases markedly during spawning (Bayne, personal observation).

The eggs are spawned in the 'germinal vesicle' stage, although the germinal vesicle is soon disrupted and development proceeds to the metaphase of the first meiotic division. The egg remains at this stage until sperm entry at which time meiosis continues (Longo & Anderson, 1969*a*). Eggs which are unfertilised for 4–6 h at 18 °C cannot subsequently develop further, and the sperm lose their motility after 1–2 h at this temperature.

Sperm penetration into the egg is facilitated by the acrosome reaction (Berg, 1950; Niijima & Dan, 1965). In *M. edulis* sperm penetration takes 3–5 min at 15 °C (Wada, 1955) and is not accompanied by any sudden change in the appearance of the vitelline coat (Humphreys, 1964) although at the site of penetration there is a loss of microvilli and formation of a fertilisation cone. Longo & Anderson (1969*a*) have discussed the block to meiosis which must be present in the unfertilised egg, and its release at fertilisation. At the time of insemination the egg assumes a spherical shape, meiotic activity is resumed and there is a dispersal of the aggregations of cytoplasmic constituents. These and other cytological aspects of fertilisation have been described by Humphreys (1964) and Longo & Anderson (1969*a*, *b*).

Young (1941) found that fertilisation of *M. californianus* was adversely affected by salinities below 25‰. Wilson (1968) recorded that fertilisation in the estuarine mussel *Xenostrobus securis* (in Australia) was possible only between 14.5 and 31.5‰, although the adults are able to tolerate salinities from 2 to 56‰. In *M. edulis* fertilisation occurs successfully over a temperature range from 5–22 °C and in all salinities from 15–40‰ (Bayne, 1965). Lough & Gonor (1971) showed that fertilisation in *Adula californiensis* could occur at temperatures from 7–20 °C and salinities from 20.4–33.2‰, except in the temperature/salinity combination of 7 °C and 20.4‰. These limits to fertilisation are often in excess of the limits to cleavage and embryogenesis.

Cleavage and development to the first shelled larva

The normal cleavage of *Mytilus* eggs has been described by Field (1922), Rattenbury & Berg (1954) and Reverberi (1971). The first cleavage is unequal and is accompanied by formation of the polar lobe. Humphreys (1964) has shown that polar-lobe formation resembles cleavage both in the appearance of a constriction of the egg surface where the polar lobe is attached to the CD cell, and also in the presence of a screen of vesicles between the polar-lobe cytoplasm and the remainder of the egg. These vesicles may be formed from the multivesicular bodies mentioned earlier; Humphreys postulated that the multivesicular bodies act as reservoirs of smaller vesicles for use in the formation of the membranes that separate the blastomeres at cleavage. The barrier of vesicles that forms between the polar lobe and the CD blastomere may isolate the polar lobe from mitosis, so helping to bring about unequal cleavage.

As the first cleavage comes to an end, the polar lobe is withdrawn into the CD blastomere. At second cleavage, a second polar lobe is formed from the CD cell, and this subsequently becomes part of the D cell, which is the largest of the first four blastomeres. The third cleavage is 'spiral, dexiotropic and equal in the A, B and C quadrants' (Rattenbury & Berg, 1954). Subsequent cleavage results in a cap of micromeres which partially envelopes the vegetal macromeres.

After 4–5 h at 18 °C, cilia first appear and the embryo begins to swim. The micromeres overgrow the macromeres, signalling the onset of gastrulation. This is completed by invagination of the macromeres within a very retricted area at the vegetal pole. The animal pole now lies opposite the blastopore, and the dorso-posterior region of the embryo is marked by a thickening of the ectoderm at the animal pole. This is the trochophore stage. An apical tuft of cilia develops at the anterior end, and a shell gland forms from the thickened dorsal ectoderm (Fig. 4.2*a*). This shell gland secretes the first larval shell, the prodissoconch I. Velar cilia develop around the apical tuft, and the anterior region enlarges to form the velum and the embryo takes on the appearance of the 'straight-hinge' larva.

Rattenbury & Berg (1954) isolated blastomeres and polar lobes at early cleavage stages of *M. edulis*, and followed their subsequent development. During the first cleavage, the factors for later differentiation of the apical tuft became localised in the CD, and later in the D blastomere, though not via the second polar lobe. Factors for the differentiation of the stomodaeum and shell gland passed into the CD blastomere. Humphreys (1964) suggested that the segregation of morphogenetic factors in the polar lobes may be caused by the partition of the cytoplasm by a sheet of vesicles at the time of cleavage. He found also, in agreement with Pucci (1961) that there is no consistent difference in the distribution of yolk granules, lipid

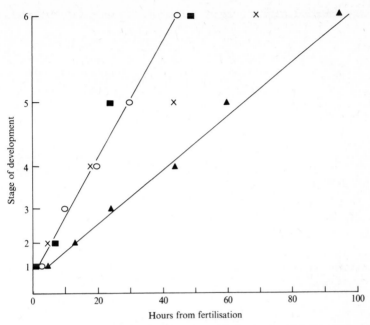

Fig. 4.4. The rates of cleavage and early development of embryos of *Mytilus edulis* at different temperatures. ○, 18 °C (Bayne, 1965); ▲, 8 °C (Bayne, 1965); ×, 20 °C (Field, 1922); ■, 19–22 °C (Rattenbury & Berg, 1954). Stages of development as follows: 1, first polar lobe; 2, first cilia; 3, trochophore; 4, appearance of velar cilia; 5, appearance of shell gland; 6, prodissoconch I.

particles or mitochondria within the polar lobe compared with the rest of the egg cytoplasm.

The rates of oxygen consumption by developing lamellibranch embryos have been measured by Balantine (1940; *Mactra, Crassostrea*), Cleland (1950; *Crassostrea*), Berg & Kutsky (1951; *Mytilus edulis*), Sclufer (1955; *Spisula*) and Black (1962*a, b; Crassostrea*). These papers are reviewed by Raven (1972). Cleland (1950) suggested that changes in metabolic rate may be associated with mitotic rate in the developing embryo, so explaining the rather complex shape of the curve of oxygen consumption with time, with its areas of reduced uptake sandwiched between areas of rapid increase in rate. Acceleration of metabolic rate was associated with ciliary activity and swimming. Black (1962*a*) measured a ninefold increase in the rate of oxygen consumption from first cleavage to the development of the trochophore. Cleland (1950) recorded a respiratory quotient (RQ) of 0.8 during cleavage in *Crassostrea commercialis*, implying the catabolism of carbohydrate and fat. In *Spisula* the RQ gradually increased from 0.7 after fertilisation to 1.0 (Sclufer, 1955).

Unlike fertilisation, cleavage and early embryogenesis have a rather

91

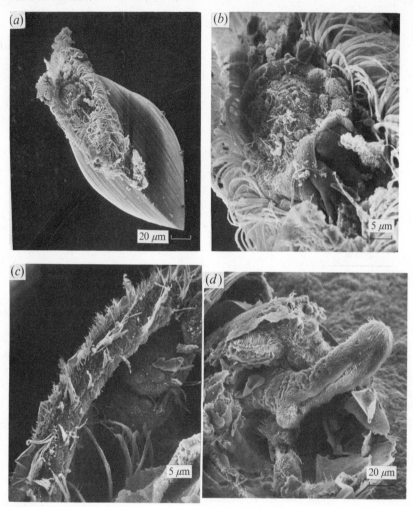

Fig. 4.5. (*a*) A scanning electron micrograph of the veliconcha larva of *Mytilus edulis* showing the partially extruded velum and, in the centre of the velum, the apical sense organ. (Courtesy of S. Cragg.) (*b*) A view at higher magnification than (*a*) of the apical sense organ of an *M. edulis* veliconcha larva. (*c*) A scanning electron micrograph of the mantle edge of a pediveliger larva of *Mytilus edulis* showing two types of cilia; within the mantle cavity are visible the distal ends of two gill buds. (Courtesy of S. Cragg.) (*d*) A scanning electron micrograph of the foot of a pediveliger larva of *Mytilus edulis*. (Courtesy of D. Lane & J. Nott, 1975.)

limited tolerance of environmental change. Bayne (1965) found that development of the trochophore of *M. edulis* occurred successfully only within a salinity of range 30–40‰ and temperatures of 8–18 °C. Within this temperature range, rate of development increased with increase of temperature (Fig. 4.4). Young (1941) discussed the salinity tolerance of *M.*

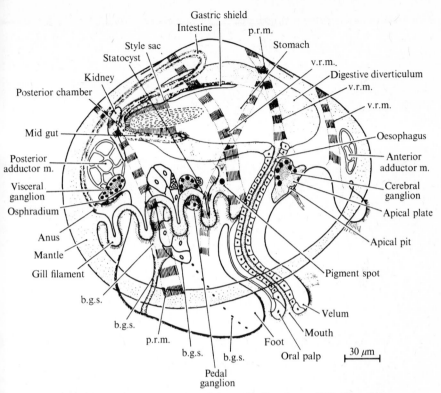

Fig. 4.6. A diagrammatic reconstruction of the pediveliger larva of *Mytilus edulis*. p.r.m.,
pedal retractor muscle; v.r.m., velar retractor muscle; b.g.s. byssal gland system. (From
Bayne, 1971*c*.)

californianus embryos, and Lough & Gonor (1971) described the temperature and salinity requirements for early development of *Adula californiensis*. Hrs-Brenko (1973) studied the development of the embryos of *M.
edulis* and *M. galloprovincialis*. Normal development of *M. edulis* (from
the east coast of USA) occurred at 15–20 °C, but not at 5 or 30 °C. The low
salinity limit for normal development was 15–20‰ at 15 °C, and 20–25‰ at
20 °C; the upper salinity was between 30 and 35‰. For *M. galloprovincialis*
from the Adriatic, optimal conditions for embryonic development were
15–20 °C and 27.5–40‰ salinity. No development to the veliger stage
occurred at 25 °C at any salinity tested, nor at 20‰ salinity at any
temperatue tested. The range of environmental conditions over which
normal embryogenesis is possible will depend on the location of the adult
population. Nevertheless, the limited data available suggest that the
environmental requirements for embryonic development may be limiting,
especially to estuarine populations of mussels.

93

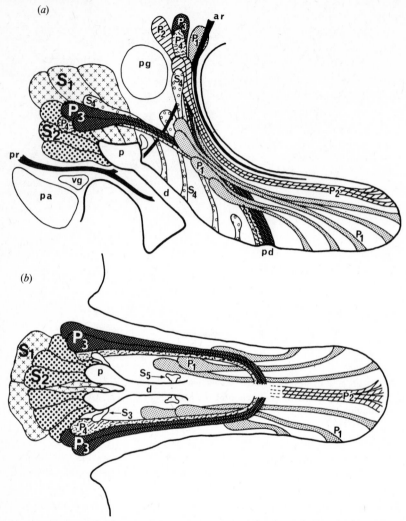

Fig. 4.7. The glands in the foot of the pediveliger larva of *Mytilus edulis*, sectioned (*a*) sagitally and (*b*) horizontally. Cells of the nine types of gland are denoted by P_1, P_2, P_3, P_4, S_1, S_2, S_3, S_4, and S_5. ar, anterior retractor muscle; d, posterior duct; p, lateral pouch; pa, posterior adductor muscle; pd, pedal depression; pg pedal ganglion; pr, posterior retractor muscle; vg, visceral ganglion. (From Lane & Nott, 1975.)

The veliconcha and pediveliger larvae

During the veliconcha larval stage (Fig. 4.5*a*) there is considerable growth in size (from 110–250 μm in shell length). Cell differentiation is limited to the velum, including its retractor muscles and the apical plate (Fig. 4.5*b*), to the alimentary system, the nerve ganglia and the cells of the mantle, which secrete the prodissoconch II shell. The gross morphology of the veliconcha of *Mytilus* closely resembles *Dreissena polymorpha*, which has been described by Meisenheimer (1901).

At a shell length of between 220 and 260 μm the larva acquires a pair of pigmented 'eye-spots' and soon afterwards develops a foot. This pediveliger stage immediately precedes settlement and metamorphosis. The gross morphology of the pediveliger larva of *M. edulis* has been described by Bayne (1971*c*), and a diagrammatic reconstruction of this larva is shown in Fig. 4.6. Lane & Nott (1975) have recently described the morphology and histochemistry of the foot of the pediveliger larva of *M. edulis* (Fig. 4.5*d*). They observed nine different kinds of gland, each of which is assigned to a specific role in crawling and in attachment. Each gland is described in detail (Fig. 4.7). This paper, together with recent studies by Cranfield (1973*a*, *b*, *c*), Gruffyd, Lane & Beaumont (1975) and S. Cragg (unpublished data) reveal the considerable complexity of the nervous, sensory and musculature systems of the lamellibranch pediveliger (Fig. 4.5*a–d*).

Metamorphosis

At metamorphosis, the gross morphological changes that occur include the secretion of byssus threads (Lane & Nott, 1975), the collapse and disappearance of the velum, the formation of the labial palps and the re-orientation of the organs in the mantle cavity (Bayne, 1971*c*). The degeneration of the velum and the formation of the labial palps occurs within one to three days of the first secretion of the byssus, the time depending, in part, on temperature. During this period the larval feeding mechanism is lost and the adult gill/palp feeding mechanism is developed; for these one to three days the larva cannot feed and relies for metabolic energy upon stored nutrients. With the degeneration of the velum, the foot migrates forwards in the mantle cavity. This allows the ctenidial filaments on each side of the posterior mantle cavity to meet ventrally, so effectively partitioning the mantle cavity between inhalant and exhalant chambers. The filtering of inhalant water by the ctenidial filaments is now possible for the first time. Simultaneously with these changes, the labial palps develop rapidly, and particles which are trapped by the ctenidia can now be directed into the mouth. At this stage (Fig. 4.8) cilia on the sides of the foot make the major contribution to causing the inhalant current of water into the mantle cavity. As the dissoconch shell develops, there are alterations in the main axes of growth (Kriaris, 1967; Bayne, 1971*c*) which result in a much enlarged posterio-ventral region of the mantle cavity. Additional ctenidial filaments are developed into this area, and their cilia gradually acquire the role of generating the feeding currents.

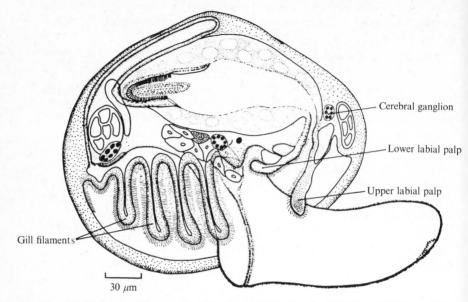

Fig. 4.8. A diagrammatic reconstruction of an early plantigrade of *Mytilus edulis*, immediately after metamorphosis and before secretion of the dissoconch shell has begun. Organ systems not labelled are as in Fig. 4.6. (From Bayne, 1971*c*.)

Physiology

Fecundity

Direct estimates of the fecundity of mussels are rare. Hancock & Simpson (1962), quoting Mateeva, suggest a fecundity of between 10^4 and 2×10^6 eggs per adult female over the size range 14–42 mm shell length. Fretter & Graham (1964) mention a figure of 10^7 eggs per female. Bayne (1975*a*) and Bayne, Gabbott & Widdows (1975) estimated that females 4–5 cm in shell length, when induced to spawn in the laboratory, released between 135 and 200 cal per female as eggs, equivalent to about 0.5×10^6 eggs per female. However, these were underestimates of the true fecundity, since autopsies showed that not all the eggs had been released. The fecundity declined when the adults suffered a stress in the laboratory (see Chapter 7). Indirect estimates of fecundity may be gained by comparing the weight of the gonad before and after spawning, and in this way size-related fecundity may also be assessed. However, insufficient data are available to make any general statements at present.

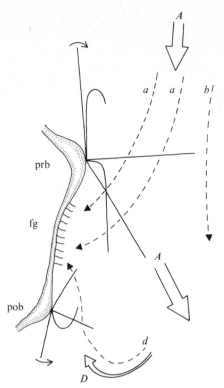

Fig. 4.9. A diagram of the 'opposed band system', as exemplified by a bdelloid rotifer. A, D; water currents created by the beat of the cilia. a, d; suspended particles carried to the food groove. b; suspended particles not captured as food. prb; pre-oral band of cilia; pob; post-oral band of cilia. fg; food groove. (Based on Strathmann, Jahn & Fonesca, 1972.)

Feeding

Lamellibranch larvae use the cilia on the velum for feeding as well as for swimming. These cilia comprise a pre-oral band of long, compound cirri and a peripherally situated band of shorter cilia, which beat towards the mouth. Whether or not a third 'post-oral' ciliated band (cf. prosobranch veligers; Fretter, 1967) is present remains uncertain. The presence or absence of post-oral cilia assumes importance in the light of studies by Strathmann, Jahn & Fonesca (1972) on feeding mechanisms of invertebrate larvae. Feeding in lamellibranch veligers is most probably by the 'opposed band system', or a modification of this (Fig. 4.9). According to Strathmann *et al.*, in this system the long cilia of the pre-oral band beat with a stiff, straight effective stroke and a curved return stroke, to produce the major water current which is used for both feeding and locomotion. Particles are carried in this current, and if they come within reach of the

97

pre-oral cilia they are swept directly onto the food-groove cilia which beat towards the mouth. The authors suggest that mucus, picked up by the pre-oral cilia in the recovery stroke, may aid in food capture. In bdelloid rotifers, on which Strathmann *et al.* (1972) based much of their description of this filtration mechanism, the post-oral cilia, which are distinct from the cilia of the food groove, beat towards the pre-oral cilia, and the water current so generated helps to capture particles for food, as well as to increase the efficiency of the pre-oral ciliary beat. Indeed, Strathmann *et al.* suggest that the opposed band system could not function effectively in the absence of a post-oral band of cilia.

In lamellibranch veligers, either previous observers have missed the post-oral cilia, or feeding does occur by the direct transfer of particles from the pre-oral cilia to the food groove, without the aid of post-oral cilia. It may be significant that some descriptions of feeding in lamellibranch veligers (e.g. Yonge, 1926*b*) mention a mucous string carrying food particles to the mouth, since, in the presence of mucus, the efficiency of a system that lacks post-oral cilia may be improved. Published values for the 'filtration rate' (or 'volume swept clear') of particles by lamellibranch veligers are lower than values quoted for other invertebrate larvae; e.g. 0.1–0.6 ml d^{-1} for *Mytilus* larvae (Table 4.1) compared with 12–25 ml d^{-1} for the larvae of the echinoderm *Luidia* (Strathmann, 1971). However, Strathmann *et al.* (1972) calculated a filtration rate by an echinoderm pluteus of 1.7 ml d^{-1} per mm of ciliated band. In the veliconcha of *M. edulis* the velum is roughly circular in outline, and the pre-oral band of cilia may be 150–200 μm in diameter, or 0.47–0.63 mm in circumference; filtration rate by the full-grown veliger is therefore in the range 0.9–1.3 ml d^{-1} per mm of ciliated band, which is in fair agreement with the pluteus.

In the type of feeding mechanism proposed for lamellibranch larvae, control over ingestion could occur either by cessation of beat in the orally-directed ciliary band, by rejection of particles over the oral palp, or by cessation of swimming. The third of these can be a temporary measure at best. The second alternative is observed to occur in dense concentrations of particles (Yonge, 1926*b*; Bayne, 1971*c*). The first alternative has not been recorded.

The feeding rates of lamellibranch larvae have been determined by measuring the rates with which they filter phytoplankton cells in laboratory culture. Some typical values are given in Table 4.1. At high cell concentrations ($>$ 200 cells μl^{-1}) feeding rate is reduced. Using algal cells labelled with radioactive phosphorus (^{32}P), Walne (1965) demonstrated that cells in the size range 3–10 μm diameter were equally well caught by the larvae of *Ostrea edulis*. There was a close correlation between the numbers of cells caught by the larvae and the larval rate of growth. At low concentrations of food cells the larvae increased the volume of water

Table 4.1. *The filtration (or clearance) rates of the larvae of* Ostrea edulis *and* Mytilus edulis

Species	Type and concentration of food; cells μl^{-1}	Filtration rate		Shell length (μm)	Authority
		$ml\ d^{-1}$ per larva	$ml\ h^{-1}$ per mg dry wt†		
Ostrea edulis (20–22 °C)	'Flagellates'; 15–26	0.65	36.2	200	Jorgensen (1952)*
Ostrea edulis (19–25 °C)	*Isochrysis galbana;* 31–54	0.43–0.49	25.6	218–280	Walne (1956)
Mytilus edulis (18 °C)	*Isochrysis galbana;* 25	0.21	23.6	210	Bayne (1965)
	60	0.24			
	365	0.14			
	Isochrysis galbana; 60	0.61	30.4	250	Bayne (1965)
	265	0.50			

* Jorgensen's calculations were based on data in Bruce *et al.* (1939).
† Calculated from values for '$ml\ d^{-1}$ per larva' using data on the flesh weight of larvae from Walne (1965).

'swept clear' of cells, either by increasing filtering activity, or by improving filtration efficiency.

Bayne (1965) recorded a Q_{10} (temperature coefficient) of 3.16 for the filtration rate of *M. edulis* larvae between 11 and 18 °C. Walne (1965) calculated a Q_{10} of 2.4 over the temperature interval 17–25 °C for the assimilation, by *Ostrea* larvae, of the flagellate *Isochrysis galbana*.

Oxygen consumption

Zeuthen (1947) measured the rate of oxygen consumption by *Mytilus edulis* larvae using a Cartesian Diver respirometer. He related metabolic rate to biomass expressed as nitrogen content, and observed that veliconcha and pediveliger larvae that were retracted within their shells had a low rate of consumption, $< 1 \times 10^{-2} \mu l\ O_2\ h^{-1}\ \mu g\ N^{-1}$. When swimming, however, the respiration rate increased to $2 \times 10^{-2} \mu l\ O_2\ h^{-1}\ \mu g\ N^{-1}$. Zeuthen also compared the rate of oxygen consumption by larvae and their isolated vela; the isolated velum consumed oxygen to about 50% of the rate of the entire swimming larva. The relationship between the rate of oxygen consumption and temperature agreed closely with the 'standard curve' of Krogh (1914). Zeuthen argued from this that, because the larvae were capable of swimming at 1.5 °C, although the metabolic rate of the velum at this temperature was only 0.17 times the rate at 16 °C, and since at 16 °C the metabolic rate of the velum was 50% the rate of the intact larva, therefore the energy that was necessary to keep the larva in 'moderate motion' in the water amounted to 8.5% of the total energy demand at 16 °C. Rapid vertical swimming, however, might account for as much as 50% of the total metabolic rate.

Zeuthen (1947) also considered the relationship between rate of oxygen consumption and body size. From his measurements of *M. edulis* larvae he concluded that it was impossible to distinguish between the metabolic rates of swimming veliconcha larvae, crawling pediveliger larvae and newly-metamorphosed plantigrades. Metamorphosis *per se* had no noticeable effect on the rate of oxygen consumption. Later (1953) Zeuthen summarised the data on *Mytilus* and recorded a regression coefficient of 0.80 for the relationship between rate of oxygen consumption ($\mu l\ h^{-1}$) and body size (total nitrogen content) for larvae from the trochophore stage ($< 0.01 \mu g\ N$) to the pediveliger (0.2 $\mu g\ N$). Zeuthen's values range from $3.2 \times 10^{-2} \mu l\ O_2\ h^{-1}\ \mu g\ N^{-1}$ for trochophores, to $1 \times 10^{-2} \mu l\ O_2\ h^{-1}\ \mu g\ N^{-1}$ for pediveligers. Bayne, Gabbott, & Widdows (1975) measured the losses of protein, lipid and carbohydrate from young *M. edulis* larvae following 48 h of starvation. When these losses are calculated as 'oxygen equivalents' using standard conversion constants, the result is an average estimate for metabolic rate of $0.18 \times 10^{-3} \mu l\ O_2\ larva^{-1}$, or $3.1 \times 10^{-2} \mu l\ O_2\ \mu g\ N^{-1}$, which is in good agreement with Zeuthen.

Table 4.2. *Rates of oxygen consumption by bivalve larvae*

Larva	Type	Dry organic weight (g)	Main body component utilised	Oxygen uptake		Authority
				(ml O_2 h^{-1})	(ml O_2 h^{-1} per g)	
				a. Estimated by respirometry		
Ostrea edulis	Newly released veliger	0.25×10^{-6}		1.6×10^{-6}	6.4	Walne (1966); Holland & Spencer (1973)
Mytilus edulis	Veliconcha	0.18×10^{-6}		$0.5-1.8 \times 10^{-6}$	2.8–10.0	Zeuthen (1947)
				b. Estimated by loss of energy reserves		
Ostrea edulis	Veliger		Lipid		5.0–6.0	Holland & Spencer (1973)
	Plantigrade		Lipid		3.0	Holland & Spencer (1973)

Adapted from Crisp (1975).

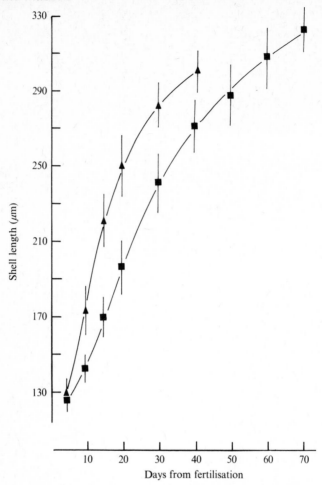

Fig. 4.10. The growth of *Mytilus edulis* larvae at two temperatures. Sizes are shown as means ±1 s.e. for *n* = 50 measurements. ■, 11 °C; ▲, 17 °C. The points were fitted with Gompertz (sigmoidal) growth curves.

Crisp (1975) has recently reviewed some of the published data on the rates of oxygen consumption by marine invertebrate larvae (Table 4.2). The available values show a consensus in the range 3–10 ml O_2 per g dry weight per hour.

Growth

The growth curve of shell length against time for lamellibranch larvae is generally sigmoidal in shape (Loosanoff, Miller & Smith, 1951; Bayne, 1965), although a linear fit to some data in the middle size ranges is not

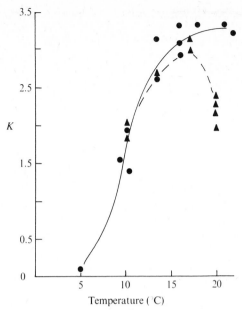

Fig. 4.11. The effect of temperature on the rates of growth (expressed as a growth constant, K) of the larvae of *Mytilus edulis* from two populations, North Wales (●) and the Oresund (▲). (From Bayne, 1965.)

unreasonable (Ansell, 1961; Yoo, 1969). Fig. 4.10 shows the growth in length of larvae of *M. edulis* in two experiments at 11 and 17 °C (30.5‰ salinity; 70 cells *Isochrysis* μl^{-1}). The calculated Gompertz (sigmoidal) growth curves gave an acceptable fit to the data.

The effects of temperature

Bayne (1965) recorded the growth of *M. edulis* larvae from 5–22 °C, as the relative growth rate (K) following Dehnel (1955);

$$K = 100 \left[\frac{\ln L2 - \ln L1}{t} \right],$$

where $L1$ is the shell length at the beginning, and $L2$ length at the end of the period t. The results for larvae from two different adult populations are plotted in Fig. 4.11. At 5 °C no growth occurred; rate of growth then increased between 10 and 16 °C, but declined at higher temperatures in the Oresund population, and remained constant in larvae from North Wales. Data from both populations indicated a range of temperatures over which the Q_{10} for growth was rather low (e.g. North Wales; $Q_{10\ (14-18\ °C)} = 1.12$). The differences in K at high temperatures were ascribed to the different temperature regimes experienced by the adults.

Marine mussels

Ursin (1963) described a method for quantifying the effect of temperature on growth, and he used Ansell's (1961) data on the growth of *Venus striatula* larvae as illustration. The procedure involves fitting the expression (due to Janisch):

$$y = y_0 \cosh p(x-x_0),$$

where y is the time to complete a specified amount of growth; x is the temperature; x_0 is the temperature at which growth is most rapid; y_0 is the corresponding minimum development time; and p is a temperature coefficient. The reciprocal of the expression describes the growth rate.

Walne (1965) arrived at a description of the time taken for larvae of *Ostrea edulis* to grow from 175 μm to 250 μm at temperatures from 17–26 °C by fitting a simple linear regression to length increment against temperature. Values from the regressions were used to construct 'Ford–Walford' plots of length on day n to length on day $n+2$, and the growth curves constructed accordingly.

A common conclusion from studies of the effects of temperature on lamellibranch larvae is that the embryos of young larval stages develop normally over a narrower temperature range than do the older larvae (Loosanoff & Davis, 1963; Bayne, 1965; Calabrese & Davis, 1970). Kennedy, Roosenburg, Castagna & Mihursky (1974) exposed cleavage stages, trochophores and prodissoconch I larvae of *Mercenaria mercenaria* to temperatures from 19–44 °C for periods from 1 to 360 min and then assessed resultant mortality. The results were analysed by multiple regression and displayed as response surfaces. The cleavage stages were most sensitive, and the shelled larvae least sensitive, to high temperatures. This type of analysis is recommended for elucidating the temperature/time relationships of development and growth.

The effects of salinity

Bayne (1965) measured the growth of *M. edulis* larvae from two populations at different salinities. Larvae from one population (North Wales) did not grow at 19‰ and showed retarded growth at 24‰; at 30–32‰ growth was normal. In larvae from a population in the Oresund, however, where the ambient salinity was lower than in North Wales, growth occurred even at 14‰. As with temperature, the growth of the larvae appeared to reflect the environmental conditions experienced by the adult.

Yoo (1969) measured the growth and survival of the larvae of *Mytilus coruscus* at salinities from 14–33‰. Greater than 80% survival occurred only at salinities more than 27‰. Some growth in length occurred at all salinities tested, but appeared to cease after 4–6 days at salinities less than 27‰.

Combined effects of salinity and temperature

In recent years, following the initiative of Costlow, Bookhout & Monroe (1960, 1962) and Kinne (1963), there has been an increase in the use of multifactorial experimental techniques to study the effects of physical variables on survival and growth of larvae in culture. Hrs-Brenko & Calabrese (1969) found that the effects of salinity and temperature on *M. edulis* larvae were significantly related only as the limits of tolerance of either factor were approached. For example, as the limits of salinity tolerance were approached (20‰) the range of temperature tolerance was markedly narrowed. Optimum larval growth occurred at 20 °C and 25–30‰, but growth declined at 25 °C and at 10 °C both at high (40‰) and low salinities. Growth occurred over a much narrower range of temperature and salinity than did survival.

This experimental approach lends itself to the quantitative description of effects in terms of multiple regression techniques as described by Alderdice (1972). Cain (1973) and Lough & Gonor (1971, 1973*a, b*) have quantified the effects of temperature and salinity on the survival and growth of embryos and larvae of *Rangia cuneata* and *Adula californiensis* respectively.

As mentioned earlier, a common feature of studies on the resistance adaptations of embryos and larvae is the greater tolerance to environmental change by shelled larvae than by the embryonic stages. However, changes in tolerance are not always predictable in this way. Lough & Gonor (1973*a*), using multiple regression techniques, found that older larvae of *Adula* became more tolerant than younger larvae to reduced salinity, but less tolerant to high temperatures. They concluded that these larvae were not adapted to estuarine conditions, although the adults commonly occur in estuaries; recruitment to the adult populations probably occurs only during the summer when 'oceanic' conditions prevail in some parts of the estuary. Wilson (1968) concluded that larval dispersal between estuaries for the mussel *Xenostrobus securis* was precluded by the relatively low upper-salinity limit of the developmental stages. Lough (1974) has used the data of Hrs-Brenko & Calabrese (1969) to carry out multiple regression analyses on percentage survival and growth, as related to temperature and salinity, of the larvae of *M. edulis*.

The effects of ration

The successful development of techniques for rearing lamellibranch larvae depended upon the realisation that their natural foods were the small cells of the nanoplankton, and upon the application of procedures for maintaining cultures of nanoplankton organisms in the laboratory (Cole, 1937;

Bruce *et al.*, 1939; Loosanoff & Davis, 1963). The nutritional qualities of a large number of algal species have now been tested for various larvae (Davis, 1953; Walne, 1963) and some algae have been found to provide a much better diet than others. Walne (1963) confirmed earlier studies that, in general, flagellates that lack a cell wall ('naked flagellates') are better as food than algae with a cell wall (e.g. *Chlorella*). In addition, Walne showed that a given species of alga may not be equally acceptable as food to different broods of larvae. There are also differences among species of lavae. Loosanoff (1954) suggested that some species can utilise for food only a few algal species, whilst others, including *Mytilus edulis*, can grow on a large variety of algal types, provided only that they are small enough to be ingested. The food value of a single species may vary with the temperature (e.g. *Chlorella* spp.; Loosanoff & Davis, 1963), with the size of the larva (Davis, 1953) and with the age of the algal culture (Bayne, 1965). Mixtures of algal species often support a faster rate of growth than single species (Davis & Guillard, 1958; Bayne, 1965) and modern rearing techniques employ different mixtures of food cells as the larvae grow; e.g. Gruffydd & Beaumont (1972) reared the larvae of *Pecten maximus* on a mixture of *Isochrysis galbana* (mean cell diameter 3.8 μm) and *Chaetoceros calcitrans* (3.7 μm) up to 14 days after fertilisation, after which *Pyramimonos obovata* (5.6 μm) and later still *Tetraselmis suecica* (6.9 μm) were added to the mixture to provide the larger larvae with larger cells in a more complex diet. In addition to the algal species already mentioned, *Monochrysis lutheri*, *Chlamydomonas* spp., *Dunaliella* spp. and *Dicrateria inornata* have been used successfully as food for lamellibranch larvae. Hidu & Ukeles (1964) reared the larvae of *Mercenaria mercenaria* on freeze-dried preparations of *Isochrysis* and *Dunaliella*, and Chanley & Normandin (1966) were successful with finely-ground fresh or frozen *Ulva lactuca*. The use of non-living food preparations presents problems of aggregation and settling-out in the larval cultures, however, and reliable techniques have yet to be developed. Until such techniques are available, the important question of the role of detritus in the nutrition of larvae cannot properly be examined. The use of microencapsulation techniques to prepare fully artificial diets should also provide a fruitful field for research.

Walne (1963) recommended the use of the 'packed cell volume' (the volume of packed cells following centrifugation of a unit volume of algal culture) as a standard measure of food concentration in comparative feeding studies; however, no standard procedure of centrifugation has been agreed. Bayne (1965) recorded rapid growth of *M. edulis* larvae on a diet of a mixture of *Isochrysis* and *Monochrysis* at a concentration of 3.2 μl (packed cell volume) l^{-1}, whereas at 0.8 μl l^{-1} growth was reduced. Walne (1965) compared the numbers of cells of different species of algae that were digested by *Ostrea edulis* larvae in 24 h. Numbers of cells digested decreased as the size of the food cell increased. The daily rates of digestion by larvae from 180–260 μm shell length were in the range 0.1–0.8 μg dry weight of cells per larva per day. When fed *Isochrysis*, the daily assimilated ration of *Ostrea* larvae declined from 56% to 29% of larval dry weight of organic matter (or 25–13% of larval carbon) over the planktonic period.

In the same series of experiments, Walne (1965) measured the numbers of cells removed from suspension by the oyster larvae. On average, 3.4 times the number of cells were removed than were actually assimilated. Some of these cells will have been rejected at the oral palp, others will have been ingested but incompletely digested; Walne's technique did not allow him to distinguish between these. Gabbott & Holland (1973) used Walne's values for numbers of cells removed from suspension, together with their own estimates of growth and metabolism, to calculate assimilation efficiencies of 41–77% for *Ostrea* larvae. Bayne (unpublished data) measured the following daily rates for larvae (150 μm length) of *M. edulis*: ingested ration 360 cal, metabolism 21.5 cal, growth 235 cal. These values are equivalent to an assimilation efficiency of 71%, and a gross growth efficiency (K_1) of 65%.

The rather scant data that are available therefore suggest that in laboratory cultures with relatively high concentrations of food cells lamellibranch larvae remove from suspension more cells than they ingest, and assimilate ingested cells with an efficiency of 40–70%. At reduced concentrations of food cells the larvae may respond by an increase in filtering activity, or an increase in filtration and/or assimilation efficiency.

Reliable determinations of the abundance of nanoplankton (2–20 μm diameter) in nature are scarce (Parsons & Takahashi, 1973). But larvae in nature will probably experience lower concentrations of living phytoplankton than are shown, by laboratory experiment, to be optimal for growth. Nevertheless, growth rates in nature and in the laboratory are similar (Walne, 1965). This discrepancy may be due, in part, to enhancement of larval growth by mixtures of algae. Also, the weight of organic carbon that represents an optimal ration in the laboratory (0.5–0.8 mg C l^{-1}) is well within estimates for inshore waters. However, the proportion of this carbon that represents utilisable food for larvae is unknown. Early studies (Davis, 1953) failed to demonstrate that oyster and clam larvae could utilise natural organic detritus to support growth, but this possibility needs re-examination.

Reports on the role of bacteria as food for lamellibranch larvae are conflicting. Davis (1953) found no evidence that bacteria were utilised by oyster larvae, but Hidu & Tubiash (1963) concluded the opposite. Walne (1963) found no difference between bacteria-free and non-bacteria-free cultures of *Isochrysis* as food for oyster larvae. On the other hand, Ukeles

& Sweeney (1969) suggested that bacteria may interfere with the normal feeding of larvae by packing the alimentary canal, and the control of bacteria in larval cultures is generally recommended (Tubiash, Chanley & Leifson, 1965) on the grounds of hygiene.

Whatever the composition of the larval diet in nature, fluctuations in food availability are likely to occur, and lamellibranch larvae are capable of surviving several days without food (Loosanoff, 1954). Bayne (1965) showed that the larvae of *M. edulis* could survive for up to 26 days without particulate food at 16 °C; for at least 10 days after the onset of starvation these larvae could begin feeding immediately any food was offered and their subsequent growth was normal. Millar & Scott (1967b) found that *Ostrea* larvae also could survive up to 8 days of starvation and still grow to full size when subsequently fed. During starvation these larvae catabolised mainly lipid (triglycerides) and protein and the 'oxygen equivalents' of the observed biochemical losses were within the range of their measured rates of oxygen consumption. The lipid reserves in lamellibranch larvae provide a useful energy store, not only during starvation (Helm, Holland & Stevenson, 1973; Gabbott & Holland, 1973; Bayne, Gabbott & Widdows, 1975), but also during metamorphosis when the larva may be unable to feed for 1–3 days. This lipid, at least in the young larva, is derived from the adult via the egg, and the amount that is stored during vitellogenesis is a sensitive link between the physiology of the adult and the survival potential of the larva.

The effect of adult 'condition'

A common observation during the rearing of lamellibranch larvae in the laboratory is the large range of size of sibling larvae cultured under identical conditions (Loosanoff & Davis, 1963). Chanley (1955) concluded that differences in the lengths of *Mercenaria mercenaria* larvae 2 days after fertilisation were caused by physiological differences between eggs, but that differences in rates of growth, including differences between siblings, could be due to inherited factors from either parent. Physiological differences between eggs from the same female could arise from unequal distribution of nutrients to different parts of the gonad during vitellogenesis. Variation between eggs from different parents could be due to differences in the physiological condition of the adults (Helm *et al.*, 1973; Bayne, 1972; Bayne, Gabbott & Widdows, 1975).

Bayne (1972) held adult *Mytilus edulis* under conditions of considerable physiological stress. The adults were then induced to spawn and the development and growth of the embryos and larvae compared, under standard conditions, with larvae from 'non-stressed' adults. Embryos from stressed adults showed increased abnormalities during development.

Helm *et al.* (1973) showed that viability was less in oyster larvae obtained from adults kept at low ration, than in larvae from adults at high ration. Bayne, Gabbott & Widdows (1975) confirmed, for *M. edulis*, that larvae derived from adults kept under the greatest stress had the lowest rates of growth.

Other determinants of growth

Loosanoff (1962) quoted experiments by Davis which showed that the growth of larvae of *Crassostrea virginica* was reduced in water carrying 0.75 g l^{-1} of silt in suspension, although *Mercenaria* larvae grew normally in a concentration of 1.0 g l^{-1}. Some of the inhibiting effect of silt may be due to changes of pH (Calabrese & Davies, 1966). The optimum pH for growth of oyster and clam larvae was between 7.5 and 8.5. Some studies of the effects of various pollutants on the growth of lamellibranch embryos and larvae are reviewed in Chapter 3.

Growth efficiency

There are few available estimates of the growth efficiency of lamellibranch larvae. Jørgensen (1952) calculated a net growth efficiency (K_2; the percentage of assimilated food that is used in growth) of 73% for larvae of *M. edulis*. From his short-term feeding experiments with *Ostrea edulis* larvae. Walne (1965) estimated that 68–80% of assimilated food was used in growth. Gabbott & Holland (1973) estimated the metabolic demand of *Ostrea* larvae from the decline in metabolic reserves during short periods of starvation. They assumed that these values represented half the normal 'swimming' demand for energy, and then used these figures to calculate net growth efficiency; their calculations indicate a decline in K_2 from 78.6% between 0 and 2 days after release from the parent, to 55.5% after 6–10 days. Bayne (unpublished data) estimated a gross growth efficiency (K_1) of 65% for *M. edulis* veliconchas. These results all suggest, therefore, a high net efficiency of growth in lamellibranch larvae.

Biochemical composition

During the development of the larva of *Ostrea edulis* organic matter increases from 17 to 26.5% of total dry weight (organic matter plus shell), and the total dry weight increases from about 1.00 µg on the day of release from the adult (shell length 179.3 µm) to 6.83 µg after 12 days (shell length 296.8 µm) (Holland & Spencer, 1973). *Mytilus edulis* larvae of 130 µm length weigh on average 0.76 µg, of which 15.9% is organic matter.

Some data on the gross biochemical composition of *M. edulis* and *Ostrea*

Table 4.3. *The biochemical composition (μg per mg dry wt) of the veliconcha larvae of Ostrea edulis and Mytilus edulis*

	Lipid			Carbohydrate			RNA	Protein	Total organic matter
	Phospholipid	Neutral lipid	Total lipid	Polysaccharide	FRS*	Total			
Ostrea edulis (Holland & Spencer, 1973)									
On day of release	14.7	15.2	29.9	3.2	11.3	14.5	9.5	118.3	172.1
Day 8	18.3	42.8	61.1	7.2	9.3	16.5	12.6	149.6	239.8
Day 12	17.8	61.8	79.6	10.0	9.1	19.1	10.3	158.6	266.6
Mytilus edulis† (Bayne et al., 1975)									
Before starvation	7.5	9.7	17.2	—	—	7.1	—	88.3	112.6
After 2 days starvation	6.9	5.8	12.7	—	—	5.1	—	69.9	87.7

* Free reducing substances.
† Mean length 130 μm.
— Not measured.

edulis larvae are listed in Table 4.3. Developing *O. edulis* larvae accumulate neutral lipid, the percentage increasing from 8.8% of total organic matter in the young veliger to 23.2% at metamorphosis. Much of this lipid is then utilised during metamorphosis when the larvae are unable to feed (Holland & Spencer, 1973). The percentage content of phospholipid, polysaccharides and free reducing substances, and the RNA:protein ratio do not change significantly during larval development (see also Chapter 8).

A striking feature of the gross biochemical data is the high proportion of lipid and relatively low proportion of carbohydrate in the larval tissues. Crisp (1975) has pointed out the great advantages of storing lipid. Lipid provides nearly twice as much energy when catabolised than does protein or carbohydrate and it also confers buoyancy on the larva, so reducing the energy cost of maintaining position in the water column. Crisp (1975) used a metabolic rate of 5 ml O_2 h^{-1} g^{-1} to calculate the potential longevity of larvae which relied entirely on their energy reserves, assuming various proportions of their body weight to be engaged as energy stores (see Table 8.2). The calculation makes it clear that a considerable store of lipid and/or protein is necessary if the larvae are to be able to survive for more than a day or two without feeding.

The large lipid stores in lamellibranch larvae are indeed used as an energy reserve during planktonic life. This was confirmed for *Ostrea edulis* by Millar & Scott (1967b) and Holland & Spencer (1973), and by Bayne, Gabbott & Widdows (1975) for *M. edulis*. Holland & Spencer showed neutral lipids to account for 41% of total organic matter lost during starvation, whereas protein accounted for 34% and carbohydrate 25%. Bayne, Gabbott & Widdows (1975) measured relatively greater losses of protein during starvation by *M. edulis* larvae; the ratios of protein:lipid:carbohydrate losses were 1.0:0.24:0.11.

The lipid that provides the energy reserve for the larva is derived in part from the female parent, as deposited in the egg during vitellogenesis, and in part from food assimilated during larval life. Helm *et al.* (1973) showed a correlation to exist between initial growth rate of oyster larvae and the lipid content of the larva on liberation from the parent. They could find no such relationship for protein, carbohydrate or RNA. Earlier Collyer (1957) had concluded that there was no correlation between larval viability and glycogen content. Bayne, Gabbott & Widdows (1975) pointed out the importance of phospholipids in the developing embryo of *M. edulis*.

Metamorphosis

The delay of metamorphosis

The ability of some invertebrate larvae to delay metamorphosis in the absence of specific environmental stimuli has been widely observed

(Thorson, 1950; Wilson, 1952). Bayne (1965) recorded that the pediveliger larva of *M. edulis* could delay metamorphosis for up to 40 days at 10 °C, and 2 days at 20 °C. The duration of this period of delay was not affected by ration, but salinity had a minor effect. During the delay of metamorphosis the rate of growth declined to zero and the velum gradually degenerated to the point where feeding was impossible and swimming was impaired. The ability to delay metamorphosis whilst at the same time being further dispersed from the parent population, might have important consequences in the ecology of the species.

Settlement and substrate selection

The act of 'settlement' consists of the descent, by the pediveliger, from the plankton to the sea bottom, followed by a pattern of swimming and crawling behaviour that ends with the secretion of the byssus threads that attach the larva to the substrate and signals the beginning of benthic existence. The morphological changes that follow have already been discussed. Cranfield (1973 *a, b, c*) has described details of settlement and attachment by the larvae of *Ostrea edulis*.

Lamellibranch larvae, in common with other invertebrate groups (Wilson, 1952; Scheltema, 1961; Thorson, 1966; Crisp, 1974), are capable of discriminating between different substrata at settlement. A common pattern characterises the behaviour of the larva at this time. This consists of a sequence of crawling movements with the foot, initiated by a hierarchy of stimuli. If the stimuli are 'positive' there is a progressive reduction in the distance moved over the substrate, culminating in attachment. If at any stage 'negative' stimuli are received, the pediveliger is capable of withdrawing the foot and swimming away from the substrate. J. Verwey (personal communication) suggested that pediveligers of *M. edulis* can secrete mucus threads while swimming, and that these facilitate contact with the substratum and possibly function in subsequent crawling behaviour. Cranfield (1973 *a, b, c*) discussed the role of mucus secretions in the settlement of oyster larvae, but Lane & Nott (1975) did not detect any 'trailing byssus' in *M. edulis* larvae.

The exact sequence of stimulus and response has not been described for any mussel larva, although some data on larval preferences for specific substrates are available. Chipperfield (1953) observed that *M. edulis* larvae attach most abundantly in areas with some shelter from water currents and wave action, and he suggested that surface discontinuities provide a stimulus for byssus secretion. De Blok & Geelen (1958) showed that *M. edulis* larvae attached most readily to various filamentous substrata, and Bayne (1965) confirmed this in laboratory experiments. The association between mussel plantigrades and filamentous substrata has been recog-

112

nised for many years (Bayne, 1964*b*; see also Chapter 2). Davies (1974) has described a synthetic filamentous material suitable for monitoring the settlement of *M. edulis* larvae.

Kiselva (1966*b*) recorded that the larvae of *M. galloprovincialis* settle on the filamentous alga *Cystoseira*, and that treating the alga with ethyl alcohol in a Soxhlet apparatus did not diminish its 'attractiveness' to the larvae. Kiselva also observed that a living bacterial and/or algal film on the substrate, and also the presence of other, recently metamorphosed plantigrades, both accelerated the settlement of *M. galloprovincialis* larvae in the laboratory. The possibility that byssus threads may encourage attachment has apparently not been examined experimentally, although various observations in nature suggest that this may be the case.

Behaviour

Observations in the laboratory

Lamellibranch larvae swim with a spiral movement propelled by the beat of the large marginal cilia on the velum. Periods of vertical rising are interspersed with periods of sinking (Isham & Tierney, 1953; Konstantinova, 1966). The larvae may sink either by withdrawing the velum between the shell valves (Carriker, 1961), by actively swimming downwards (Isham & Tierney, 1953; Lough & Gonor, 1971) or, more slowly, with the velum uppermost and the velar cilia beating to retard the rate of fall (Cragg & Gruffydd, 1975). The vertical pattern of upward swimming permits a directional response to environmental stimuli, (i) by varying the time spent in each horizontal component of the spiral, resulting in a change in horizontal direction of travel, and (ii) by varying either the diameter of the spiral component or the number of spirals per unit depth of water, or both, resulting in a change in the rate of vertical travel. The normal orientation during vertical swimming is probably the direct result of the uneven distribution of mass in the larval body, with the heavy shell lowermost and the velum uppermost.

Cragg & Gruffydd (1975) measured the swimming paths of *Ostrea edulis* larvae and were able to distinguish between mean vertical velocity, mean actual velocity, the path length per unit vertical rise, the diameter of the spiral and the amount of vertical rise per complete spiral. This type of study allows a quantitative analysis of the locomotory response to various stimuli. In a qualitative study of the swimming of *M. edulis* larvae, Bayne (1963, and unpublished observations) observed the responses to changes in hydrostatic pressure. At 20 °C, under low-intensity and diffuse light, at atmospheric pressure, veliconcha larvae swam slowly, in broad spirals with a small vertical component. Immediately the pressure was raised, the vertical component was increased, and the horizontal component decreased, resulting in more rapid movement upwards through the water column. The mean vertical velocity could be varied between 0 and 4 mm s⁻¹. Cragg & Gruffydd (1975) showed that increased upward swimming by oyster larvae in response to a pressure increase of 2 bars was partly due to a greater velocity component in the spiral path and partly to a greater actual swimming velocity. The pressure threshold for young *Ostrea* veligers was < 0.1 bar, and this increased to between 0.1 and 0.2 bar after 11 days. This compares with a threshold of < 0.1 bar for the veligers of *Mercenaria mercenaria* (Haskin, 1964) and < 0.54 bar for *M. edulis* veliconchas (Bayne, 1963). Bayne reported that eyed veligers of *M. edulis* responded more slowly to increase in pressure than did early veliconchas, and the pediveligers did not respond to pressure. In other experiments, using directional light stimuli, Bayne (1964*a*) observed that mussel larvae could respond by varying both horizontal and vertical directions of travel.

Some responses by different stages in the larval development of *M. edulis* to light, gravity and pressure are summarised in Table 4.4. Bayne (1964*a*) suggested that all these responses had different threshold intensities in each larval stage and that temperature had a marked effect on both the sign and the intensity of the responses. Taken together, these behaviour patterns would tend to keep the shelled larvae near the surface until development of the pediveliger, when negative phototaxis and a failure to respond to pressure stimuli would encourage the larvae towards the bottom.

Observations in the field

Recorded speeds for the vertical movement of lamellibranch larvae are in the range 0.15–10 mm s⁻¹. Mileikovsky (1973) compared these speeds with published values for holoplankton organisms and concluded that veliger larvae are probably able to control their vertical distribution in estuarine and nearshore waters, even in the presence of tidal currents. Wood & Hargis (1971) noted that even in an estuary with horizontal tidal currents that may exceed 800 mm s⁻¹, the rate of vertical water movement may be as low as 0.1 mm s⁻¹, which is low enough to allow an oyster larva to negotiate 10 m in about 15 min. Verwey (1966) suggested a diurnal pattern of vertical distribution of *M. edulis* larvae in the Oosterschelde (Holland). In a thorough analysis of the distribution of the larvae of *Mercenaria mercenaria*, Carriker (1961) concluded that the veligers were able to control their vertical distribution to some extent and were 'more than passive buoyant particles in the water'. Carriker suggested that the larvae were stimulated to swim by turbulence.

Carriker (1961) and Wood & Hargis (1971) have reviewed the literature on the vertical distribution of lamellibranch larvae in estuaries. Much of

Table 4.4. *A summary of the sign and relative intensity of the behavioural responses of different stages in the larval development of* Mytilus edulis *to light (directional stimulus), gravity and pressure*

| | Environmental stimulus | | |
Larval stage	Light	Gravity	Pressure
Trochophore	0	0	0
Veliger	–	0	0
Veliconcha	+++	0	+++
Eyed veliger	++	– –	+
Pediveliger (swimming)	– –	+	0
Pediveliger (crawling)	– – –	– –	0

From data in Bayne (1963, 1964*a*).
0. no response; –, negative response; +, positive response.

this literature is concerned with the retention of larvae within the estuary, or with the supply to the up-river adult populations of larvae that are ready to settle. Larval retention within estuaries does occur, and Wood & Hargis (1971) point out that the important question is whether evolved patterns of larval behaviour contribute to this process of retention. These authors compared the distribution of oyster larvae in the James River estuary (USA) with the distribution of coal particles of similar size and density. Larvae and coal were distributed differently in time and in space. Maximum concentrations of larvae coincided in most cases with salinity increases at flood tide, whereas maximum concentrations of coal particles coincided with current speed maxima, regardless of whether the tide was in flood or ebb. There was a net transport of larvae up-river, and Wood & Hargis conclude that 'selective swimming' by the larvae did occur, confirming the earlier hypothesis of Nelson (1912). The sensory bases of the observed patterns of selective swimming remain to be elucidated (Haskin, 1964), but one possibility which has not been sufficiently examined is the role of pressure changes, combined with salinity change as the tide floods, in initiating vertical swimming by the larvae.

Ecology

Growth, mortality and recruitment

Mussel larvae require between 15 and 35 days to grow from fertilisation to the stage (the pediveliger) when settlement and metamorphosis first become possible. As discussed earlier, the duration of the larval life depends on the available ration and on temperature, salinity and other variables, but evidence from samples taken in the field suggests that a

115

larval period of 3 weeks is a reasonable approximation. During this time the weight of the larva increases from about 0.1 μg of organic matter to about 1.0 μg. The efficiency of this growth is high (50–70% net efficiency) and the larva probably requires a ration of between 30 and 60% of its own weight per day. The larvae feed on the nanoplankton and their efficiency of assimilation is also high, about 40–70%.

Mortality during the free-swimming larval period is considerable, possibly approaching 99% (Thorson, 1946, 1950; Mileikovsky, 1971). The main mortality factors are predation, excessive dispersal (dispersal to areas where no suitable sites for post-larval survival exist) and death due to extreme physical factors. Many vertebrate (including herring; Lebour, 1933; Legaré & Maclellan, 1960) and invertebrate (Thorson, 1946) predators feed on the free-swimming larvae. Mileikovsky (1959, 1971) estimated that an average of 3% of the standing stock of lamellibranch larvae in Velikaya Salma Bay (White Sea) were eaten daily by the larvae of the polychaete *Nepthys ciliata*. One source of larval mortality that has received some attention is the ingestion by filter-feeding adult invertebrates of the free-swimming larvae of the same or different species (Thorson, 1946, 1966; Bayne, 1964b; Hancock, 1973; Mileikovsky, 1974). Mileikovsky concluded that although larvae ingested in this way will not normally survive, the quantitative importance to the larvae, and to the benthic communities, remains to be elucidated.

Excessive dispersal, although not strictly affecting larval survival directly, can represent a significant mortality to the species population as a whole. The large numbers of *M. edulis* larvae recorded by Rees (1954) in the North Sea probably represented a considerable wastage to the coastal parental stocks. Ayers (1956) applied the term 'dilution' to those processes, such as flushing-out to sea, which produce a decrease in population density without any of the population dying. Ayers considered the combined effects of mortality (by predation and all other causes of death) and dilution, using the equation:

$$P_t = P_0(1-f)^t e^{-kt},$$

where P_0 is the initial population size, P_t the size of the population which both survives and remains in the spawning area, t is time in tidal cycles, k is the coefficient of mortality and f is the fraction of water lost from an embayment or estuary during one complete tidal cycle. This model does not consider the possiblity of larval retention within the estuary by control of their own vertical movement, and it therefore represents an extreme case. Nevertheless it can provide a useful first approximation in some studies of the dynamics of bivalve populations. Ayers (1956) considered that forty plantigrades per adult breeding pair per year represented a stable condition for populations of *Mya arenaria*. In Fig. 4.12 this stable condition is related to flushing rate (f) and mortality (k).

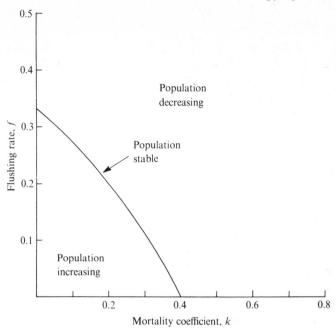

Fig. 4.12. The stability of an estuarine bivalve population relative to a mortality coefficient (*k*) and the flushing rate of the estuary (*f*). The 'population stable' curve represents a recruitment of forty plantigrades per year per breeding pair of adults. (Based on a model of the population dynamics of *Mya arenaria* by Ayers, 1956.)

The mortality of larvae due to extremes of temperature, salinity or ration is probably of little importance when compared with other factors such as predation (Thorson, 1950). The temperature and salinity tolerances of mussel larvae are relatively broad (p. 105) and probably reflect the environmental conditions experienced by the parent population. Equally, the ability of larvae to survive for long periods with no apparent food (p. 108) suggests that death due to starvation may be rare. Variations in these conditions probably have their greatest effect in prolonging the free-swimming period and increasing the chances of losses due to predation or dilution (Thorson, 1946, 1950).

Thorson (1966) identified three stages of 'selection' likely to affect the recruitment of pelagic larvae to benthic stocks. The first is due to the temperature and, primarily, the food characteristics of the particular water mass; a change in either will affect the chances of successful metamorphosis by a particular species. The second selection is associated with stratification of the water mass in coastal or estuarine areas, which might restrict the larvae to a certain stratum and render unavailable particular areas of the sea-bottom. Thorson's third selection occurs at the time of settlement, when the larvae select particular substrates for metamorphosis. Laboratory experiments (p. 112) confirm Wilson's (1952) hypothesis of substrate-selection for bivalves. Thorson (1966) suggests that under natural conditions selection by larvae will be more coarse than it is in laboratory experiments, but nevertheless such selection in nature has been confirmed (Muus, 1973).

There is, therefore, a considerable qualitative understanding of the factors likely to determine the success with which mussel larvae complete their pelagic development. Hancock (1973) has summarised these as follows:

(*a*) suitable environmental conditions, notably temperature;

(*b*) adequate food supply;

(*c*) predation;

(*d*) accidental ingestion by benthic filter-feeding adults;

(*e*) contact with areas and conditions favourable for settlement.

However, the quantitative expression of these factors for any particular stock is not yet possible. Hancock (1973) reviewed twelve years' data on the dynamics of a population of the bivalve *Cardium edule*, and he showed a negative correlation to exist between number of recruits and the size of the total stock. Hancock concluded that a tendency towards overpopulation is accompanied by various density-regulating mechanisms, such as reduced fecundity, poor recruitment and reduced growth rate, all of which tend to favour the better survival of individuals already established in the population, and act against further additions. The relationship between stock and recruitment in mussels is not so apparent, due at least in part to the lack of sufficient data.

The reproductive 'strategy' of mussels

In recent years papers by Vance (1973*a*, *b*), Strathmann (1974) and Crisp (1975) have sought to quantify and generalise discussions initiated by Thorson (1946, 1950) and developed by Mileikovsky (1971, 1972) on the relative advantages of the various types of larval development in marine invertebrates. Three types of development are recognised – planktotrophic, lecithotrophic and direct development (Thorson, 1946; Ockelmann, 1965) – and Mileikovsky (1971) adds a fourth, the demersal larva. Mussels are planktotrophic and their reproductive 'strategy' is one of high fecundity, small eggs, external fertilisation and a pelagic larva that feeds on the phytoplankton.

The advantages of planktotrophy over lecithotrophy are usually considered to include; (i) better possibilities for wide dispersal and the invasion of new habitats; (ii) more rapid recovery from damage to a population, as larvae recolonise the area from outside; (iii) the possibility of rapid population expansion under favourable conditions; and (iv) a trophic advantage due to the larvae feeding upon the phytoplankton; this in turn allows a reduction in egg size and increased fecundity. This last advantage is usually considered to be balanced by a high larval mortality.

Vance (1973*a*, *b*) considered, in particular, the possible trophic advantages of planktotrophy. He assumed that a certain minimum amount of energy was required for successful metamorphosis, and he viewed larval development as consisting of two successive stages of varying length, viz. a non-feeding stage during which nutrient reserves support metabolic activity, followed by a feeding stage during which nutrient is gained from the plankton. Planktotrophy and lecithotrophy then, represent the extremes of the transition between non-feeding and feeding stages. Vance suggested that the maximum reproductive efficiency (the number of metamorphosing larvae per unit of energy devoted to the eggs) will result either from complete planktotrophy, with egg size as small as is consistent with viability, or complete lecithotrophy; all intermediate patterns conferred reduced efficiency. Further, when planktonic predation is low and planktonic food is abundant, planktotrophic development is favoured. Vance acknowledged that his model did not consider the possible advantages accruing from dispersal.

Crisp (1975) argued that it is still unresolved whether there is any net gain in energy as a result of planktonic feeding, i.e. whether gain in larval growth balances predation losses. Crisp examined with a simple model the advantages of dispersal that might accrue from planktotrophy in an heterogeneous environment. He concluded that even when the trophic advantage is nil, there remains a benefit in having a pelagic larva whenever the population density is unevenly distributed among isolated parts of the habitat. Since the effects of dispersal are to even-out recruitment over a wide range of habitats, the effects of density-independent mortality are also evened, allowing the maintenance of a more uniform adult population.

In a closely reasoned paper, Strathmann (1974) concludes that the short-term advantages of large-scale dispersal are due primarily to the spread of sibling larvae rather than to the general spread of larvae of the entire population. Strathmann suggests that larvae settling in opportune areas nevertheless subsequently disperse their offspring, and cannot take advantage of the favourable local conditions by limiting their dispersal. Nor does large-scale dispersal increase in response to deteriorating conditions. In addition, because of gregarious settlement behaviour, dispersal often does not result in an escape from crowding. Dispersed larvae may derive some benefit from their ability to choose suitable sites for settlement, but since larvae are unable to take such advantage of suitable sites during much of the larval period, substrate selection cannot be the only advantage of a long pelagic larval stage. On the other hand, dispersal of offspring is favoured when the probability of survival, or of reproductive success, varies independently in time and in space (see also

Crisp, 1975). 'When the parental organism cannot predict whether its present area is likely to provide better opportunities for survival and reproduction than some other area, then dispersal each generation is the optimal strategy' (Strathmann, 1974; p. 32). Parents that spread their offspring even out the variation in survival and reproduction among areas, and therefore have the most uniform success in each generation. Strathmann (1974) ends by considering some of the ways in which the spread of sibling larvae may be maximised.

These papers provide a contemporary framework for the assessment of marine invertebrate reproductive adaptations. Future studies should be directed towards the measurement of fecundity, physiological and genetical differences, and the survival of recruits in many populations over time, in order to determine the flexibility of the reproductive strategy of the genus.

5. Physiology: I

B. L. BAYNE, R. J. THOMPSON & J. WIDDOWS

Physiological ecology is the study of how an animal is adapted to function in its particular environment. Such a study must begin with knowledge of the natural conditions which are normally experienced by the organism. It is necessary to know, for example, that *Mytilus edulis* is a benthic (littoral and near sublittoral) semi-sessile species; that it may experience wide fluctuations in salinity and temperature; that it is a particulate suspension-feeder, relying for its food on a seasonally variable supply of particles within a limited size range; and that, in its intertidal habit, it is exposed to air for varying periods of time with all the resulting problems of potential desiccation, thermal shock and lack of oxygen. In addition, the organism's tolerance limits must be appreciated, preferably as a result both of observations 'in the field' and of experiments in the laboratory. Fortunately, there is a considerable literature that deals with these aspects of the ecology of mussels. Finally, in a study of the physiological ecology of a species, the flexibility of the individual's response to changes in the environment must be determined. All species are capable of some degree of compensation for environmental change, and the limits of this capacity for physiological adaptation must be appreciated if we wish to achieve an understanding of the species' ecology. Fry (1947) has clarified the distinction to be made between an organism's 'zone of resistance' and its 'zone of tolerance'.

When assessing an organism's capacity to compensate physiologically for changes in its environment, it is important to establish the relevant steady-state values for particular physiological rate functions, for it is these that will describe the basic pattern of the response. This may require long-term experiments with frequent measurement of the particular physiological processes. But in addition, shorter-term experiments are necessary, in which the dynamic nature of the physiological mechanisms are examined by observing their responses to sudden environmental changes. Kinne (1964b) has distinguished between the immediate response to an environmental change, the stabilisation of this response, and the new steady-state. Taken together, information on all of these states is likely to yield the most complete understanding and possibilities for prediction of how the animal functions in its normal world.

In this and the following chapter, we discuss some of the considerable amount of published information on the physiological ecology of mussels with these principles in mind. We have chosen to consider separately each of the main physiological systems although this has, inevitably, led to some duplication since, when emphasising the performance of the 'whole organism', it is often impossible to make absolute distinctions between certain of the physiological processes involved. In this chapter we consider the processes of feeding and digestion, and the effects of environmental factors on respiration and metabolic rate. In Chapter 6, the circulation of the blood, excretion, ionic and osmotic regulation and, briefly, the mechanisms of chemical and nervous co-ordination, are discussed.

Feeding and digestion

Gill structure in some bivalves, especially the Ostreidae, has been examined in considerable detail (see reviews by Owen, 1966a, 1974b; Jørgensen, 1966; Purchon, 1968), but the gross morphology of the gill in the Mytilidae is not particularly well described. Paradoxically, there are several excellent accounts of the fine structure of the gill filaments in *Mytilus edulis*, including details of their ciliation and innervation.

The gills consist of four pairs of demibranchs which separate the pallial cavity into inhalant and exhalant (or suprabranchial) chambers throughout its length (Fig. 5.1). Each demibranch comprises two lamellae, one attached to the gill axis (descending limb), the other lying free (ascending limb). The two lamellae forming each demibranch are held together by connective tissue junctions (interlamellar junctions) which link the ascending and descending limbs. Individual lamellae are formed by rows of ciliated filaments. Lateral cilia are responsible for moving water through the ostia (the spaces between pairs of adjacent filaments), latero-frontal cilia remove particles from the inflowing water and frontal cilia transport the particles to acceptance or rejection tracts at the margins of the lamella (Fig. 5.2). In the Filibranchia (which includes the Mytilidae), individual filaments are discrete, although adjacent ones are loosely joined by ciliated discs (Murakami, 1962).

Water enters the pallial cavity through the inhalant siphon (which is continuous along the entire length of the ventral surface) before passing through the gill ostia into the suprabranchial chamber, and is expelled through the exhalant siphon, which is narrower than the inhalant and lies at the posterior edge of the mantle, immediately dorsal to the inhalant margin. Both inhalant and exhalant siphons possess a velum (Dodgson, 1928) which can regulate the current flows. The inhalant velum serves another important function, since it can deflect the incoming current to prevent it striking the gill directly.

Food particles are bound into mucus strings on the gill lamellae and are carried to the labial palps via ciliated grooves on the lamellar margins. The palps regulate the amount of food which enters the mouth, and direct surplus material on to the rejection tracts of the mantle surface. From the

122

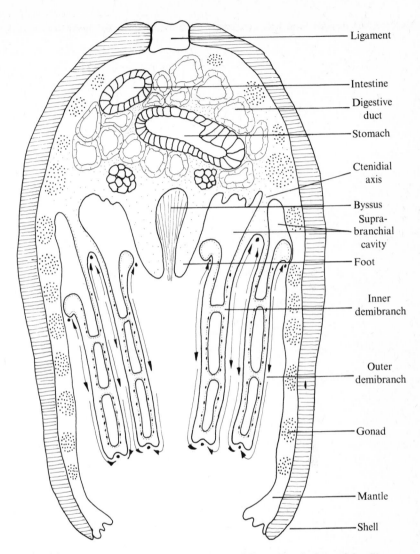

Fig. 5.1. Diagrammatic transverse section through *Mytilus edulis* to show the form of the gills and the direction of the main ciliary currents, represented as arrows.

palps, food is directed into the mouth, passes through the oesophagus and enters the stomach, which is a complex structure containing a system of ciliary tracts which sort the particulate material and direct it either towards the digestive tubule duct openings or towards the intestine. Intracellular and extracellular digestion occurs in the digestive gland, which also stores nutrient reserves and regulates their transfer to other tissues.

123

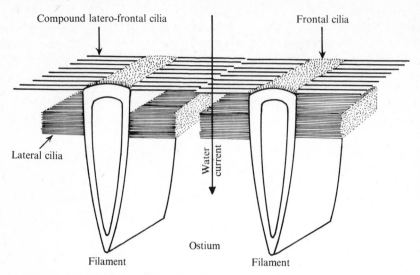

Fig. 5.2. A schematic representation of two filaments on the gill of *Mytilus edulis*. The cilia are at rest. (Redrawn from Dral, 1967.)

The digestive gland envelops the stomach and part of the intestine; the intestine passes through the heart and pericardial cavity, over the posterior adductor muscle and opens into the suprabranchial chamber, adjacent to the exhalant siphon.

Function of ctenidia in feeding

The mechanism by which the lamellibranch gill removes suspended particles from the water pumped through the mantle cavity aroused considerable interest when it became apparent, in the early nineteenth century, that the ctenidia play a role in feeding as well as in gas exchange. Alder & Hancock (1851) suggested that the gill acts as a sieve, the latero-frontal cilia forming the meshwork. This concept provided the basis of the 'classical' view of filtration by the bivalve gill, a view which has only recently been modified (Dral, 1967; Moore, 1971; Owen, 1974a). Observations of filtering behaviour and measurements of filtration rates revealed two possible anomalies in the classical theory. Firstly, considerable variations were observed in filtration rates recorded from mussels of the same species under similar conditions (Willemsen, 1952; Tammes & Dral, 1955; Jørgensen, 1960; Theede, 1963; Chappuis & Lubet, 1966). In some cases, fluctuations in filtration rates were observed in short-term experiments with individual animals. Although some of this variation was attributable to differences in experimental method, and in some instances to poor technique, it became evident that bivalves are able to control the rate at which suspended matter is removed from the surrounding water (Jørgensen, 1966; Owen, 1966a).

Jørgensen (1949, 1959, 1960) and Jørgensen & Goldberg (1953) demonstrated retention by *M. edulis* of graphite particles 1–2 μm in diameter. Zobell & Feltham (1938) found that *M. californianus* can remove bacteria from suspension, and according to Duff (1967) *M. edulis* is capable of filtering viruses (diameter 25 nm) from the surrounding medium. These observations, however, do not provide unequivocal evidence that individual small particles are retained by the gill, since larger particles may have been formed by aggregation of micro-organisms or small graphite particles in the static system used in these experiments. Nevertheless, Vahl (1972) has demonstrated high retention of small motile flagellates (1–2 μm diameter) by undisturbed *M. edulis* in flowing water. These observations led to a second difficulty with the classical model. The distance between the latero-frontal cilia in *M. edulis* is 2–3 μm (Dral, 1967), so retention of particles smaller than 2 μm appears to be inconsistent with a concept of the gill as a sieve. An alternative possibility, first suggested by Wallengren (1905), is that particles may be removed by adhesion to the latero-frontal cilia rather than by retention in a meshwork. Foster-Smith (1975a) has observed that alumina particles (5 μm diameter) often adhere to gill cilia in *M. edulis*. Although this mechanism appears to play a role in feeding, it is unlikely to account for the high retention efficiencies recorded for small particles by Vahl (1972).

An alternative mechanism for the retention of particles by the lamellibranch gill was proposed by MacGinitie (1941), who maintained that a mucus sheet was secreted over the gill during feeding. Such a mechanism would account for removal of very fine particles from suspension, and from this point of view the hypothesis was not unattractive. Jørgensen (1949, 1966) attempted to reconcile MacGinitie's observations with the conventional 'sieve' concept by envisaging the mucus sheet in operation only under conditions of poor food supply, whereas the latero-frontal meshwork would function at higher particle concentrations. Nevertheless, it is difficult to visualise latero-frontal ciliary function in the presence of a sheet of mucus, so that the two proposed mechanisms appear to be mutually exclusive. Other objections to the mucus net theory are discussed by Owen (1966a). Subsequent detailed analyses of the function of latero-frontal cilia in filtration by *Mytilus* (Dral, 1967; Moore, 1971; Owen, 1974a) have shown how small particles are efficiently retained without the formation of a mucus net, and stereoscan electron microscope studies have failed to demonstrate any mucus sheets on the gill (Moore, 1971; Owen, 1974a).

The controversies regarding the mechanism of particle retention could not be resolved without a critical evaluation of the behaviour and function

125

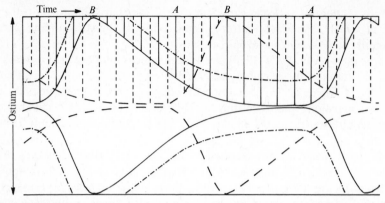

Fig. 5.3. Projections of the latero-frontal cilia on the ostium during the execution of the beat (*Mytilus edulis*). The horizontal axis represents time. The solid and the dashed lines are the projections of two adjacent latero-frontal cilia in water with a low concentration of suspended matter. The dash–dot line is the projection of a latero-frontal cilium in water with a high concentration of suspended matter. For further discussion, see the text. (Redrawn from Dral, 1967.)

of the latero-frontal cirri. Dral (1967) analysed the latero-frontal cirral beat from observations (using stroboscopic illumination) on young, intact *Mytilus edulis* possessing transparent shells and negligible mantle tissue. The cirral beat has two important features. Firstly, alternating cirri on any given filament (Fig. 5.3) beat synchronously, differing by approximately half a phase from the two adjacent ones. Secondly, the velocity of the beat varies during the cycle. The resulting projections of the latero-frontal cirri on the ostium are shown in Fig. 5.3. At *A*, all cirri are lying across the ostium and the mesh is equal to the intercirral distance (3 μm). Alternate cirri then execute the slow part of the cycle and remain straightened over the ostium, whilst the adjacent cirri enter the fast phase, opening the ostium (*B*) before moving across it again (*B* → *A*). Thus at *B*, alternate cirri effectively close the ostium and their neighbours open it, so that the mesh size is twice the intercirral distance, i.e. 5–6 μm. In dense suspensions, the amplitude of the cirral beat is reduced so that the ostium is only partly closed and retention efficiency is decreased. At any given time the degree of the amplitude reduction may vary from one part of the gill to another, indicating that latero-frontal cilia as well as lateral cilia may be locally controlled (see section on control of ciliary beat, p. 131). *Mytilus* therefore appears to have the capacity for local and independent control of both filtration and ventilation.

Whereas Dral's analysis demonstrated the considerable degree to which *Mytilus* can control the filtration mechanism, it did not account for the high retention of small particles observed by Vahl (1972) and others. This problem was resolved when Moore (1971) and then Owen (1974*a*)

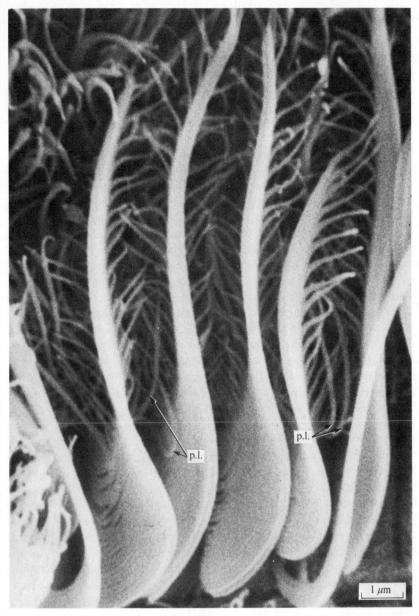

Fig. 5.4. A scanning electron micrograph of the latero-frontal cirri of *Mytilus edulis* as viewed from the abfrontal aspect and showing the pro-latero-frontal cilia (p.l.) projecting between the bases of the cirri. (From Owen, 1974*a*.)

examined the structure of the latero-frontal cirri using scanning and transmission electron microscopy. In *Mytilus edulis* each cirrus comprises twenty-two to twenty-six pairs of cilia arranged in two parallel but alternating rows (Fig. 5.4). Individual cilia leave the shaft of the cirrus at regular intervals on each side to extend across the intercirrus space, producing a complete meshwork of lateral branches between adjacent cirri in the same phase. The limiting dimension of the meshwork is in fact the distance between adjacent cilia on the same side of the cirrus shaft (0.6 μm), not the distance between alternate cirri (4.8 μm) as Dral supposed. The mesh therefore has the dimensions 0.6 μm by either 2.4 μm or 4.8 μm, depending on the relative position of adjacent cirri. It should be remembered that the mussel possesses the capacity to vary the dimensions of the ostium (Dral, 1967) but the meshwork formed by the cirri on each filament appears to be more or less constant (Owen, 1974a).

Retention of fine particles, therefore, is accomplished by the meshwork of cilia on the latero-frontal cirri. There is no evidence to suggest that individual cilia themselves undergo a beat cycle independently of the cirrus. In fact, Owen (1974a) found that in *M. edulis* each cilium contains an electron-dense structure which appears to stiffen the cilium, so that the free lateral regions of the cilia essentially maintain a constant relationship to the cirrus shaft throughout the latter's cycle. Control of retention efficiency may be achieved by regulating the amplitude of the cirrus beat, thus varying the free or non-filtered space in the ostium. There are indications that this regulation of cirrus beat amplitude is local, adding another level of complexity to the control mechanism.

The control of particle retention efficiency is independent of the ventilation current produced by the lateral cilia. Were this not so, regulation of retention efficiency could not be achieved without an alteration in the parameters of gas exchange, which could impose serious limitations on the effectiveness of the gill for this purpose. A filtration control mechanism which is not coupled to the respiratory pump contributes substantially to the efficiency of the gill in its dual function of particle retention and gas exchange. Thus in suspensions sufficiently dense to saturate the gill filter, the ostia may be opened to permit the through-passage of a proportion of the particles entering; there is no need for the mussel to reduce the ventilation current, thereby impairing gas exchange, or to close the valves together, although this latter does happen in extremely dense suspensions. Conversely, in very dilute suspensions filtration activity may cease altogether, although oxygen uptake, and presumably ventilation rate, remain normal (Thompson & Bayne, 1972). The overall efficiency of the gill is demonstrated by the fact that under favourable conditions the efficiency of oxygen utilisation is low (approximately 5%; see Chapter 7) and filtration activity optimal (Bayne, 1967;

Thompson & Bayne, 1972), providing the animal with a good food supply whilst the gill retains the capacity for increasing the utilisation efficiency for oxygen under unfavourable conditions (e.g., low P_{O_2} (oxygen tension), Bayne, 1971*b*).

Movements of demibranchs or of individual lamellae may also be used to control the efficiency of the gill filter in *Mytilus edulis* (Foster-Smith, 1974). The mussel is able to close the ostia over the entire lamellar surface or parts of it. The outer demibranch can be moved towards the mantle (especially in the posterior part of the mantle cavity) and the inner demibranch can be moved towards the visceral mass, although this occurs only in dense suspensions. Such adjustments of the ostia and the position of the lamellae reduce the water flow through the gill, thereby decreasing the amount of food reaching the ctenidia and also causing a decrease in the oxygen available to the animal. To some extent this may be counterbalanced by a reduced oxygen demand, but the mussel does possess the capacity to maintain a steady rate of oxygen uptake under these conditions, partly by increasing heart rate (Chapter 7).

In dense suspensions, therefore, the mussel controls the amount of particulate matter coming into contact with the gill either by reducing the effective area of lamellar contact with the water, or by changing the beat frequency and pattern of the latero-frontal cirri. Both mechanisms have been observed in high particle concentrations (Dral, 1967, 1968; Foster-Smith, 1974) but there is no information regarding the relative importance of each under specified conditions. It is clear, nevertheless, that the gill is physiologically an extremely flexible structure with a considerable degree of local control.

Particles removed from suspension by the latero-frontal cirri are transferred to the frontal cilia. Some material may also impinge directly onto the frontal cilia, with the exception of the narrow zones adjacent to the dorsal grooves. Both dorsal and ventral grooves transport material anteriorly, i.e. towards the mouth.

The mucus-bound particles on the frontal cilia are conveyed to the marginal groove (which is very deep in mytilids) by one or more of three ciliary tracts bordering the groove (Foster-Smith, 1974). These appear to control the route taken by food strings when the mussel is feeding in dense suspensions. Fine material is channelled into the deepest part of the marginal groove, whereas coarser strings are directed to its superficial regions, from which rejection is more likely. In *Modiolaria marmorata* (= *Musculus marmoratus*), long cilia are present on the lobes of the marginal groove (Atkins, 1937*a*), allowing only small particles to pass into the groove.

Function of the cilia

The lateral and latero-frontal cilia of the gill of *Mytilus* have been studied extensively. Sleigh (1969, 1974) provides up-to-date and comprehensive reviews of this literature. We wish to comment here on one aspect only of recent studies by Aiello & Sleigh (1972) and Sleigh & Aiello (1972), i.e. on metachronism in the beat of the lateral cilia.

The presence of a metachronal wave in the beat of the lateral cilia of *Mytilus* has been recognised for some time. Aiello & Sleigh (1972) re-examined this subject using interference-contrast and scanning electron microscopy. They found that the beat of the lateral cilia is not planar, as was previously assumed, but that the cilia rotate in a clockwise sideways movement in the recovery stroke; the effective stroke is normal to the surface of the gill epithelium, but the recovery stroke is inclined to the surface. Aiello & Sleigh suggest that the form of this recovery stroke determines the pattern of metachronism in the cilia.

Sleigh & Aiello (1972) examined also the movement of isolated lateral cilia. They reported that the maximum velocity of particles transported by these cilia was approximately 1.2 mm s^{-1}. Since the lateral cilia are about 15 μm long and beat with an amplitude of 150°, each tip describes an arc of about 40 μm. At a frequency of 20 beats s^{-1}, and with the effective stroke completed in one-fifth the time of the entire cycle, the velocity at the tip of the cilium would be approximately 4 mm s^{-1}. Sleigh & Aiello (1972) estimated that water flows through approximately one-fifth of the total gill surface (an area of 3.4 cm^2). At a velocity of 1.2 mm s^{-1} this would constitute a total flow of 1.5 l h^{-1}, which is within the published range for ventilation rate.

Sleigh & Aiello (1972) also note that the dimensions of the interfilamentar space, the length of the lateral cilia and the length of the metachronal wave are such that no part of the water between the lateral epithelia is more than about 7 μm away from a lateral cilium in the effective stroke; this is well within the range of water movement caused by a rapidly beating cilium. In addition they note that the metachronal waves in the two groups of lateral cilia that face each other across the ostium move in opposite directions; this probably minimises the interference, between cilia, of their effective strokes, and means also that the frequency of ciliary beat is effectively doubled, since both rows of lateral cilia act on the same body of water. The metachronal rhythm in the lateral cilia has the result that the water flow through the gill is continuous, and this probably represents the main contribution of metachronism to functional efficiency.

Innervation of the gill and control of gill cilia

The possible role of the nervous system in regulating ciliary activity on the bivalve gill was first examined in *M. edulis* by Lucas (1931*a, b*), who confirmed Field's (1922) earlier observation that the gill is supplied by the branchial nerve, which originates in the visceral ganglion. Lucas, however, was unable either to demonstrate any ciliary response to stimulation of the branchial nerve or to trace any branches of the branchial nerve into the gill filaments proper. He therefore concluded that the branchial nerve is sensory and that the gill cilia are not under nervous control.

More recent studies on the mytilid gill have established that the epithelial cells are supplied by the branchial nerve, and also that the activity of frontal, latero-frontal and lateral cilia is regulated via the nervous system (Aiello, 1960, 1970; Aiello & Guideri, 1965; Paparo & Aiello, 1970; Paparo, 1972). There appears to be an excitatory mediator, probably 5-hydroxytryptamine, and an inhibitory mediator, possibly dopamine (Paparo & Aiello, 1970). The action of these mediators is discussed more fully in Chapter 6. A significant aspect of this work is the morphological and physiological evidence for independent control of the gill filaments. According to Aiello & Guideri (1965), approximately ten adjacent filaments are innervated by fibres from a single bundle. Such local control of filaments must clearly facilitate the regulation of ventilation and clearance rates which is known to occur, and is consistent with Dral's (1968) observation that ciliary activity can vary in different regions of the gill at any given time; in dense suspensions, for example, the mussel can 'shut-off' parts of the gill.

Stimulation of particular areas of the visceral ganglion gives rise to cilio-excitation on individual filaments, indicating that discrete control is facilitated by the distribution of post-ganglionic fibres. In the case of the ciliated lateral cells of the filament, only one of a series of such cells may be innervated (Paparo, 1972), but the presence of septate junctions connecting homologous ciliated cells suggests that the remainder are excited indirectly, by passive spread of current; the innervated cell possibly acts as a pacemaker.

Although there has been much discussion in the past of the possibility that metachronism in the lateral cilia was controlled by nervous mechanisms, Aiello & Sleigh (1972) and Aiello (1974) suggest that metachronal co-ordination is probably mechanical.

Filtration rate

The most commonly used measurement of filtering activity is filtration rate (clearance rate), which is defined as that volume of water completely cleared of particles in unit time. Filtration rate is frequently confused with ventilation rate (pumping rate), which is the volume of water flowing through the gills in unit time. If all particles present are removed by the gills, i.e. retention efficiency is 100%, filtration rate and ventilation rate have the same numerical value. There is evidence for complete or near-complete retention by mussels feeding undisturbed in suspensions of particles larger than 2 μm diameter (Vahl, 1972). Where retention efficiency is significantly below 100%, filtration rate is less than ventilation rate; in such circumstances the concept of filtration rate, as defined above, becomes somewhat abstract, but nevertheless provides a useful index of feeding activity.

Estimation of feeding rate in bivalves may be done either directly, by separating the inhalant and exhalant currents and measuring the water flow in the latter; or indirectly, by determining the rate of removal of particles from the water. Whereas the indirect technique provides an estimate of clearance rate, direct methods measure ventilation rate, which may exceed filtration rate (see reviews by Owen (1966a, 1974b) and Jørgensen (1966)). The following discussion will be confined to methods used in feeding studies on mytilid species.

Davids (1964) measured ventilation rate in *Mytilus edulis*, using a direct method (Drinnan, 1964) in which the animal was placed in a constant-level tank and the outflow measured from a tube inserted into the exhalant siphon. The same technique was employed by Chappuis & Lubet (1966) with *M. edulis* and *M. galloprovincialis*, and is similar to an earlier method used for *M. edulis* by Tammes & Dral (1955). (Tammes & Dral also determined particle densities in the separated inhalant and exhalant currents, thereby estimating filtration rate as well as pumping rate by a combination of the direct and indirect approaches.) In such procedures, care must be taken to ensure that the mussel is not pumping against a back-pressure and that water is not siphoning across the gills. A further disadvantage is the possibility that normal pumping activity may be disturbed by the constraints imposed by a tube in the exhalant siphon. Such factors may account for the considerable variation observed in the data from all these experiments, and for the very low ventilation rate values recorded by Chappuis & Lubet (30 ml h^{-1} g (wet wt)$^{-1}$ for a mussel 6 cm in length). An alternative direct method (Coughlan & Ansell, 1964), originally suitable for siphonate bivalves only, has been successfully modified by White (1968) for use with *M. edulis*. This technique, which depends on the introduction of a metered current of dye into the inhalant siphon, has proved more satisfactory than the more conventional direct methods previously described.

The indirect approach has proved the more popular, however, especially in its simplest form, namely a static system in which the mussel is confined in an appropriate volume of water containing a suspension of suitable particles (usually graphite or unicellular algae). Filtration results in a decrease in particle concentration which is monitored, either photometrically (Fox, Sverdrup & Cunningham, 1937; Jørgensen, 1949, 1960; Willemsen, 1952; Rao, 1953; Segal, Rao & James, 1953; Theede, 1963; Ward & Aiello, 1973) or by measuring the rate of removal of radionuclide-labelled plant cells from suspension (Kuenzler, 1961b; Allen, 1970). In such systems, filtration rate is calculated from the exponential decrease in particle concentration which occurs during the feeding period. The theoretical basis of the method is discussed by Coughlan (1969), who points out that the different forms of the exponential relationship which have been used by various workers are essentially the same.

The static system has certain disadvantages. Accumulation of ammonia and other excreted compounds may inhibit normal filtration behaviour. Reduction in P_{O_2} of the surrounding water may result in an increased ventilation rate and possibly an increase in filtration rate. The significance of these factors will depend on the duration of the experiment, the geometry of the vessel, the volume of water in relation to the size of the animal, and the animal's metabolic rate. The method also assumes that filtration rate is constant throughout the period of the experiment. The amount of particulate matter available to the mussel is continually decreasing, which introduces an uncontrolled variable into the filtration rate measurement and also limits the duration of the experiment, since the particle concentration ultimately approaches the minimum detectable level. In addition, the animal must do work against the inertia of static water, particularly at the beginning of an experimental period, and this is likely to reduce the energy available for passing water across the gills. The very low filtration rate values recorded in some experiments (Tammes & Dral, 1955; Allen, 1970; Ward & Aiello, 1973) may well be attributable to this factor, which, it should be noted, is also a feature of most direct measurement experimental systems. However, other values recorded in static systems by indirect methods (Fox *et al.*, 1937; Jørgensen, 1949, 1960; Willemsen, 1952; Rao, 1953; Segal *et al.*, 1953; Theede, 1963) are comparable with measurements subsequently made in flow systems. The problem of maintaining the animal in a constant particle concentration may be circumvented by the addition of fresh cells to replace those removed by the animal. Filtration rate may then be calculated from the number of additional particles required to maintain equilibrium. This technique has been used very successfully by Winter (1969, 1970, 1973).

Marine mussels

The introduction of the Coulter counter to marine science (Sheldon & Parsons, 1967) greatly facilitated rapid and accurate determinations of the density of particles suspended in seawater. Thus it became feasible to measure filtration rate in a flow-through system in which the concentration of particles in the inflowing water is kept constant and the particle concentration in the outflow monitored for a suitable period (Bayne, 1971*b*; Widdows & Bayne, 1971; Thompson & Bayne, 1972, 1974; Vahl, 1972, 1973; Walne, 1972; Widdows, 1973*a*, *b*; Bayne, Thompson & Widdows, 1973). Flow systems have certain advantages not shared by static systems: (*a*) a steady-state may be maintained over a relatively long period of time (several hours), minimising any variation in filtration rate estimates; (*b*) the situation may be manipulated for experimental purposes, e.g. using the repeated establishment and maintenance of new steady-states to examine the mussel's response to fluctuations in particle concentration; (*c*) long-term experiments may be carried out without disturbing the animal (Thompson & Bayne, 1972); (*d*) problems of low P_{O_2} and the accumulation of excreted metabolites are avoided; (*e*) it is possible to detect periods of non-activity. There arises the problem of selecting a suitable flow rate for use in the experimental procedure, since Walne (1972) maintains that in *Mytilus edulis* (and other bivalves) clearance rate is an increasing (power) function of flow rate. However, there is a danger here of a potential artefact. Filtration rates in flowing systems are usually determined by counting the particles flowing into (I) and out of (O) an experimental chamber. The difference between I and O, as a proportion of I, is then multiplied by the flow rate of the water to arrive at an estimate of filtration rate as volume×time^{-1}. But the animal is actually filtering in a particle concentration (C) within the experimental chamber. I and C may not have the same numerical value, and indeed the difference between them will increase at slower flow rates. This introduces an error into the calculation of filtration rate which is greatest at low flow rates and least at higher flow rates where I and C approach the same value. In practice, we have found (Bayne, Thompson & Widdows, unpublished data) that a flow rate of 60 ml min^{-1} is sufficient to reduce the error to a negligible value, but this will vary with the size and geometry of the experimental flask used. Clarification of this potential artefact should precede any study of the filtration rates of bivalves using flow-through techniques.

Filtration rate values for lamellibranchs have been reviewed by Owen (1966*a*, 1974*b*), Jørgensen (1966) and Ali (1970). In the following account, filtration rates by mussels will be discussed in relation to environmental and other factors.

Size

Many investigations have been carried out to determine the relationship between filtration rate and the length or body weight of the mussel. Table 5.1 summarises data available for various species. In some instances, e.g. Bayne, Bayne, Carefoot & Thompson (1975a) for *Mytilus californianus*, the relationship between filtration rate and dry flesh weight is adequately described by a power function for a wide range of body weights. In other cases, however, a power curve fits the data for small animals only. According to Thompson & Bayne (1974) the weight exponent is 0.39 for *M. edulis* less than 1 g dry weight, but decreases for larger animals. Similar observations were made by Theede (1963) and Winter (1973). Measurements of filtration rate in *M. californianus* made by Rao (1953) and Segal *et al.* (1953) illustrate the point very well (Table 5.1); the size class below 2 g dry weight has a greater weight exponent, *b*, than that from 2 to 12 g. (Bayne *et al.* (1975a) studied a weight range of only 0.3–5.0 g.) The explanation probably lies in the relationship between the gill area, or possibly the area of the gill ostia, and the dry weight of the animal, the curves in Fig. 5.5 demonstrating a reduction in the growth rate of the gill in larger (= heavier) animals. Since an increase in the area of the gill is achieved only by lengthening the filaments and by the addition of more filaments, one might predict an increase in filtration rate which is directly proportional to the additional gill area (Hughes, 1969). Vahl (1973) determined gill area and filtration rate in *M. edulis* (animals 0.01 to 1.0 g dry flesh weight) and found little difference in their weight exponents. Foster-Smith (1975a) related filtration rates to the area of gill ostia (the latter measured *in vivo* by observation through a window in the shell) in three bivalves, including *M. edulis*; although absolute filtration rates varied, when related to area of ostia all three species varied between 235 and 245 ml h^{-1} cm^{-2}. Winter (1973) comments, however, on the reduced activity of gill cilia in old specimens of *Mytilus edulis* and *Modiolus modiolus* noted by Schlieper, Kowalski & Erman (1958), which may in part account for the reduction in the rate of increase in filtration activity observed for larger animals. Whatever its cause, this reduction is a significant factor in the decreased growth efficiencies of larger mussels, to be discussed in Chapter 7.

Weight exponents for filtration rate determinations vary from negative values for large *Mytilus californianus* to values of 0.74 to 0.76 for *M. edulis* and *Modiolus* spp. (Table 5.1). It is difficult to make a definitive statement regarding the significance of these differences, except to account for the low exponents for large animals in terms previously discussed and to note that most of the weight exponents recorded for filtration rate are smaller

Table 5.1. *Relationships between filtration rate $C(l\,h^{-1})$ and dry flesh weight $W(g)$ as described by the equation $C = aW^b$*

Species	Authors	Suspended particles	Flow rate (ml min⁻¹)	Temperature (°C)	Weight (W) of dry flesh (g)	a	b
Mytilus edulis	Willemsen (1952)	Silt	Static system	12–15	0.70–3.10	1.273	0.38
	Theede (1963)	Graphite	Static system	15	0.06–3.28	1.656	0.49
	White (1968)		?	?	0.28–2.90	2.214	0.27
	Walne (1972)	Natural seawater	200	10	0.50–4.00	3.846	0.25
			300	10	0.50–4.00	4.788	0.26
			200–299	20–22	0.50–2.00	4.212	0.11
			200	20–22	0.19–3.28	3.046	0.34
	Winter (1973)	*Dunaliella marina* 20×10⁶ cells l⁻¹	3000	12	0.003–1.19	2.410	0.74
		40×10⁶ cells l⁻¹	3000	12	0.003–1.19	2.410	0.74
	Vahl (1973)	*Isochrysis galbana* and *Monochrysis lutheri*	150	10	0.01–1.00	3.90	0.60
	Thompson & Bayne (1974)	*Tetraselmis suecica*	50	15	0.05–1.00	1.944	0.39
Mytilus californianus (Los Angeles)	Rao (1953)			20 (ambient)	3.00–12.00	2.270	0.29
				16	2.00–12.00	1.162	0.37
				10	0.30–12.00	0.599	0.21
(Friday Harbor)	Rao (1953); Segal et al. (1953)	Graphite	Static system	10	3.00–12.00	1.835	−0.39
				16 (ambient)	2.00–12.00	2.069	0.27
				21	0.50–12.00	2.138	0.48
				21	3.00–12.00	2.156	0.80
				21	0.50–12.00	3.962	0.13
				8	0.50–12.00	0.527	0.27
				8	0.50–12.00	0.527	0.27
(Friday Harbor)	Bayne et al. (1975a)			8	3.00–12.00	0.940	−0.06
		Algae (various) High ration		15	0.30–5.00	1.610	0.46
		Low ration		15	0.30–5.00	1.018	0.42
Modiolus modiolus	Winter (1969)	*Chlamydomonas*	3000	12	0.43–3.95	0.929	0.74
Modiolus demissus	Kuenzler (1961a)	³²P-labelled *Nitzschia*	Static system	10	0.01–0.81	3.444	0.76

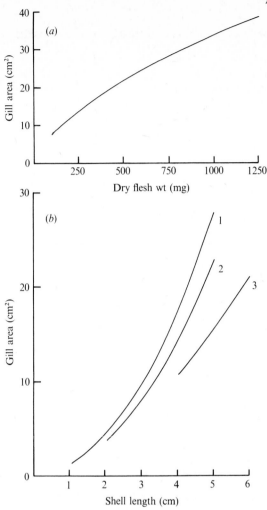

Fig. 5.5. *Mytilus edulis*: the relationship between gill area (cm²) and (*a*) body weight (Vahl, 1973); (*b*) shell length (1, Hughes, 1969; 2, Foster-Smith, 1975*a*; 3, Dral, 1967).

than those for oxygen uptake (p. 161), a fact which is important in the determination of relationships between growth efficiency and size.

Filtration rate data may be used to calculate daily ingested ration, as a percentage of dry weight, which may then be related to the dry weight of the animal, thus providing an ecologically relevant index of feeding activity. Fig. 5.6 is derived from a series of such transformations, which in this case are based on the assumption that mussels are feeding continuously in a suspension of 10 000 cells ml⁻¹ of a flagellate containing 90% organic matter. With the exception of Winter's (1973) data, the curves for

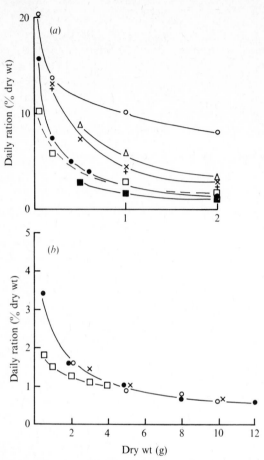

Fig. 5.6. The relationship between ingested ration (as % of dry body weight per day) and dry body weight (g), as calculated for a concentration of 10000 cells of *Tetraselmis suecica* per ml. (*a*) *Mytilus edulis* (■, Willemsen, 1952; □, Theede, 1963; ×, White, 1968; △, Walne, 1972; ○, Winter, 1973; ●, Thompson & Bayne, 1974; +, Thompson, unpublished data). (*b*) *Modiolus modiolus* (□, Winter, 1969) and *Mytilus californianus* (○, Rao, 1953 (animals from Friday Harbor); ×, Rao, 1953 (animals from Los Angeles); ●, Bayne *et al.*, 1975*a*).

M. edulis are similar to one another, daily ration decreasing from 6–13% of body weight for a 0.2 g mussel to 1–3% for a 2.0 g animal. Values are slightly lower for *M. californianus* and *Modiolus modiolus* of comparable size to *M. edulis*.

Particle size

Some filter-feeders, notably copepods, can discriminate between particles on the basis of size (Poulet, 1974), but mussels do not apparently exhibit such selective behaviour. Vahl (1972) demonstrated 80–100% retention by *Mytilus edulis* of all particles in the range 2–8 μm diameter. Observations of mussels filtering in natural seawater (particle size range 2–100 μm) have confirmed that selective removal does not occur (R. Conover & W. MacKay, personal communication; Bayne *et al.*, unpublished observations). Foster-Smith (1975*b*) offered *M. edulis* (and two other bivalve species) mixed suspensions of inorganic particles and algae; these were equally accepted by the pallial organs, but the inorganic material was rejected by the alimentary canal. The effect of inorganic material in suspension was chiefly to limit, by 'dilution', the amount of algae ingested rather than the total amount of material filtered. The filtration mechanism is such that the mussel can regulate the lower size limit of particles retained by controlling the width of the ostium, but increasing this lower limit has only been observed in dense suspensions and apparently functions to prevent clogging of the gills and palps. There appears also to be little qualitative sorting on the palps, which are adapted to deal with the large quantities of material retained by the gills when food is abundant. Nevertheless, observations are lacking of filtration activity in dense, mixed suspensions. To the extent that particle size, when multiplied by concentration, represents an estimate of the mass of particulate matter, it may be important in determining the concentrations at which feeding efficiency is reduced and production of pseudofaeces begins (Winter, 1969).

Particle concentration

Mytilus edulis does not filter in very dilute suspensions. Filtration is initiated at a critical particle concentration which presumably corresponds to the threshold of the receptor concerned (Theede, 1963; Thompson & Bayne, 1972). Schlieper & Kowalski (1958) have suggested that the frontal cilia on the gills of *M. edulis* only beat at maximal velocity if stimulated to do so by food particles or by dissolved organic compounds in the water. Irrespective of the particle concentration, work must be done by the lateral cilia in ventilation and by the latero-frontal cirri in filtration. It is unlikely that the energy expended is significantly affected by increased particle densities, a view consistent with the observation of Thompson & Bayne (1974) that filtration rate is independent of a wide range of particle concentration, provided that the threshold is exceeded. Levinton (1972) comments that suspension-feeders which create a ventilation current are making some effort regardless of the food concentration. In very dense suspensions (20000–40000 *Dunaliella* cells ml^{-1}), Winter (1973) observed a decrease in filtration rate in *M. edulis*, which supports Dral's (1967) observations on activity of the latero-frontal cirri under these conditions. Davids (1964) found that pumping rate of *M. edulis* was reduced in a

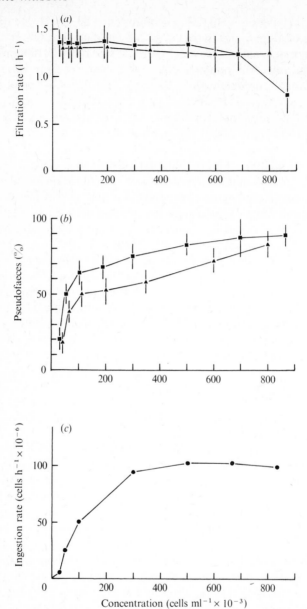

Fig. 5.7. (*a*) *Mytilus edulis*: filtration rate with increasing concentrations of ▲, *Isochrysis galbana*; and ■, *Phaeodactylum tricornutum*. (*b*) Percentage of material rejected before ingestion, by *Mytilus edulis*, with increasing concentrations of ▲, *I. galbana* and ■, *P. tricornutum*. (Values in (*a*) and (*b*) ±1 S.E.) (*c*) Rates of ingestion by *Mytilus edulis* at increasing concentrations of *P. tricornutum*. (All redrawn from Foster-Smith, 1975*a*.)

suspension of 610×10^3 cells of *Nitzschia* ml^{-1}. Foster-Smith (1975a) has recently recorded remarkably constant rates of filtration by *M. edulis* over a concentration range from $< 50 \times 10^3$ to $> 800 \times 10^3$ cells ml^{-1} (Fig. 5.7a).

In considering the effects of particle concentration on ingestion rate, as opposed to filtration rate, it is important to identify the rate of production of pseudofaeces. Winter (1969) found that *Modiolus modiolus* reduced filtration rate in the proportion $2.7 : 2 : 1$ in concentrations of *Dunaliella* of 10×10^6, 20×10^6 and 40×10^6 cells l^{-1}, respectively; no pseudofaeces were produced, even at 40×10^6 cells l^{-1}, at 20 °C, though pseudofaeces were found at this concentration at 12 °C. Thompson & Bayne (1972, 1974) observed no production of pseudofaeces by *M. edulis* at particle concentrations up to 25×10^6 cells l^{-1}. Foster-Smith (1974), using a rather different procedure, recorded pseudofaeces production by *M. edulis* as an increasing proportion of cells filtered above a concentration of 2×10^8 cells l^{-1}. Subsequently (Foster-Smith, 1975a), he quantified the production of pseudofaeces by *M. edulis*, as a percentage of material rejected before ingestion (Fig. 5.7b); this rose rapidly with increasing concentration from 50×10^3 to 100×10^3 cells ml^{-1}, then increased more gradually at higher concentrations. By subtracting the rate of production of pseudofaeces from rate of filtration, Foster-Smith (1975a) determined a true ingestion rate (Fig. 5.7c). This increased gradually up to a concentration of 300×10^3 cells ml^{-1}, then remained constant to 800×10^3 cells ml^{-1}.

Winter (1969) offered a scheme of seven 'concentration stages' for visualising the interactions between ventilation rate, filtration rate, retention efficiency and particle concentration. This scheme could not consider the most recent understanding of control of filtration rate, and it suffers also from implying a more rigid behaviour than is realistic. Nevertheless, the scheme is useful for general comparative purposes.

During prolonged starvation, the oxygen uptake of *M. edulis* decreases to reach a standard metabolic rate (Bayne *et al.*, 1973). This reduction in oxygen consumption (see p. 163), also recorded in *M. californianus* by Bayne *et al.* (1975a), is associated with a reduction in ventilation rate (see also Flügel & Schlieper, 1962; Theede, 1963). The amount of food reaching the gills must presumably also decrease, causing a reduction in filtration rate, which is consistent with the low values recorded for starved *M. californianus* (see Table 5.1).

Temperature

Acute and long-term responses of *Mytilus edulis* to instantaneous temperature change have been described by Widdows & Bayne (1971) and Widdows (1973a, b). Between 5 and 20 °C the immediate response to a reduction in temperature is a decrease in filtration rate (Fig. 5.8), whereas

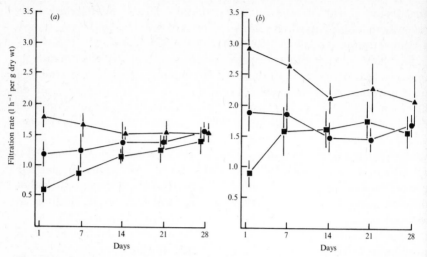

Fig. 5.8. The acclimation of filtration rates by *Mytilus edulis*. (*a*) Food concentration 5500 cells ml^{-1} (November). ■, 5 °C; ●, 10 °C (ambient); ▲, 15 °C. (*b*) Food concentration 3000 cells ml^{-1} (July). ■, 10 °C; ●, 15 °C (ambient); ▲, 20 °C. The vertical bars represent one standard deviation on either side of the mean. (From Widdows & Bayne, 1971.)

increased temperature results in a corresponding rise in filtration rate (data for mussels acclimated to 15 °C). Above 20 °C, however, an animal acclimated to 15 °C responds to elevated temperature by reducing filtration rate. Between 5 and 20 °C there is a complete acclimation of filtration rate (see p. 172), but only partial acclimation at 25 °C (see also Flügel & Schlieper, 1962; Theede, 1963). Rao (1953) observed higher Q_{10} values for filtration rate by large than by small *M. californianus* over the temperature range 10–20 °C.

Rao (1953) investigated three populations of *Mytilus californianus* from different latitudes on the west coast of the United States, and found filtration rate to be greater at any given temperature in mussels from higher latitudes than in animals of corresponding weight from lower latitudes. Weight-specific filtration rates of these *M. californianus* were identical at the minimum temperatures recorded for the respective latitudes, providing a good example of temperature acclimation. Effects of temperature are discussed further in a later section (p. 166).

Position in the intertidal zone

According to Segal *et al.* (1953), filtration rate of intertidal *Mytilus californianus* is partly determined by the vertical distribution of the population on the shore. At any given temperature, mussels from the lower intertidal exhibit a consistently higher clearance rate than those further up the shore. Animals from lower down in the intertidal zone appear therefore to be cold-adapted (Rao, 1953) relative to those from higher levels only a few feet away. However, it is not known whether the higher intertidal animals in this particular population were in fact exposed to greater average temperatures, e.g. as expressed in day-degrees.

Reduced oxygen tension (P_{O_2})

Bayne (1971*b*) showed that *Mytilus edulis* may regulate oxygen uptake under conditions of reduced P_{O_2}. This is achieved by a decrease in ventilation rate (= filtration rate in this case), a reduced ventilation:relative perfusion ratio, and an improved efficiency of oxygen utilisation (Chapter 7). The bifunctional gill is inefficient under these conditions, since the physiological adjustments necessary for adequate gas exchange lead to a reduction in food intake (Bayne, 1975*b*). The converse does not necessarily apply; filtration rate may be reduced without affecting the parameters of gas exchange (see previous discussion). The gill appears to afford an efficient compromise between the requirements of gas exchange and those of food-particle retention. Under normal circumstances P_{O_2} is high but, teleologically speaking, food input must be maximised, so the strategy adopted is the development of a large surface area with a high ventilation rate and a low oxygen utilisation efficiency. During regulation to low P_{O_2}, adjustments to provide an adequate oxygen supply are essential, and a reduction in food intake is inevitable.

Salinity

Theede (1963) measured the filtration rates of *Mytilus edulis* from the North Sea and the Baltic, in relation to salinity. At their respective ambient salinities (15‰ for the Baltic, 30‰ for the North Sea) filtration rates were similar, but on transfer to either higher or lower salinities filtration rate was markedly reduced. After a period of 7 to 10 days, however, filtration rates increased at salinities between 5 and 10‰ higher, or lower, than the environmental ambient values, but acclimation to the new salinity is not completed in this time. Many studies, reviewed by Schlieper (1971), have recorded similar effects of salinity on isolated gill cilia. This subject is discussed later in this chapter (p. 200).

The palps

The contents of the food grooves on the gills are transferred to the inner surfaces of the labial palps, which are folded into complex ciliary tracts. These sort the incoming material and convey it either to the mouth or to the

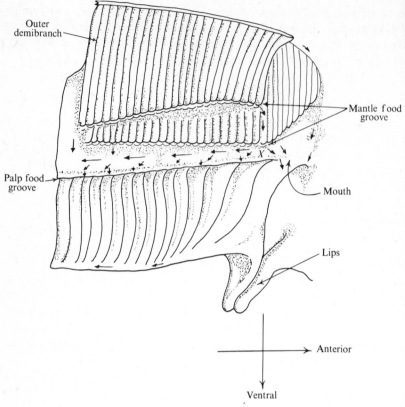

Fig. 5.9. The junction between demibranch and palp in *Mytilus edulis*, viewed from the inner surface of the right inner palp. *X*, see text. (Redrawn from Foster-Smith, 1974.)

rejection tracts which remove unsuitable or excess particles as pseudofaeces. The following account of palp function in *Mytilus edulis* is based on the observations of Foster-Smith (1974).

Fig. 5.9 illustrates the junction between demibranch and palp. The lips are formed by a continuation of the outer and inner palps on each side as low folds which meet in the midline. Posterior to the lips lies a series of ciliated ridges, the four most anterior being larger than the remainder. The dorsal part of the palp inner surface is not folded, and forms the roof of the palp food groove. Gilmour (1974) has suggested that the lips function to prevent water being taken into the mouth.

Material from the marginal food groove of the outer demibranch crosses the face of the inner demibranch and merges with the mucus string in the inner demibranch marginal groove before joining the palp surface at a point (*X*) dorsal to the first two palp ridges and posterior to the mouth. Cilia anterior to *(X)* beat orally, those posterior to *(X)* beat aborally. Food entering this aboral tract is directed posteriorly, then transferred to the sorting system on the ridges and folds of the palp inner surface. In general, the greater the amount of material entering the dorsal aboral tract from the gill lamellae, the further its posterior displacement before transfer to the sorting currents. Thus as the palp receives more food, an increasing number of the available ridges and folds is utilised, beginning with the first (most anterior) and progressing posteriorly.

Re-sorting tracts on the aboral surfaces of the folds of the palps direct fine material dorsally into the palp food groove, which is a direct pathway to the mouth. Larger strings do not reach the re-sorting tracts but are passed oralwards across the crests by means of local contractions on the aboral side. A crest rejection tract, which serves primarily to remove surplus material from the palp food groove, is recessed into the ridge and does not interfere with the passage of mucus strings from crest to crest. Deep rejection tracts are also present at the base of the palp folds.

The origin of the crest rejection tract is marked by an elevation of the palp folds at the ventral margin of the food groove. This is most prominent when the mussel is feeding in dense suspensions and the food groove is full; it brings the crest rejection tract into contact with the food groove margin, thus removing excess material from the groove. Under these conditions the food groove is also restricted by contractions of the dorsal margin of the palp, which forms its roof. When there is relatively little food material on the gills, the mucus strings in the marginal groove of the lamella may enter the palp food groove without coming into contact with the palp ridges. Conversely, high food concentrations lead to a constriction of the palp food groove, the activation of crest rejection tracts to direct surplus material away from the groove, and to an increase in the frequency of palp movements across the lamella surface to remove loose food strings directly from the gill. Disposal of the excess material which results from these processes is effected by the rejection currents, of which the crest tracts are the most important, especially at intermediate concentrations. The deep rejection tracts appear to be utilised only in situations in which the palp is saturated with material. All material to be removed is transferred to the rejection tracts of the mantle edge. These tracts convey the material to the posterior margin of the inhalant siphon, adjacent to the exhalant opening, where it is deposited as pseudofaeces and is carried away by the exhalant current. Pseudofaeces do not accumulate on the palps but are transferred to the mantle as soon as they are formed.

Available evidence indicates that when *Mytilus* is feeding in dense suspensions, the ctenidia and palps, which form the feeding apparatus, regulate the quantity of food material which reaches the mouth. The lower size limit of particles efficiently retained by the gill filter may be controlled by varying the amplitude of the latero-frontal cirral beat, which tends to be maximal in dilute suspensions and minimal in dense ones, and by control of the ostia. It seems probable that the primary purpose of these mechanisms is to regulate the amount of material retained on the gill by the exclusion of smaller particles, rather than the selection *per se* of larger particles. The function of the labial palps is the continuous removal of material from the lamellar food tracts in order to prevent saturation of the gill, which would inhibit both gas exchange and further filtration. In dense suspensions, the sorting and rejection tracts channel much of the filtered material away from the mouth and dispose of it as pseudofaeces, so that the animal may continue to filter and ingest at an optimum rate. No selection of particles by size has been observed on the palps, and it is difficult to envisage any such mechanism, since the particles are bound in mucus strings.

The stomach

The stomachs of three species of mussels have been examined in detail: *Mytilus edulis* (Graham, 1949; Reid, 1965), *Modiolus modiolus* (Reid, 1965) and *Lithophaga nasuta* (Purchon, 1957). These are all very similar morphologically. The following account is based on Reid's description of the stomach of *Mytilus edulis*.

The stomach is a dorso-ventrally flattened sac, into which the oesophagus opens antero-ventrally and the mid-gut leaves posteriorly, together with a style sac, which communicates with the mid-gut. The crystalline style, a hyaline rod of mucoprotein, protrudes from the style sac into the stomach proper and rests against the gastric shield, a chitinous, thickened region of the stomach wall. The ducts of the digestive diverticula open into the stomach wall. Extending from the duct openings are the right and left duct ciliary tracts, which traverse the floor of the stomach to the intestinal groove. These tracts are bounded by two low, narrow ridges, the minor and major typhlosoles.

Details of the ciliary tracts are shown in Fig. 5.10. Food enters the stomach from the oesophagus in the form of a mucus string, which is carried by the cilia of the buttress to the gastric shield. The string is wound slowly round the style, which rotates against the gastric shield and appears to draw in more material from the oesophagus. It is often assumed that style rotation is continuous, but Purchon (1971) suggests that style formation and rotation occur intermittently at certain stages in the digestive process.

The mechanical action of the style breaks up the food material into fragments which fall onto the typhlosole tongue and the duct tracts. The latter lead into a sorting caecum, which is attached to the ventral wall of the stomach. The narrow opening to the caecum restricts passage to small

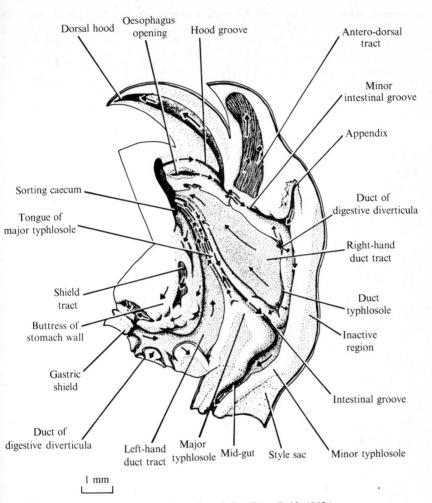

Fig. 5.10. Interior of the stomach of *Mytilus edulis*. (From Reid, 1965.)

aggregations of material, the large masses being carried back to the tip of the style for further maceration by the shield and buttress tracts. Within the caecum, material in the duct tracts is drawn into the strong current along the typhlosole tongue. During this process, small, dense particles (e.g. sand grains) fall into the intestinal groove and are rejected via the mid-gut. Smaller and less dense particles are either re-circulated into the stomach sac or enter the dorsal hood groove. Some rejection occurs from the hood groove to the minor intestinal groove, but most of the material taking this route accumulates in the dorsal hood and eventually reaches the style.

The cilia of the duct tracts and typhlosole tongue set up currents in the

147

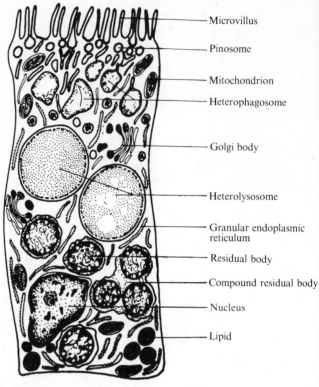

— Microvillus

— Pinosome

— Mitochondrion

— Heterophagosome

— Golgi body

— Heterolysosome

— Granular endoplasmic reticulum

— Residual body

— Compound residual body

— Nucleus

— Lipid

Fig. 5.11. Diagram of a bivalve digestive cell. (From Owen, 1972.)

stomach sac, keeping small, light particles in suspension and directing them towards the digestive gland duct openings. These ducts possess inhalant currents which channel this material into the digestive diverticula (Owen, 1955).

The fine structure of the cells lining the style sac and intestine was investigated in *Mytilus galloprovincialis* by Giusti (1970), who found that the crystalline style is secreted by the cells on the upper sides of the typhlosoles which lie between the style sac and the intestinal furrow.

It has been noted that when *M. edulis* is feeding in dense suspensions of algae, two components are readily identifiable in the faeces (Van Weel, 1961; Widdows & Bayne, 1971; Thompson & Bayne, 1972), one containing well-digested material, the other comprised of undigested cells, some of which may still be viable after passing through the gut. The undigested fraction is presumably rejected during the sorting process and channelled into the intestinal tracts without entering the digestive tubules. This rejection may be initiated when the digestive tubules are full. Reid (1965) noted irregular, spontaneous contractions of the stomach wall in *M. edulis*,

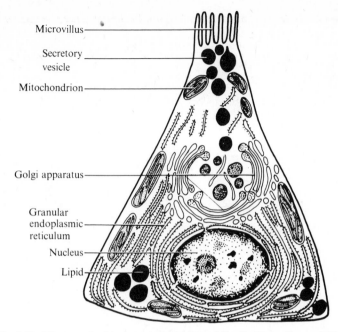

Fig. 5.12. Diagram of a bivalve pyramidal basophil cell. (From Owen, 1972.)

associated with widening of the intestinal rejection tracts. He suggested that these movements may represent a rejection reflex to remove excess food which would otherwise interfere with normal sorting processes in the stomach. If true, this would provide a third stage at which excess material may be removed to prevent saturation of the food processing mechanisms, the others being the gill lamellae and the labial palps.

The digestive gland

The digestive diverticula consist of blindly ending tubules which are linked with the stomach by a system of branched, partially ciliated ducts. The tubule epithelium contains two cell types recognisable in the light microscope, one acidophilic, columnar and vacuolated (Fig. 5.11), the other pyramidal and basophilic (Fig. 5.12). The acidophilic cells are responsible for intracellular digestion of food, and are usually referred to as digestive cells. The literature on the digestive gland of bivalve molluscs is extensive. A recent review by Owen (1972) is particularly useful here because it illustrates the fine structure of the tubule cells of *Mytilus edulis* and discusses the role of the digestive cell in terms of the formation and breakdown of lysosomes, a concept previously applied to the bivalve digestive tubules by Sumner (1969).

149

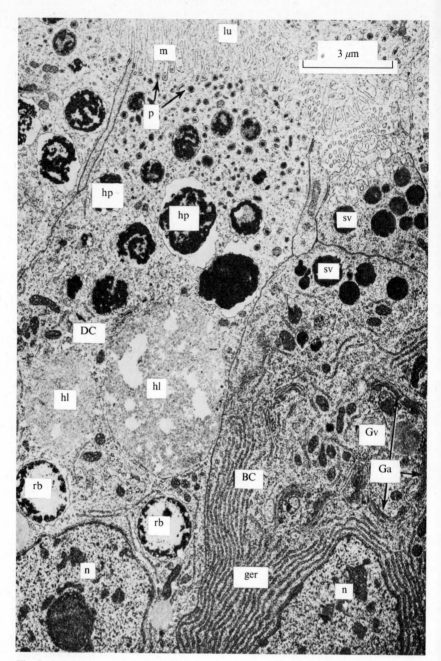

Fig. 5.13. An electron micrograph of a portion of the digestive tubule of *Mytilus edulis* to show the essential features of a digestive cell (DC) and a pyramidal basophilic cell (BC). The section is taken from an animal which had been fed whole blood of pigeon 1 h earlier. The heterophagosomes (hp) contain electron-dense material derived from the blood. Ga, Golgi apparatus; ger, granular endoplasmic reticulum; Gv, Golgi vesicle; hl, heterolysosome; lu, lumen of the tubule; m, microvilli; n, nucleus; p, pinosome; rb, ribosome; sv, secretory vesicle. (From Owen, 1972.)

150

The fine structure of the bivalve digestive cell is illustrated in Fig. 5.13 (*M. edulis*) and Fig. 5.11 (generalised diagrammatic representation). The cell is characterised by the presence of numerous cytoplasmic vesicles and by microvilli which project from the cell apex into the tubule lumen. Particulate material from the lumen is taken up at the base of the microvilli by pinocytosis, and the pinocytotic vacuoles so formed appear to fuse (Thompson, Ratcliffe & Bayne, 1974) giving rise to heterophagosomes (Owen, 1972). The roles of the various types of vesicle in the heterophagic activity of the cell are described by Owen (1972). It is not yet established whether heterophagosomes are formed continuously as ingestion proceeds, a typical sequence in many cell types, or whether they are permanent structures which persist for the life of the cell. The origin and identification of the primary lysosomes which contain the hydrolytic enzymes is also uncertain. The fate of the residual bodies varies from species to species, but in *M. edulis* they migrate to the apical region of the cell and are rejected into the tubule lumen.

The main features of the basophil cell (Figs. 5.12 and 5.13) include an extensive granular endoplasmic reticulum, numerous free ribosomes, an active Golgi body, granules at the cell apex and microvilli at the interface between cell apex and tubule lumen. All available evidence indicates that it is equipped for extensive protein synthesis. The function of the basophil cell is not known, but it has been regarded by various authors either as an immature digestive cell or as an enzyme-secreting cell (see discussion by Owen, 1972). In some bivalves, including *M. edulis* (Thompson *et al.*, 1974), the basophil cell occurs in two forms, distinguishable primarily by the presence of a flagellum at the apex in one of the types. Sumner (1966) distinguished between two forms of the basophil cell of *Anodonta* by histochemical methods. One type has a low RNA content, no apical granules and undergoes mitosis; Sumner considered this to be an immature form of the second type, which is rich in RNA and has apical granules, and which he regarded as a mature secretory cell. Thompson *et al.* (1974) examined the fine structure of the degenerating tubules of starved *M. edulis*, and observed that the autophagic breakdown of the digestive cells appeared to disturb the normal balance in relative numbers of the flagellate and non-flagellate basophil cells, the former type occurring more frequently. It is possible that the flagellate form of the basophil cell may differentiate into a mature digestive cell under suitable conditions, because recovery of normal structure was observed in tubules from mussels which were subjected to less severe stress. Clarification of the function or functions of the basophil cell types would greatly improve our present understanding of digestive physiology in the bivalve mollusc.

In addition to its role in the digestion of food, the digestive gland of *Mytilus edulis* serves as a site for the storage of metabolic reserves which

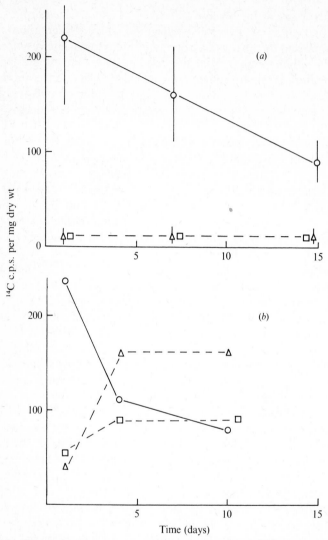

Fig. 5.14. Changes in the amount of radioactive label (^{14}C) in the digestive gland (○), adductor muscle (△) and mantle (□) of *Mytilus edulis*, following administration of the label in the diet. (*a*) Experiment in January; values ± 1 s.D. (*b*) Experiment in June. (From Thompson, 1972.)

provide a source of energy utilised during gametogenesis and during periods of physiological stress (Thompson *et al.*, 1974). Sastry (1966) and Sastry & Blake (1971) postulated a flow of reserves from the digestive gland to the gonad during gametogenesis in the bivalve *Aequipecten irradians*, and Vassalo (1973) supported this with data from another scallop, *Chlamys hericia*. Thompson (1972) administered ^{14}C to *M. edulis* in

January and in June, in the form of labelled algal cells. The label was rapidly taken up into the digestive gland, and then gradually lost over the next 10 to 30 days (Fig. 5.14). In June, when the animals normally build up their energy reserves (Chapter 8), loss of label from the digestive gland was matched by accumulation of the label in the mantle and adductor muscle. In the winter (January), however, loss of label from the digestive gland did not lead to accumulation in the other tissues; presumably, in the winter, this material was metabolised. Thompson *et al.* (1974) confirmed this storage/distribution function of the digestive gland, and they concluded that nutrients stored during the summer are used to 'fuel' gametogenesis during the autumn and winter, when the reserves in the mantle tissues have been depleted. As in so many other aspects of the physiology of mussels, the seasonal gametogenic cycle plays a key role.

Phasic activity in the digestive gland

In a critique of the so-called 'classical' theory of digestion in bivalves, Purchon (1971) remarks that the digestive diverticuli in some genera exhibit a well-defined rhythmical cycle, and are not in a steady-state, as previously supposed. These phases reflect a sequence of events in the digestive cells, involving absorption, intracellular digestion, disintegration and regeneration. In some cases the phasic activity in the tubules is associated with changes in other components of the digestion process, such as pH of the stomach contents and protein content of the style (Langton & Gabbot, 1974).

Such a digestive rhythm has been claimed by Morton (1970, 1971) for a variety of intertidal bivalves. The important element in Morton's hypothesis (Owen, 1974b) is that extracellular digestion in the stomach and intracellular digestion in the digestive diverticuli occur in alternating phases. As Owen (1974b) points out, rhythmic patterns of feeding are imposed on intertidally distributed bivalves, but is the observed digestive synchrony an endogenous one, or is it co-ordinated with natural feeding cycles, and imposed by tidal periodicity?

Owen (1972) proposed that at low tide, when food is not available, the tubules assume a common appearance which he termed a 'holding phase'. When the animals are immersed by the tide, and feeding commences, food material rapidly reaches the tubules to be endocytosed and undergo intracellular digestion within the lysosomal system. The appearance of the digestive cells now changes and, due to unequal 'delivery' of food to different parts of the diverticulum, the tubules become asynchronous. Owen (1972) suggested that variations in the digestive cycle will depend on the periodicity of food availability, the nature of the food, and whether or not individual digestive cells break down completely when releasing

153

residual bodies. Langton (1975) has recently confirmed Owen's basic scheme for *Mytilus edulis*, and he concluded that different 'phasing' of the cycle may be due simply to different times of exposure or to different levels of feeding.

Digestive enzymes

Molluscs possess a wide variety of digestive enzymes, particularly carbohydrases, and according to van Weel (1961) there are no major differences between the mollusc groups in terms of their digestive enzyme complements. In bivalves, digestive enzymes are produced by the style, by the cells of the digestive diverticula and, in some genera examined, in the mid-gut (Purchon, 1971). There is an extensive literature on the subject of digestive enzymes in lamellibranchs (Owen, 1966*b*, 1974*b*; Purchon, 1971), in which a wide range of enzymes is recorded. Unfortunately, very few quantitative measurements of enzyme activity have been made. Most reports are qualitative or at best semi-quantitative, and are often based on methods which are not very specific.

Kristensen (1972) examined carbohydrases from the crystalline styles of a number of bivalves, including *Mytilus edulis*, and found less activity than in the digestive gland. Wojtowicz (1972) measured specific activities for a number of carbohydrases from the style and digestive gland of the scallop *Placopecten magellanicus*. The activity of α-amylase was nearly forty times greater in the style than in the digestive gland, which confirms earlier semi-quantitative observations on other bivalves, but the other carbohydrases (α-glycosidase, β-glucosidase, β-galactosidase, laminarase and chitobiase) were confined to the digestive tissue. The presence of laminarase, probably a multi-component enzyme, is particularly interesting. This enzyme has been reported in a number of bivalves (Sova, Elyakova & Vaskovsky, 1970) and includes exo- and endohydrolytic β-1,3-glucanases and β-glucosidases.

Crosby & Reid (1971) recorded endogenous cellulases in the style and the digestive gland of *Mytilus californianus*, confirming earlier work by Fox & Marks (1936) and Newell (1953). Very high activity in stomach contents indicated that a bacterial cellulase may be present. Cellulolytic activity was higher in the mussel and other filter-feeders than in deposit-feeders, presumably because the former group have to break down cellulose cell walls in plant detritus. It is not known whether cellulase content can be increased by induction in response to food of high cellulose content, or whether increased activity is an evolutionary function under genetic control.

Lipase activity in the style and digestive gland of *Modiolus demissus* was recorded by George (1952), who also noted that fatty acids were absorbed by the stomach epithelium and the epithelium of the ducts in the digestive gland, not by the digestive cells in the tubules.

A detailed account of the distribution of hydrolytic enzymes in the digestive gland cell of *Mytilus edulis* is provided by Sumner (1969). Most activity is in the digestive cell, especially within the cytoplasmic granules which form part of the lysosome complex involved in the cycle of heterophagic digestion within the cell. Lloyd & Lloyd (1963) describe the occurrence of a digestive glycosulphatase in *M. edulis* which hydrolyses sulphated polysaccharides from algal tissue, but the site of production of this enzyme is not known.

Reid (1968) examined digestive enzymes in a variety of bivalves (including *Modiolus modiolus*) in order to determine the relationships between stomach morphology,* diet, enzyme complement and evolutionary trends. He describes three main patterns of enzyme distribution, corresponding to the three polysyringian stomach types. The gastrotetartikan condition is the most primitive, exhibiting mainly intracellular digestion of proteins and fats. In the Gastrotriteia (which includes the Mytilidae) and the Gastropempta, there is increased extracellular digestion of these components of the diet. Gastrotriteians, however, retain a strong endopeptidase in the digestive gland. This enzyme is weaker in the Gastropempta. The gastropemptan condition is also associated with stronger gastric esterases, stronger extracellular endopeptidases, and an increased role of the mid-gut as an organ of digestion and assimilation. The tendency therefore is towards the evolution of extracellular digestion, which is believed to provide more efficient utilisation of food.

Assimilation efficiency

Because it is often difficult to recover all the faeces produced by an aquatic filter-feeder, direct estimation of assimilation efficiency by comparing the organic or energy contents of food and faeces is frequently impractical. Conover (1966) proposed a method by which assimilation efficiency may be determined from the ash-free dry weight to dry weight ratios of food and faecal samples. The method is based on the premise that only the organic component of the food is significantly affected by digestion, but according

---

* Purchon (1960, 1963) has suggested a classification of the Bivalvia based on stomach morphology. Whereas details of this classification are irrelevant in the present context, a brief indication of the proposed taxa is useful, since they are now commonly referred to in the literature. The class Bivalvia is divided into two subclasses, the Oligosyringia (bivalves having few ducts of the digestive diverticula opening into the stomach) and the Polysyringia (those with many ducts opening into the stomach). The Oligosyringia contains two orders, the Gastroproteia (Protobranchia) and the Gastrodeuteia (Septibranchia). The Polysyringia comprises three orders, the Gastrotriteia (most of the Filibranchia, including the Mytilidae), Gastrotetartika (the remaining filibranchs plus some of the Eulamellibranchia), and Gastropempta (the remaining eulamellibranchs).

to Forster & Gabbott (1971) this assumption may not always be valid. They estimated assimilation efficiency in prawns by incorporating into the food a physiologically inert reference compound, chromic oxide, measuring the ratio of the concentration of the reference material in the food to that of a given nutrient, and then determining the corresponding ratio in the faeces. Calow & Fletcher (1972) have modified the latter method, incorporating a ^{51}Cr reference marker into ^{14}C-labelled food. This procedure gives an estimate of the assimilation efficiency of the ^{14}C-labelled component of the diet.

According to Johannes & Satomi (1967), some unassimilated food may be released in soluble form, so that assimilation-efficiency estimates based on analyses of food and faeces may be overestimates. This criticism is difficult to meet, because it would involve not only the measurement of dissolved organic carbon but also the distinction between material released from the gut or faeces (i.e. unassimilated) and that originating from other surfaces (i.e. assimilated, metabolised and released).

Measurements of assimilation efficiency in *Mytilus edulis* feeding on flagellates have been made by Widdows & Bayne (1971) and Thompson & Bayne (1972, 1974), and by Vahl (1973) for *M. edulis* in suspensions of natural seawater. In each case Conover's method was used. Foster-Smith (1975*b*) fed three species of bivalve, including *M. edulis*, with *Phaeodactylum* labelled with ^{32}P, and estimated assimilation efficiency by counting the label in the food, faeces and pseudofaeces. Assimilation efficiency is a decreasing linear function of food concentration for mussels fed on *Tetraselmis* (Fig. 5.15; Widdows & Bayne, 1971; Thompson & Bayne, 1972). There appears to be a limit to the quantity of food which the digestive gland can deal with in any given time, but the filtration mechanism can provide food in excess of this quantity if the suspension is sufficiently dense. As a result, a proportion of the ingested material does not enter the digestive gland but is channelled directly into the intestine. Van Weel (1961) observed two components in faeces produced by bivalves, one which he termed 'glandular', the other 'intestinal'. They are easily distinguishable in faeces from *M. edulis* feeding in high densities of *Tetraselmis*, the intestinal component being packed with undigested and often live algal cells (p. 148). Similar observations were made by Fox (1936) on *M. californianus*. If the food intake is very high, the intestinal fraction comprises the bulk of the faecal material, but when little food is available it decreases significantly. These changes are reflected in assimilation efficiency. Foster-Smith (1975*b*) concluded that assimilation efficiency was inversely related to the amount of food ingested. Foster-Smith compared his values for assimilation efficiency with values published by Thompson & Bayne (1972) and found good agreement (Table 5.2) up to a ration of 468×10^6 cells ingested per day. The composite picture of assimilation

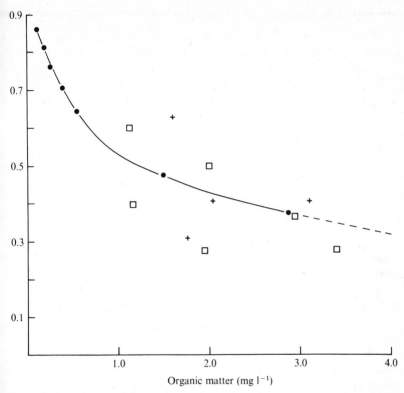

Fig. 5.15. Assimilation efficiency by *Mytilus edulis* as a function of the concentration of suspended organic matter. ●, from laboratory experiments; +, □, from measurements in the field, on two different populations. All values are for animals of 1.0–1.5 g dry flesh weight, at 10–12 °C. Efficiency determined by the method of Conover (1966).

efficiency given in Fig. 5.15 also suggests a minimum value at very high levels of ration (2.0 to 3.0 mg organic matter l^{-1}).

Winter (1969) found that the efficiency of protein assimilation by *Modiolus modiolus* was high (80–90%) and was independent of the food concentration, but it is difficult to compare these observations with other published values, which are for assimilation of organic matter as a whole. Whereas temperature did not affect protein assimilation in *Modiolus* (Winter, 1969), Widdows & Bayne (1971) demonstrated an inverse relationship between assimilation efficiency and temperature in *Mytilus*. However, more recent studies (Widdows, unpublished data) have shown this effect of temperature on assimilation efficiency of *M. edulis* to be slight.

Vahl (1973) expressed assimilation efficiency by *Mytilus edulis* as a power function of dry weight and obtained an exponent not significantly

Table 5.2. *A comparison of assimilation efficiencies by* Mytilus edulis *calculated by Thompson & Bayne (1972) and Foster-Smith (1975b)*

Concentration of suspension (cells ml⁻¹)	Cells ingested per day ($\times 10^6$)	Cells assimilated per day ($\times 10^6$)	Assimilation efficiency Thompson & Bayne	Assimilation efficiency Foster-Smith
1 000	31	27.8	0.89	0.83
2 500	78	62.4	0.80	0.78
10 000	312	187.2	0.60	0.45
15 000	468	187.2	0.40	0.40
20 000	624	124.8	0.20	—
25 000	780	0	0	—

From Foster-Smith (1975b).

different from zero. Similar observations were made by Thompson & Bayne (1974), also for *Mytilus edulis*, and by Winter (1969) for *Modiolus modiolus*. Our more recent studies, however, suggest that there is a slight, but significant effect of body size on assimilation efficiency (Chapter 7), which is most noticeable at medium concentrations of suspended matter.

Direct absorption

Several authors (e.g. Stephens, 1967, 1968) have demonstrated that marine invertebrates can absorb from solution organic compounds such as amino acids and sugars. Péquignat (1973) found that *Mytilus edulis* removed dissolved, labelled amino acids and glucose, most of which were absorbed onto the gills before transfer to the mantle and digestive gland. Acid phosphatase activity in the gill epithelium of *M. edulis* has been demonstrated by Pasteels (1968, 1969), who claims that the enzyme is located in the goblet cells of the epithelium and is exuded with the mucus into the pallial cavity. In other words, Pasteels considers that the mucus secreted onto the gill has a digestive as well as a mechanical function. The absorptive mechanism is chemically selective, but small particles are removed non-selectively by pinocytosis. Amoebocytes located in the gill tissue also show acid phosphatase activity, and are responsible for translocating material to other sites (Pasteels, 1967).

Although there is little doubt regarding the mussel's ability to absorb dissolved organic compounds, it is difficult to assess the nutritional significance of such material. Available information suggests that the concentration of dissolved organic matter in seawater is normally very low (Jørgensen, 1966) so that its contribution to the energy input in invertebrates may only be a fraction of the total.

It is not possible to make a definitive statement at the present time, but the subject clearly deserves further research.

Respiration

In marine mussels, the extraction of oxygen from the water occurs primarily at the gill surface. The gill therefore serves a dual function, in feeding and in respiration. The structure of the lamellibranch gill is well-suited to function in gas exchange, by virtue of its considerable surface area and rich supply of blood. Large volumes of water are passed through the mantle cavity, ensuring a supply of oxygen, and the diffusion distance for oxygen from water to blood is small. Normally, the efficiency with which the oxygen made available to the gill is utilised is rather low (3–10%, see Chapter 7). Under certain circumstances this efficiency may be increased considerably, up to 30% or more (p. 264) and this provides one way in which a mussel may vary its rate of consumption of oxygen in order to meet a variable metabolic demand. However, calculations suggest that a high proportion of the total oxygen consumed may be used to support the 'respiratory pump'; this, and the high energetic costs of increasing ventilation rate (p. 171) impose an upper limit to the operation of the respiratory system.

Knowledge of the limits of respiratory function are important for understanding the physiological adaptation of a species, since many features of aerobic metabolism can be studied indirectly by measurement of the rate of oxygen consumption by intact animals. This has led to a considerable number of investigations into the rates of respiration of mussels. As with many species, the rate of oxygen consumption has been found to vary with change in virtually any environmental variable. In this section, some of the factors known to effect changes in rates of oxygen consumption by mussels are reviewed. In addition, since it is often impossible to discuss the intensity of respiration without at the same time considering the activity that contributes to the metabolic demand, we shall also review here some aspects of the activity of the individual.

Size

The relationship between body size (usually expressed as weight) and the rate of oxygen consumption by animals has received considerable attention (see reviews by Zeuthen, 1947, 1953; Von Bertalanffy, 1957; Hemmingsen, 1960) including many studies on bivalve molluscs. The relationship is most commonly expressed in the form of an allometric equation:

$$Y = aX^b,$$

Marine mussels

where Y is the rate of oxygen consumption, X is the body weight and a (the 'intercept') and b (the 'slope') are fitted parameters. The equation may be transformed to a linear form by taking the logarithm of both sides:

$$\log Y = \log a + b \log X.$$

In many of the published accounts of oxygen consumption and animal size, oxygen uptake is expressed as a rate per unit body weight:

$$\frac{Y}{X} = a \cdot X^{b-1}.$$

Davies (1966) has suggested that, in order to avoid confusion in the use of the notation b, this expression for the 'weight-specific' rate of oxygen consumption should be written:

$$Y^1 = a \cdot X^{b^1}.$$

Some confusion also exists over the use of short-hand notations for the terms 'rate of oxygen consumption by the whole animal' and 'weight-specific oxygen consumption', i.e. for Y and Y^1 respectively. In this chapter we will refer to Y as V_{O_2}, usually expressed as ml O_2 h^{-1}, and to Y^1 as Q_{O_2}, usually expressed as ml O_2 h per g dry flesh weight. Finally, in making comparisons between rates of respiration measured in different experiments, it is sometimes convenient to refer the data to an animal of standard body size. This can be done by dividing V_{O_2} by W^b, when it has been shown that the value for b does not vary significantly between experiments.

It has been standard practice to fit the allometric equation to data on oxygen consumption rate and body size by taking logarithmic transformations and applying linear regression analysis by the method of least squares. Analysis of covariance is then used to assess the goodness of fit to the data, and to compare separate regressions for determining the significance of the differences between the fitted parameters (Snedecor & Cochran, 1972). Recently, however, Glass (1969) has argued that linear regression analysis on logarithmically transformed data results in erroneous estimation of the parameters a and b in situations where the random-error component in the measurement of Y is additive. Glass suggested that non-linear, iterative least squares techniques should be employed, and he illustrated his argument with data relating oxygen consumption to body size in fish. Jolicoeur & Heusner (1971) have also challenged the assumption, implicit in normal regression analysis, that X is determined virtually without error. They propose that a and b should be determined with no distinction made between an independent and dependent variable; the line of best fit is then the major axis of an equiprobability ellipse. Since neither of these recommendations have to date been

Table 5.3. *Some relationships between the rate of oxygen consumption and body weight in mussels*

Species	Range of dry flesh weights (g)	Temp. (°C)	No. of observations (n)	a	b	b'	Authority and comments
Modiolus demissus	0.25–2.60	22	12	0.629	0.798	−0.202	Read (1962a)
Modiolus demissus	0.40–1.30	14	4	0.260	0.690	−0.310	Kuenzler (1961a)
Mytilus edulis	0.11–1.68	16	24	0.698	0.660	−0.340	Rotthauwe (1958); 16‰ salinity
Mytilus edulis	0.01–1.75	15	14	0.525	0.930	−0.070	Krüger (1960); Mar. 1959
Mytilus edulis	0.01–1.75	15	14	0.320	0.700	−0.300	Krüger (1960); Aug. 1959
Mytilus edulis	0.20–4.00	12	25	0.559	0.595	−0.405	Read (1962a)
Mytilus edulis	0.07–3.00	15	44	0.263	0.724	−0.276	Bayne et al. (1973); Winter, standard
Mytilus edulis	0.07–3.00	15	41	0.549	0.774	−0.226	Bayne et al. (1973); Winter, routine
Mytilus edulis	0.07–3.00	15	42	0.164	0.670	−0.330	Bayne et al. (1973); Summer, standard
Mytilus edulis	0.07–3.00	15	68	0.339	0.702	−0.298	Bayne et al. (1973); Summer, routine
Mytilus edulis	0.007–1.02	10	20	0.370	0.750	−0.250	Vahl (1973); 22‰ salinity
Perna perna	0.06–3.46	20	18	0.342	0.625	−0.375	Bayne (1967)
Mytilus californianus	5.00–52.00	17–21	17	0.625	0.837	−0.163	Whedon & Sommer (1937)
Mytilus californianus	0.35–5.01	13	22	0.542	0.648	−0.352	Bayne et al. (1975a); routine
Mytilus californianus	0.35–5.01	13	19	0.233	0.648	−0.352	Bayne et al. (1975a); standard

a and *b* are fitted parameters in the allometric equation $Y = a \cdot X^b$; *b'* is *b* − 1.
Units are ml O_2 h^{-1} (*Y*) and g dry flesh weight (*X*).

considered in studies on bivalve respiration, they will not be discussed further.

Some of the published values for oxygen consumption and body size in mussels are listed in Table 5.3; where data were available for a range of temperatures, values from the temperature closest to the presumed ambient environmental condition were chosen. Rotthauwe's (1958) measurements were taken in water of 16‰ salinity, and Vahl's (1973) in water of 22‰ salinity. Krüger (1960) measured oxygen consumption rate monthly (except June and September) between March 1959 and February 1960; his two extreme valves for *b* are quoted in the table, although he concluded that *b* = 0.80 was probably a fair estimate. Bayne *et al.* (1973) recorded different values for *a* and *b*, in summer and winter, for both 'routine' and 'standard' rates of oxygen uptake (see p. 163). Whedon & Sommer (1937) did not fit an equation to their data; the values quoted in Table 5.3 were calculated by Bayne *et al.* (1975*a*), from data in Table 1 of Whedon & Sommer's paper. Bayne *et al.* (1975*a*) recorded separate values for routine and standard rates of oxygen consumption by *M. californianus*.

The data in Table 5.3 suggest that there is less variability in estimates of the coefficient *b* than in the 'constant' *a*. Paloheimo & Dickie (1966*a*) reached a similar conclusion from their review of published data on fish respiration. A simple arithmetic mean of the values for *b* in Table 5.3, weighted according to the number of observations, *n*, is 0.710. The meaning of such a general value is obscure (Hemmingsen, 1960) and its use in the application of respiratory data to the study of ecological problems should be limited to the most general cases only.

The variability to be expected in the numerical value of *b* is still uncertain. Some studies suggest that *b* may be variable in a predictable way. For example, Zeuthen (1953) generalised from a large amount of data on oxygen consumption by *Mytilus edulis*, from eggs to adults, around three values for *b* (0.80, 0.95 and 0.65) which he suggested were characteristic of particular stages in the life history. Kuenzler's (1961*a*) data on *Modiolus demissus* suggested that *b* varied with temperature. However, other studies on a variety of bivalves (e.g. Read (1962*a*) on *Mytilus edulis* and *Modiolus demissus*, Pamatmat (1969) on *Transennella* and Ansell (1973) on *Donax*) indicate that over a range of temperatures, at different levels of ration and at different times of the year, there is no significant variation in the value of *b*. Ansell (1973) reported that a value of *b* of 0.703 was common to sixteen species of bivalves in Scottish waters.

Some of this uncertainty over the expected value for *b* is due both to the fact that the rate of oxygen consumption by one individual is very variable and dependent on environmental conditions, and also to some insensitivity in the statistical procedures used to distinguish differences in the regression coefficients. Krüger (1960) recommended that, at the least, a range in size of 1:50 between the smallest and largest individuals should be measured for oxygen uptake in order for estimates of *a* and *b* to be reliable. Much of the variation in the literature is probably due to the measurement of respiration rate over too small a size range of animals. Multiple regression procedures that allow separation of variation in respiration rate due to size from variation due to other factors (Newell & Roy, 1973) may result in more precision in the estimation of *b* and consequently more confidence in generalising inter- and intraspecific comparisons.

The value of *a* in the allometric equation is very variable, and in what follows we shall examine the way in which this parameter (or respiration rate per unit of body size) changes in response to environmental conditions.

Ration

Three different levels of oxygen consumption rate by *Mytilus edulis* have been identified by Thompson & Bayne (1972) and Bayne *et al.* (1973). The description of these three levels was empirically based on changes in ration, as follows. In the absence of particulate food, the rate of oxygen consumption declines to a steady-state condition (called the 'standard rate') which is typical of an animal with shell valves open and with the mantle edge partially extruded but showing minimal filtration activity (or clearance rate). When such an animal is fed, filtration rate increases to a maximum (see also Widdows, 1973*a*) coincidentally with a marked increase in the rate of oxygen consumption to the 'active rate'. Between the limits of the active and standard rates, the mussels can show a variety of 'routine rates' of oxygen consumption; these are related to spontaneous variation in filtration or ventilation activity, and they are predictable according to animal size, season and to ration level.

The rate of decline of oxygen uptake from routine to standard varies with animal size (being more rapid in smaller individuals) and with season (Bayne, 1973*a*). In the summer and autumn, when the routine rate is low (p. 191), decline to the standard rate may occur within 10 days. In the winter, when gametogenesis is active and the routine rate is high, decline to the standard rate may require 25–30 days (Gabbott & Bayne, 1973). In the spring, however, with gametogenesis virtually complete and the energy reserves of the body at a minimum, starvation results in a very rapid decline to the standard rate. This standard rate of oxygen consumption could also be called a 'starvation rate', since it is empirically defined as a result of starvation. The decline from routine to standard is due, in part, to reduced activity, and is also equated with the reduction of body reserves and other conditions characteristic of starvation (Bayne, 1973*b*). However, prediction of the standard rate from extrapolation to zero activity on

163

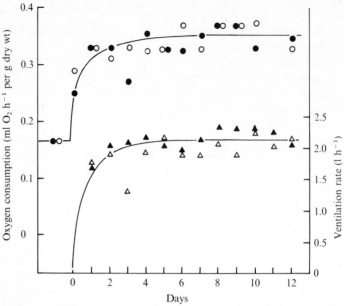

Fig. 5.16. The rates of oxygen consumption (circles) and ventilation (triangles) of two starved mussels (*Mytilus edulis*) in response to the provision of food on day 0. (From Widdows, 1973*a*.)

a graph relating filtration rate to rate of oxygen consumption (Thompson & Bayne, 1972) yields estimates that are not statistically distinguishable from measurements of the standard rate for similar-sized animals.

The increase from standard to active rates of oxygen consumption, resulting from the offer of food to animals previously starved, has been described by Thompson & Bayne (1972) and Widdows (1973*a*). Such an increase is almost instantaneous, and is associated with a greatly enhanced level of activity (Fig. 5.16). If feeding at a high ration level is then continued, the rate of oxygen consumption may decline (at the time of year when routine rate is significantly less than active rate; see p. 191) or it may remain close to the active rate. The routine rate of oxygen uptake is established spontaneously by the animal once the ration level exceeds the maintenance requirement (Chapter 7). At low ration levels, only slightly in excess of maintenance, the rate of oxygen consumption increases rapidly with any increase in the concentration of food particles (Fig. 5.17) reaching an asymptotic value which is then maintained over a considerable range of further ration increase (Thompson & Bayne, 1974).

The relationships between respiration rate and ration are therefore complex, involving the metabolic demands, not only of activity (primarily the action of gill cilia in bringing about the movement of water and the filtration of particles in the mantle cavity), but also of processes related to

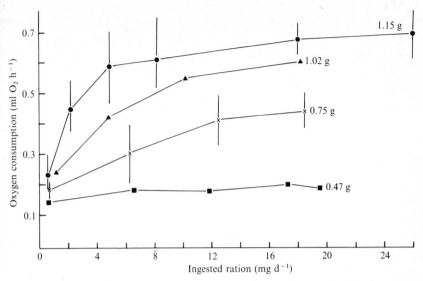

Fig. 5.17. The relationship between rates of oxygen consumption and ingested ration in *Mytilus edulis* of different dry flesh weight. Values are plotted as means of three to seven observations, with the range of measurements shown for two size groups only. (From Thompson & Bayne, 1974.)

the assimilation of food (the metabolic costs of digestion, excretion and growth). The former of these comprise the 'mechanical costs' associated with feeding, and the latter comprise the 'physiological costs' (Bayne *et al.*, 1973). Bayne *et al.* (1975*a*) estimated these various metabolic demands in *Mytilus californianus* by relating the rate of oxygen consumption to ventilation rate. Fig. 5.22 is a plot of this relationship in both *M. edulis* and *M. californianus*; in both cases the form of the curve is exponential. Bayne *et al.* calculated from these data the calorific costs of increments of $1 \, l \, h^{-1}$ in ventilation rate; for an animal of 1 g dry flesh weight at 13 °C this cost increased from 0.95 cal h^{-1} for an increase in ventilation from 1 to $2 \, l \, h^{-1}$, to 2.7 cal h^{-1} for an increase from 3 to $4 \, l \, h^{-1}$. Bayne *et al.* (1975*a*) also estimated the physiological and mechanical costs associated with an ingested ration of 6.0 cal h^{-1}. The physiological cost was 7.7%, and the mechanical cost 25%, of the ration.

The physiological costs of feeding include energy utilised in digestion and assimilation, and the 'non-utilised energy freed through deamination and other processes' (Warren & Davis, 1967). In an attempt to estimate this last component in *M. edulis*, Bayne & Scullard (unpublished data) measured the increase and subsequent decline in oxygen consumption by mussels starved for some days and then given a suspension of algal cells for one hour. The decline in oxygen consumption after feeding could be assigned to two components (Fig. 5.18) one of which was associated with

165

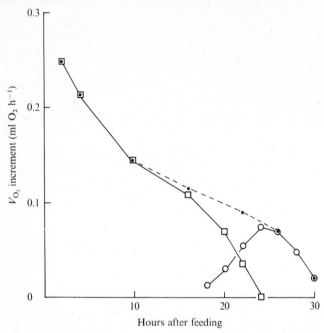

Fig. 5.18. Increase in the rate of oxygen consumption by *Mytilus edulis* after feeding for 1 h following a prolonged period of starvation. ●, total increment in oxygen uptake; □, increment due to increased feeding activity; ○, increment due to the 'specific dynamic action' of the ration. (From Bayne & Scullard, unpublished data.)

the decline in feeding activity (the 'activity component'), and the other associated with digestion and assimilation (the 'specific dynamic action' or s.d.a. component). This latter component coincided in time with a pulse of increased nitrogen excretion, and amounted to between 5 and 30% of the total increment in oxygen uptake following feeding.

The measurement of three different levels of oxygen consumption rate allows the calculation of 'scope for activity', as described by Fry (1957), by subtracting the standard from the active rate of oxygen uptake. In *M. edulis* the total scope for activity, although variable according to season and body size, is approximately twice the standard rate of oxygen consumption.

Temperature

Temperature is recognised as one of the major environmental determinants of the rate of metabolism and the level of activity of poikilothermic organisms. However, mussels, like many other littoral invertebrates, although apparently unable to regulate their rate of heat loss or gain from the environment, are able to vary their respiratory and feeding rates in such

166

a way as to maintain them relatively independent of the environmental temperature.

The subject of thermal compensation in poikilotherms has been reviewed by Bullock (1955), Prosser (1955), Gunter (1957), Fry (1958), Precht (1958), Prosser & Brown (1961), Segal (1961), Vernberg (1962), Kinne (1963, 1964b, 1970c), Newell (1970, 1973), Vernberg & Vernberg (1972) and Wieser (1973). The general concept of thermal adaptation includes both genetic and non-genetic aspects. Genetic adaptation takes place over many generations and sets the upper and lower limits of thermal tolerance. Non-genetic adaptations are induced directly by the environment, and can be divided into two categories on the basis of the time course of the physiological compensation. Firstly, there is 'immediate compensation' for change in temperature, occurring over hours rather than days (the 'acute' response). Secondly, there are compensatory adjustments which take place over days or weeks. These are commonly referred to as acclimatisation, when occurring within a complex of environmental variability (normally the animal's natural habitat), or acclimation, when occurring in response to a maintained deviation of a single environmental factor (in this case temperature), normally imposed in the laboratory. The term acclimation has often been misused, as when applied to animals kept under specific laboratory conditions for an arbitrary period of time, irrespective of whether or not any compensatory adjustments have been shown to have taken place. The term should only be used when a measured change in a physiological rate has occurred.

The effects of temperature on the respiration and activity of mussels are examined in this section in terms of short-term, or acute effects, and more long-term, or acclimatory responses. The biochemical mechanisms responsible for metabolic compensation for temperature change (Hochachka & Somero, 1973; Hazel & Prosser, 1974) are discussed in Chapter 8.

The acute effects of temperature

The effects of a change in temperaure on the rate of oxygen consumption are often expressed as rate/temperature or RT curves (Fig. 5.19). The form of the curve may conveniently be described as values for Q_{10} over specified increments of temperature, when Q_{10} is:

$$Q_{10} = \left[\frac{V_1}{V_2}\right]^{\frac{10}{t_1 - t_2}},$$

where V_1 and V_2 are the rates of oxygen consumption at temperatures t_1 and t_2, respectively. The Q_{10} provides an index of the dependence of the physiological rate on temperature.

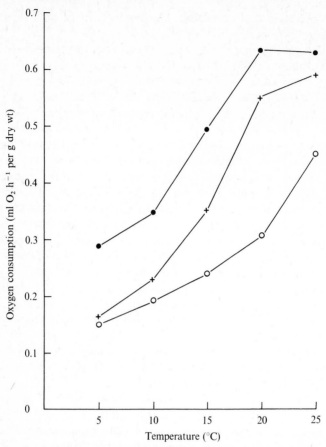

Fig. 5.19. The 'acute' responses of standard (O), routine (+) and active (●) rates of oxygen consumption by *Mytilus edulis* to changes in temperature. Mussels were acclimated at 15 °C.

The acute temperature response of the rate of oxygen consumption by mussels is modified by the thermal history of the animal, by its body size, its reproductive condition and by its level of metabolism. Fig. 5.19 illustrates the acute response of standard, routine and active rates of oxygen consumption by *M. edulis* (animal weighing 1 g dry flesh weight, with a mature gonad, and acclimated to 15 °C). The Q_{10} for these standard, routine and active rates between 10 and 20 °C are 1.6, 2.4 and 1.9 respectively (Widdows, 1973*b*). Lower Q_{10} values of 1.0–1.2 for the standard rate of oxygen consumption by *M. edulis* were recorded by Newell & Pye (1970*a, b*) and supported with evidence of temperature-independent metabolic rates of cell-free homogenates. Newell & Pye studied small, sexually immature mussels (about 30 mg dry flesh weight). Widdows (1973*b*) and Bayne *et al.* (1973) studied larger, mature individuals

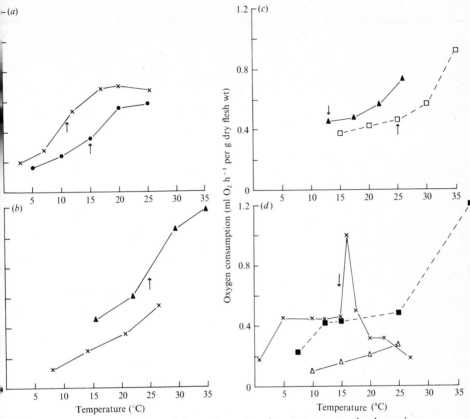

Fig. 5.20. The 'acute' responses of the routine rates of oxygen consumption by various mussels to changes in temperature. Wherever the field-ambient temperatures (acclimation temperature) were recorded, they are arrowed. (a) *Mytilus edulis* (×, Read, 1962a; ●, Widdows, 1973a). (b) *Modiolus demissus* (▲, Read, 1962a; ×, Kuenzler, 1961a). (c) *Mytilus californianus* (▲, Bayne et al., 1975a) and *Perna perna* (□, Bayne, 1967). (d) *Mytilus edulis*, tissue preparations (■, retractor muscle×10 (Glaister & Kelly, 1936); △, gill (Schlieper, Kowalski & Erman, 1958); ×, cell-free homogenate (Newell & Pye, 1970a)).

(about 1 g dry flesh weight). The presence of gametes may impose a greater dependence of total metabolic rate on temperature change. Rao & Bullock (1954) found that the ventilation rate of *Mytilus californianus* became less dependent on temperature with decreasing body size. Bayne et al. (1975a) recorded a Q_{10} of 1.13 between 12 and 21 °C for oxygen consumption by *M. californianus* in the summer, after the animals had spawned.

The ecological significance of a standard metabolic rate that is relatively independent of temperature is apparent in situations of environmental stress, such as extremes of temperature, during starvation, or when exposed to air at low tide, when insensitivity to temperature change may help to conserve metabolic energy reserves.

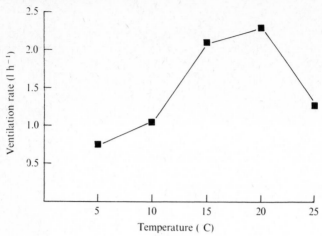

Fig. 5.21. The 'acute' response of ventilation rate by *Mytilus edulis* to changes in temperature. Mussels acclimated to 15 °C.

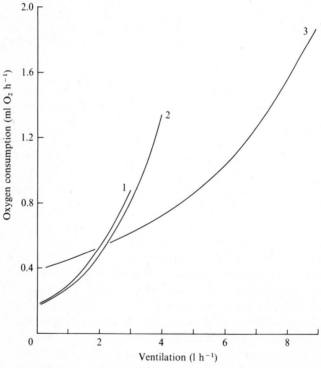

Fig. 5.22. Relationships between ventilation rate and rate of oxygen consumption for *Mytilus edulis* (1, Widdows, 1973*b*), *Mytilus californianus* (2, Bayne *et al.*, 1975*a*) and *Mercenaria mercenaria* (3, Verduin, 1969).

Table 5.4. *The Q_{10} for rates of oxygen consumption by four species of mussel over temperature increments of 5 deg C*

Species	Q_{10} for temperature increments (deg C)					
	5–10	10–15	15–20	20–25	25–30	30–35
Mytilus edulis (15 °C)*	2.25	2.13	2.47	1.10	—	—
Modiolus demissus (25 °C)	—	—	1.55	1.89	1.86	1.24
Mytilus californianus (13 °C)	—	—	1.30	1.69	—	—
Perna perna (25 °C)	—	—	1.25	1.17	1.54	2.59

Values calculated from data in Fig. 5.20.
* Values in brackets are the acclimation temperature.

In contrast to standard and active rates, the routine rates of oxygen consumption by *M. edulis* (Bruce, 1926; Read, 1962a; Widdows, 1972; Bayne *et al.*, 1973) and *Modiolus demissus* (Read, 1962a) are more clearly dependent on changes in temperature, with Q_{10} values greater than 2 over much of the normal temperature range, declining only as the upper limit of each species' thermal tolerance (25 °C for *M. edulis*; 35 °C for *M. demissus*) is approached (Fig. 5.20a and b). The rotation of the routine relative to the standard and active RT curves (Fig. 5.19), is due in part to the effects of temperature on spontaneous activity, or on ventilation rate (Fig. 5.21). The cost of ventilation probably accounts for a significant proportion of the oxygen consumption, and this cost increases exponentially (Fig. 5.22). In response to a rise in temperature the ventilation rate increases, leading to a rise in oxygen consumption towards the active rate. Between 20 and 25 °C a decline in ventilation rate is coincident with a reduction in the rate of increase of routine oxygen consumption. Newell & Pye (1970a, b) recorded a temperature-dependent rate of oxygen consumption by *M. edulis* which is thought to correspond to the routine rate described here (Newell & Bayne, 1973).

In Fig. 5.20(c) the rate/temperature relationships for oxygen consumption by *M. californianus* and *Perna perna* are plotted. In these species the routine V_{O_2} is apparently rather less temperature-dependent than in *M. edulis* or *Modiolus demissus* (Fig. 5.20 a, b), at least over part of the normal temperature range. These differences are made more apparent when the Q_{10} values for routine oxygen consumption (acute response) are compared (Table 5.4). These differences may be real species attributes, or they may be due to differences in the physiological and/or reproductive condition of the animals at the time of measurement; only further comparative studies will clarify this.

Isolated tissue preparations may also show a relative independence of temperature. In Fig. 5.20(d) some values for oxygen consumption by slices of pedal retractor muscle, isolated gill tissues and a whole-animal cell-free homogenate (all *M. edulis*) have been plotted. Percy & Aldrich (1971), studying tissue respiration in the American oyster, *Crassostrea virgincia*, recorded low Q_{10} values for the acute temperature response of adductor muscle tissue, whereas mantle and gill tissue did not have such a marked independence of temperature. Newell & Pye (1970 *a*) have demonstrated that homogenates of *M. edulis* show seasonal variation in the range of temperatures over which metabolic compensation occurred. They have interpreted these results (see also Newell, 1967, 1969) to indicate that acute metabolic compensation may be controlled at the subcellular level. Possible mechanisms for this are reviewed by Hazel & Prosser (1974) and discussed in Chapter 8.

Acclimation to change in temperature

The possibility that compensatory mechanisms might exist in the physiology of poikilotherms from cold and warm seas was suggested in 1916 by Krogh, but it was Spärck (1936) and Thorson (1936) who first demonstrated thermal acclimation in marine lamellibranchs, including *M. edulis*. Spärck (1936) examined latitudinally separated populations of *M. edulis* and showed that mussels from high latitudes had higher rates of oxygen consumption than those from lower latitudes, when measured at the same temperature. A similar adaptation of the ventilation rate of latitudinally separated populations of *M. californianus* has been demonstrated by Rao (1953).

The responses of the three levels of oxygen consumption rate by *M. edulis* to maintained changes in temperature are shown in Fig. 5.23. The active rate shows partial acclimation whereas the standard rate does not acclimate. In contrast, the routine V_{O_2} shows a distinct acclimatory response over 14 days (Widdows & Bayne, 1971).

Fig. 5.23 also illustrates the response of filtration rate to a maintained temperature change. Following an increase in temperature, filtration declines from the elevated rate which marks the acute response to the acclimated, steady-state level after 14 days. Acclimation to a decrease in temperature shows the reverse trend. As with the acute response to temperature change, changes in filtration rate are reflected in altered routine rates of oxygen consumption. The exponential relationship between ventilation rate and oxygen consumption signifies that, as a result of a small reduction in ventilation, there can be a large 'saving' in total metabolic cost. This may explain, at least in part, the observed acclimation of V_{O_2} to temperature change.

The acclimation of routine respiration rate occurs at temperatures from < 5 to 20 °C (Widdows, 1972). At the limits of this range there are critical

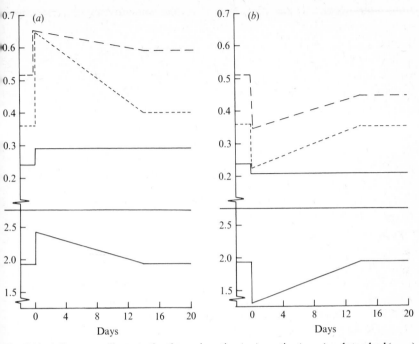

Fig. 5.23. A diagram to illustrate the change in active (– –), routine (- - - -) and standard (——) rates of oxygen consumption, and in the filtration rate, of *Mytilus edulis* following a maintained change in temperature. (*a*) Increase in temperature from 15 to 20 °C. (*b*) Decrease in temperature from 15 to 10 °C. (From Widdows, 1973*b*; Bayne *et al.*, 1973.)

temperatures at which a breakdown of the compensatory mechanisms occurs and the mussel is no longer able to acclimate. But within a wide range of temperatures the animal can maintain its activity relatively independent of seasonal changes; thus filtration rate is kept fairly constant throughout the year (Bayne & Widdows, unpublished data). However, measurement of the routine rate of oxygen consumption by *M. edulis* from a North Sea population showed a marked seasonal cycle, with high rates in winter and low rates in the summer (Bayne, 1973*a*). These changes were apparently related to the seasonal gametogenic cycle, high rates of oxygen consumption coinciding with active gametogenesis, and low rates occurring after spawning in the summer (p. 191). These seasonal changes may not be as marked in more northerly populations where very low winter temperatures inhibit gametogenesis, or in more southerly populations, where gametogenesis may occur over a large portion of the year (personal observations).

Three components of aerobic metabolism may be recognised, therefore; maintenance metabolism, which may be estimated as the standard rate of oxygen consumption (or the 'rate of fasting catabolism'; Kleiber, 1961) metabolic processes associated with ventilation, movement and feeding and gametogenesis. The last two of these contribute to the routine rate of oxygen consumption. Fig. 5.24 shows a simple schematic model that may help to account for some of the observed effects of temperature on the respiration rate of *M. edulis*. Gametogenesis, feeding/activity and the processes of maintenance contribute towards the observed rate of oxygen uptake. Temperature acts directly on these processes as a 'disturbance' (Wieser, 1973) and also on a 'control unit' or temperature transducer as a 'stimulus'. Gametogenesis, feeding/activity and maintenance are influenced by reference inputs such as season, body size and age; these act on activity via the control unit. The disturbance effects of a rise in temperature are: (i) to increase gametogenesis along a hyperbolic curve and (ii) to increase ventilation rate and maintenance metabolism in the manner shown (Figs. 5.19 and 5.21). We postulate that there is a negative feedback component that results from changes in metabolic rate and that acts, together with the temperature stimulus, on the control unit to regulate the level of activity; this is observed as the acclimation of the ventilation rate. In a starved animal, at a time when gametogenesis is not proceeding, ventilation activity is reduced and the total metabolic rate largely reflects the costs of maintenance metabolism. An increase in temperature at this time acts directly to inflate the maintenance costs and, to a slight degree, to increase the cost of ventilation (Fig. 5.22); the result is a slight increase in metabolic rate (a low Q_{10} for the acute response) and an insignificant capacity to acclimate. In a well-fed animal, also at a time when gametogenesis is proceeding at a slow rate, but when ventilation and filtration rates are high, a temperature increase markedly inflates the total metabolic rate, due largely to an exponential increase in activity cost. Feedback, acting (in an unknown manner) through the control unit, results in reduced ventilation rate and the concomitant acclimation of the routine metabolic rate. When gametogenesis is active, a temperature increase accelerates this, and limits the animal's capacity to acclimate.

A schematic diagram of the sort featured in Fig. 5.24 can do no more than to suggest some ways in which the various components of the respiratory response to temperature may be integrated and controlled. It says nothing of the mechanisms involved. Of the elements featured in Fig. 5.24, the feedback, the nature of the control unit, the identity of the signals and the processes underlying the form of the various response curves are all unknown at this time, although a variety of suggestions have been made in the literature (Wieser, 1973; Hazel & Prosser, 1974; see Chapter 8 of this volume).

In attempts to systematise the considerable literature that has developed on the non-genetic adaptation, or acclimation, of metabolic rate to changes

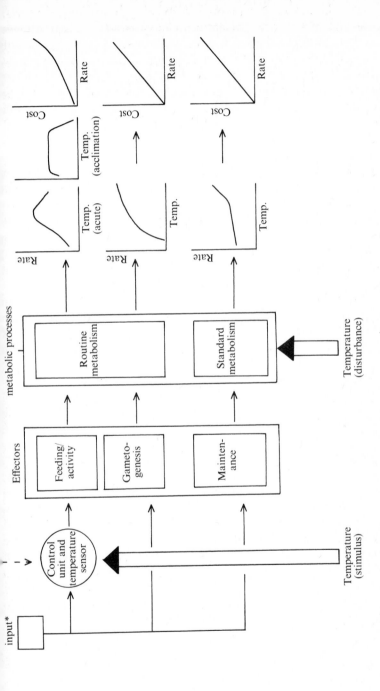

*e.g. season, size, ration

Fig. 5.24. A scheme (based on Wieser, 1973) to illustrate some of the effects of temperature on respiration and activity of *Mytilus edulis*. The form of the temperature response is shown as rate/temperature curves and, more speculatively, as curves of the 'energy cost' against rate. See text for further details.

7-2

in temperature, Precht (1958) and Prosser (1958) each suggested a scheme of classification. Prosser's scheme distinguishes four basic patterns: (1) little or no compensation, indicated by negligible change in the position of the rate/temperature (RT) curves; (2) a shift in position of the RT curves (translation) without any change in Q_{10}; (3) change in the Q_{10} of the RT curves (rotation) without any translation; and (4) translation combined with rotation. Identifying the particular pattern of temperature response may be useful in making generalised comparisons between species. However, the plasticity of the adaptive responses to temperature by some species (including mussels) brought about by seasonal differences, differences due to body size and by changes in metabolic intensity, sometimes render the application of rigid schemes of classification difficult, and somewhat arbitrary. In Table 5.5 we have attempted to identify types of temperature response reported for feeding (or ventilation) rates, and for rates of oxygen consumption, by some mytilid species. The impression gained is of the general occurrence of complete acclimation (Precht type 2) by translation of the RT curves, but the treatment is of limited usefulness at this time.

In a somewhat different approach, Kinne (1964a) suggested that in assessing the quantitative aspects of non-genetic adaptation distinction should be made between the amount, the stability and the velocity of acclimation; meaningful inter- and intraspecies comparisons may then be made. However, perhaps the greatest need at present is for a synthesis between the biochemical or cellular mechanisms of temperature adaptation (Hazel & Prosser, 1974) and the very varied processes of physiological compensation by the whole organism. In addition, the problem of the phenotypic versus the genotypic basis for interpopulation differences in physiological response requires further study (Bullock, 1955; Prosser, 1957; Segal, 1961). Loosanoff & Nomejko (1951) and Korringa (1957) discussed the possibility that 'physiological races' of oysters may exist. Reshöft (1961) suggested that the analysis of cellular tolerance limits to heat provides a useful tool for assessing genetic differences between populations; he concluded that *M. edulis* from the Baltic and the North Sea do not represent distinct races. Recent studies (see Chapter 9) have considerably advanced understanding of the population genetics of mussels and just as there is a requirement that information on biochemical and physiological processes should be integrated, it is also important that both physiological and genetical mechanisms be considered together to provide further understanding of thermal adaptation.

Table 5.5. *Patterns of acclimation to change in temperature in various species of mussels*

Species	Function	Acclimation pattern	Reference	Comments
Mytilus edulis	Oxygen consumption	Prosser 2	Spärck (1936)	Latitudinally separate populations
Mytilus edulis	Oxygen consumption	Prosser 2	Lagerspetz & Sirkka (1959)	
Mytilus edulis	Filtration rate	Precht 2	Theede (1963)	$Q_{10}(10–20\ °C) = 1.13$
Mytilus edulis	Oxygen consumption (standard)	Precht 2; Prosser 2	Newell & Pye (1970a)	
Mytilus edulis	Oxygen consumption (active)	Precht 4; Prosser 1	Newell & Pye (1970a)	
Mytilus edulis	Oxygen consumption (routine)	Precht 2	Widdows & Bayne (1971)	
	Filtration rate	Precht 2	Widdows & Bayne (1971)	
Mytilus californianus	Pumping rate	Precht 2	Rao (1953)	Latitudinally separate populations $Q_{10}(4–14\ °C) = 2.33$ $Q_{10}(10–20\ °C) = 1.63$
Modiolus modiolus	Filtration rate	Precht 3	Winter (1969)	

For details of the various patterns, see text, Precht (1958) and Prosser (1958).

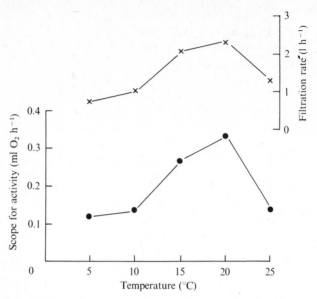

Fig. 5.25. The scope for activity (●, ml O$_2$ h^{-1}) of *Mytilus edulis* during acute response to change in temperature. The filtration rate (×, l h^{-1}) is also plotted. (From Bayne *et al.*, 1973.)

The scope for activity

As discussed in the previous section, the scope for activity is calculated as the numerical difference between the active and the standard rates of oxygen consumption, and it provides an index of the amount of energy available for activity and other physiological processes. During the acute response to increases in temperature, the scope for activity increases between 5 and 20 °C and then declines at 25 °C (Fig. 5.25); there is a close agreement between the form of this curve and the RT curve for ventilation (or filtration) rate. During temperature acclimation there is no marked change in the scope for activity (Fig. 5.23). Newell (1973) concluded from a review of the literature that temperature acclimation in some species involves the maintenance of a maximum scope for activity to coincide with the ambient thermal conditions. *M. edulis* is able to maintain a maximum scope over an impressive range of temperature, and during the course of a year in the North Sea shows little variation in scope.

It is also possible to calculate a 'routine scope for activity' as the difference between routine and standard rates of oxygen consumption. Changes in the routine scope illustrate the extent to which the animal varies its activity, and its metabolism, in compensating for environmental change. During temperature acclimation, for example, the routine scope is either increased (cold acclimation) or decreased (warm acclimation). The result is to establish a more or less constant routine metabolic rate whilst maintaining a maximum total scope for activity.

Fry (1947), in introducing the concept of the scope for activity, identified the temperature optimum of a physiological function as that point on a scale of temperature where an activity curve reaches its maximum, coincident with the maximum scope for activity. For *M. edulis* a discrete temperature optimum can be identified for its acute responses (Fig. 5.25). However, due to acclimation, the physiological functions have been adapted to express 'optimum' activity over the very wide range of temperatures normally encountered in the environment.

Species comparisons

Mytilus edulis shows a well-developed capacity to compensate for changes in temperature by acclimation of the metabolic demand. However, metabolic demand could equally conform to a change in temperature so that respiration rate is augmented in response to increased temperature. Ansell & Sivadas (1973) recorded an exaggerated enhancement of the respiration rate of the clam *Donax vittatus* in response to temperature increases within the normal physiological range. Although some temperature acclimation occurred (Ansell, 1973) this was minimal, and unlikely to have much practical significance for the species. Kennedy & Mihursky (1972) studied the effects of temperature on the respiration of three lamellibranch species; *Mya arenaria*, *Macoma balthica* and *Mulinia lateralis*. The three species all differed in their responses. *Macoma* showed a marked ability to compensate for temperature change; *Mulinia* largely conformed to environmental temperature and the responses of *Mya* were intermediate. Pamatmat (1969) found that *Transennella* showed only partial acclimation to temperature.

Whereas an acclimatory response of metabolic rate to temperature change signifies energy conservation by the species, conformity to temperature change represents an exploitation of the environment, suggesting an 'opportunistic' as opposed to a 'conservative' physiological strategy. Opportunism can result not only in deriving benefit from certain benign environmental changes, but also in an increased vulnerability to environmental stress. Differences between conservatism and opportunism may apply not only between species, but also within a single species, as between populations and possibly also between individuals of different age. Elucidating the inter- and intraspecific significance of these differences should provide a fruitful area for future research.

179

Fluctuating temperatures

When considering the effects of temperature as an ecological factor, a distinction has to be made between long-term seasonal changes and shorter-term diel fluctuations. Surface seawater temperature in some temperate regions, although having a large annual range, can be regarded as relatively constant in the short-term, with only small diurnal fluctuations. Under these conditions, studies in physiological adaptation to constant temperatures are generally meaningful. Further, the rate of acclimation of physiological functions to changes in temperature may be more rapid than the rate of change in the mean environmental temperature, so that seasonal constancy of the physiological function is possible. For example, the filtration rate of *Mytilus edulis* can compensate completely (Precht type 2) for gradual temperature changes between 2 and 20 °C, thereby enabling feeding activity to be maintained at a constant level throughout the year independent of the annual cycle of temperature.

Short-term environmental fluctuations are those which occur over a tidal or diel cycle, and may be regular or irregular in nature. Marked temperature fluctuations may occur in estuarine and coastal environments as a result of solar radiation in shallow water; exposure of the animals to both air and water, which may be markedly different in temperature; and thermal additions from industrial sources. There are several components of a fluctuating variable to which an organism may respond: (*a*) the amplitude of the change; (*b*) the frequency of the oscillation; (*c*) the rate of change between maxima and minima; and (*d*) the average value about which the oscillations occur.

The only experiments on the effects of fluctuating temperatures on mussels known to us are due to Widdows (1976) working with *M. edulis*. He adapted the mussels to diel temperature changes both within and above the normal environmental temperature range. The animals acclimated their rates of filtration and oxygen consumption to fluctuating temperatures between 11 and 19 °C by reducing the amplitude of response over 14–16 days (Fig. 5.26), and thereby increasing their independence of temperature. In these experiments the metabolic RT curves for two different mussel populations reflected their different thermal environments. An estuarine population which normally experienced a small diurnal temperature fluctuation only (mean ambient ± 1 deg C) had a metabolic rate that was relatively temperature-dependent up to 20 °C. A different population, sited in the outflow of a cooling water system, experienced a fluctuation of ambient ±8 deg C, and had a corresponding region of temperature independence between 15 and 25 °C. When subjected to fluctuating temperatures which extended above their normal environmental maxima (21–29 °C; Fig. 5.27) the mussels showed a temperature-independent

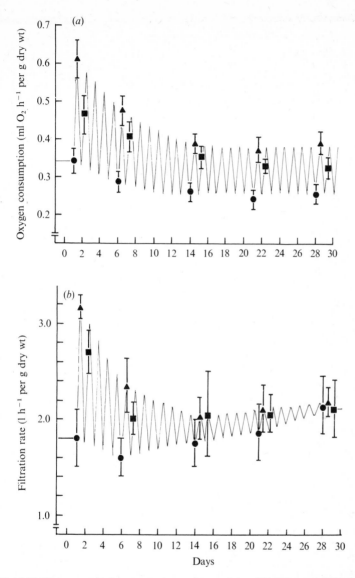

Fig. 5.26. The response of oxygen consumption rate (*a*) and filtration rate (*b*) by *Mytilus edulis* to fluctuating temperatures over 30 days. The temperature regime was a fluctuation from 11 to 19 to 11 °C in 24 h. Measurements were made at 11 (●), 15 (■) and 19 °C (▲) at appropriate times in the cycle. (From Widdows, 1976.)

181

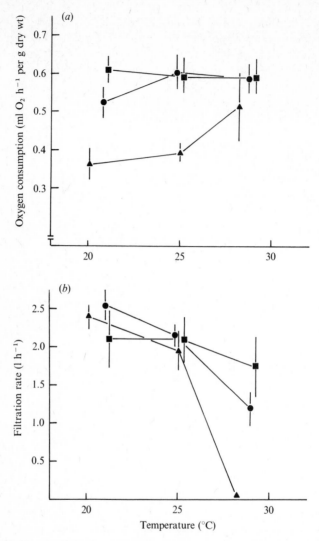

Fig. 5.27. The response of rate of oxygen consumption (*a*) and filtration rate (*b*) by *Mytilus edulis* following acclimation to fluctuating and constant temperatures. ●, an estuarine population, fluctuating temperatures 21–29 °C; ■, a population normally subjected to heated water (ambient+8 deg C), fluctuating temperatures 21–29 °C; ▲, an estuarine population, maintained at constant temperatures of 20, 25 and 28 °C. Mean ±1 s.D. (From Widdows, 1976.)

respiration rate over the entire range, and an extension of the zone of activity and temperature tolerance. These studies suggest that the physiological response to a specific temperature in a fluctuating cycle is not significantly different from the acclimation response to equivalent constant temperature. However, in a fluctuating thermal environment respiratory adaptation may occur at higher temperatures than is possible in a constant thermal environment, and there is an increase also in activity and thermal tolerance.

Temperature as a lethal factor

Fry (1947) described lethal factors as those which act to destroy the integration of the organism, and he distinguished between two components of the lethal effect: (*a*) the incipient lethal level, or 'that level...beyond which the organism can no longer live for an indefinite period of time'; and (*b*) the effective time, which is the time necessary to cause death by a given level of the environmental factor beyond the incipient lethal level. Fry stressed that both these components should be considered within the context of the animal's previous experience of the factor concerned. In addition, subsequent studies have shown that the lethal limit may depend upon sex, age (Ushakov, 1965), nutritive conditions and season (Newell & Pye, 1970*a*).

The simplest experimental procedure used to determine lethal temperatures has been to warm, or cool, the organism at a constant rate until it dies. As Fry (1947) points out, this does not allow the two variables, time and temperature, to be distinguished. Nevertheless, most of the recorded lethal limits in the literature have been determined in this way. Using this procedure, Henderson (1929) recorded a lethal temperature of 36.3 °C for *Modiolus modiolus* and 40.8 °C for *Mytilus edulis*. Read & Cumming (1967) determined an upper lethal temperature of 27 °C for *M. edulis* by raising the temperature 1 deg C every 3.5 days. This appeared to provide an ecologically meaningful estimate, since Wells & Gray (1960) have stated that the southern distribution limit for *M. edulis* occurs where the mean summer water temperature is 26.7 °C. Read (1964) has attempted to correlate heat tolerance of some tropical and boreal bivalves with their intertidal habits. Lent (1968) subjected *Modiolus demissus* to 10 h at temperatures between 32 and 44 °C, when exposed to air at 80% relative humidity; 50% of the mussels died at between 36.6 and 37.8 °C. Although reports differ, the consensus view appears to be that increasing size is accompanied by decreasing heat tolerance (Lent, 1968; Kennedy & Mihursky, 1971). When desiccation accompanies thermal shock, however, the relationship between size and mortality is more complex (Kinne, 1970*c*). Recently V. S. Kennedy (personal communication) determined the lethal temperatures for *Mytilus edulis aoteanus, Perna canaliculus* and *Aulacomya maoriana* in New Zealand; at an acclimation temperature of 20 °C, the lethal temperatures varied from 30–33 °C for *A. maoriana* to 36–37 °C for *M. e. aoteanus*. However, as Kennedy points out, lethal temperature limits alone could not explain differences in intertidal distribution.

A more elaborate and reliable experimental method of determining lethal temperature limits is described by Fry (1947) and Speakman & Krunkel (1972). This method allows the definition, not only of the zone of tolerance, bounded by the upper and lower incipient lethal levels, but also of the zone of resistance, beyond the tolerance zone, in which the animal will ultimately succumb to the effects of heat, but in which it can resist this lethal effect for a period of time. Speakman & Krunkel (1972) describe the incorporation into this treatment of an assessment of the rate of temperature change, resulting in a three-dimensional surface with axes representing acclimation temperature, lethal temperature and the time taken to complete the temperature change. A complete experimental treatment of this sort has not been attempted for any mytilid, but Waugh & Garside (1971) and Waugh (1972) did include different acclimation regimes in their studies of the upper lethal temperatures of *Modiolus demissus* (see also Kennedy & Mihursky (1971) for a similar study on three other bivalve species). The upper lethal temperature for *M. demissus* over periods of 24-h exposure ranged from 38.4 to 40.2 °C and increased with increase in the acclimation temperature. Waugh (1972) recorded that as the environmental temperature declined, with season, from 25 to −1.1 °C, the upper lethal temperature also declined, from 39.5 to 37.7 °C. For any acclimation temperature, upper lethal temperature declined with departure from the appropriate acclimation salinity (Waugh & Garside, 1971). However, in all these experiments, the upper lethal temperature was independent of size (as shell length) and sex, and never differed by more than 3 deg C in any of the experimental treatments.

Considerable work has been done on the upper and lower temperature limits for isolated tissue preparations, notably gill tissue, of mussels, and these studies have been comprehensively reviewed by Kinne (1970c) and Theede (1973). Ushakov (1965) distinguished between obligatory and situational changes in heat tolerance; the former are due to endogenous factors, and the latter to exogenous, environmental changes. Resistance of the gill cilia of *M. edulis* to freezing or to high temperatures varies seasonally (Theede, 1965) and can be altered by salinity (Theede, 1965) or by the acclimation temperature (Schlieper, Flugel & Rudolf, 1960; Schlieper, 1966, 1967). Both tidal and geographical differences in distribution affect the thermal tolerance of isolated tissues (Reshöft, 1961; Vernberg, Schlieper & Schneider, 1963). Vernberg *et al.* (1963) recorded

184

thermal acclimation by gill tissue of *Modiolus demissus* within 17 days; the thermal tolerance of this tissue was also influenced by photoperiod.

Mytilus edulis appears, from these studies, to be considerably eurythermal (see p. 186 for discussion of freezing tolerance) and, indeed, euryoecious in general. Reshöft (1961) recorded a survival period of 100 min for isolated gill pieces of *M. edulis* at 36 °C, whereas *Modiolus modiolus* gill pieces survived for less than 20 min at this temperature.

Exposure to air

An intertidal distribution exposes a marine animal to a variety of possible environmental stressors, including the abrasive action of waves and ice; discontinuous availability of food; wide variations and extremes of salinity and temperature; desiccation; and the necessity either to withstand periods without a supply of oxygen, or to gain oxygen from a medium to which the respiratory apparatus is ill-adapted. Nevertheless, certain marine invertebrates, including some species of mussels, tolerate and show certain adaptations to these conditions in a way that allows their successful colonisation of the shore. For many such species the shore represents a refuge, with the lower limits to their distributions set by biotic factors of competition and predation (Paine, 1974) and their upper limits determined by such stressors as temperature, desiccation and shortage of food. Harger (1970a) has indicated some of the ways in which mussels are adapted to wave exposure, including their shell-shape, strong byssal attachments and clumped distribution. The relevance to an intertidal habit of rhythms of feeding and digestion has been discussed by Morton (1970, 1971), Owen (1974b) and, with particular reference to *Mytilus edulis*, by Langton (1975; see also p. 153). In this section we shall discuss the effects of temperature, desiccation and oxygen supply on mussels during periods of exposure to air. These problems have been examined particularly by Kuenzler (1961a) and Lent (1968, 1969) in *Modiolus demissus*; Coleman & Trueman (1971) and Coleman (1973a, b) in *Mytilus edulis*, and Moon & Pritchard (1970) and Bayne et al. (1975b) in *Mytilus californianus*. In addition, Boyden (1972a, b) studied exposure to air in two species of *Cardium* (*Cerastoderma*), the common cockle.

It is during periods of aerial exposure that extremes of temperature are likely to occur and thus, not surprisingly, intertidal mussels are markedly eurythermal. However, behavioural adaptations may help to reduce heating. *Modiolus demissus* occurs in salt marshes buried in mud with only the posterior extremity of the shell, and the siphons, extended. As Lent (1969) indicates, two benefits accrue from this; evaporative cooling of the mud reduces the microhabitat temperature by 5–10 deg C, and the relative humidity at the mud/air interface is nearly 100%, so

185

reducing desiccation. *Mytilus edulis* also often occurs attached in clumps which retain water at low tide and ensure high local humidity. *M. californianus* may occur in clefts and gullies on the shore, protected from direct sunlight and the drying effects of wind. Lent (1968) measured the extent of evaporative cooling in *M. demissus;* temperature in the mantle cavity was reduced by 2.2 deg C within 1 h of emersion and then increased to ambient air temperature over the following 20 h. Lent suggested that evaporative cooling was of benefit to individual mussels during brief exposures on extremely hot days.

Kanwisher (1955) pointed out that on a high-latitude shore in winter, mussels may be exposed to temperatures of −20 °C or lower; a sudden drop in temperature of 20 deg C can occur as the animals are emersed by the ebbing tide. Under these conditions body temperatures of *M. demissus* and *M. edulis* equilibrated to air temperature within a few minutes and at −15 °C 62–65% of the body water was frozen, but the animals survived. Williams (1970) concluded that −10 °C was the minimum tolerable tissue temperature of the *M. edulis* that he studied. Temporary survival at lower ambient temperatures is effected by a variety of factors which influence the time required for tissue temperature to reach −10 °C. The most important of these are 'extrinsic sources of heat', such as the thermal reserve of the rock substrate. However, when temperatures do fall to very low values, the latent heat of freezing of the water in the mantle cavity greatly retards the cooling of the tissues, and this delay helps in survival until immersion by the next tidal flood. Williams (1970) found that tissue death occurs when 64% of tissue water has been removed as ice, and he concluded that this represents the minimum tolerated cell volume, regardless of the absolute concentration of solutes involved. The tolerance of freezing of *Mytilus* is apparently due to the presence of a constant 20% of total tissue water which is osmotically inactive and possibly retained by high molecular weight compounds in the cells. Changes in intracellular free amino acid concentration alter the freezing point of the tissues, and so effect a gain or loss in freezing tolerance on a simple colligative basis without influencing the basic cryoprotective mechanism.

Pamatmat (1969) discussed an interesting aspect of the influence of aerial exposure on the physiology of the bivalve *Transennella tantilla*. The rate of oxygen consumption showed evidence of cold acclimation in the autumn, but not in the winter. In the autumn in the area studied (False Bay, Washington, USA) the tidal pattern is such that low tide occurs progressively later at night, at which time there may be a sudden drop in temperature in the microhabitat of the clam. By late winter, however, low tide occurs progressively earlier in the day, and temperatures may then increase. Pamatmat concludes that cold acclimatisation was induced by the periodic exposure to lower temperatures during low tide in the autumn, and warm acclimatisation resulted from periodic warming in the winter at low tide. This not only suggests an influence of the thermal regime experienced during emersion on the metabolism of the animal when subsequently re-immersed, but also provides a strong argument for careful ecological analysis before laboratory-based physiological data are interpreted in terms of the animals' responses in nature.

Lent (1968, 1969) and Boyden (1972*a*) reported that groups of *M. demissus* and *Cardium* spp., respectively, tolerated between 33 and 42% loss of weight as water before mortality reached 50%. The absolute loss of water is greater in larger than in smaller mussels, but the percentage loss is greater in the smaller individuals (see also Coleman, 1973*b*). V. S. Kennedy (personal communication) recorded that whereas *Perna canaliculus* and *Mytilus edulis aoteanus* began to suffer significant mortality after 20% or more of body weight was lost through desiccation, *Aulacomya maoriana* did so after only 15% weight loss. Small *Perna* desiccated faster than either *Mytilus* or *Aulacomya*. Kennedy suggested that these differences help to explain the distribution of the three species on the shore, since *Perna* is less common high on the shore than *Mytilus*, and *Aulacomya* tends to occur in shade, in tide pools or under clumps of *Mytilus*, where there is some protection from desiccation.

Whether the animals ever actually experience desiccation of the order of 30–40% in their natural habitats seems unlikely. P. S. Davies (1969) showed that limpets seldom lose more than 10% of their body weight from dessication. Bayne *et al.* (1975*b*) did not record greater than 10% desiccation in *M. californianus* which were directly exposed to sunlight on a horizontal rock platform. Coleman (1973*b*) recorded considerable loss of water from *Modiolus modiolus* exposed to air (this species only occurs intertidally either in rock pools or at extreme low water of spring tides) and less loss from *Mytilus edulis*. Much of the water loss from *Modiolus* occurred very rapidly, probably by spillage due to shell gape rather than evaporation, and there followed a period of desiccation. Coleman stressed that wind was a major factor in accelerating water loss, and an important cause of mortality. V. S. Kennedy (personal communication) observed that small *Perna canaliculus* in New Zealand gaped excessively during air exposure, and so aggravated their desiccation.

The ability to endure desiccation is due, at least in part, to the mussel's tolerance of increased osmoconcentration of the tissues. During the desiccation of *Modiolus demissus*, the rate of rise in the osmoconcentration of the body fluids (recorded as depression of the freezing point) suggested that the animals were freely permeable to water at low desiccation rates, but less permeable as percentage water loss increased; alternatively, as the blood osmoconcentration increased, the intracellular concentration of organic molecules also increased (Lent, 1969). Recent experiments by Pierce & Greenberg (1972, 1973) indicate that both mechanisms probably occur (see section on osmotic control, Chapter 6).

Much interest has concerned the possible utilisation of oxygen in the air during aerial exposure. Kuenzler (1961*a*) and Lent (1968) reported that *Modiolus demissus* opens its valves during exposure to air. There are similar reports for two tropical bivalves, *Isognomon alatus* (Trueman & Lowe, 1971) and *Geloina ceylonica* (Bayne, unpublished data) and for the cockle, *Cardium edule* (Boyden, 1972*a*). In some other species the pattern is not as clear. *Mytilus edulis* (Coleman & Trueman, 1971; Coleman, 1973*a*) and *M. californianus* (Bayne *et al.*, 1975*b*) sometimes show a slight degree of shell gape, but this behaviour is not consistent. Subtidal species, such as *Modiolus modiolus* (Coleman & Trueman, 1971; Coleman, 1973*b*) not surprisingly show a variable and apparently uncontrolled response to air exposure. The shell movements that immediately follow emersion vary in these species. *M. demissus* adducts the valves to a position representing half the maximum gape (about 2 mm) and remains in this position during exposure (Lent, 1968). *Cardium edule* exhibits rapid valve movements on emersion, associated with the expulsion of some water from the mantle cavity and its replacement with a bubble of air; the valves then remain between 50 and 100% agape during exposure (Boyden, 1972*a*). This behavioural dissimilarity probably reflects a morphological difference between the two species. *M. demissus* does not have discrete siphons, and attaches in clumps with the anterior end pointing downwards and the posterior mantle margin exposed at the surface of the mud (Lent, 1968); gaping allows direct access by the air to a relatively large surface area of water trapped in the mantle cavity. *Cardium* on the other hand, possesses small but distinct siphons, and it is possible that a small bubble of air in the mantle cavity increases the surface area of contact between the water in the mantle cavity and the air.

Table 5.6 summarises the recorded relationships between size (as flesh weight) and the rate of oxygen consumption in air and water, for three species. The rate of oxygen uptake in the air is slightly less than the minimum recorded values in water. In spite of an apparently negligible degree of shell gape, *M. californianus* consumes oxygen from air at high relative humidity equivalent to 0.74 times the immersed standard rate at the same temperature; there was no significant difference between the regression coefficients relating oxygen consumption to weight in air and water (Bayne *et al.*, 1975*b*). Moon & Pritchard (1970) had earlier deduced indirectly that *M. californianus* is able to extract oxygen from the air.

Coleman (1973*a*) has recorded very low Q_{10} values for the rate of oxygen consumption in air by *M. edulis* between 10 and 21 °C ($Q_{10} = 1.29$ between 10 and 16 °C and 1.09 between 16 and 21 °C). He equated this rate with the standard rate of oxygen uptake in water. Boyden (1972*b*) also likened the

Table 5.6. *Equations relating the rates of oxygen consumption* (V_{O_2}) *by three species of mussel, in air and water, to their weights*

Species (and authority)	Allometric equation*	Temp. (°C)	Rate in air as % of rate in water†
Cardium edule (Boyden, 1972*b*)			
Air	$V_{O_2} = 0.131\,W^{0.440}$	15	65.5
Water	$V_{O_2} = 0.200\,W^{).438}$	15	
Modiolus demissus (Kuenzler, 1961*a*)			
Air	$V_{O_2} = 0.240\,W^{0.890}$	20	63.2
Water	$V_{O_2} = 0.380\,W^{).380}$	20	
Mytilus californianus (Bayne *et al.*, 1975*b*)			
Air	$V_{O_2} = 0.172\,W^{0.648}$	13	73.8
Water	$V_{O_2} = 0.233\,W^{0.648}$	13	

* V_{O_2}, ml O_2 consumed per animal per hour; W, g dry flesh weight.
† For animals of 1 g dry flesh weight.

respiration of *C. edule* in air to the standard or 'quiescent' rate in water; $Q_{10} = 1.88$ between 10 and 22.5 °C for the aerial rate. *M. demissus* had a Q_{10} (29–25 °C) of 1.52 (Kuenzler, 1961*a*). These values for Q_{10} are all rather less than the equivalents for routine respiration rate in water, and suggest some degree of metabolic conservation during aerial exposure at low tide (Newell, 1970).

Patterns of activity during air exposure also vary between species. In the cockle *C. edule*, there is a slight reduction in the rate of heart beat during exposure and a reduced, but still marked, pattern of shell adduction and abduction (Boyden, 1972*a*). In *M. demissus* and in *M. edulis* there is little valve movement during exposure (Lent, 1969; Coleman & Trueman, 1971) and, in *M. edulis* and *M. californianus* heart beat is reduced to low values (see Chapter 6) and gill cilia apparently cease to beat (Schlieper, 1955*b*; Schlieper & Kowalski, 1958). Trueman & Lowe (1971) recorded continued heart beat in *Isognomon alatus* during exposure to air, whereas the heart beat of *Modiolus modiolus* is erratic under these conditions (Coleman & Trueman, 1971). These differences probably reflect the varied degrees of adaptation to aerial conditions. In species relatively well-adapted to gas exchange in air, ciliary activity on the gills, and heart beat, probably continue during air exposure to facilitate gaseous diffusion. Species less well-adapted probably reduce their levels of activity, as reflected in bradycardia and minimal shell movement. Coleman (1973*b*) has suggested that *M. edulis* controls its physiological reactions during air exposure (maintained heart rhythm and controlled gape) whereas *Modiolus modiolus* shows no such control (erratic heart beat and irregular shell gape).

Marine mussels

The ability to respire aerobically when exposed to air has survival value. The median survival time of *M. demissus* in air is proportional to the amount of oxygen present (Lent, 1968), and Boyden (1972*a*) showed that, by preventing *C. edule* from gaping, its survival in air was significantly reduced. However, intertidal species also have a high tolerance to anoxia when exposed. Lent (1968) showed that *M. demissus* survived 5 days in an atmosphere of nitrogen, and *Cardium edule* and *C. glaucum* (the latter of which does not show shell gape, and has a very small consumption of oxygen in air) both survived 75–85 h in a nitrogen environment (Boyden, 1972*a*). This apparent capacity for anaerobiosis has been equated with an inflated rate of oxygen consumption during re-immersion after a period of exposure to air. Moon & Pritchard (1970) recorded increased oxygen uptake by *M. californianus* after 6 or 12 h exposure, and Bayne *et al.* (1975*b*) calculated an apparent cumulated 'oxygen debt' in *M. californianus* exposed to air, as follows:

Time of exposure (h)	Oxygen debt (ml O_2)
2	0.283
5	0.406
14	0.606

Such data are difficult to interpret without a more complete understanding of the processes of anaerobic glycolysis that are likely to accompany air exposure (Chapter 8). Boyden (1972*a*) suggested that inflated rates of oxygen uptake following air exposure may be caused by hyperactivity associated with excretion and the 'flushing out' of ammonia from the body as the animal is re-immersed. But this is unlikely to account for the entire apparent oxygen debt.

The height at which individuals are located on the shore can affect their respiration rates when immersed. Both *M. edulis* and *M. californianus* from high on the shore show lower ventilation rates than mussels lower down (Bullock, 1955). Moon & Pritchard (1970) found that the rate of oxygen consumption by *M. californianus* high on the shore was greater by 32% than by individuals lower on the shore; high-shore animals also showed a greater increase in respiration rate after periods of exposure to air than did the low-shore individuals. This suggests an 'inverse compensation' (Precht, 1958) to temperature, in contrast with observations by Segal *et al.* (1953) on ventilation rate. However, interpreting these data in terms of temperature-response alone is probably misleading.

The results of these various studies all indicate that mussels have very different capacities for survival in the intertidal. In *Modiolus modiolus* lack of sufficient control over its physiological functions when placed in air limits its distribution on the shore (Coleman, 1973*b*). Two species of *Mytilus* exhibit a controlled response to air exposure, which results in reduced metabolic demand and the possibility for some consumption of oxygen from the air at high humidity. *Modiolus demissus* shows relatively advanced physiological adaptation to the intertidal. In all species, however, tidal exposure to air probably results in metabolic stress through the inability to feed, a limit to gas exchange, and exposure to desiccation. These act to reduce scope for growth (see Chapter 7) and ultimately impose the upper limit to intertidal distribution. To understand these adaptations further we need more information on the biochemical processes that accompany aerial exposure.

Season

In the light of the many, varied effects of environmental change on respiration rate, it is not surprising to find a regular seasonal pattern in the rate of oxygen consumption by some species. However, even when the effects of temperature, for example, or body size, are excluded experimentally from determinations of oxygen consumption, a residual seasonal pattern often remains. Bruce (1926) reduced his data on oxygen consumption by *Mytilus edulis* to standard values at 10 °C, 160 mm Hg P_{O_2}, and 10 g wet tissue weight (Fig. 5.28 *a*) and deduced a seasonal pattern of high values in the winter and spring and low values in the summer. Krüger (1960) measured oxygen consumption by *M. edulis* throughout the year at a constant temperature (15 °C) and recorded a seasonal pattern of high rates in the spring, declining to minimum rates after spawning in the summer, and increasing again during the autumn (Fig. 5.28*b*).

Both Bruce (1926) and Krüger (1960) correlated this seasonal pattern with the cycles of gametogenesis and the storage and utilisation of nutrient reserves. Bruce concluded that the increasing proportion of gonad material in the body through the late summer and autumn increased the oxygen demand of the animal. Krüger was impressed by the parallel changes in the rate of oxygen uptake and the concentration of protein in the tissues; a ratio of oxygen consumed : protein concentration showed little change over the year.

This relationship between gametogenesis, body reserves and routine rate of oxygen consumption has also been documented by Bayne & Thompson (1970), Widdows & Bayne (1971) and Bayne (1973*a*), (Fig. 5.28*c*). In the summer, gametogenesis is in the 'resting stage' (Chapter 2) and reserves of glycogen are high; the high proportion of metabolically inert material results in a low rate of oxygen consumption per unit weight. During the winter gametogenesis is active, the glycogen stores are being utilised and the metabolic demand is increased. In the spring, a large mass of developing gametes continues to impose a high demand for oxygen which is only reduced after spawning. Throughout the year there is

191

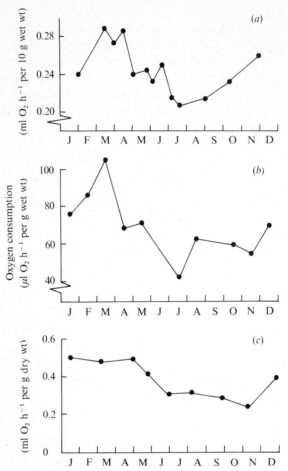

Fig. 5.28. Seasonal changes in the rates of oxygen consumption by *Mytilus edulis*. (*a*) From Bruce (1926); (*b*) from Krüger (1960); (*c*) from Bayne (1973*a*).

relatively little change in the standard rate of oxygen consumption, however, and the total scope for activity also remains relatively unchanged (Bayne, 1973*a*).

In spite of these marked seasonal changes in oxygen demand, which are, at least in part, independent of temperature, Bruce (1926) estimated that in the natural populations temperature would be more important in controlling the rate of oxygen consumption than innate physiological changes. Bruce was not aware, however, of temperature acclimation. Ansell (1973) noted that increased glycogen reserves depressed the weight-specific respiration rate of *Donax*. Nevertheless, only 14% of the variation in oxygen consumption could be accounted for by changes in body 'con-

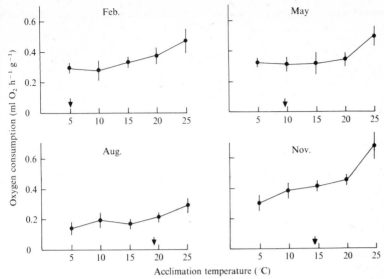

Fig. 5.29. The rates of oxygen consumption by *Mytilus edulis* acclimated to different temperatures during four months of the year. The arrows indicate the field-ambient temperatures at the time of the measurements. (From Bayne *et al.*, 1973.)

dition'. However, *Donax*, unlike *Mytilus*, shows little capacity for acclimation to temperature change (Ansell, 1973). Fig. 5.29 shows acclimated rate/temperature curves for oxygen uptake by *M. edulis* during four months of the year, with the relevant field-ambient temperatures indicated by arrows. The rates of consumption are clearly independent of temperature over a considerable range on each occasion, and the level of routine respiration is largely a function of innate physiological events.

Seasonal changes in the metabolic rate of isolated tissues have been demonstrated by Percy, Aldrich & Marcus (1971) working with the American oyster (*Crassostrea virginica*). They found that the oxygen consumption rates of gill and mantle tissues (but not adductor muscle) were depressed in winter, when the rate/temperature curves showed evidence of 'inverse acclimation'. There was a negative correlation between a condition index (which they suggest was a measure of glycogen content) and the respiration rate of mantle tissue. However, Hopkins (1946) who studied *Mercenaria mercenaria* (the American hard clam) found tissue respiration to be higher in winter than in summer, i.e. there was a positive seasonal acclimatisation. These differences may be a function of the different gametogenic cycles of the species (Percy *et al.*, 1971). *Mercenaria*, like *Mytilus*, has active gametogenesis in the early winter and retains ripe gametes throughout the winter to the spring. *Crassostrea*, on the other hand, has a gametogenic 'resting stage' in the winter. These studies might suggest, therefore, that the gametogenic cycle, which is under hormonal control (Chapter 8) is a determinant, directly or indirectly, of the seasonal metabolic pattern. However, Van Winkle (1968) found that respiration by isolated gill slices of *Mercenaria* was higher in summer than in winter; in *Modiolus demissus* the picture was reversed, and in *Mytilus* and *Crassostrea* there was no significant difference between oxygen uptake rates in winter and summer. Extrapolation from tissue to whole-animal respiration is not justified at the present time, until more information is available on systemic and cellular control mechanisms. In mussels, the metabolic demands of feeding and ventilation are relatively steady throughout the year (at least in temperate boreal regions) due to seasonal acclimation. Innate physiological demands, resulting from gametogenesis and nutrient cycles possibly determine the seasonal pattern of oxygen consumption, but further research is needed into the identity and character of 'innate physiological demands'.

Oxygen tension

Mussels are found in a variety of coastal habitats that may be subjected to fluctuations in oxygen tension (P_{O_2}); for example, in lagoons and enclosed areas of brackish water (Muus, 1967), in muddy areas and estuaries (Theede, Ponat, Hiroki & Schlieper, 1969), in salt marshes, and in areas exposed to pollution (Chapter 3). In addition, some intertidal forms regularly experience periods of low P_{O_2} during shell closure at low tide (p. 185). But there is a physiological, as well as an ecological, interest in examining the respiratory response to reduced P_{O_2}, as a way of understanding the mechanisms and control of gas exchange. In this section we discuss some of the effects of reduced oxygen tension on respiration. Discussion of the integrated response of the 'whole animal' to reduced P_{O_2} is left to Chapter 7, and cellular adaptations to altered oxygen tension and to anoxia are examined in Chapter 8.

A distinction should be made between environmental hypoxia and tissue hypoxia. The former refers to any reduction in the environmental P_{O_2}, and consequently in the P_{O_2} of the water taken into the mantle cavity (the inspired water). The term tissue hypoxia refers to oxygen limitation in the tissues and cells, and its use should be restricted to conditions that result in a shift of cellular oxidation–reduction systems to a more reduced state (Miller, 1966). Hypoxia might also result from hypoventilation of the gas exchange surfaces, from an increase in the diffusion distance for oxygen between water and blood, or from a reduction in the perfusion of blood to the gills (Hughes, 1973). However, these conditions have not been studied adequately in bivalves, and our discussion here will be limited to environmental hypoxia.

194

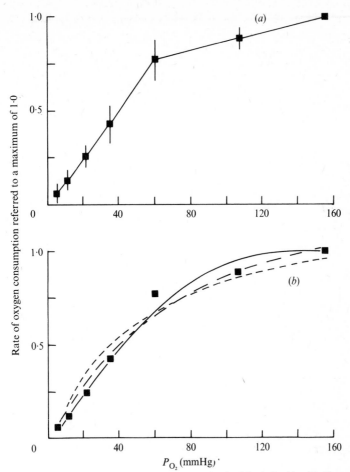

Fig. 5.30. (*a*) Rates of oxygen consumption (means ± standard deviation) by *Mytilus edulis* at different oxygen tensions (P_{O_2}), standardised with maximum consumption = 1.0. (*b*) Data from (*a*) fitted by quadratic (——), hyperbolic (– –) and semilogarithmic (- - - -) models.

Mussels, and bivalves in general, have been shown to be capable of some regulation of their rates of oxygen consumption under hypoxic conditions, so that the relationship between oxygen uptake and P_{O_2} takes the form shown in Fig. 5.30(*a*). This relationship has been described for *Mytilus edulis* (Rotthauwe, 1958; Bayne, 1971*a*), *Perna perna* (Bayne, 1967), *Mytilus californianus* (Moon & Pritchard, 1970; Bayne *et al.*, 1975*b*), *Modiolus demissus* (Mangum & Van Winkle, 1973), and for other bivalve species by Van Dam (1938, 1954), Gaarder & Eliassen (1954), Bayne (1971*a*, 1973*c*), Mangum & Van Winkle (1973), Taylor (1974) and Mangum & Burnett (1975).

Marine mussels

Tang (1933) proposed that the relationship between oxygen consumption P_{O_2} and oxygen tension V_{O_2} in many organisms was hyperbolic, and could be described by the expression:

$$V_{O_2} = \frac{P_{O_2}}{K_1 + K_2 \cdot P_{O_2}},$$

(5.1)

where K_1 and K_2 are fitted parameters. A linear form of this expression is:

$$\frac{P_{O_2}}{V_{O_2}} = K_1 + K_2 \cdot P_{O_2}.$$

(5.2)

As K_1 increases relative to K_2, V_{O_2} becomes more directly proportional to P_{O_2}, whereas as K_2 increases relative to K_1, V_{O_2} becomes more independent of P_{O_2}. Bayne (1967, 1971a) suggested that the value for K_1/K_2 could provide an index of the degree of dependence of oxygen consumption on oxygen tension. This index was subsequently used by Bayne (1973c) to compare the response of different species to declining P_{O_2} at reduced salinity, and by Taylor (1974) in assessing the behaviour of the bivalve *Arctica islandica*.

Mangum & Van Winkle (1973) examined a number of statistical models, including the hyperbolic equation, in attempts to relate oxygen consumption to oxygen tension. They concluded that the hyperbolic model had little general value, because it predicted (a) that the animals 'exhaust their oxygen supply before completing the shift to anaerobic pathways' (and they reported this not to be the case for a number of different species), and (b) that oxygen consumption has an upper asymptote (whereas they suggest that the 'perfect regulation of V_{O_2}' that this implies seldom occurs). Mangum & Van Winkle (1973) propose instead the use of a polynomial (quadratic) equation of the form:

$$V_{O_2} = K_0 + K_1 \cdot P_{O_2} + K_2 \cdot (P_{O_2})^2.$$

(5.3)

Alternatively, Sassaman & Mangum (1972) have earlier used the semilogarithmic expression:

$$V_{O_2} = K_1 + K_2 \cdot (\log_{10} P_{O_2}).$$

(5.4)

Mangum & Van Winkle argue for the quadratic expression on the basis of its generality, and they applied it to a variety of invertebrates, including *Modiolus demissus*.

Fig. 5.30(b) compares the fit of (5.1), (5.3) and (5.4) to the data plotted in Fig. 5.30(a) (Bayne & Livingstone, unpublished data). The quadratic expression gives the best fit (sums of squares of deviations = 0.022) followed by the hyperbolic (sums of squares = 0.036) and the semilogarithmic (sums of squares = 0.064) models. Mangum & Van Winkle

(1973) suggested that the second order coefficient in the quadratic expression indexes the capacity to regulate V_{O_2} at reduced P_{O_2}. Clearly, each unique set of experimental values must be examined for its best quantitative expression, and in a complex physiological response such as respiration rate as affected by oxygen tension it is not surprising that no single expression proves to be universally valid. This hardly matters where the aim is simply to describe the relationship accurately. On the other hand, if the aim is to apply biological meaning to statistical constants, then we need to gain more understanding of the component physiological processes.

Small (= young?) mussels show less capacity to regulate their rates of oxygen consumption at reduced P_{O_2} than larger (= older?) animals; Bayne (1971a) recorded that the K_1/K_2 index was related to 'weight-specific' oxygen consumption (or Q_{O_2}) by a power function in three bivalve species. The equation for *M. edulis* was:

$$K_1/K_2 = 15.5 \cdot Q_{O_2}^{0.818}.$$

Differences between species suggested different capacities for regulation. At reduced salinity, capacity for regulation of V_{O_2} was reduced, but the reduction was slight in the case of *M. edulis*. Capacity to regulate V_{O_2} at reduced oxygen tension is also reduced at high temperatures (Bayne & Livingstone, unpublished data). Taylor (1974) confirmed, for the bivalve *Arctica islandica*, that capacity to regulate V_{O_2} is size-dependent.

The capacity for *M. edulis* to regulate its rate of oxygen consumption during environmental hypoxia varies also with the physiological condition of the animal. When mussels were starved in the laboratory their oxygen consumption rates became linearly dependent on P_{O_2} (Bayne, 1971a). During starvation, V_{O_2} at normal saturation conditions declines from a routine to a standard level (p. 163). The curves in Fig. 5.31 showing 'regulation' and 'conformity' of V_{O_2} may therefore represent the responses of the routine and the standard respiration rates respectively. The numerical differences between these curves is a measure of the routine scope for activity (p. 178), which increases in a slightly hypoxic environment, reaches a maximum at about 80 mm Hg P_{O_2}, and then declines to the point where the routine (= 'regulating') and standard (= 'conforming') curves coincide. Fry (1947) identified an 'incipient limiting tension' (sometimes referred to as the 'critical tension', or P_C), at which the scope for activity begins to decline, and an 'incipient lethal tension' at which both routine and standard metabolic rate have the same numerical value. Between the incipient limiting tension (about 80 mm Hg in Fig. 5.31) and 160 mm Hg P_{O_2} (or air-saturation), the limiting effect of changes in the oxygen supply to the mussel is minimal, because scope for activity can increase. Below the incipient limiting tension, however, activity is limited by the decline in the scope. Within the range of the limiting and lethal

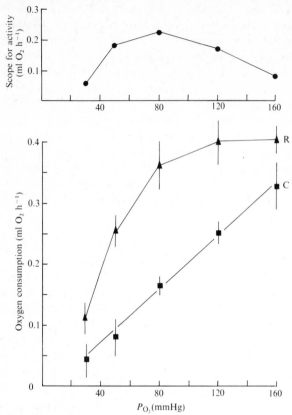

Fig. 5.31. Rates of oxygen consumption by *Mytilus edulis* (mean±standard deviation) at different oxygen tensions (P_{O_2}). R, individuals 'regulating'; C, individuals 'conforming' (see text). A scope for activity, calculated as the numerical difference between regulators and conformers, is also shown.

tensions, the animal continues to make compensatory adjustments in its physiological processes (see Chapter 7), but, to use Fry's (1947) words, the lower the level of oxygen 'the greater the difficulty of repaying any metabolic deficit that may be incurred'.

In Fig. 5.31 the incipient lethal tension would be, by extrapolation, approximately 8 mm Hg P_{O_2}. Mussels are known to have a considerable tolerance of anoxia, and bivalves in general have been called 'facultative anaerobes' (Maloeuf, 1937; Hochachka & Somero, 1973; see also Chapter 8). However, a proper understanding of the metabolic processes that occur under hypoxia and anoxia require further study. Systemic adjustments may continue to very low P_{O_2} values, but the point at which cellular adaptive processes occur is not known, although D. R. Livingstone (unpublished data) has shown that a pathway of anaerobic glycolysis

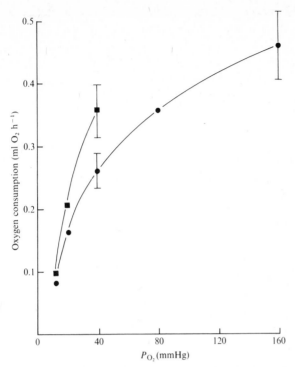

Fig. 5.32. Partial acclimation of oxygen consumption rate to reduced oxygen tension (P_{O_2}) by *Mytilus edulis*. ●, mussels held at 160 mm Hg; ■, mussels held at 40 mm Hg. Values are the means of measurements on eight animals (± 1 standard deviation shown for three points) transformed to a standard body size of $W^{0.7}$. (From Bayne, 1975*b*.)

(Chapter 8) is utilised at 80 mm Hg P_{O_2} and, to a greater extent, at 40 mm Hg.

Recently Bayne (1975*b*) and Bayne & Livingstone (unpublished data) have described the acclimation by *M. edulis* to low P_{O_2}. Individuals were kept at 40, 80 and 160 mm Hg P_{O_2} and their rates of oxygen consumption and ventilation measured as they were subjected to further reduction in oxygen tension. Mussels held at 80 mm Hg fully acclimated within 5 days, i.e. their rates of oxygen consumption at this oxygen tension increased to the same value as the control animals at 160 mm Hg P_{O_2}. This resulted in an increase in the scope for activity at tensions < 80 mm Hg, and a lowering of the incipient limiting tension from about 80 to about 40 mm Hg P_{O_2}. Mussels which were held at 40 mm Hg were unable to acclimate fully, although even at this P_{O_2} some increase in oxygen consumption rate over the control values did occur (Fig. 5.32). The mechanisms underlying this acclimation are discussed in Chapter 7.

Mussels are relatively tolerant of totally oxygen-deficient (anoxic)

199

water, to which they respond by closing their shell valves. Thamdrup (1935) kept *M. edulis* for 7 days in anoxic water at 10 °C and recorded an 80% survival. Collip (1921) found that tolerance of anoxia by *Mya arenaria* (American soft-shelled clam) was temperature-dependent. Lent (1968) observed that *Modiolus demissus* survived 5 days in an atmosphere of nitrogen. Theede *et al.* (1969) point out that resistance to hydrogen sulphide may be very important for marine invertebrates that are exposed to anoxic environments. They found that *M. edulis* from the North Sea survived 35 days in water at 10 °C containing 0.15 ml O_2 l^{-1}; this period was reduced to 25 days in the presence of hydrogen sulphide. When compared with other bivalve species, *M. edulis* gill tissue had an intermediate tolerance to anoxic water with and without hydrogen sulphide; *Modiolus modiolus* was generally less tolerant than *M. edulis*. Torres & Mangum (1974) recently recorded that *Modiolus demissus* was not visibly affected after 9 days under hyperoxic conditions (27–35 ml O_2 l^{-1}).

When mussels that have been held for some hours in water of reduced oxygen content are recovered to air-saturated P_{O_2}, the rate of oxygen consumption may be greater than before hypoxic exposure (the apparent 'repayment of an oxygen debt'), it may be the same, or it may be less. The factors controlling these responses are complex (Bayne & Livingstone, unpublished data). For example, short exposure to water < 20 mm Hg P_{O_2} may cause no change in V_{O_2} on recovery. Longer exposure, of 3–6 h, may result in an inflated V_{O_2} on recovery. But if exposure to low P_{O_2} is for much longer (2–3 days), there may be a more fundamental shift to pathways of anaerobic glycolysis, and when these animals are recovered to a higher P_{O_2}, oxygen consumption may remain extremely low, suggesting that the animal has become 'locked in' to anaerobic metabolism, requiring some minutes at high P_{O_2} before metabolism is 'unlocked' and V_{O_2} returns to normal. These responses require further study in the context of biochemical processes involved (Chapter 8).

Mussels, therefore, respond to environmental hypoxia by a variety of physiological compensations that constitute systemic adaptation. These serve to increase the scope for activity and possibly to maintain the optimum delivery of oxygen to the cells. At what point oxygen becomes truly limiting to metabolism, however, in the sense of affecting ATP production and the normal ratio of $NAD/NADH_2$, remains a subject for further research.

Salinity

The relationships between the salinity of the environment and the respiration (and general physiology) of marine invertebrates have been reviewed in detail by Kinne (1964a, b) and by Remane & Schlieper (1971).

Many mussels are considered to be euryhaline (i.e. able to survive in a wide range of salinities). For example, *Mytilus edulis* has an extremely wide estuarine and marine distribution, from salinities of 4–5‰ (The Bay of Finland in the inner Baltic) to fully marine conditions. *Modiolus demissus* occurs in waters of salinity as low as 8–9‰ (Vernberg, Schlieper & Schneider, 1963). *Mytilus galloprovincialis* does not tolerate as wide a range of salinities as *M. edulis*; Mars (1950) gives the salinity range for *M. galloprovincialis* as 12–38‰. *M. californianus* may occur at salinities down to 17‰. These species are all euryhaline, therefore, although to differing degrees.

The effects of salinity on the respiration and activity of mussels have been examined most extensively in *Mytilus edulis*. The results indicate the necessity for very careful consideration of the time-course of the metabolic adaptations to change in salinity. When the respiration rates of mussels from a range of environments have been measured at their field-ambient salinities, they have been found to be similar in extreme brackish water (5–6‰), in intermediate (15‰) and in fully marine (30‰) conditions (Remane & Schlieper, 1971). Theede (1963) recorded similar filtration rates by *M. edulis* from the North Sea (30‰) and western Baltic (15‰). However, when mussels are experimentally exposed to altered salinity over a relatively short period, the rate of oxygen consumption is highest in the salinity to which the animal has become adapted in nature (Bouxin, 1931; Remane & Schlieper, 1971). Bouxin (1931) exposed *M. galloprovincialis* which were adapted to normal seawater, to changes in salinity, and measured their rates of oxygen consumption. Over a range from 20 to 36‰ there was little change, but with further increase or decrease in salinity oxygen consumption was depressed. Lubet & Chappuis (1966) recorded decline in ventilation rate by *M. galloprovin-cialis* below 19‰. Lagerspetz & Sirkka (1959) exposed *M. edulis* from the coast of Finland (mean salinity 5.5‰) to 15‰ and 30‰ and measured their rates of oxygen consumption immediately and after 2–6 weeks (Table 5.7). At each change of salinity the oxygen consumption rate was reduced and remained low over the experimental period. Theede (1963) found that immediately after transfer to above- or below-ambient salinities, filtration rate by *M. edulis* was reduced; after 7–10 days this reduction was diminished, but the difference between North Sea and Baltic populations was not completely eliminated.

The apparent discrepancy between results of laboratory experiments and observations in nature may partly be explained by the long period needed for full acclimation to change in salinity, far exceeding the 2 weeks that is typical of the time required for temperature acclimation (p. 172). Bøhle (1972) kept *M. edulis* at 34‰, 26‰ and 18‰ and measured ' filtration activity ' over the ensuing 7 weeks. Filtration was initially depressed below

201

Table 5.7. *The rates of oxygen consumption by* Mytilus edulis *at different salinities*

Acclimation salinity (‰)	Experimental salinity (‰)	Oxygen consumption (ml h^{-1} per g dry wt)±1 S.E.
5	5	0.470±0.040
5	15	0.400±0.040
5	30	0.355±0.045
15	15	0.390±0.035
30	30	0.345±0.015

From Lagerspetz & Sirkka (1959).

the control values at both 26‰ and 18‰. At 26‰ the mussels required 4 weeks for activity to increase to the control value, and at 18‰ recovery was not quite complete even after 7 weeks. Schlieper (1955*a*) had suggested much earlier that 4–7 weeks was required for complete metabolic compensation for a salinity change, the time taken depending on temperature.

The respiratory response by mussels to a marked change in salinity may therefore be generalised as follows. The immediate response of the animal is to close the shell valves (Motwani, 1955; Gilles, 1972*b*). Pierce (1971*b*) recorded that *Modiolus demissus* may remain with the shell valves shut for up to 7–10 days, depending on the extent of the salinity change. During this time the shell valves may open infrequently, and for short periods of time, and eventually the animal either resumes ventilating or it dies. During this initial period of 'shock', respiration rate is reduced (Bayne, 1973*c*). There follows a period of physiological stabilisation during which respiration rate recovers to an apparent new steady-state. It is during this time that the processes of isosmotic intracellular regulation (Chapter 6) probably take place. If the salinity change has been large (e.g. 5–15‰, or 15–30‰) this new steady-state may be lower than the previously acclimatised value (Lagerspetz & Sirkka, 1959). This period in the salinity response may last for some weeks, during which fundamental biochemical changes probably take place, finally resulting in a recovery of respiration and filtration rates to control values. Kinne (1970*a*) concluded that the rate of acclimation tends to increase with rise in temperature, and to be variable according to animal size and the extent of the salinity change. Lange, Staaland & Mostad (1972) summarised the rather complex nature of salinity adaptation in *M. edulis* as a 'primary inhibition followed in the course of some weeks by a secondary activation' of oxygen consumption.

In view of the long periods of time involved in successful salinity adaptation by mussels, it is not surprising that experimentally based assessments of salinity-tolerance limits have been very varied. Dodgson (1928) reported the survival of *M. edulis* from North Wales from 8.75 to 31‰, but he observed defective byssus formation below 16‰ and irregular ventilation below 12‰. The formation of byssus threads can be a sensitive index of the effects of salinity (and other environmental factors) on mussels. Glaus (1968) observed that *M. edulis* produces fewer byssus threads at 15.9‰ than at 31.3‰, and Van Winkle (1970) did not observe any byssus production by this species at 16‰. Reish & Ayers (1968) reported a sharp decline in byssus production by *M. edulis* between 25.3 and 28.9‰. Pierce (1970) used normal ventilation behaviour and byssus production as criteria to establish the following salinity limits for species of *Modiolus*: *M. demissus granosissimus*, 3–48‰; *M. demissus demissus*, 8–48‰; *M. squamosus*, 22–41‰; *M. modiolus*, 27–41‰. Castagna & Chanley (1973) observed that the salinity tolerance of a number of bivalve species varied with season, temperature and previous salinity experience; feeding and reproduction occurred at nearly all salinities at which survival in nature is possible, but byssus formation required a higher salinity than was necessary for other activities. Castagna & Chanley found that in Virginia (USA) *M. edulis* did not occur over the entire salinity range that it can tolerate in the laboratory, probably because the species is near its geographic limit and its local distribution is controlled by temperature.

There have been a large number of investigations into the effects of salinity on isolated gill preparations of mussels (see Remane & Schlieper, 1971, for a review). In some of these ciliary activity has been measured (Tomita, 1955*a*, *b*; Schlieper, 1955*a*; Schlieper & Kowalski, 1956; Vernberg, Schlieper & Schneider, 1963; Theede, 1965; Van Winkle, 1972); in other studies the rates of oxygen consumption by the isolated gill pieces were measured (Lagerspetz & Sirkka, 1959; Erman, 1961; Lange, 1968; Van Winkle, 1968).

Schlieper & Kowalski (1956) recorded that mussels from the eastern Baltic (5–6‰ salinity), western Baltic (15‰) and North Sea (30‰) had mean gill activity indices (the rate of travel of silver foil over the surface of an exposed gill lamella) of 14, 34 and 41 mm min^{-1}; ciliary activity increased from 14 to 25 mm min^{-1}, when the salinity was raised from 5‰ to 16‰ for *M. edulis* adapted to 5‰ at Tvarmine in the Baltic. Van Winkle (1972) found that ciliary activity was inhibited at first, if the salinity change was marked. Within certain limits, recovery followed after 0.5–2.5 h, and both *M. edulis* and *Modiolus demissus* then showed a plateau of maximal activity that extended down to about 15‰ for animals adapted to 29–30‰. This approximates the range within which respiratory compensation for dilution can occur within a few hours by *M. edulis* acclimated to full salinity (Bayne, 1973*c*).

When mussels which were adapted to full salinity in the North Sea were

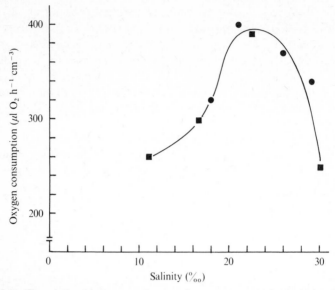

Fig. 5.33. The relationship between the rate of oxygen consumption by isolated gill tissue of *Mytilus edulis* and the salinity of the medium. ■, gas phase pure oxygen, shaking rate 75 strokes min^{-1}; ●, gas phase air, shaking rate 115 strokes min^{-1}. All experiments at 10 °C. (From Lange, 1968.)

transferred to the Baltic (15‰) oxygen consumption by isolated gill apparently increased to 20% within a few hours and gradually rose to 170% of the initial value within subsequent weeks (Schlieper, 1955*a*). Lange (1968) re-examined these responses, and argued for expressing the results as oxygen consumption per unit volume of tissue, rather than per unit weight, because of solute changes in the cells (Chapter 6). At first, Lange's results seemed to confirm Schlieper's (1955*a*) observations that oxygen consumption by the gill increases with decreasing salinity, but Lange suspected that this might be due to inadequate diffusion of oxygen in the experimental preparation. He therefore increased the shaking rate of the reaction flasks in the Warburg apparatus (from 75 to 115 strokes min^{-1}) and then recorded a 'bell-shaped', or typical optimum response curve for oxygen uptake at different salinities (Fig. 5.33). Lange (1968) concluded that the metabolic rate of gill tissues, like that of the whole animal, was maximal at the salinity to which the animal was adapted.

Theede (1965) recorded acclimatisation to salinity by the gill cilia of *M. edulis* which he transplanted between the Baltic (15‰) and the North Sea (30‰). Prior to transplantation, gill cilia from Baltic mussels were more tolerant of low salinity but less tolerant of high salinity than mussels from the North Sea. During acclimatisation following transplantation, these

204

differences disappeared, and the original tolerances were reversed, after 30 days. Van Winkle (1972) also showed a shift towards greater tolerance of low salinity in the gills of mussels acclimated to 12–15‰, when compared with mussels acclimated to 29–30‰. The general principle that acclimation to low salinity enhances low-salinity tolerance is now well-established (Theede & Lassig, 1967; Remane & Schlieper, 1971). Acclimation temperature has also a marked effect on the salinity tolerance of gill cilia (Vernberg *et al.*, 1963).

The gill tissue of bivalves has been a popular preparation for investigating tolerance limits (see p. 184 for a discussion of temperature tolerance). A positive correlation between the distribution and salinity tolerance of the whole organism and its gill tissue has been claimed by Pilgrim (1953a, b), Schlieper & Kowalski (1956), Reshöft (1961), Vernberg *et al.* (1963), Theede (1965), Schlieper, Flugel & Theede (1967), Theede & Lassig (1967) and Van Winkle (1972). Schlieper *et al.* (1967) recorded that the tropical mussel *Modiolus auriculatus* had a higher tolerance to high salinities and a lower tolerance to low salinities than the temperate *Mytilus edulis*.

Although considerable information has now been gathered on the respiratory responses of some mussels to salinity change, there is as yet little real understanding of the processes involved. The apparent similarity between the responses of gill tissue and the whole organism suggest that similar mechanisms are involved at the cellular and systemic levels, from the initial inhibition of physiological processes to their gradual acclimation. As with so many aspects of 'physiological' adaptation, there is a need for more fundamental biochemical information (e.g. the proposed enzyme inductions and 'salt sensitivity' of Lange (1968), and the two-stage adaptation to salinity proposed by Remane & Schlieper (1971)), especially in view of the apparent separation in time between the processes of intracellular isosmotic regulation (Chapter 6) and metabolic compensation. In addition, the role of fluctuating, as opposed to constant, salinities, and the effects of multiple environmental variables acting together in time, remain to be examined.

Other factors

Other environmental variables are known, or may be expected, to affect the respiration rate of mussels. Various pollutants, notably 'heavy metals' (Brown & Newell, 1972; Delhaye & Cornet, 1975) depress the respiration rate, possibly by inhibiting ciliary activity as well as by direct metabolic inhibition; the literature is reviewed in Chapter 3. In this section we deal briefly with two other aspects; the presence of exogenous or endogenous rhythms and the effects of water flow rate.

There have been reports of rhythmic activity of mussels correlated with

205

tidal phenomena. Rao (1954) claimed tidal rhythmicity in filtration rate in *Mytilus edulis* and *M. californianus*. The rhythm was independent of temperature between 9 and 20 °C, and independent also of light. *M. edulis* transplanted over a distance of 5000 km gradually shifted their filtration rhythm over 3 weeks to correspond with the tidal pattern at the transplant site. However, neither Jørgensen (1960), Theede (1963) nor Davids (1964) could demonstrate the tidal rhythmicity in *M. edulis*, and Jørgensen (1960) was critical of Rao's (1954) technique for measuring filtration rate. Winter (1969) observed two maxima per day in the feeding rate of *Modiolus modiolus*; he suggested that this may indicate a tidal rhythm, but the data were insufficient to draw a firm conclusion. Later, Winter (1973) could find no evidence of a tidal feeding rhythm in *M. edulis*. Salanki (1966) observed an adduction/abduction rhythm in *Lithophaga lithophaga* in the Mediterranean, and equated this with diurnal changes in light intensity.

However, there has been no firm demonstration of rhythmic changes in the respiration rate of mussels. Thompson & Bayne (1972) recorded continuous feeding by *M. edulis* whenever food cells above a certain concentration were present; Bayne, Thompson & Widdows (unpublished data) have been unable to detect a respiratory rhythm in *M. edulis*. Rhythmic activity in the digestive gland is discussed earlier in this chapter (p. 153).

The effects of the rate of water flow on the respiration of marine bivalves has been neglected in most experimental studies. Kerswill (1949) recorded an increased growth rate in clams and oysters in flowing water, and Walne (1972) also observed that small oysters had a higher growth rate when kept at 183 ml min^{-1} flow than at 74 ml min^{-1}. Kirby-Smith (1972) on the other hand, found that growth of the scallop *Aequipecten* was inversely related to current speed until, at low current velocities, food became limiting. We have discussed the effects of water flow on filtration rate on p. 134.

Nixon, Oviatt, Rogers & Taylor (1971) examined the influence of current speed on the 'metabolism' of a mussel bed *in vivo*. Dissolved oxygen was measured upstream and downstream of the bed, and the results corrected for diffusion to give an estimate of community respiration. Nixon *et al.* (1971) also recalculated some earlier data of Kuenen's (1942) and they concluded that the relationship between respiration rate (R; g O$_2$ h^{-1} m^{-2}) and current speed (C; m s^{-1}) was described by:

$$R = \frac{C}{0.006 + 0.35C},$$

over a range of current speeds from 0 to 0.6 m s^{-1}. The mechanism responsible for such an effect of current speed is unclear, and the relationships between current speed, filtration rate, particle concentration and respiration rate in bivalves remain to be elucidated.

206

6. Physiology: II

B. L. BAYNE, J. WIDDOWS & R. J. THOMPSON

Circulation

The blood vascular system of *Mytilus edulis* has been described by Field
(1922) and White (1937). The heart is situated on the mid-dorsal line behind
the posterior termination of the hinge, where it is enclosed in the
pericardial cavity. The heart is composed of a median ventricle and two
lateral auricles. Blood leaves the ventricle by the single anterior aorta,
from which it is distributed to the body through five channels: (*a*) pallial
arteries, which supply the mantle and the pallial muscles; (*b*) the
gastro-intestinal arteries which supply the stomach, intestine, posterior
retractor muscles, posterior adductor muscles and mesosoma; (*c*) the
pericardial artery, which carries blood to the pericardium, intestine and
nearby genital glands; (*d*) hepatic arteries which go to the digestive gland;
and (*e*) The terminal arteries which distribute blood to the anterior parts of
the body.

The venous system collects blood into three main sinuses; the pallial, the
pedal and the median ventral sinus. From these sinuses the blood is carried
to the kidneys and thence to the gills and finally back to the heart via the
oblique vein. It should be noted that there is not a complete branchial
circulation in mussels. Blood can be returned from various organs to the
heart, without passing through the gills.

The haemolymph

The blood, or haemolymph (the terms will be used synonymously),
consists of cells within a colourless plasma. There is no respiratory
pigment and the oxygen-carrying capacity of the haemolymph of *Mytilus
edulis* is equivalent to that of seawater (p. 266). The blood volume of
bivalves is large; Martin, Harrison, Huston & Stewart (1958) have
reviewed this topic, and they recorded a blood volume for *Mytilus
californianus* of 50.8% of the wet body weight excluding the shell.

Blood cells

The blood cells of molluscs have been variously referred to as haemocytes,
amoebocytes, leucocytes, lymphocytes and phagocytes. Cheng & Rifkin
(1970) consider these terms to be synonymous; we shall use the generic
term haemocytes. The haemocytes are cells capable of independent

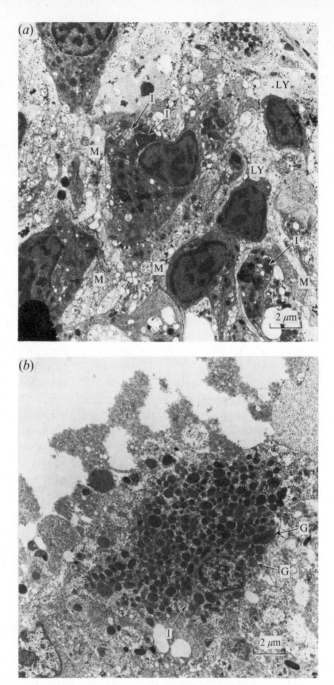

Fig. 6.1. (*a*) Haemocytes from the mantle connective tissue of *Mytilus edulis*. This group of cells comprises two agranular lymphocytes (LY) with large nuclei surrounded by a small volume of cytoplasm; and a number of larger macrophages (M) with irregular nuclei, and containing numerous cytoplasmic inclusions (I) which are believed to be phagosomes and phagolysosomes. Uranyl acetate and lead citrate staining. (*b*) A granular haemocyte (granulocyte) from the mantle connective tissue of *Mytilus edulis*. The small spherical nucleus differs from the nucleus of an agranular cell in the distribution of the intensely stained heterochromatin. The cytoplasm is packed with electron-dense membrane-bound spheroidal granules (G). There are numerous pseudopod-like processes and some evidence for pinocytotic inclusions (I). Uranyl acetate and lead citrate staining. (Both photographs by courtesy of M. Moore & D. Lowe.)

movement, and they play an important role in digestion, nutrient transport, excretion, regeneration and repair, and internal defence. The haemocytes are distributed throughout the vascular system, and because this system is an open one, they are also found within the tissues.

The classification of cell types present in molluscan blood has been the subject of much speculation, but there is still no general agreement (for reviews see Tripp, 1963; Andrew, 1965; Feng, 1967; Cheng & Rifkin, 1970). Bruyne (1896) recognised seven cell types in *M. edulis* and other bivalves, based on the presence or absence of cytoplasmic granules, and on the cells' different staining affinities. However, Galtsoff (1964) stated that there are only two cell types in the oyster *Crassostrea virginica*, viz. granular and hyaline cells. The granulocytes are recognised by the large numbers of granules present in the cytoplasm, although the hyaline cells are not completely devoid of cytoplasmic granules. Moore & Lowe (1975) have recently described three morphological types of haemocyte in *M. edulis*; lymphocytes, macrophages and granulocytes. The agranular lymphocytes and the macrophages are believed to form a developmental series, maturing from small, RNA-rich lymphocytes to the larger, phagocytic macrophages. The granulocytes probably represent a separate cell line.

Moore & Lowe (1975) suggest that the lymphocytes represent circulating stem cells; they are small (4–6 μm), spherical, with little cytoplasm and a spherical nucleus (Fig. 6.1*a*). As the cytoplasm increases in volume the cells assume an irregular appearance, with pseudopod-like formations and vacuoles in the cytoplasm. These macrophages (7–10 μm in diameter; Fig. 6.1*b*) are phagocytic, and Moore & Lowe observed high activity of lysosomal hydrolase enzymes, associated with vacuoles, suggesting the presence of an intracellular vacuolar digestive system. The cytoplasm of the granular leucocytes (Fig. 6.1*b*) is filled with spherical granules (0.5–1.0 μm) which range from neutrophilic to acidophilic in staining properties. The degree of acidophilia of these granules apparently increases with increasing size, and Moore & Lowe were unable to use the staining properties of the granules to make any functional distinction between different granulocytes.

Haemopoiesis. The origin of the haemocytes in molluscs is unknown, although Wagge (1955) has suggested that they originate in the epithelial and connective tissues of the mantle and the digestive gland.

Digestion and storage. Haemocytes play an important role in the digestion and transport of material in the alimentary canal (Wagge, 1955). They ingest directly particles of nutritional value which are too large to enter the cells of the digestive diverticulum (Yonge, 1926*a*). These haemocytes are enzymatically equipped with lipases (Yonge, 1926*b*; George, 1952; Zachs

209

& Welsh, 1953), carbohydrases (Takatsuki, 1934) and proteases (Yonge, 1926*b*; Takatsuki, 1934). Moore & Lowe (1975) found positive staining reactivity for lysosomal enzymes (acid phosphatase, β-glucuronidase) in the cytoplasmic inclusions of the macrophages of *M. edulis*. When individuals were injected with carbon, the macrophages showed evidence of phagocytosis and the presence of phagolysosomes.

Glycogen and lipid are present in the haemocytes of *M. edulis* (Bubel, 1973*a*) but it seems unlikely that these provide a significant energy reserve for the animal; their carbohydrate content is less than 2% of the total tissue carbohydrate reserve (M. Moore, unpublished observations). Granulo-cytes are found in close association with the basal regions of the digestive cells, the Leydig cells of the mantle (the main glycogen storage cells; Chapter 8) and with maturing and mature gametes which are rich in phospholipid (Moore & Lowe, 1975). Although the function of these cells remains obscure, present evidence indicates a role in the transport of nutrient reserves around the body.

Excretion and internal defence. The haemocytes are able to recognise, and then to phagocytose, foreign particulate matter such as micro-organisms, carbon particles and foreign protein molecules (Wagge, 1955; Feng, 1965; Pauley & Heaton, 1969; Cheng & Rifkin, 1970; Feng, Feng, Burke & Khairallah, 1971; Sparks, 1972; Moore & Lowe, 1975). There appear to be three stages in an 'inflammatory response' by *M. edulis* to mantle injury and invasion by micro-organisms (Mikhailova & Prazduikar, 1961, 1962). The first stage develops during the first 12 h, and involves a migration of haemocytes to the damaged area followed by encapsulation of foreign material. The second stage, occurring from 12 h to 5–10 days after injury, is characterised by intensive phagocytosis and both intra- and extracellular digestion of foreign particles. The third stage is the migration of haemo-cytes containing engulfed particles to the blood and thence to the digestive gland, kidney and other epithelial linings where they are discharged to the exterior.

Reade & Reade (1972) have shown that only the macrophages are involved in the phagocytic clearance of injected carbon particles by *Tridacna maxima*. Moore & Lowe (1975) also associated phagocytosis mainly with the macrophages. Nakahara & Bevelander (1969), working with the clam *Pinctada radiata*, produced evidence of ingestion of colloidal thorium dioxide by agranular haemocytes (macrophages?) through minute pinocytotic channels which terminated in vesicles. However, Galtsoff (1964) and Cheng & Rifkin (1970) have reported that the granulocytes are more phagocytic than the agranular cells in *Crassostrea virginica*. Further experimental studies are clearly needed to resolve the phagocytic func-tions of the various cell types.

A pigmented haemocyte, the 'brown cell' (Takatsuki, 1934) is also thought to have an excretory function in bivalves. The removal of degradation products of dead parasites, and the metabolic by-products of successful parasites (Cheng & Rifkin, 1970), as well as the storage of fats and the secretion of shell-forming materials, are other proposed functions of these brown cells.

Bivalves respond to parasites that are too large to be phagocytosed by encapsulation, and this represents the main line of defence against helminths (Cheng, Shuster & Anderson, 1966; Cheng & Rifkin, 1970).

Tissue repair. The outcome of wound repair in oysters is a reconstruction of the original tissue, rather than the formation of scar tissue (see review by Sparks, 1972). Ruddell (1971) showed that in *Crassostrea virginica* a wound is repaired by haemocytes concentrating in the damaged area to form a plug, while the epithelium is regenerated. Haemocytes in *M. edulis* are also concerned with the early stages of shell repair, the transport of shell precursor material and the laying down of proto-ostracum (Bubel, 1973*b*).

Disease. In recent years naturally occurring tissue abnormalities and proliferative diseases have been recorded in *M. edulis*. Farley (1969) and Farley & Sparks (1970) recognised a disease characterised by enlarged amoebocytes with polyploid nuclei two or four times the normal size and often irregular in shape. Advanced stages of the disease were fatal.

Blood plasma

The inorganic composition of the haemolymph of mussels is variable, according to the composition of the medium. At any given time, the ionic composition of the body fluids will be determined by the permeability of the cell membranes, by the diffusion properties of the individual ions, and by the degree of control exerted by the animal on the individual ionic species. The osmotic concentration of the haemolymph is mainly determined by the inorganic ions (Remane & Schlieper, 1971). Although extracellular osmotic regulation is not achieved by mussels (see p. 236), some degree of ionic control is effected. In particular, the concentration of potassium ions in the blood is maintained at 30–40% higher than that in the medium (Robertson, 1953, 1964; Potts, 1954). Some values for the ionic composition of the blood of *M. edulis* are listed in Table 6.1. The pH of the blood of bivalves varies between 7.2 and 7.7; Watabe & Kobayashi (1961) recorded pH 7.3 ± 0.09 for the blood of *Modiolus demissus*.

Bayne (1973*a*) reported values for the organic composition of the haemolymph of *M. edulis* over an annual cycle, and as affected by changes in temperature and ration (Table 6.2).

211

Table 6.1. *The ionic composition of the blood of* Mytilus edulis *at two salinities*

	Ionic concentrations (mM per kg H$_2$O)					
	Na$^+$	K$^+$	Ca^{2+}	Mg^{2+}	Cl$^-$	SO$_4$$^{2-}$
Seawater (37.5‰ salinity)	507	10.6	11.3	57.9	594	30.5
Blood (pH 7.7)	502	12.7	12.6	55.8	586	30.7
Blood as % of seawater	99.0	120.0	111.5	96.4	98.8	100.5
Brackish water (16.0‰ salinity)	215	5.1	5.7	24.2	253	13.2
Blood	213	7.5	5.8	24.5	253	13.2
Blood as % of brackish water	99.3	147.8	101.4	101.1	100.0	100.0

Seawater values from Potts (1954), brackish water values from Seck (1958), quoted by Schlieper (1971).

Table 6.2. *Organic constituents in the blood of* Mytilus edulis

	Concentration range (mg 100 ml^{-1})
Protein	115–282
Non-protein-nitrogen	3.5–23.4
Ammonia	0.31–1.79
Sugars	9.8–35.7
Lipid	20.4–84.3

From Bayne (1973*a*).

Heart beat

The physiology and pharmacology of the hearts of bivalve molluscs have been studied for many years; recent reviews include Krijgsman & Divaris (1955), Hill & Welsh (1966), Pujol (1968) and articles in the volume edited by McCann (1969). Four neurotransmitter substances have been identified; acetylcholine (ACh), 5-hydroxytryptamine (5-HT), catecholamines, and an unknown 'substance X'. Only ACh and 5-HT have been shown conclusively to play a natural role in regulating the heart beat in bivalves. However, as Greenberg (1969) makes clear, much of this work has been done with the clam *Mercenaria mercenaria*, and comparative studies with other species indicate a variety of detailed differences that have proved difficult to generalise. For example, Greenberg summarises work on the effects of 5-HT on the hearts of mussels as follows: *Mytilus* spp., *Modiolus americanus* and *Modiolus modiolus* hearts are excited by 5-HT; *Lithophaga bisulcata* heart is depressed by 5-HT, but this depression can be blocked by the drug benzoquinonium which is known to block the action of ACh; and *Modiolus demissus* heart is also depressed by 5-HT, but this depression is virtually unaffected by benzoquinonium. Clearly there is a need for much more comparative study of this sort.

Ramsay (1952) and Krijgsman & Divaris (1955) have proposed the so-called 'constant-volume' hypothesis to explain the filling of the heart following ventricular systole. They state that the pericardium can be considered to be a closed chamber. A regular alteration of beat between the auricles and ventricles would be assisted by the rigidity of the pericardial wall, which is well-supported by the shell. During systole, pressure in the pericardial cavity and in the auricles would be reduced, leading to flow of blood into the heart. The hypothesis explained earlier observations that the discreteness of the pericardial wall is a prerequisite for the maintenance of the heart rhythm. Studies by Trueman (1966) and Brand (1972) have strongly supported the constant-volume hypothesis in bivalves. The

213

general haemodynamics of bivalve molluscs has been studied by Smith & Davis (1965), Trueman (1966) and Brand (1972).

The reader is referred to the reviews mentioned above for further details of the general physiology and pharmacology of bivalve hearts. We shall confine our discussion here to recent studies on the relationships between heart beat and environmental change, and the role of various environmental factors in affecting the heart rhythm in mussels.

Two methods have been used for measuring the heart beat in bivalves; (*a*) by cutting a window in the shell and observing the beat of the heart directly, and (*b*) by passing a small oscillating electrical current between two electrodes implanted alongside the pericardium and recording the changes in impedance that result from pulsatile changes in the heart (Hoggarth & Trueman, 1967; Trueman, 1967). This latter method has the considerable advantage that disturbance to the animal is minimal and heart beat can be monitored remotely. The pen-recorder trace that results from the (amplified) impedance changes affords a direct read-out of heart rate; the trace has also been used to estimate the amplitude of the heart beat by relating the distance of pen travel to an internal calibration signal in the impedance pneumograph. The assumption that this 'amplitude' of pen trace (or the product of rate and amplitude) is proportional to heart output is as yet untested (Bayne, 1971*b*; Brand & Roberts, 1973), but until more direct techniques for recording stroke volume have been employed, this indirect measure is probably acceptable as an index of changes in heart output in a single preparation.

Temperature

The frequency of the heart beat in *Mytilus edulis, M. californianus, M. galloprovincialis* and *Perna perna* increases linearly with increase in temperature (Figs. 6.2 and 6.3; Pickens, 1965; Lubet & Chappuis, 1967; Widdows, 1973*a*; Bayne *et al.*, 1975*a*). For *M. edulis* and *M. californianus* there is an apparent upper thermal limit to heart activity at 25–27 °C, which coincides with the temperatures at which other physiological processes are disrupted. In *M. galloprovincialis* the heart beats rhythmically up to 27 °C, and in *Perna perna* up to 32 °C (Lubet & Chappuis, 1967). These latter authors remark on the presence of an inflection point ('un point de rupture') at 19 °C when the rate/temperature (R/T) curves for *M. edulis* and *M. galloprovincialis* heart beats are plotted on logarithmic co-ordinates, but the significance of this is obscure. Lubet & Chappuis (1967) showed a shift to the left (translation) in the R/T curves for *M. edulis* and *M. galloprovincialis* (temperate species) relative to *Perna perna* (tropical), which they equate with the concept of latitudinal temperature compensation as discussed by Bullock (1955) and Prosser (1955). Unlike the rates

214

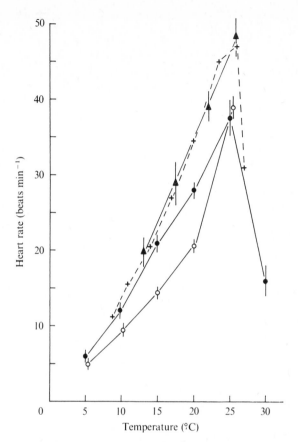

Fig. 6.2. The frequency of the heart beat of *Mytilus edulis* (●, fed animals; ○, starved animals) and *M. californianus* (+, from Pickens, 1965; ▲, from Bayne *et al.*, 1975*a*) as affected by temperature. Values ± 1 s.D.

of oxygen consumption and ventilation, heart beat frequency does not acclimate in response to temperature change (Widdows, 1973*a*).

The heart beat of other bivalves also increases with rise in temperature with a Q_{10} of about 2 (Galtsoff, 1964). Trueman & Lowe (1971) showed that the heart rate of the tropical bivalve *Isognomon alatus* immediately increased or decreased with rise or fall in temperature, and they suggested that temperature-sensitive receptors on the mantle might be responsible. Lowe (1974) equated changes in heart rate in *Mya arenaria* and *Crassostrea gigas* with the temperature of the water in the mantle cavity. Lowe & Trueman (1972) depressed the heart beat frequency of *Mya arenaria* by lowering the oxygen tension of the water and recorded that this depressed rate was less temperature-dependent (Q_{10} (increment in temperature not

215

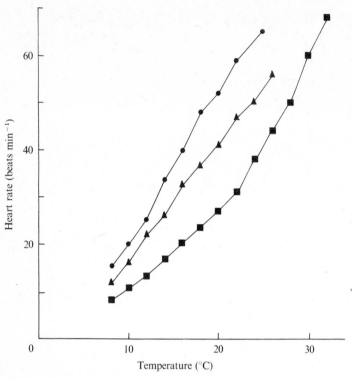

Fig. 6.3. The frequency of the heart beat of *Mytilus edulis* (●), *M. galloprovincialis* (▲), and *Perna perna* (■), as affected by temperature. (From Lubet & Chappuis, 1967.)

stated) = 1.84) as compared with the normal heart beat (Q_{10} = 2.16). The significance of this finding is uncertain, however.

Size

Pickens (1965) recorded that heart beat frequency (as deduced from counting the pulsations in the plicate membranes) was related to body size in *M. californianus* by the expression:

$$R = aV^b,$$

where R is beats per minute and V is the inner shell volume; values for b varied between −0.10 and −0.24 for all the populations he studied. Lubet & Chappuis (1967) record that heart beat in small *M. edulis* is faster than in larger individuals, but they did not quantify the difference.

216

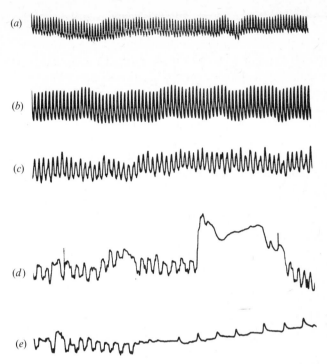

Fig. 6.4. Examples of the recorded heart trace of *Mytilus edulis* in one experiment, with the animal regulating oxygen uptake. (*a*) 150 mm Hg P_{O_2}; (*b*) 60 mm Hg; (*c*) 30 mm Hg; (*d*) 20 mm Hg; (*e*) 18 mm Hg. (From Bayne, 1971*b*.)

Oxygen tension (P_{O_2})

Bayne (1971*b*) studied the effects of oxygen tension of the medium on the heart rhythm of *M. edulis*. Heart rate was maintained relatively constant down to 40–60 mm Hg P_{O_2}, followed by a slight increase (tachycardia) and then a decline (bradycardia) that eventually resulted in cardiac arrest at < 20 mm Hg P_{O_2} (Fig. 6.4). During 'cardiac arrest' the heart sometimes showed short periods of near-normal beat (see also p. 223). Bayne (1971*b*) also recorded the amplitude of the heart beat, measured relative to the signal amplitude from the impedance pneumograph. When the mussels regulated their rates of oxygen consumption during decline in P_{O_2}, the amplitude of the heart trace gradually increased, with a more marked increase just prior to cardiac arrest. When the oxygen consumption was not regulated (see p. 194), amplitude of the heart beat altered less at low P_{O_2}. In all of these experiments the P_{O_2} at which frequency and amplitude of heart beat began to decline (the ' P_c ' for heart beat) was lower than the ' P_c ' for oxygen uptake. Similar results have been obtained by Brand & Roberts

(1973) working with *Pecten maximus* and Taylor (1974) on *Arctica islandica*.

On recovery from hypoxia, the amplitude of the heart beat often increases from cardiac arrest to exceed normal values, remaining high for some minutes following recovery to full oxygen saturation (Bayne, 1971*b*). An 'overshoot' of amplitude, and occasionally also of frequency, is a common feature of recordings of heart beat in bivalves. Brand & Roberts (1973) suggested three explanations for such overshoot, viz. that it is an indication of the repayment of an oxygen debt; that it is the result of a temporary imbalance in the system controlling the heart beat (Segal, 1962); or that it is due to increased activity that accompanies the 'flushing out' of excretory products from the body tissues (Boyden, 1972*a*). Taylor (1974) has shown that, on recovery from periods of anaerobiosis, *Arctica* exhibits an increased rate of oxygen consumption, and also an increased heart rate, for 20–25 h. In addition, there is an increased level of alanine (a presumed end-product of anaerobic metabolism; Chapter 8) in the blood, and this also declines to normal over 20–25 h. This evidence, together with the observations of Hers (1943), Schlieper (1955*b*), Karandeeva (1959) and Bayne (1971*b*), which indicate a general relationship between increased heart frequency and the duration of increased oxygen uptake, all suggest that, following periods of respiratory stress, there is a general enhancement of respiratory rate, accompanied by increased heart beat, which are both related in some way to anaerobic metabolism. More experiments are needed to carry this explanation further.

These changes in heart rate resulting from decreased environmental P_{O_2} are similar in many ways to changes that follow shell adduction. After shell closure, heart rate often increases slightly (Trueman, 1967; Brand, 1968; Taylor, 1974; Bayne *et al.*, 1975*b*) before the more frequently observed decline, so mimicking the slight tachycardia followed by bradycardia that occurs during decline in ambient P_{O_2}. Taylor (1974) has shown that both rapid and more gradually induced hypoxia elicit identical cardiac responses in *Arctica*. Brand (1968, working with *Anodonta anatina*, a freshwater bivalve) and Taylor (1974, with *Arctica*) canulated the mantle cavity and so were able to perfuse water of different P_{O_2} through the mantle cavity when the animals were fully adducted. In each case the heart beat responded to the P_{O_2} of the perfusate (Fig. 6.5), indicating that the heart was responding to the oxygen tension of the fluid in the mantle cavity rather than to any mechanical effect of shell closure. Also, the slight delay (about 2 min) between perfusing the cavity with water of reduced P_{O_2} and tachycardia suggests that the heart responds to the oxygen tension of the blood (Taylor, 1974). These studies strongly implicate P_{O_2} as affecting heart rate directly, in spite of the contradictory evidence of Baskin & Allen (1963) on isolated ventricles of *Tivella stultorum*.

218

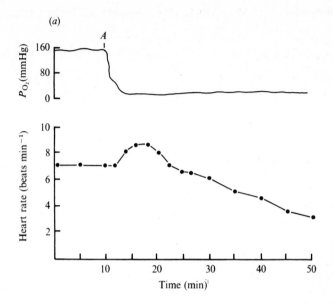

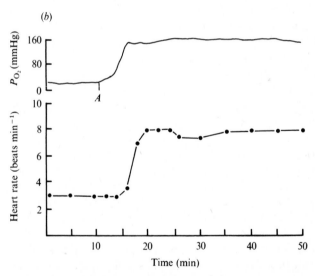

Fig. 6.5. Heart rate and the oxygen tension (P_{O_2}) of the water perfusing the mantle cavity of *Arctica islandica*. (a) Changing to water of low oxygen tension (at A); (b) changing to water of high oxygen tension (at A). (From Taylor, 1974.)

219

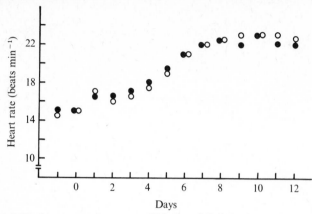

Fig. 6.6. The heart rates of two starved *Mytilus edulis* following the provision of food at day 0. (From Widdows, 1973*a*).

Ration

During prolonged starvation heart beat frequency declines from a routine to a standard level over at least 28 days (Widdows, 1973*a*). If food is then made available to the starved mussel, heart rate does not increase immediately, although both oxygen consumption rate and ventilation rate do increase to their active levels (Thompson & Bayne, 1972; see p. 163) However, if feeding is maintained, heart rate increases to the routine rate over a period of 10 days (Widdows, 1973*a*; Fig. 6.6). Pickens (1965) also found that heart rate in *M. californianus* was related to ration level, and increased with the addition of food. Pickens concluded that the apparently rather complex relationship he had recorded between heart rate and temperature was largely due to adjustments in heart rate in response to changes in the amount of available food.

The significance of this delayed heart response to ration level is the apparent absence of nervous coupling between ventilation, general metabolism and heart function. This is also apparent in the response to temperature, when heart beat does not acclimate in spite of acclimation of routine oxygen uptake. To say that heart rate reflects general metabolic rate (Trueman & Lowe, 1971; Walne, 1972) is therefore an unacceptable simplification. Changes in metabolic rate have to be considerable before a change in heart rate necessarily results. As Coleman (1974) points out, the relationship recorded by Walne (1972) between filtration rate and heart rate in the European oyster (*Ostrea edulis*) represents a 17% increase in heart rate for a 300% increase in filtration rate. Nevertheless, further studies of the relationships between respiration, ventilation and heart beat are clearly needed. We will discuss later (Chapter 7) the ways in which changes in the perfusion of the gill with blood may affect the efficiency of gas exchange.

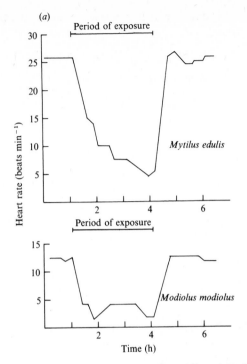

Fig. 6.7 (*a*) The frequency of the heart beat of *Mytilus edulis* and *Modiolus modiolus* during, and after, periods of exposure to air. (From Coleman & Trueman, 1971.)

Aerial exposure

Changes in the heart rhythm of bivalves that result from exposure to air have been studied by Trueman (1967; *Donax* and *Cardium*), Helm & Trueman (1967; *Mytilus edulis*), Trueman & Lowe (1971; *Isognomon alatus*), Coleman & Trueman (1971) and Coleman (1973*b*; *M. edulis* and *Modiolus modiolus*), Brand & Roberts (1973; *Pecten maximus*) and Bayne et al. (1975*b*; *M. californianus*).

Aerial exposure of *M. edulis* results in partial or complete valve closure (p. 185); this is accompanied by reduced heart beat frequency (bradycardia) and occasionally leads to complete suppression of the heart beat (Schlieper, 1955*b*; Helm & Trueman, 1967), (Fig. 6.7*a*). Prolonged valve closure during immersion produces the same effects on heart rate (Coleman, 1973*b*) and this bradycardia is usually interpreted as a response to the reduction in the oxygen tension of the water in the mantle cavity (Trueman & Lowe, 1971). Trueman & Lowe (1971) found that the heart beat of *Isognomon alatus* was not normally reduced during aerial exposure, because the shell valves were usually not fully adducted. They

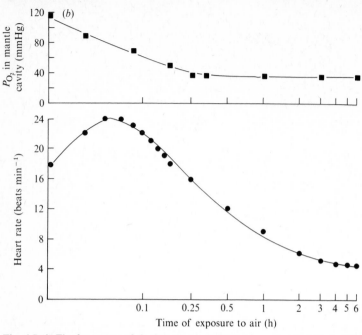

Fig. 6.7 (*b*) The frequency of the heart beat, and the oxygen tension (P_{O_2}) of the fluid in the mantle cavity, of *Mytilus californianus* during exposure to air. (From Bayne *et al.*, 1975*b*.)

concluded that gas exchange between the mantle cavity and the atmosphere resulted in a high oxygen tension in the mantle cavity (aided by continued ciliary activity on the gills) and an active heart rhythm was maintained as a result.

Bayne *et al.* (1975*b*) measured the heart frequency and the P_{O_2} in the mantle cavity of *Mytilus californianus* during exposure to air (Fig. 6.7*b*). Within 2–5 min of exposure, heart rate increased from nineteen to twenty-three beats min^{-1}, coinciding with a decline in mantle cavity P_{O_2} to 80–90 mm Hg. With more prolonged exposure, mantle cavity P_{O_2} declined to 40 mm Hg and frequency of heart beat to 4–8 beats min^{-1}. On re-immersion, heart frequency increased with 'overshoot', coinciding with a short period of increased oxygen consumption, and then declined to the normal rate. There was a delay of 2–5 min following re-immersion before heart rate increased from 4–8 to > 24 beats min^{-1}. Bayne *et al.* (1975*b*) point out the marked similarities between the relationship of heart beat to P_{O_2} during aerial exposure and during the reduction in dissolved oxygen tension, and they suggest that some effect of mantle cavity P_{O_2} on heart beat is indicated.

However, the decline in the P_{O_2} in the mantle cavity is probably not the only factor causing bradycardia during aerial exposure. Both *Modiolus*

222

modiolus (Coleman & Trueman, 1971) and *Pecten maximus* (Brand & Roberts, 1973) gape widely during exposure to air, and yet both species show gradual bradycardia at this time. Brand & Roberts (1973) suggest that this may be due, in part, to the loss of hydrostatic support for the tissues, and in particular the pericardium. Many recordings of heart beat during aerial exposure show periods of heart rhythm, with full amplitude and high frequency, interspersed between periods of cardiac arrest. These periods of regained heart rhythm may be correlated with movement of the animal in the shell causing increased venous return, and initiating heart beat.

On re-immersion following a period of aerial exposure, the heart beat often shows increased frequency and amplitude (e.g. Helm & Trueman, 1967) similar to the behaviour following anoxia while immersed. The possible relationship between such tachycardia and the respiratory consequences of anaerobiosis have been discussed earlier.

Salinity

There have been no detailed studies of the effects of salinity on the heart beat of bivalves. Bayne (1973c) indicated that salinities down to 16‰ had no marked effect on the heart beat of *M. edulis*. More recently, Feng & Van Winkle (1975) found that the heart rate of the oyster *Crassostrea virginica* was sensitive to salinity at temperatures between 20 and 30 °C, but not between 5 and 10 °C. Heart beat acclimated to salinity change over several days.

Heart beat as an index of metabolic activity

Coleman (1974) has reviewed some of the recent literature on the heart rate and activity of bivalves in their natural habitats, and particularly addressed the question of the extent to which heart rate may be used to monitor activity in nature. He concludes that whereas the measurement of heart rate does provide an index of 'extreme levels of activity and shows responses to drastic changes in environmental conditions or internal physiological states', it does not provide a 'comprehensive document of molluscan activity' under normal conditions. We agree with this assessment, and would add only that when heart rate is measured in parallel with other physiological functions, such as ventilation rate, more insight is gained into physiological condition than when heart beat alone is monitored (Chapter 7).

Excretion

Regulation of the chemical composition of the body fluids by the elimination of waste and the conservation of useful metabolites is the function of the mechanisms of ionic and osmotic control, and of excretion. In this section we shall discuss excretion in the context of the loss from the body of nitrogenous wastes; this includes the processes of production, concentration and final voiding of the end-products of nitrogen metabolism. Excretion in the Mollusca has been reviewed by Martin & Harrison (1966), Potts (1967, 1968) and Kirschner (1967). Florkin (1966), Campbell & Bishop (1970) and Florkin & Bricteux-Grégoire (1972) have comprehensively reviewed more general aspects of nitrogen metabolism in molluscs.

Organs of excretion

The primitive excretory organ of the Mollusca is the protonephridium. Paired protonephridia are present in the trochophore larva of bivalves (Meisenheimer, 1901) but are lost during subsequent larval development, when the excretory function is presumably met by the coelomic renal organs. The relationship between the coelomic cavities in ontogeny and phylogeny are discussed by Goodrich (1945). In mussels, the renal coelom communicates with the pericardium through the reno-pericardial canals, and opens to the exterior through the renal ducts on the common reno-genital papilla.

The excretory system of the adult mussel consists of paired kidneys and pericardial glands (White, 1937, 1942). The reddish-brown kidneys lie ventral to the pericardium and dorsal to the gill axis. They each take the form of a U-shaped tube, extending from the labial palps anteriorly to the posterior adductor muscle; at one end they open into the pericardium via the ciliated reno-pericardial canals, whilst at the other end they open to the exterior. The proximal portions of the kidneys are glandular, the distal portions are more thin-walled. The pericardial glands develop from the posterior wall of the pericardium and eventually come to invest the walls of the auricles. White (1942) could find no 'apertures for the discharge of the secretion' of these pericardial glands, but presumed that secretions from the gland entered the pericardial cavity. The kidney is supplied with blood via the renal sinus. The glandular part of the kidney is sometimes called the organ of Bojanus (Martin & Harrison, 1966); the pericardial glands are sometimes called Keber's organ. Whereas this kidney complex almost certainly functions in excretion, excretory products are probably also lost across the general body wall, and particularly across the gills.

Function of the excretory organs

Urine formation is usually considered to require three processes which are separate in time, and often also in space, viz. filtration, secretion and re-absorption. The long-held view of urine formation in bivalves (see Martin & Harrison, 1966) is that filtration of the blood occurs in the pericardium, either across the wall of the ventricle, or elsewhere within the pericardial coelom, and that the fluid so produced flows to the glandular part of the kidney where secretion and re-absorption of ions occurs.

Florkin & Duchâteau (1948) recorded that the pericardial fluid and the blood of the freshwater bivalve *Anodonta* contained equal concentrations of chloride, calcium and inorganic phosphate, but that these ions were much less concentrated in fluid from the organ of Bojanus. Martin & Harrison (1966) review the evidence for suggesting that secretion into the urine as well as re-absorption takes place in the organ of Bojanus. They also discuss earlier studies that point to the accumulation of waste material in certain cells of the pericardial glands. White (1942) quotes Grobben as recording the accumulation of 'accretions' in the cells of the pericardial glands which, on reaching a certain concentration, are discharged into the pericardial cavity and thence eliminated through the kidneys. Martin & Harrison (1966) conclude that the accumulation of waste materials and the filtration of the blood might both occur in the pericardial glands, the former in the glandular cells and the latter through flattened epithelial cells.

Pickens (1937) suggested that the pericardial fluid in *Anodonta* (and, by implication, in other molluscs) was a filtrate of the blood through the wall of the ventricle. This conclusion was based on three pieces of evidence; (*a*) the pericardial fluid was isotonic with the blood; (*b*) the pericardial fluid contained less 'non-mineral substance' than the blood, as based on measurements of refractive index; and (*c*) the blood had a high hydrostatic pressure, in the range 3 to 8 cm of water. The identification of the ventricle wall as the site of ultrafiltration of the blood received indirect support from Smith & Davis (1965) who measured high systolic pressures in the ventricles of a number of bivalves (including *Mytilus californianus*; pressures recorded in the range 0.4 to 3.0 mm Hg) and suggested that the morphology of the walls of the ventricle was suited to a function in filtration. Similar ventricular pressures have since been recorded by Trueman (1966) and Brand (1972).

However, Pierce (1970), when investigating water balance in *Modiolus* spp., found the blood to be significantly hyperosmotic to the pericardial fluid. In such a situation a gradient of osmotic pressure would result between the pericardial fluid and the blood, and in order for filtration to occur across the ventricle wall, the systolic blood pressures would have to exceed this osmotic pressure difference. Pierce (1970) used Smith & Davis'

Table 6.3. *Comparison of the osmotic pressure difference between blood and pericardial fluid* (Π) *with the effective filtration pressure (EFP) in three marine bivalves*

Species	Π (atmos.)	EFP (atmos.)	Π/EFP
Dinocardium robustum	1.96×10^{-1}	8.77×10^{-3}	22
Macrocallista nimosa	2.20×10^{-1}	5.65×10^{-3}	39
Mercenaria mercenaria	2.20×10^{-1}	5.65×10^{-3}	39

From Tiffany (1972).

(1965) values for the hydrostatic pressure of the blood, compared these with the observed osmotic differential, and concluded that ultrafiltration across the wall of the heart, from the ventricle to the pericardial cavity, could not occur. Tiffany (1972) confirmed that an osmotic pressure gradient occurs between the blood and the pericardial fluid in five species of bivalve. He also measured ventricular pressures, and confirmed Pierce's (1970) conclusion that the site of ultrafiltration in bivalves cannot be the wall of the heart, since the effective filtration pressure (ventricular systolic pressure – pericardial cavity pressure) was always less than the osmotic pressure differences between blood and pericardial fluid (Table 6.3). Martin & Harrison (1966) had earlier concluded that, in all probability, filtration occurs in the pericardium, but the precise location of the site of filtration in bivalves remains an open question. Kirschner (1967), in reviewing the processes of urine formation in invertebrate kidneys, suggested that filtration may well be the main mechanism for forming the primary urine in molluscs, but stressed that the evidence was rather meagre.

Pickens (1937) estimated the rate of filtration of the blood in the freshwater bivalve *Andonta* by opening the pericardial cavity and collecting the fluid as it was formed. Kirschner (1967) points out that by so abolishing the pericardial back-pressure the result was probably an overestimate of the filtration rate. Measurements based on the rate of inulin clearance (Martin *et al.*, 1958) suggest a filtration rate of 23 ml kg^{-1} h^{-1} for *Mytilus californianus* and Kirschner (1967) has suggested that, for most invertebrates, filtration rates in the kidney lie in the range of 1–10% of body weight per hour.

ble 6.4. *Nitrogenous excretory products of some bivalve molluscs*

Species	Excreted components as % of total measured nitrogen (N)				Authority
	NH$_4$-N	Urea-N	amino-N	uric acid-N	
diolus demissus	62–75	0	25–38	—	Lum & Hammen (1964)
	66	0	34	—	Hammen (1968)
assostrea virginica	65	13	5	—	Hammen et al. (1966)
	68	8	21	3	Hammen (1968)
rcenaria mercenaria	66	0	30	4	Hammen (1968)
lemya velum	70	0	27	3	Hammen (1968)
nax variabilis	75	0	24	1	Hammen (1968)
gellus plebius	50	0	31	19	Hammen (1968)
a arenaria	94	6	—	—	Allen & Garrett (1971b)
tilus edulis	41	4	55	—	Bayne (1973a): winter
tilus edulis	67	5	28	—	Bayne (1973a): summer

signifies that this component was not measured.

Products of excretion

The identification of the excretory products of bivalves has been undertaken entirely by analysis of the water in which individuals have been kept
for between 2–3 h (Bayne, 1973a) and one or more days (Hammen, Miller
& Geer, 1966; Allen & Garrett, 1971b). The inadequacies of this technique
are generally appreciated (Florkin, 1966), viz. the possibility of bacterial
contamination and the presence of dissolved material voided in the faeces,
but suitable microtechniques for analysis of fluid in the kidneys have not
been employed with bivalves. The available information on the excretory
products of mussels is also unsatisfactory in the virtual disregard of
purines in the analysis of excreted nitrogen. Uric acid has been identified in
certain tissues of *Mytilus edulis*, notably the byssus gland (Kasuga &
Ishida, 1957) and the kidney, but Hammen (1968) could find, with one
exception, only very small amounts excreted by seven species of bivalve.
The possibility that concretions rich in uric acid are shed from the pericardial gland and voided via the kidney has not been properly examined.

With these reservations in mind, some recent data on the nitrogenous
excretory products of some bivalves are listed in Table 6.4. Although
ammonia is the dominant product, large amounts of amino-nitrogen are
lost, and there is a small but significant amount of urea also excreted by
some species. The proportions of excreted ammonia to amino acid can
vary with season and as a result of environmental stress (Hammen, 1968;
Bayne, 1973a). Hammen et al. (1966) could not account for 17% of total
excreted nitrogen in *Crassostrea virginica*, and this may represent the

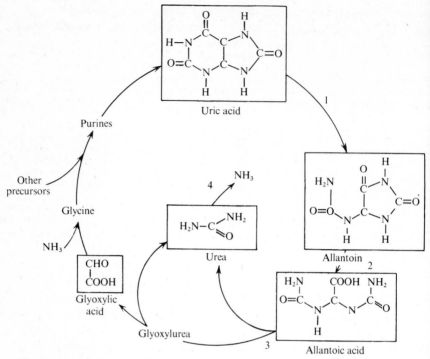

Fig. 6.8. Diagrammatic representation of the purinolytic cycle. Enzymes are: 1, uricase; 2, allantoinase; 3, allantoicase; 4, urease. (From Andrews & Reid, 1972.)

contribution of purines. Bayne (1973*a*) recorded a positive correlation between non-protein-nitrogen concentration in the haemolymph of *M. edulis* and the rate of loss of amino-nitrogen, but did not find a similar correlation between haemolymph ammonia concentration and ammonia excretion.

The ornithine and purine cycles

It has been known for some time that *Mytilus* has the complete complement of uricolytic enzymes (Florkin & Duchâteau, 1943). Xanthine oxidase, which is active in the conversion of xanthine to uric acid, has also been demonstrated in *M. edulis* by Tsuzuki (1957). More recently, Andrews & Reid (1972) have demonstrated that all the uricolytic enzymes are present in the digestive diverticula of four bivalve species including *M. californianus*, though they could find no activity in extracts of kidney or mantle tissue. Andrews & Reid confirmed earlier studies (Florkin & Duchâteau, 1943; Hammen, Hanlon & Lum, 1962) on the presence of urease in the

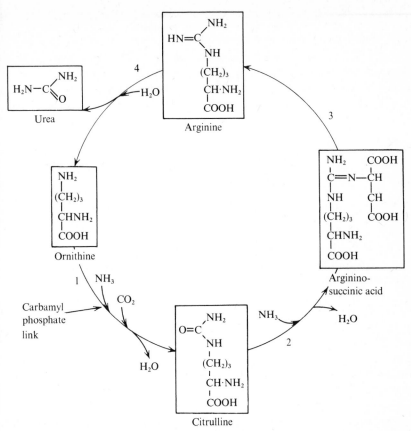

Fig. 6.9. Diagrammatic representation of the ornithine cycle. Enzymes are: 1, ornithine transcarbamylase; 2, arginosuccinate synthetase; 3, arginosuccinate lyase; 4, arginase. (From Andrews & Reid, 1972.)

digestive gland. It appears therefore, that the classical pathway for purine catabolism, or purinolysis (Fig. 6.8) is present in mussels.

Andrews & Reid (1972) also recorded that all of the enzymes of the ornithine cycle (Fig. 6.9), with the exception of carbamylphosphate synthetase, were present in digestive gland tissue, though none was present in kidney or mantle. They suggested that their inability to demonstrate carbamylphosphate synthetase might be due to a lack of sensitivity in the analytical technique. The absence of urease in the kidney may be significant in explaining the excretion of small quantities of urea by some bivalves (Campbell & Bishop, 1970).

All of the enzymes of the ornithine cycle and purinolysis are therefore probably present in at least some bivalves, including *Mytilus*. But other possibilities for the degradation of arginine also exist. Baret, Mourgue,

Broc & Charmoit (1965) did not detect arginase activity in *M. edulis*, but they did observe activity of the enzyme γ-guanidinobutyrate ureohydrolase. Blaschko & Hope (1956) confirmed the existence of an L-amino acid oxidase in *M. edulis* which oxidised arginine. Campbell & Bishop (1970) suggest that molluscan L-amino acid oxidase may be inhibited *in vivo* by high concentrations of alanine, glycine and proline, and by low tissue oxygen tension. Nevertheless, as Florkin (1966) points out, three mechanisms may be involved in the catabolism of arginine; (i) the action of arginase to form ornithine and urea; (ii) oxidative decarboxylation, and (iii) oxidative deamination. The oxidative deamination of arginine may lead to the production of γ-aminobutyrate and urea via the action guanidinobutyrate ureohydrolase. Gaston & Campbell (1966) recorded relatively low arginase activity in estuarine bivalves, and could find no activity in marine species. In view of this, Andrews & Reid (1972) suggest that much of the catabolism of arginine *in vivo* may occur by oxidative deamination.

The physiological significance of the ornithine and purinolytic cycles in bivalves is not known, although certain possibilities have been suggested. Andrews & Reid (1972) suggest that during prolonged periods of valve closure, the capacity to detoxify ammonia by conversion to urea would have adaptive value. De Zwaan & Van Marrewijk (1973a) have postulated a different mechanism, with the same adaptive significance in mind, namely the fixing of ammonia into alanine (involving an alanine dehydrogenase). Andrews & Reid (1972) further suggest that urease may be functional in degrading to ammonia any excess urea that accumulates. Speeg & Campbell (1969) have postulated an alternative role for urease; they suggest that free ammonia might react with hydrogen ions formed by the action of carbonic anhydrase on bicarbonate, so releasing carbonate ions for deposition in the shell as calcium carbonate.

Campbell & Bishop (1970) have pointed out that the primitive role of the ornithine cycle was probably in the synthesis of arginine, which in turn is the precursor for phosphoarginine, the muscle phosphagen of the molluscs. Phosphoarginine was demonstrated by paper chromatography in *Mytilus californianus* by Seraydarian & Kalvaitis (1964). Arginine may serve another physiological role, as the precursor of octopine. The biosynthesis of octopine involves a condensation of arginine with pyruvic acid, followed by a reduction step to yield octopine (Florkin & Bricteux-Grégoire, 1972). This reaction has been demonstrated in bivalve muscle tissue by Thoai & Robin (1959) who later (1961) found an inverse correlation in several invertebrates between lactate dehydrogenase activity and the capacity to synthesise octopine. Gäde & Zebe (1973) have recently confirmed a ratio of octopine dehydrogenase:lactate dehydrogenase of 2.9 in *M. edulis* adductor muscle. Robin & Thoai (1961) suggest that the reaction serves to regenerate oxidised coenzyme during anaerobic glycolysis (Chapter 8), but, as Campbell & Bishop (1970) point out, this would remove arginine from the phosphagen pool, and so require a source of arginine.

Clearly, the physiological role of the enzymes associated with the ornithine cycle and the biosynthesis of arginine and urea, as related not only to nitrogen excretion but also to anaerobic metabolism, requires further study. The catabolism of nitrogen-containing compounds almost invariably leads to the production of ammonia, which is highly soluble and easily voided from the animal when immersed. But ammonia is also highly toxic, and a variety of metabolic sequences is available for 'detoxifying' the ammonia, should this be necessary. The significance of these metabolic sequences in the normal life of mussels warrants further investigation.

Loss of amino acids

The data in Table 6.4 emphasise the large amounts of amino acids that are normally lost from bivalves, amounting to 5–55% of total measured excreted nitrogen. Bayne & Scullard (unpublished data) have recently demonstrated a linear relationship between ammonia excretion and amino acid loss in *M. edulis*. The proportion of amino-nitrogen to total excreted nitrogen may increase considerably above normal, however, when the animals are placed under environmental stress. For example, Hammen (1968) records a change in the proportions of excreted nitrogen in *Tagelus plebius* over 20 h (presumably starved) in the laboratory; amino-nitrogen increased from 31 to 67% of total excreted nitrogen. Bayne (1973a) found that the proportion of amino-nitrogen increased from 55 to > 75% during stress from temperature and starvation in *M. edulis*. The loss of amino acids may constitute a considerable energy (calorie) 'drain' for the animal. Bayne (1973a) calculated for *M. edulis* that amino-nitrogen loss normally represented 11% of the routine metabolic rate, when expressed as calories, and this could rise, during stress, to as high as 63%.

The question arises whether this loss of amino acids represents simple leakage or an active excretion process. Hammen (1968) emphasised the enormous difference in concentration between the animals' tissues and the medium, giving an internal:external ratio of 10^5. He suggested (see also Potts, 1958) that this concentration gradient caused leakage of amino acids across the body membranes, and in order to replace this loss bivalves have very active aminotransferase enzymes. As recognised by Hammen (1968), for such replacement to occur, transamination must be coupled with the fixing of ammonia, and this is traditionally viewed as a possible role for glutamate dehydrogenase (GDH). However, the activity of this enzyme in bivalve tissue appears to be rather low. Thornberger, Oliver & Scutt (1968) could not demonstrate GDH activity by starch-gel electrophoresis in

Marine mussels

molluscan tissue. De Zwaan & Van Marrewijk (1973b), however, refer to unpublished data of Addink as demonstrating a cytoplasmic GDH in *Mytilus* tissue, and Livingstone (1975) recorded a maximum activity in the adductor muscle of *M. edulis* of 0.03 μmol min^{-1} per mg protein. This is equivalent to about 30 μmol h^{-1} per g wet tissue, which compares with average values of 360 μmol h^{-1} per g wet tissue for aspartate aminotransferase in bivalve whole tissue (see Table IX in Campbell & Bishop, 1970). Further study of GDH in bivalve tissues is urgently needed (see also Chapter 8), but present data do not suggest a very active enzyme in the cytosol.

That the loss of amino acids may be a process of active excretion rather than of simple diffusion has been suggested by Lange (1970, 1972). Lange (1970) considered that the intracellular concentration of taurine, at least, may be controlled by 'variation in the permeability of the cell membranes for this substance'. Loss of other amino acids might be similarly controlled. Bayne (1975b) presented some preliminary data on the loss of five amino acids (including taurine) from *Mytilus* to the medium; the rates of loss of three of these (taurine, glutamate and glycine) suggested an active excretion not directly proportional to their tissue concentrations. Lange (1972) has raised the further possibility that since certain amino acids may be formed as end-products of energy metabolism (Chapter 8) they may be expected to be excreted. Studies on the responses of mussels to reduced salinity (p. 239) indicate that, in the control of intracellular amino acid concentrations, at least three processes may be involved. One of these involves degradation to ammonia; another involves loss of amino acids. Just as the production of urea may be considered, under certain circumstances, a means for the detoxification of ammonia, so perhaps may amino acid loss meet a similar objective.

Rates of excretion

Rates of nitrogen excretion by mussels are extremely variable, which is not surprising in view of the marked seasonal changes in nutrient storage and utilisation of reserves (Chapter 8). However, there is considerably less information available on the factors that influence the rate of excretion, than there is, for example, on rate of oxygen uptake.

Lum & Hammen (1964; for *Modiolus demissus*) and Hammen, Miller & Geer (1966; for *Crassostrea virginica*) published average values for excretion rate of 42 and 25 μg ammonia-nitrogen d^{-1} per g fresh weight, respectively. Allen & Garrett (1971b) recorded an average rate for *Mya arenaria* at full salinity of 94.7 μg ammonia-nitrogen d^{-1} per g fresh weight. Bayne (1973a) reported rates of ammonia-nitrogen excretion by *M. edulis* between 9.6 and 67.2 μg d^{-1} per g fresh weight. Values for the excretion

232

Table 6.5. *Relationships between the rate of excretion of ammonia-nitrogen* (V_N; *mg* d^{-1}) *and body size* (W; *g dry flesh wt*) *in* Mytilus edulis

Month	Ration	Temperature (°C)	Equation
January	Fed	5.0	$V_N = 0.074 \cdot W^{1.20}$
	Starved	5.0	$= 0.566 \cdot W^{0.35}$
April	Fed	7.0	$= 0.170 \cdot W^{1.20}$
	Starved	7.0	$= 1.939 \cdot W^{0.35}$
July	Fed	17.0	$= 0.024 \cdot W^{0.72}$
	Starved	17.0	$= 0.024 \cdot W^{0.72}$
October	Fed	12.5	$= 0.037 \cdot W^{0.72}$
	Starved	12.5	$= 0.037 \cdot W^{0.72}$

rate by *M. edulis* varied considerably, depending on animal size, temperature, ration level and salinity. We will consider some of these factors, with special reference to *Mytilus*.

Bayne & Scullard (unpublished data) and Widdows (unpublished data) found that an allometric relationship held, for all conditions of season, temperature and ration, between excretion rate (μg ammonia-nitrogen d^{-1}) and body size (g dry flesh weight) in *M. edulis*. However, the numerical values for the parameters of the allometric equation (a and b in the expression: ammonia-nitrogen excretion = $a \cdot$ body size^b) showed considerable variability. Regression equations for mussels fed and starved, at field-ambient temperatures, in the laboratory in four months of the year, are listed in Table 6.5. The regression coefficient (b) varied between 0.35 (starved mussels in the winter) and 1.20 (fed mussels in winter); both fed and starved animals in summer had a similar coefficient of 0.72. Rates of excretion were high in the winter and low in the summer, and they increased with starvation in the winter but did not change appreciably in the summer. Bayne (1973a), working with a different population of mussels, also recorded high excretion rates in the winter and lower rates in the summer. In that study, starvation in the winter caused a sharp rise in excretion of ammonia over the first month followed by a decline, whereas in the summer starvation induced a drop in excretion rate.

The changes in the relationship between excretion rate and body size may be explained, in part, by seasonal changes in the synthesis and utilisation of nitrogenous compounds as substrates for energy metabolism. The effects of starvation interact with these seasonal changes; small individuals, with a relatively small glycogen reserve, increase considerably their protein catabolism during starvation, whereas larger individuals rely to a greater extent on their relatively large glycogen stores.

Emerson (1969) observed that ammonia excretion by *Mya arenaria* was dependent on body size, but he gave no further details. Ansell & Sivadas

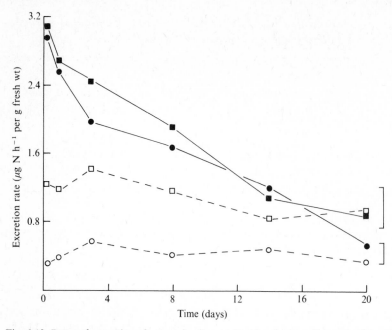

Fig. 6.10. Rates of excretion of ammonia-nitrogen (○, ●) and rates of loss of amino-nitrogen (□, ■) by *Mytilus edulis* at 32.5‰ salinity (open symbols joined by dashed lines) and 14.5‰ salinity (closed symbols joined by continuous lines). The vertical brackets indicate the least significant differences between values for ammonia-nitrogen (top bracket) and amino-nitrogen (bottom bracket). (From Bayne, 1975*b*.)

(1973) recorded a regression coefficient of 0.639 in relating ammonia excretion to body size in the clam *Donax vitatus*. Bayne *et al.* (1975*a*) measured ammonia excretion by *Mytilus californianus* at 13 °C in the summer and recorded the relationship with body size as: Ammonia-nitrogen excretion (μg h^{-1}) = 23.9 · dry weight (g)$^{0.82}$.

The rate of ammonia excretion by *M. edulis* increases with rise in temperature and does not then acclimate (Bayne & Scullard, unpublished data). The coefficient relating excretion rate to body size (over a weight range of 0.3 to 1.3 g dry flesh) is not altered by temperature change, i.e. Q_{10} for excretion rate is not size-dependent. However, Q_{10} may vary with season. Ansell & Sivadas (1973) recorded a mean Q_{10} for ammonia excretion by *Donax* between 10 and 20 °C of 2.07.

The rate of excretion of ammonia by *M. edulis* is increased at reduced salinity, as is the rate of loss of amino-nitrogen (Fig. 6.10). Both rates decline over a period of 20 days at 14.5‰ salinity, eventually returning to values typical of animals at full salinity (Bayne, 1975*b*). Emerson (1969) and Allen & Garrett (1971*b*) observed that rates of ammonia excretion by *Mya arenaria* increased at reduced salinity. Allen & Garrett recorded an increase from 3.22 mg NH_4-N d^{-1} at 34‰ salinity to a maximum of 64.4 mg NH_4-N d^{-1} at 17‰ salinity. Within 6 to 7 days at 17.5‰ salinity, excretion rates returned to normal. Allen & Garrett (1971b) did not observe an enhanced excretion of urea at reduced salinity.

During exposure to air, *Mytilus californianus* continues to produce ammonia, which accumulates in the fluid trapped in the mantle cavity (Bayne *et al.*, 1975b). However, the rate of production of ammonia at this time is only about 5% of the immersed rate, whereas the rate of oxygen consumption in air is only reduced to 75% of the standard value recorded during immersion (p. 188). This difference could be due to an increase in glycolysis relative to protein degradation, to an increased production of urea (as postulated by Andrews & Reid, 1972), a 'fixing' of ammonia as alanine (De Zwaan & Van Marrewijk, 1973a), or to an increased production of amino-nitrogen. The relationships between nitrogen metabolism and physiological adaptations to aerial exposure require further study.

Conclusions

Just as a knowledge of the energy metabolism of a species may be useful in understanding its ecological energetics, so information on nutrient relationships, including nitrogen metabolism, may be important for an appreciation of a species' role in the nutrient cycles and budgets of the natural community. Kuenzler's (1961b) study of a population of *Modiolus demissus* in a salt marsh ecosystem is a vivid illustration of this. Dense populations of mussels may be expected to contribute significantly to the nutrient budgets of inshore waters, both in terms of particulate and dissolved nutrients. The loss of amino acids may represent a significant input of nitrogenous substrate for microbial production; for example, based on an amino acid loss of 20 μg N h^{-1}, and a population biomass of 1.2 kg fresh wt m^{-2} (see Chapter 2), a simple calculation shows a yield to the environment of 41.5 g organic N yr^{-1} m^{-2} which is not insignificant. There is a need for investigations into the role of mussel populations in the nutrient cycles of estuaries and near-shore waters.

Other areas of ignorance have been identified in the relevant sections of the text. Two that have not been mentioned specifically are the clarification of the relationships between individual excretory products and the sites, within the kidney or elsewhere, of filtration and re-absorption in the production of urine. And there is a need for investigation of the ultrastructure of the kidneys and pericardial glands. Finally, as in all aspects of physiological ecology of bivalves, there is a need for much more biochemical understanding of the animals' adaptive processes.

Ionic and osmotic relationships

Amongst the variety of mussels there is a wide spectrum of salinity tolerance, from the markedly euryhaline *Xenostrobus securis* (Wilson, 1968) and *Modiolus demissus* (Pierce, 1970) to the more stenohaline *Modiolus modiolus*. It has been known for many years, however, that in spite of the euryhalinity of some species (*Mytilus edulis* in particular), these animals are incapable of 'osmoregulation', at least over much of their normal salinity range. For example, Maloeuf (1937) forcibly kept the shell valves of *M. edulis* open with a glass tube and placed the mussels in distilled water; he determined that the animals then increased in weight exponentially. Equally it has been realised for some time that mussels are capable of some regulation of the internal (haemolymph) concentrations of some ions (Robertson, 1953, 1964; Schlieper, 1971). More recently, mechanisms for the control of intracellular osmotic concentration have been described (Duchâteau & Florkin, 1956). In considering physiological adaptation to varying salinity, therefore, three different degrees of control of the composition of the body fluids need to be recognised; control of the concentrations of various ions, extracellular anisosmotic control, and intracellular isosmotic regulation. These three are all interrelated and are themselves components of a fourth process, the control of cell volume. These topics have been reviewed recently by Robertson (1964), Florkin & Schoffeniels (1965), Schlieper (1971), Lange (1972) and Schoffeniels & Gilles (1972). Ionic regulation (Robertson, 1964) is the maintenance of ionic concentrations in the body fluids which are different from the concentrations to be expected should passive equilibrium occur between the internal and external media. Extracellular anisosmotic control (Florkin, 1962) is the regulation of the osmotic pressure of the extracellular body fluids. Isosmotic intracellular regulation (Duchâteau & Florkin, 1956) is the term applied to the control of intracellular osmotic pressure.

Extracellular anisosmotic regulation

Mussels have generally been considered incapable of extracellular osmotic control, and to have extracellular fluids (haemolymph and pericardial fluid) isosmotic with seawater (Schlieper, 1971). However, Wilson (1968) reported that the pericardial fluid of the estuarine mussel *Xenostrobus securis* (in Australia) was slightly hyperosmotic to the environmental medium, and Remmert (1969), who reviewed similar data for a variety of invertebrates, concluded that a hyperosmotic body fluid may be characteristic of all poikilosmotic animals. Pierce (1970) examined this problem in four species of *Modiolus*, two of which are poikilosmotic and euryhaline, and two of which are poikilosmotic but stenohaline. In all four cases the extracellular fluids were slightly, but significantly, hyperosmotic to the medium. The average difference between seawater and the blood of *M. demissus*, for example, was about 20 mOsm kg(H$_2$O)$^{-1}$, representing a difference in freezing-point depression of 0.036 deg C. Pierce (1970) suggested that some techniques used in the past to estimate freezing-point depression may not have been sensitive enough to detect this difference. Schlieper (1971) quotes values due to Beliav, for 'internal medium' freezing-point depression for *M. edulis* which were between 0.03 and 0.14 deg C lower than the external medium. Pierce (1970) concluded that this hyperosmoticity of body fluids is not a function of the species' habitat nor is it an active process, but is due to a passive Gibbs–Donnan equilibrium caused by proteins in solution in the blood. This osmotic difference will result in the influx of water into the animal, unless it is opposed by a hydrostatic pressure in the body fluids.

There is some evidence, however, that at very low salinities the difference between the osmoconcentrations of the medium and the blood of some bivalves is increased, indicative of an active extracellular osmotic control. Wilson (1968) suggested that this might be so in *Xenostrobus*, and Bedford & Anderson (1972) demonstrated the phenomenon in the clam *Rangia cuneata*, which maintained an osmotic differential between blood and medium of 55–65 mOsm l^{-1} in water at 16–100 mOsm l^{-1}.

The mechanisms that provide for this apparent osmoregulation in *Rangia* have not been described. However, osmoregulation requires the expenditure of energy. Potts (1954), in a calculation that yielded minimum estimates, suggested that the energy required to maintain a differential between haemolymph and water of 36 mOsm l^{-1} in the freshwater bivalve *Anodonta* represented about 1.2% of the energy equivalent of the total metabolic rate. Bedford & Anderson (1972), in assuming Potts' model to hold also for *Rangia*, suggest that 2.4% of the available energy could be used for osmoregulation at salinities of 3‰ or less. Unfortunately, there are insufficient data for the rate of urine production by marine mussels, and Potts' (1954) model is sensitive to changes in this variable; for example, a change in the rate of urine production from 1 to 5 μl h^{-1} (see p. 226) can represent an increase in osmotic work, done at the surface of the animal, from 1.5 to 7% of the estimated total metabolic expenditure at a salinity of 3‰ (Fig. 6.11). Nor are there data available on the osmotic concentration of the urine of estuarine bivalves; Potts (1954) demonstrates the considerable energetic advantages to be gained in low salinity water from producing a urine hyposmotic to the blood. Potts also emphasises that the permeability characteristics of the body are of crucial importance to survival at low salinities. Bivalves, with their very large surface area (mantle, convoluted gills etc.) exposed directly to the medium, will have high permeabilities which will impose an upper limit on the extent to which their blood can be

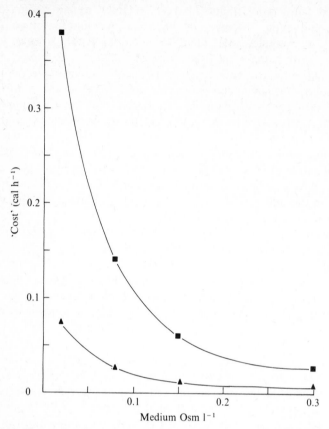

Fig. 6.11. The predicted 'cost', as calories per hour, of maintaining an osmotic differential between blood and the medium, calculated according to Potts (1954), using values for the clam *Rangia cuneata* as given by Bedford & Anderson (1972). ■, rate of urine production set at 5 μl h^{-1}; ▲ rate of urine production set at 1 μl h^{-1}.

maintained hyperosmotic to the medium without incurring a very large metabolic cost.

Anderson (1975) recorded that although the concentrations of both ninhydrin-positive-substances (NPS) and chloride ion in the haemolymph of *Rangia* decreased as the salinity of the medium was reduced, the NPS was apparently regulated whereas chloride ion was not; NPS was reduced from 5.5 to 1.3 mM l^{-1} over a salinity range from 17 to 0.5‰. Pierce (1971*a*) concluded that the hyperosmoticity of the extracellular fluids of *Modiolus* relative to the medium was due in approximately equal portion to potassium ion, free amino acids and proteins. NPS are involved in intracellular isosmotic regulation (p. 239) and also in anaerobic metabolism (Stokes & Awapara, 1968; Chapter 8), and the possibility that they may function in anisosmotic regulation as well suggests a rewarding area for future research.

Amongst the bivalves in general, behavioural responses probably contribute most to their adaptation to fluctuations in salinity. Maloeuf (1937) quotes Bendant as being the first authority (in 1816) to note that bivalves can respond to sudden changes in salinity by closing their shells. In doing so, they isolate their tissues from osmotic changes in the external medium, and expose them instead to the fluid now trapped in the mantle cavity. Milne (1940) described how, by virtue of this response, the salinity within the mantle cavity of *Mytilus edulis* can be as high as 24‰ when the salinity of the water is only 7‰. Pierce (1971b) recorded valve movements by different species of *Modiolus* (*M. demissus, M. squamosus* and *M. modiolus*) during exposure to dilute and to concentrated seawater, and during recovery from salinity stress; he also measured tissue weight changes as an index of tissue volume. The animals closed their valves on first experiencing a change in salinity, and the period they remained closed could last for some days, with only short, intermittent periods of opening. Pierce also determined that *Modiolus* can control its volume in diluted seawater by loss of solutes from the tissues. However, this volume control response was unidirectional, in the sense that, on subsequently returning to full salinity, the body volume did not recover to the previous value, but was stabilised at the reduced level. That is, the loss of solutes that facilitated volume control at reduced salinity 'committed' the animal to this salinity regime and, concomitantly, to a reduced upper lethal salinity limit. Pierce (1971b) concluded that valve closure affords the mussel a 'period of grace' following reduction in salinity. If this reduction is transient, the animal is not committed to a full acclimation of tissue volume, but instead must tolerate a period of valve closure. If the reduction in salinity is more lasting, volume regulation by solute loss is forced upon the animal by its demand for food and oxygen.

Gilles (1972b) has described some of the consequences of shell-closing behaviour for three molluscs, including *Mytilus edulis*. In *Mytilus*, the haemolymph remains isosmotic to the fluid in the mantle cavity. When the animals were subjected to a marked reduction in salinity (75% dilution), closure of the shell valves resulted in the mantle cavity fluid remaining hyperosmotic to the medium for at least 96 h. Ultimately, as full acclimation is forced upon the animals, the processes of intracellular isosmotic regulation are invoked, resulting in regulation of cell volume.

Intracellular isosmotic regulation

Early studies (reviewed by Florkin & Schoffeniels, 1965, 1969) showed that the osmotic pressure of the cells of marine invertebrates was due, in large

Table 6.6. *The concentrations and osmotic functions of certain inorgani*
and organic solutes in the adductor muscle of Mytilus edulis

Solute	Concentration (μmol per g wet wt)
Inorganic ions	
Sodium	158–91
Potassium	125–92
Calcium	8.5–9.6
Magnesium	38
Chloride	187–307
Sulphate	13
Phosphate	13
Organic solutes	
L-amino-nitrogen	155–389
Taurine	89–91
Betaine	97
Arginine phosphate	13.4
ATP	4.9
Acid-soluble phosphate	7.4
Osmotic pressure (mOsm per kg H_2O) due to:	
Inorganic ions	560.5–851.9
Organic solutes	421.9–569.2
Total	982.4–1411.1

Data compiled from Potts (1958) and Bricteux-Gregoire *et al.* (1964).

part, to small-molecular-weight organic compounds (Table 6.6) which were
present at high concentration within the cells but at much lower concentrations in the extracellular fluid. These organic compounds largely comprised
ninhydrin-positive-substances (NPS) and of these, in the muscle tissues
of *M. edulis* (Table 6.7), alanine, arginine, aspartic and glutamic acids,
glycine, taurine and betaine were particularly important. Some of these
early studies also demonstrated that, on reduction of the ambient salinity,
there was a marked decline in the concentration of certain NPS which
could not be explained solely as a result of the slightly increased hydration
of the tissues (Table 6.7). That is, an active regulation of the concentration
of the NPS within the cells occurs. Florkin (1962) termed this process
isosmotic intracellular regulation, and it has since become accepted as a
major control mechanism in marine and brackish-water organisms. By
means of this control over the concentration of intracellular solutes, an
animal is able to maintain its cells isosmotic with the extracellular fluids
whilst at the same time regulating their volume.

Lange & Mostad (1967) and Lange (1970, 1972) have discussed theoretical aspects of volume regulation in marine bivalves, and presented a means

Table 6.7. *The concentration of amino acids in adductor muscle of* Mytilus edulis *at 100 % and 50 % seawater*

| | Concentration (μmoles per g wet wt) | | | |
| | 100% seawater | | 50% seawater | |
Amino acid	A*	B†	A*	B†
Alanine	18.40	28.07	13.0	8.77
Arginine	11.94	—	8.38	—
Aspartic acid	9.77	11.25	1.73	2.01
Glutamic acid	9.79	5.00	10.61	1.03
Glycine	61.87	132.69	18.40	58.92
Histidine	1.80	0.58	1.22	0.72
Isoleucine	0.31	0.90	0.39	0.35
Leucine	0.42	1.57	0.60	0.53
Lysine	2.81	1.09	1.71	0.72
Phenylalanine	0.51	0.60	0.28	0.40
Proline	6.29	45.65	2.80	10.95
Serine	5.99	2.86	3.05	3.17
Threonine	4.37	3.07	1.93	1.21
Tyrosine	0.77	0.90	0.48	0.58
Valine	0.94	2.24	0.50	0.75
Taurine	58.09	20.50	44.51	9.21
Betaine	62.99	—	40.66	—
Water content	69.80	71.62	80.20	77.26
Osmotic pressure *(mOsm per kg H₂O)* *due to:*				
Amino acids	216.54	366.49	90.19	129.46
Taurine	92.47	31.77	61.66	13.23
Betaine	100.27	—	56.23	—
Osmotic pressure *of the medium* *(mOsm)*	1180		573	

* Data from Bricteux-Grégoire *et al.* (1964).
† Data from Gilles (1972*b*).

of quantifying the capability of a species for control of cell volume by means of intracellular isosmotic regulation. Water content should be assessed as the weight of water per unit volume of tissue, over a range of salinities. The concentration of osmotically active solutes in the tissues needs also to be measured, and the rate of change in water content that would be necessary to accomplish complete volume regulation at reduced salinities can then be calculated (Lange & Mostad, 1967). These theoretical expectations may then be compared with experimentally observed values to arrive at a true estimate of efficiency of volume regulation. Lange (1970)

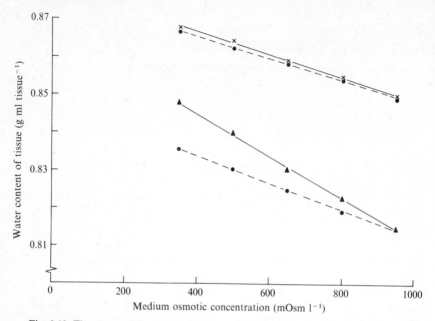

Fig. 6.12. The observed ($\times$, $\blacktriangle$) and the predicted ($\bullet$) water content of tissues of *Cardium edule* ($\times$) and *Modiolus modiolus* ($\blacktriangle$) at different osmotic concentrations of the medium, (Adapted from Lange, 1970.)

carried out such an analysis for six bivalve species, including *Modiolus modiolus* (Fig. 6.12), and showed a correlation to exist with the presumed steno- or euryhalinity of the species. These studies provide a quantitative basis for the hypothesis of Florkin & Schoffeniels (1969) that the degree of euryhalinity exhibited by a species is dependent on the extent to which cellular volume can be regulated in response to change in salinity. However, Lange (1972) has commented on the general paucity of data on the influence of other environmental variables on the control of cell volume, and also on the time-course of the processes of isosmotic regulation.

The involvement of 'free amino acids (FAA)'* in intracellular isosmotic regulation by bivalves has now been confirmed for *Mytilus* (Potts, 1958; Lange, 1963; Bricteux-Grégoire *et al.*, 1964), *Rangia* (Allen, 1961), *Crassostrea* (Lynch & Wood, 1966), *Macoma inconspicua* (Emerson, 1969), *Mya arenaria* (Virkar & Webb, 1970) and *Modiolus* (Pierce, 1971a). In all cases, there is an increase in the concentration of total FAA with

---

* The terms 'free amino acids' or 'free ninhydrin-positive-substances' have not been rigidly defined in the literature. Various authors have extracted the tissues either with 80% ethanol, or 5% trichloroacetic acid, or with perchloric acid, and taken the supernatant to include the FAA. At the least, until the differences between 'free' and 'bound' amino acids is more clearly defined, full details of extraction procedures should always be given.

increase in salinity (or a decrease with reduced salinity), although Allen (1961) reported that in *Rangia*, concentrations declined with increase in salinity between 17 and 25‰. Equally, in all cases recorded so far, the individual amino acids do not respond in a simple way to changes in salinity, some increasing in concentration while others decrease or remain constant (Table 6.7). Lange (1963) recorded a linear change in both NPS and taurine in *Mytilus edulis* with change in salinity, but the rate of change was greater for taurine. Virkar & Webb (1970) found that alanine, glycine and, to a lesser extent, taurine and glutamic acid were the most important in reducing the FAA 'pool' at reduced salinities (see also Lynch & Wood, 1966). These observations suggest that the changes in concentration are subject to metabolic control. Campbell & Bishop (1970) argue that the amino acids which are lost are generally those which are metabolic 'end-products', such as taurine, betaine and alanine, whereas the apparent retention of others, such as aspartate and glutamate, may be due to increased biosynthesis. Lange (1964) suggested that taurine loss may have a 'sparing effect' on other, more metabolically active, amino acids.

Pierce (1971*a, b*) demonstrated volume regulation at reduced salinity by loss of solutes in species of *Modiolus*, and identified the active solutes as being, in large part, FAA. In *M. demissus* four compounds were important (taurine, alanine, glycine, and proline) whereas in *M. modiolus* only taurine, alanine and glycine were significantly effective in volume regulation. Pierce also found that the total FAA pool in the subtidal *M. modiolus* was smaller than in the intertidal *M. demissus*. *M. demissus* therefore has more 'expendable intracellular solutes' than the subtidal species, and can maintain its cellular volume over a wider salinity range.

Virkar & Webb (1970), working with *Mya arenaria*, concluded that the NPS reached equilibrium concentration by the second day of exposure to reduced salinity; decline in concentration was linear in the first 24 h, and although a significant part of this early decline was due to increased tissue hydration, this effect was transient. Pierce (1971*a*) showed that valve movements took some days to return to normal after a salinity change (7 days at 3‰, 5 days at 18‰ for *Modiolus demissus*) but volume control in diluted water was complete within 24 h at 27‰ salinity and 48 h at 3‰. Bayne (1975*b*) recorded that although the concentration of free amino-nitrogen in *M. edulis* declined very rapidly at reduced salinity (14.5‰), equilibrium concentrations were achieved within 3 h for taurine, 24 h for glycine and proline and 3 days for alanine; total free amino-nitrogen was at a new steady-state concentration after 2 days. As Lange (1972) points out, the fact of amino acid involvement in the regulation of cell volume by some marine bivalves is no longer in doubt (but see Gilles (1972*b*) on *Glycymeris*). The mechanism of control of the amino acid pool, however, remains to be clarified. Fig. 6.13 depicts the main pathways by which FAA ·

243

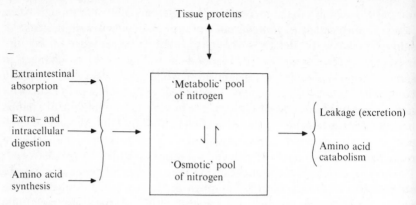

Fig. 6.13. Possible mechanisms for the increase or decrease in the concentration of free amino acids (FAA) in the tissues. (From Lange, 1972.)

may be increased or decreased. Control of FAA might operate at one (or all) of four points; assimilation from the ration or from the medium; the balance between synthesis and catabolism; the equilibrium between 'bound' (= protein) and 'free' amino acids; and by control of excretion.

Assimilation

The possibility that the assimilation efficiency for amino acids may be varied in response to a change in salinity has not, to our knowledge, been examined. However, Anderson & Bedford (1973) and Anderson (1975) demonstrated that the uptake of ^{14}C-labelled glycine from the medium by *Rangia* was suppressed at salinities below 6‰. Uptake by isolated gill tissue was not suppressed at low salinity (2‰), but over 90% of the accumulated glycine was incorporated into alcohol-insoluble, osmotically inactive compounds. These experiments have more relevance to our later discussion (p. 246) but they do support the suggestion of Stephens (1967) that 'the processes that underlie osmotic regulation...[may be]...incompatible with the rapid accumulation of amino acids from the ambient medium'.

Synthesis and catabolism of amino acids

Schoffeniels (1964) and Florkin & Schoffeniels (1969) have suggested that isosmotic intracellular regulation in the Crustacea proceeds primarily by variations in the balance between amino acid synthesis and degradation, brought about by the sensitivity of the enzymes of amino acid metabolism to changes in ionic concentrations in the cell. In particular, the enzyme glutamate dehydrogenase (GDH) is considered to be sensitive to both

244

anions and cations in the incubation medium. Decreased salinity results in a reduction in the ionic concentration of the cell, which reduces the activity of GDH leading to a decline in the intracellular amino acid concentration.

As discussed earlier, GDH activity in mussels appears to be relatively low. Gilles, Hogue & Kearney (1971) demonstrated that increased concentrations of various cations within the physiological range caused a reduction in the activity of succinic dehydrogenase extracted from mantle tissue of *Mytilus californianus*. However, this was a preliminary study only, and much more work is needed to elucidate the ionic effects on enzymes of amino acid metabolism.

Du Paul & Webb (1970) examined changes in the FAA and NPS in the adductor muscle of *Mya arenaria* following a salinity increase from 20 to 30‰. They distinguished three components in the accumulation of FAA; a fast component, active during the first 24 h; a slow component effective after 35 h; and a 'taurine' component that occurred between the fast and slow elements. They suggested that the fast component consists of the release of osmotically active amino acids from an 'inactive' form. These results with *Mya* suggest a complex sequencing of events leading to an increase in the FAA pool.

Some of the evidence for increased catabolism of amino acids during isosmotic regulation at low salinity stems from the observation that the rate of ammonia excretion may increase markedly at this time (Emerson, 1969 (for *Macoma inconspicua*); Allen & Garrett, 1971*b* (for *Mya arenaria*); Bayne, 1975*b* (for *Mytilus edulis*)). Emerson (1969) recorded that the rate of ammonia excretion more than doubled on first exposing *Macoma* to 50% seawater; it then declined to normal over 16 days, and this period coincided with the decline in concentration of NPS in these animals. Bayne (1975*b*) recorded an increase in the rate of ammonia excretion by *Mytilus* from 1.3 to nearly 3.0 μg NH_4-N h^{-1} per g fresh weight when the ambient salinity was reduced from 32.5 to 14.5‰; the excretion rate then declined towards the 'control' value at full salinity after 20 days. In these experiments, however, even the sum of ammonia-nitrogen and amino-nitrogen loss accounted for < 10% of the change in FAA concentration in the mantle during the first day of the salinity change. Other processes leading to reduced FAA were presumed to occur during the first few hours, and one possibility is that free amino acids are 'bound' to osmotically inactive macromolecules.

Equilibrium between 'bound' and 'free' amino acids

Shaw (1958) was the first to suggest that variations in the intracellular amino acid pool might result from an altered equilibrium between FAA and proteins. Bedford (1971) suggested that such a new equilibrium could be established within 5 hours in the gastropod *Melanopsis trifasciata*. Anderson & Bedford (1973) came to a similar conclusion from their work with *Rangia*. At salinities of 5‰ or less, the uptake of glycine by *Rangia* was reduced and the glycine that was accumulated was very rapidly (within 3 h) incorporated into alcohol-insoluble compounds, which were probably protein. As Anderson (1975) indicates, such a response would have adaptive value by enabling the animal to conserve chemical energy while at the same time reducing the osmotic concentration of the cell. Presumably the 'bound' molecules could be made available subsequently, either as osmotic effectors or for energy metabolism. But the time-course of such a process needs to be studied further, especially in view of Pierce's (1971*a*) conclusion that solutes, once 'lost' to the intracellular pool cannot readily be replaced.

Excretion of amino acids

The loss of amino acids to the medium by bivalves is now well-documented (p. 231). A certain loss appears to be a normal component of nitrogenous excretion, but during exposure to reduced salinity this loss is considerably increased. Bayne (1975*b*) recorded rates of loss of amino-nitrogen (from *M. edulis*) which increased from 0.42 μg amino-N h^{-1} per g wet weight at 32.5‰, to 2.95 μg amino-N h^{-1} per g wet weight within 3 h of transfer to water at 14.5‰. Specifically, taurine, glutamate and glycine showed marked increases in the rate of loss. During the 20 days following dilution of the medium the rates of loss of amino-nitrogen at reduced salinity declined. Anderson & Bedford (1973) found that *Rangia* which had been labelled with [^{14}C]glycine released labelled material (unidentified) into the medium when exposed to various dilutions of seawater. Interestingly, this loss of radioactivity occurred rapidly from animals subjected to slight reductions in salinity, but more slowly from animals exposed to extreme dilutions (2 and 4‰ salinity). The possibility of 'binding', followed by gradual release, is apparent.

Pierce & Greenberg (1972) reported experiments on isolated ventricles from *Modiolus* exposed to dilutions of the medium. The ventricles were capable of cell volume regulation, and taurine, alanine, glycine and proline were involved in amino acid efflux. Pierce & Greenberg (1973) then examined the effects of particular ions on this efflux and concluded that salinity-induced free amino acid regulation is initiated by a decrease in external osmotic pressure, but that the time-course of the efflux is dependent on external divalent ion concentration. They propose that the cell membrane contains specific sites permeable to amino acids. The sites exist in either an open (Sp) or closed (Sn) conformation and are normally in steady-state equilibrium. In the open conformation, amino acid efflux is

246

possible; in the closed conformation it is not. The Sp conformation is favoured by any decline in extracellular osmotic pressure ($\Delta\pi$), whereas Sn is maintained by a cellular process (unspecified) that requires Ca^{2+} and Mg^{2+}:

$$Sn \xrightarrow[-\Delta\pi]{[Ca^{2+}+Mg^{2+}]} Sp.$$

Normally, some sites are open, explaining the slight efflux of amino acids that occurs in full-strength seawater. Pierce & Greenberg (1973) indicate some of the cellular mechanisms that may be involved in the divalent ion control, and they suggest that lowered external osmotic pressure operates on the membrane via an increase in turgor pressure as water enters the cell.

Taurine and betaine

The importance of taurine in isosmotic regulation in bivalves is often referred to (Campbell & Bishop, 1970; Lange, 1972; Florkin & Schoffeniels, 1965). *Mytilus* is rich in taurine, and its metabolism in this genus has been studied by Allen & Awapara (1960), Yoneda (1967, 1968), Girard, Huguet & Solère (1969) and reviewed by Allen & Garrett (1971a). Allen & Awapara (1960) observed that whereas both *Rangia* and *Mytilus* are capable of the synthesis of taurine, *Rangia* loses the taurine as rapidly as it is formed, whereas *Mytilus* retains the taurine against a large concentration gradient. This, and other information reviewed by Allen & Garrett (1971a), has led to the view that taurine is a metabolic end-product which is utilised by some species, but not others, in isosmotic regulation. However, Allen & Garrett (1971a) suggest that taurine may have other functions, in particular in the formation of sulphonated polysaccharides, which play a large role in the physiology of bivalves. Taurine clearly warrants further study in mussels.

Equally, the physiological role of betaine merits further investigation. Bricteux-Grégoire *et al.* (1964) found high concentrations in the adductor muscle of *M. edulis*, higher even than taurine (Table 6.7), and they suggested its involvement in isosmotic regulation. Apart from this account, and the suggestion by Florkin & Bricteux-Grégoire (1972) that it may be involved as a 'methyl donor' in the synthesis of methionine, nothing is known of the physiological role of betaine in bivalves.

Chemical control

The co-ordination and integration of the functions of the body are performed by the nervous system and by hormones. The nervous and hormonal systems in bivalves are closely inter-related both structurally and functionally, and many neurones play a dual role in the propagation of nerve impulses and in the manufacture and discharge of secretions. There are two types of secretory activity. In the first, substances known as neurohumours, such as acetylcholine (ACh) and 5-hydroxytryptamine (5-HT), are synthesised and act 'locally'. In the second, synthesis of neurohormones occurs in the cell body and the substances pass down the axon to nerve endings for storage or release into the haemolymph; these neurohormones act 'at a distance' from the site of synthesis. Two areas of research into chemical control in bivalves have been particularly active; the neurosecretory activity in nerve ganglia, and the role of neurohumours in the excitation and inhibition of gill cilia.

Neurosecretion

Neurosecretion in the mollusca has been reviewed by Welsh (1961), Gabe (1966) and Martoja (1972). Development of the subject has been hampered by the presence of the shell, which makes microsurgery difficult, by the diffuse distribution of the neurosecretory cells, and by ignorance of the chemical nature of the neurohormones. Umiji (1969) has suggested that a neurohaemal area exists on the cerebral commissure of *Perna perna*, but the major neurosecretory function in bivalves is centred on certain cells in the nerve ganglia. Gabe (1955), Lubet (1955*a*, 1956, 1957, 1959, 1965, 1966) and Umiji (1969) have demonstrated such neurosecretory cells in the cerebral and visceral ganglia of *Mytilus* species; the number and exact location of these cells differ from species to species. The presence of neurosecretory cells in the pedal ganglia is controversial, however. Gabe (1955) and Lubet (1955*a*) conclude that they are absent in *M. edulis* or *M. galloprovincialis*, but Umiji (1969) has recorded them as present on the pedal ganglia of *Perna perna*.

Two types of neurosecretory cell have been identified in *Mytilus*. Some cells are pear-shaped, unipolar and up to 25 μm in length; others are small and multipolar (Lubet, 1956). The transport, and the fate of neurosecretory products is largely unknown. Lubet (1956) and Umiji (1969) have reported the movement of neurosecretory substances by axons, intermediate cells and possibly by glial cells.

Neurosecretion in lamellibranchs has been associated with reproduction, growth, metabolism and responses to environmental stress.

Reproduction

Investigations by Lubet (1956, 1959) demonstrated an annual neurosecretory cycle in the pear-shaped neurosecretory cells of the cerebral ganglia in *M. edulis* and in *M. galloprovincialis*. The small multipolar neurones, and the neurosecretory cells of the visceral ganglion showed continuous activity throughout the year. These observations were confirmed for other bivalves by Nagabhushanam (1963, 1964).

The annual neurosecretory cycle and the gametogenic cycle appear to be closely correlated. Secretory material is accumulated in the cerebral ganglia during gametogenesis, and then evacuated from the cells when the gametes become fully mature (Lubet, 1956, 1959). Removal (ablation) of the cerebral ganglia during the resting phase of the gametogenic cycle (Chapter 2) and at the beginning of active gametogenesis delays maturation in *M. edulis*, with many oocytes undergoing lysis before spawning (Lubet, 1965). Ablation of the cerebral ganglion at the end of gametogenesis, but before spawning, hastens maturation and gamete release. However, Lubet remarks that it is impossible to decide if the mechanism is nervous or hormonal in nature. If the activities of the neurohormones are important, removal of an internal inhibition, such as the neurosecretory products of the cerebral ganglia, may allow the animal to become receptive to external stimuli which then induce the release of the gametes. However, Antheunisse (1963), who worked with the fresh-water mussel *Dreissena polymorpha*, has suggested that premature spawning may be a function of the intensity of the operative shock when ablating the ganglia. Antheunisse concluded that, in spite of a parallelism between the neurosecretory and reproductive cycles in this species, a direct causative relationship did not operate.

Lubet (1956) and Lubet & Choquet (1971) have associated a 'hormone', which is linked with the spermatozoa, with spawning in *M. edulis*. Galtsoff (1938*a*, *b*, 1940) and Nelson & Allison (1940) had earlier demonstrated the presence of a hormone active in inducing spawning in *Crassostrea virginica*.

Growth and metabolism

There is a close parallelism between a seasonal cycle of neurosecretory activity and the rate of oxygen consumption by *Mytilus edulis* but, as with spawning and the gametogenic cycle, whether there is a causative relationship is in doubt. Removal of the cerebral ganglion of *M. edulis* (Lubet, 1966) or *Perna perna* (Umiji, 1969) has little or no effect on shell or body growth, or on glycogen metabolism and storage. However, ablation of the cerebral ganglion does result in disorders of lipid metabolism, particularly a reduction in lipid accumulation (Lubet, 1965). The ablation of the visceral ganglion has more severe consequences, and the mussels do not survive for more than 3 or 4 months (Lubet, 1966; Umiji, 1969). Some of these points are discussed further in Chapter 8, and by Gabbott (1975).

Environmental stress

Lubet & Pujol (1965) demonstrated the effects of salinity and temperature change on neurosecretory activity within the cerebral ganglia of *Mytilus galloprovincialis*. A sudden rise in temperature (10 deg C), maintained for 1 h, or a sudden reduction in salinity to 20‰ resulted in an emptying of the neurosecretory cells of the cerebral ganglia. A fall in temperature, or an increase in salinity (to 45‰) were followed by an accumulation of neurosecretory products. There was no variation in neurosecretory activity in the visceral ganglia following similar experimental treatments (Lubet & Pujol, 1963, 1965). Ablation of the cerebral ganglia impaired the mussels' capacity for isosmotic intracellular regulation at reduced salinity resulting in an increase in amino-nitrogen content in the tissues.

These experiments do suggest that neurohormones play a significant role in the physiology of mussels, but the limited range of techniques available (ablation of whole ganglia, indirect correlation of neurosecretory activity with other physiological events) has constrained real progress in understanding the processes involved. Further insight will probably depend on knowledge, lacking so far, of the chemical and pharmacological nature of the neurohormones themselves.

Neurohumours

Control of ciliary activity

The ciliated gill epithelium of lamellibranchs provides a favourable system for investigating the mechanisms and the control of ciliary motion and a considerable literature has developed (most recently reviewed in a volume edited by Sleigh, 1974) on the excitation and inhibition of gill ciliary activity in *Mytilus*.

Aiello (1957) demonstrated that ciliary activity on the gill of *M. edulis* was affected by stimulation of the branchial nerve. Later, the single branchial nerve was traced from the visceral ganglion to the ciliated epithelium, where it branched to small groups of adjacent filaments (Aiello & Guideri, 1965). More recently, Paparo (1972) has shown that fibres from the branchial nerve lie adjacent to the gill filaments and penetrate the fibrous basal lamina under the epithelium.

The pattern of ciliary activation following electrical stimulation of the visceral ganglion and branchial nerve indicates that individual filaments, or small groups of adjacent filaments, are independently innervated, allowing for discrete control of ciliary activity on different parts of the gill (Aiello & Guideri, 1965).

Aiello & Guideri (1966) showed that electrical stimulation of the branchial nerve in an isolated ganglion/nerve/gill preparation activated quiescent cilia and increased the beating of active cilia. They suggested that cilio-excitation was mediated by 5-hydroxytryptamine (Aiello, 1957, 1962, 1965, 1970; Gosselin, 1961; Gosselin, Moore & Milton, 1962). Other studies have revealed measurable quantities of 5-HT in the gill of *Mytilus* (Paparo, 1972) and also the enzymes of 5-HT synthesis (Milton & Gosselin, 1960) and degradation (Blaschko & Milton, 1960). 5-HT has no effect on the musculature of the gill (Aiello, 1970).

The gill cilia of bivalve molluscs may continue to beat for several days after excision from the animal (Gray, 1928; Usuki, 1962*a*), which suggests that an energy-rich substrate is stored endogenously within the cells. Moore *et al.* (1961), Moore & Gosselin (1962) and Usuki (1962*b*) have shown this substrate to be glycogen. The endogenous formation of ATP from glycogen may be linked to an anaerobic pathway or to oxidative phosphorylation (Chapter 8). 5-HT has been shown to stimulate both anaerobic glycolysis and the rate of oxygen consumption by gills of *M. edulis* and *Modiolus demissus.* (Moore *et al.*, 1961; Moore & Gosselin, 1962). Earlier studies by Gray (1928) and Aiello (1960) had led to the conclusion that oxygen was required for prolonged ciliary beating. However, Malanga & Aiello (1971) used a bicarbonate/carbonic acid buffer system to prevent organic acids, which might accumulate as anaerobic end-products, from lowering the pH, and they showed that lateral cilia of *Modiolus* and *Mytilus* are capable of prolonged activity without oxygen, with the addition of micromolar concentrations of 5-HT. Anaerobic cilio-excitation can be blocked by inhibitors of glycolysis, indicating that 5-HT may exert its effect through a rate-limiting step in glycolysis. Malanga & Aiello (1971) suggest that 5-HT may activate the enzyme phosphoglucomutase.

Spontaneous and momentary cessation of ciliary beat on the gill of *M. edulis* has been recorded by Lucus (1931*a*, *b*), Aiello & Guideri (1965) and Dral (1967). Takahashi & Murakami (1968) and Aiello & Paparo (1968) demonstrated that stimulation of the branchial nerve may result either in excitation (10 pulses s^{-1}) or inhibition (25 or 50 pulses s^{-1}) of the lateral cilia of *M. edulis*. The existence of cilio-inhibitory fibres suggested that activity and inhibition were mediated by different neurohumours.

Acetylcholine is a widely distributed inhibitory neurotransmitter in invertebrates, but it fails to stop ciliary beat when applied to the gill of *M. edulis* (Takahashi, 1971). Paparo & Aiello (1970) suggested that the catecholamine, dopamine, was the natural inhibitory neurotransmitter, especially since Sweeney (1963) had earlier found dopamine in the nerve ganglia of *M. edulis* and other bivalves. More recently, Paparo & Finch (1972) localised catecholamines in the branchial nerve within the gill epithelium of *M. edulis*; they have also shown that the gill has the capacity to metabolise dopamine. Gosselin (1966) reported that other catechol-

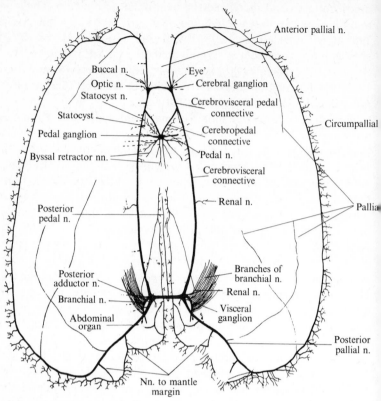

Fig. 6.14. A schematic representation of the nervous system in *Mytilus galloprovincialis*. n, nerve; nn., nerves. (From List, 1902.)

amines, such as norepinephrine or epinephrine had no significant effect on ciliary beating in *M. edulis*.

These experiments all strongly support the argument (Paparo & Aiello, 1970) that 5-HT and dopamine are the two neurohumours, for cilio-excitation and cilio-inhibition respectively, in mussels. Further work is now needed to show how this control is integrated into the regulation of ventilation and filtration in the life of the individual mussel.

Nervous system and receptors

The nervous systems of several mytilid species have been described by Purdie (1887), List (1902), Field (1922), Clasing (1923) and White (1937). List's monograph is the most detailed and probably the most useful of these accounts, which agree closely with one another apart from details attributable to minor differences between species. The following descrip-

252

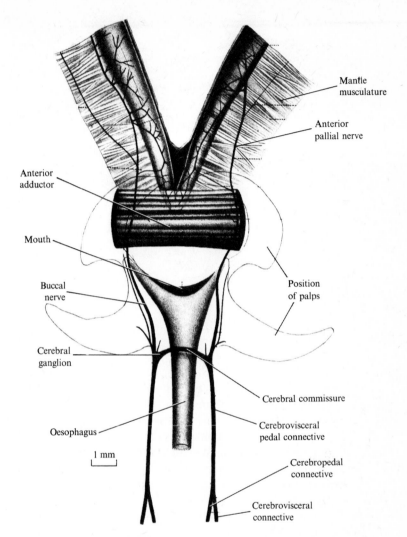

Fig. 6.15. The nervous system in the oral region of *Mytilus galloprovincialis.* (From List, 1902.)

tion of the mytilid nervous system is based on these five publications and on the more recent treatise of Bullock & Horridge (1965), which reviews nervous systems in the invertebrates.

Most bivalves, including the Mytilidae, possess a nervous system comprising three pairs of ganglia joined by paired connectives (Fig. 6.14). The cerebral, pleural and buccal ganglia of the primitive molluscan system are fused into a single structure, the cerebro-pleuro-buccal ganglion, for

253

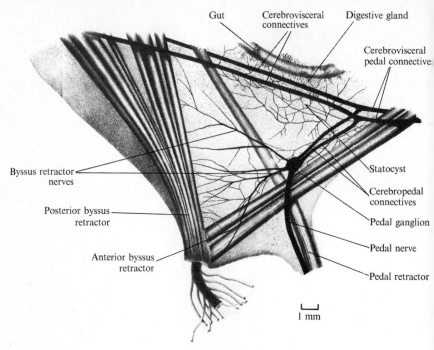

Fig. 6.16. The pedal ganglia in *Mytilus galloprovincialis*. (From List, 1902.)

convenience often termed the cerebral or cerebropleural ganglion. The buccal ganglion is, however, sometimes visible as a discrete structure (e.g. in *Musculus*). In *Mytilus* the buccal ganglion is not distinct, but a buccal commissure ventral to the oesophagus is recognisable, as in *Musculus*. The visceral, intestinal and parietal ganglia are represented by a pair of visceroparietal ganglia, commonly referred to as the visceral ganglia. The paired pedal ganglia remain discrete and have not migrated to fuse with other elements of the nervous system.

Whereas there is considerable species variation in the distance between the two visceral ganglia, and hence in the length of the visceral commissures, the cerebral ganglia are always well-separated and are joined by a commissure which passes over the oesophagus (Fig. 6.15). In contrast, the pedal ganglia are fused in the midline, although the right and left components are distinguishable (Fig. 6.16).

The cerebral and visceral ganglia are linked by the cerebrovisceral connectives, and the pedal and cerebral ganglia by the cerebropedal connectives. In some species the latter arise directly from the cerebral ganglia but in others they separate from cerebrovisceral pedal connectives posterior to the cerebral ganglia (Fig. 6.14).

Most of the anterior region is supplied from the cerebral ganglia via:

(1) anterior pallial nerves, which form anastomoses with branches of the posterior pallial nerves to form circumpallial nerves;

(2) nerves to the anterior adductor muscle;

(3) branches of the buccal nerves to supply the labial palps (innervation of the palps is described for *Mytilus edulis* by Welsch & Storch, 1969);

(4) statocyst nerves;

(5) optic nerves to the larval eyes, which persist in adult mytilids;

(6) a plexus overlying the oesophagus and stomach (discussed by Clasing, 1923).

The pedal ganglia give rise to:

(1) pedal nerves, which penetrate the posterior pedal retractor muscles and enter the foot;

(2) dorsal retractor byssus nerves, supplying part of the posterior retractor byssus musculature;

(3) ventral retractor byssus nerves, which supply the byssus organ and parts of the anterior and posterior retractor byssus muscles.

The posterior region and most of the visceral mass are supplied from the visceral ganglia by means of:

(1) posterior pallial nerves, which have dorsal and ventral branches;

(2) siphonal nerves, arising from the posterior ventral pallial nerves and supplying the exhalant siphon;

(3) dorsal pallial nerves to the edge of the mantle;

(4) posterior renal nerves (the kidney is also innervated by branches of the cerebrovisceral connectives);

(5) posterior pedal nerves;

(6) nerves to the posterior adductor muscle;

(7) nerves supplying the digestive gland and the gonad;

(8) nerves reaching the heart via cardiac ganglia (Carlson, 1905, for *Mytilus californianus*);

(9) osphradial nerves, arising from the branchial nerves and the visceral ganglion (Lucas, 1931*a*, *b*, for *Mytilus edulis*);

(10) branchial nerves, supplying the musculature and epithelium of the gills. Lucas (1931*a*, *b*) described the course of the branchial nerves in *M. edulis* but could not trace any branches into the gill filaments. However, it has subsequently been demonstrated that fibres from the branchial nerves penetrate the basal lamina of the gill epithelium and permit local control of ciliary activity (Aiello & Guideri, 1965; Paparo, 1972; see p. 250).

The histology of the ganglia has been described by Rawitz (1887, 1890), List (1902) and White (1937). Rawitz's accounts are the most detailed, but have been criticised on several grounds, particularly the use of non-specific staining techniques (Dakin, 1910; Bullock & Horridge, 1965).

Ganglion extirpation studies using *Mytilus edulis* (Woortmann, 1926) demonstrated that foot retraction is controlled almost entirely by reflex arcs in the pedal ganglia, although in other species there appears to be cerebral involvement. The transection of the cerebropedal connectives does not inhibit byssus secretion but the placement of byssal threads is probably under cerebral control. The visceral ganglia control the heart beat and movements of the shell, mantle, siphons and gills. The literature on ganglion function in bivalves is reviewed by Bullock & Horridge (1965).

Whereas the mytilid nervous system has been described in detail, there is much less information available concerning receptor structure and function. Some histology has been done (List, 1902; Field, 1922; Clasing, 1923) but illustrations are poor. Laverack (1968) pointed out that despite the existence of a large literature on animal receptors, comparatively little is known about their characteristics in marine invertebrates. Some of the molluscan sensory organs (e.g. the osphradium and the statocyst) have been studied in gastropods or cephalopods but not in bivalves. This is especially true when one considers physiological work. Most studies of bivalve receptors have concerned dermal photoreception, mechanoreceptors on the mantle and siphons, and the complex eyes found in some genera (e.g. *Pecten, Cerastoderma, Lima*). None of these accounts feature mytilid species. For our present purposes it will therefore be necessary to rely extensively on information obtained from other bivalves and from marine gastropods, but receptors which obviously have no relevance to the Mytilidae (e.g. pallial eyes) will not be discussed.

Tactile receptors. Horridge (1958) noted that the mantle edge and siphon tip of *Mya arenaria* are sensitive to touch, a slight stimulus eliciting a local contraction of the musculature but a strong stimulus producing a co-ordinated retraction of the entire animal. Similar observations were made by Prior (1972*a, b*) on the surf clam, *Spisula solidissima*. In *Spisula* the local contractions produced in response to weak stimulation are not mediated by the central nervous system but result from reflexes in the siphon wall, whereas the contraction of the siphon retractors which results from strong stimulation is under central control. The local reflex is mediated by nerve cell bodies located at the peripheral branching points of the siphonal nerves; these cells are efferents to the siphon wall musculature and receive synaptic input from touch-sensitive afferents originating in the siphons and mantle. The central-mediated response involves touch-sensitive neurones whose somata lie in the visceral ganglion (Mellon, 1965, 1972). Most of the sensory neurones possess discrete, single receptive fields on the inner walls of the siphons, although some have multiple fields. The cells have morphologically unmodified end organs which may be exposed terminal axonal arborisations, possibly stimulated directly. Alternatively, these touch-sensitive nerve cells may be second-order neurones excited by touch-sensitive epithelial cells.

Prior (1972*b*) considers that the behavioural discrimination between weak and strong tactile stimuli is adaptive – complete siphon retraction or valve adduction in response to minor disturbances could severely restrict essential processes of feeding and gas exchange. This suggestion is consistent with observations on *Mytilus edulis*, which often partially withdraws the mantle for a very brief period in response to impingement of large particles on the mantle margin.

Eyes. In the Mytilidae, the larval eyes are retained in the adult. Each takes the form of an invaginated eyecup (Fig. 6.17) lined with pigmented epithelial cells and filled with a mucoid substance often termed the 'lens'. The eye is innervated from the cerebral ganglion and is presumably functional, but it has not been a subject of electrophysiological studies.

Dermal light sense. Millott (1968) defined the dermal light response as a sensitivity to electromagnetic radiation of wavelengths 390–760 nm not mediated by eyespots or eyes and in which light does not act directly on the effector. Dermal photoreception may function to detect the photoperiod or sudden changes in light intensity, e.g. due to shadows cast by approaching predators. Responses of the latter type are common in bivalves, including *Mytilus*. Hecht (1934) used the response to light by the siphons of *Mya arenaria* to elucidate some fundamental principles of photoreceptor physiology. He also discovered that *Mya* shows light and dark adaptation; after a period of darkness, the clam responds strongly to light by retracting the siphon; continued exposure to light leads to expansion of the siphon. Further responses are then elicited only by very high light intensities.

Some electrophysiological studies have been made of dermal photoreceptors in bivalves. Kennedy (1960) recorded responses from such a receptor in the pallial nerves of *Spisula solidissima*. Each nerve contains a single afferent fibre which responds directly to light intensity, apparently mediating the 'shadow response' of siphon retraction. During darkness, spontaneous activity of constant frequency occurs in this afferent, but the activity is inhibited by light. The onset of darkness results in a prolonged burst of impulses at high frequency – the so-called 'off' response. There is no structural specialisation in the photosensitive neurone, and it is not known whether the latter is first- or second-order. Wiederhold, MacNichol & Bell (1973) recorded activity in single axons of the siphonal nerve of *Mercenaria mercenaria* and found a response to reduced illumination only, i.e. an 'off' response.

The photoreceptor sites in the *Mercenaria* siphon are physically and functionally isolated from one another; the effective region over which a stimulus triggers a single axon is about 85 μm in diameter. Such sharp localisation implies that the receptors are superficial and may be the

257

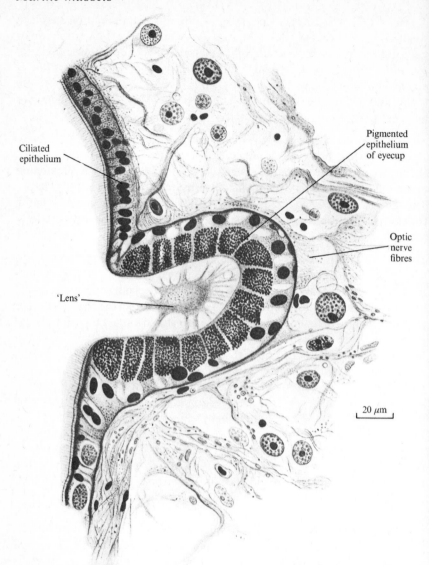

Ciliated
epithelium

Pigmented
epithelium
of eyecup

Optic
nerve
fibres

'Lens'

20 μm

Fig. 6.17. The eye of *Mytilus galloprovincialis*. (From List, 1902.)

neurones themselves. There are no photoreceptor cells distal to the axons, so the neurones appear to be first-order in this case.

There is some doubt regarding the nature and number of the photosensitive pigments present. *Mercenaria* contains a rhodopsin-type pigment with a maximum sensitivity at 510 nm. According to Kennedy (1960), more than one pigment may be involved in photoreception by *Spisula*. Mpitsos (1973) obtained only an 'off' response in the dermal photoreceptors of *Lima scabra*, and found some evidence for the involvement of two separate pigments.

Osphradia. The paired osphradia appear to be patches of specialised epithelium, richly innervated, which lie ventral and lateral to the visceral ganglia. Descriptions of the mytilid osphradium (List, 1902; Field, 1922) are poor. The structure is represented as a single layer of non-ciliated cells. Electron microscopy of the prosobranch osphradium (Crisp, 1973) has revealed a more complex organisation in which there are two discrete regions of epithelium; one a secretory area, one cell thick, and another so-called sensory area containing several cell types in a layer five to seven cells thick. Unfortunately, there is no comparable information for the bivalve osphradium, which is not necessarily homologous with the prosobranch organ.

The modality to which the osphradium responds has yet to be established unequivocally in any mollusc. Bailey & Laverack (1966) applied a variety of discrete chemical and mechanical stimuli to the osphradium of *Buccinum undatum* and recorded activity in the osphradial nerve. Responses were elicited by many of these chemical stimuli, but not by mechanical ones. It appears that the osphradium of the whelk is a chemoreceptor, probably playing a role in prey detection. Yonge (1947) noted the presence of osphradia in a wide variety of gastropods, including predators, filter-feeders and grazing herbivores, and suggested that the organs were mechanoreceptors, sensing the quantity of sediment in the water current. Bailey & Benjamin (1968) recorded from the osphradial ganglion cells of a herbivore, *Planorbis*, which has a small osphradium. No response was obtained either to mechanical or to chemical stimulation. Other modalities, e.g. pH, temperature and ions, also failed to stimulate the osphradium, implying that it may be an effector, possibly secretory. This view is consistent with Crisp's (1973) observations of a secretory region in the prosobranch osphradium. Stinnakre & Tauc (1969) obtained evidence for osmoreception by the osphradium of *Aplysia*. One particular neurone in the visceral ganglion is especially responsive to osmotic changes in the medium surrounding the osphradium. This neurone is easily identified and does not respond to stimuli of other modalities.

Evidence from electrophysiological experiments therefore tends to support the conclusion that osphradia from different gastropods are not necessarily homologous. There seems to be no information regarding the physiology of the bivalve osphradium.

Abdominal sense organs. These paired structures lie on the ventral posterior side of the posterior adductor muscle. Each consists of a ciliated

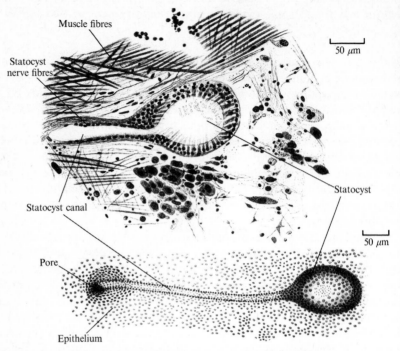

Fig. 6.18. The statocyst of *Mytilus galloprovincialis*. (From List, 1902.)

epithelium several cells thick, innervated from the posterior pallial nerve. Nothing is known of their function.

Statocysts. A pair of statocysts (incorrectly termed otocysts in much of the literature) lie near the pedal ganglia, beneath the epithelium between the cerebropedal and cerebrovisceral connectives. They are innervated from the cerebral ganglia. In the Mytilidae, the connection between the statocyst and the exterior is retained by means of the statocyst canal (often termed the excretory canal), which terminates in an invagination of the body epithelium (Fig. 6.18). According to Barber (1968) there are no studies on the physiology of statocysts in molluscs other than cephalopods, apart from a few extirpation experiments showing statocyst function in orientation (e.g. *Pecten*).

7. Physiological integrations

B. L. BAYNE, J. WIDDOWS & R. J. THOMPSON

In Chapters 5 and 6 we discussed the different physiological systems of mussels with the aim of describing the scope of their adaptive responses to environmental change. In this chapter our objective is to bring together some of this information in an integrated way (in the sense of the combination of parts into a whole) in order to try to gain more insight into the physiology of the individual. Integration of this kind has been a traditional tool of the physiologist. For example, the calculation of the respiratory quotient (RQ) has for long been a means of bringing together data on respiratory gas exchange to provide increased understanding of the underlying metabolic processes. With a more ecological orientation, the 'basic energy equation' of Winberg (1956) is an attempt to further understanding of the energy relationships of organisms by summation of the component parts of the 'energy budget'.

Integrations of this kind have an additional importance in ecology, for they suggest quantitative indices which may be used to describe, and possibly to predict, the physiological 'condition' of the whole organism as it responds to changes in its environment. In this chapter we shall discuss a few such integrations which promise to provide some useful ecological understanding.

Simple integrations

Gas exchange

Mussels effect the uptake of oxygen by ventilating water over their gill filaments, which are simultaneously perfused with blood. The arrangement of these flows of blood and water is probably similar in many respects to the multicapillary arrangement (Fig. 7.1) proposed by Piiper & Schumann (1967) for elasmobranch fishes and discussed earlier by Bartels & Moll (1964) in the context of exchange in the human placenta. Some general properties of gas exchange under these conditions are:

(1) The volume of blood in a single gill filament is small in contrast with the ventilation volume, so that each filament is surrounded by a volume of water which is virtually homogeneous as regards its partial pressure of oxygen.

(2) It is therefore immaterial whether the flows of blood and water in and around the individual filament are con-current or counter-current.

(3) The partial pressure of oxygen in the water is gradually reduced as

261

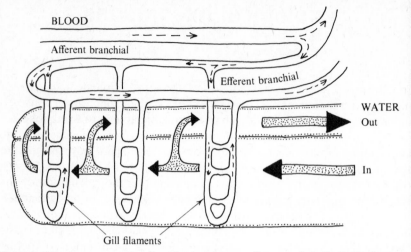

Fig. 7.1. Diagram representing the blood and water flows at the gill of *Mytilus* as a 'multicapillary' arrangement (see text). Possibilities exist for both blood and water to be shunted past individual gill filaments without participating in gas exchange.

the ventilation current passes anteriorly from the posterior edge of the gill; different filaments are therefore surrounded by water of different oxygen tensions, although these differences will be small under normal conditions, due to the low utilisation efficiency (see later).

The degree of contact between water and blood may be varied by shunting some of the blood flow away from the gills (Fig. 7.1), or by failing to ventilate all of the gill filaments, or both. Mechanisms for both of these options have been described for bivalves. Dral (1967) and Foster-Smith (1974) have demonstrated how the surface area of the gill which is exposed to the ventilation current may be varied (discussed in Chapter 5). The lack of a complete branchial circulation (see Chapter 6) permits the shunting of blood away from the respiratory surfaces. On the other hand, Aiello & Guideri (1965) describe how contraction and relaxation of individual filaments may aid branchial circulation which, under normal circumstances, may be rather sluggish. The rate of perfusion of the gills may be varied by regulation of heart output. It is evident, therefore, that a variety of responses is available to the mussel in order to regulate gas exchange at the gill.

Unfortunately, a thorough quantitative analysis of gas exchange in bivalves is limited by the difficulties of measuring directly the various parameters that are known to be important in effecting oxygen uptake in water (Hughes & Shelton, 1962; Rahn, 1966; Dejours, Garey & Rahn, 1970; Randall, 1970; Rahn, Wangensteen & Farhi, 1971; Dejours, 1972). For

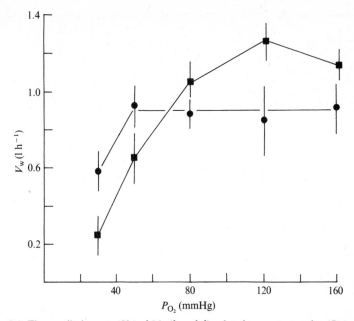

Fig. 7.2. The ventilation rate (V_w) of *Mytilus edulis* related to oxygen tension (P_{O_2}). ■, animals 'regulating' their rates of oxygen consumption; ●, animals 'conforming' in their rates of oxygen consumption. Values are means ± 1 s.e. for between five and eight measurements.

example, none of the parameters of perfusion, such as the stroke volume of the heart, or the oxygen tension of the blood, have been measured. Nevertheless, by determining only the rate of oxygen consumption, the ventilation rate and the rate of heart beat, some understanding of gas exchange in these animals is possible.

When *Mytilus edulis* is 'regulating' its rate of oxygen consumption during environmental hypoxia (see p. 195), ventilation rate (or V_w)* increases slightly between oxygen tensions (P_{O_2}) 160 and 120 mm Hg, and then declines sharply (Fig. 7.2). When the rate of oxygen consumption 'conforms' to changes in P_{O_2}, however (p. 195), V_w, which is initially somewhat low, remains constant to low values of P_{O_2} (Fig. 7.2). Knowing the rate of oxygen consumption (V_{O_2}) and the ventilation rate (V_w), the oxygen utilisation efficiency (E_w) may be calculated as:

$$\left[\frac{Ci_{O_2} - Co_{O_2}}{Co_{O_2}} \right] \cdot 100,$$

* In this discussion, 'ventilation rate', or the flow of water across the gills, has been calculated from the clearance rate of particles in suspension, where the retention efficiency for these particles is known to exceed 85%. The assumption is made that particle retention is not dependent on the oxygen tension. As with so many other aspects of the analysis of gas exchange in bivalves, more direct methods of measurement are needed.

263

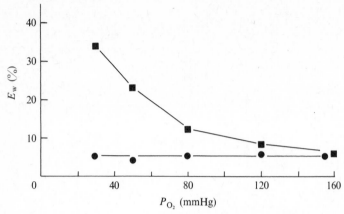

Fig. 7.3. The oxygen utilisation efficiency (E_w) by *Mytilus edulis* 'regulating' (■) and 'conforming' (●) their rates of oxygen consumption at reduced oxygen tension (P_{O_2}).

where Ci_{O_2} and Co_{O_2} are the concentrations of oxygen in the inhalant and exhalant water (or the water before and after passing through the mantle cavity), respectively. E_w increases at reduced P_{O_2} during the regulation of the rate of oxygen uptake, but remains unchanged at a low value during conformity of V_{O_2} (Fig. 7.3). Low values for E_w are typical for bivalves respiring at 160 mm Hg P_{O_2} (Table 7.1).

The term:

$$\left[\frac{V_w}{V_{O_2}} \right] \cdot E_w \cdot Ci_{O_2} = 1,$$

is a fundamental expression of gas exchange in water (Dejours, 1972). When Ci_{O_2} is reduced, for example during hypoxia, either V_w/V_{O_2} (the 'convection requirement'), or E_w, or both, must increase. Under normal, 'unstressed' conditions, *Mytilus* increases utilisation efficiency, while holding the convection requirement constant, or nearly so (Fig. 7.4). Under conditions of starvation, however, E_w remains constant and the convection requirement increases.

The curves for convection requirement reproduced in Fig. 7.4 probably represent extremes for the general response of *M. edulis* to hypoxia. Bayne & Livingstone (unpublished data) recorded acclimation by *M. edulis* to low oxygen tensions (p. 199); this was accompanied by a reduction in the convection requirement and an increase in utilisation efficiency (Fig. 7.5a). They also observed that the ability to regulate V_{O_2} during declining P_{O_2} was reduced at high temperature (22 °C), and this was correlated with an increase in the convection requirement and a reduced utilisation efficiency (Fig. 7.5b).

Until the perfusion characteristics of gas exchange in bivalves can be

Table 7.1. *The efficiency with which various bivalve species utilise oxygen in the medium at air saturation (160 mm Hg P_{O_2})*

Species	Utilisation efficiency for oxygen (%)	Authority
Mya arenaria	3–10	Van Dam (1938)
Cardium tuberculum	6–10	Hazelhoff (1938)
Pinna nobilis	3–8	Hazelhoff (1938)
Solen siliqua	7–12	Hazelhoff (1938)
Pecten grandis	0.5–9	Van Dam (1954)
Pecten irradians	1–13	Van Dam (1954)
Mytilus edulis	3–15	Rotthauwe (1958)
Mytilus edulis	5–10	Bayne (1971*b*)
Arctica islandica	7–12	Taylor (1974)
Mytilus californianus	3–10	Bayne *et al.* (1975*a*)
Anadara ovalis	5 (±1.0 s.e.)	Mangum & Burnett (1975)
Modiolus demissus	6.3 (±4.0 s.e.)	Mangum & Burnett (1975)
Mytilus edulis	8.5 (±1.0 s.e.)	Mangum & Burnett (1975)

The experimental conditions and the methods of measurement vary from authority to authority, but all values are of a similar order.

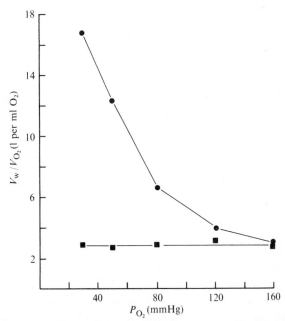

Fig. 7.4. The 'convection requirement' (ventilation rate (V_w)/rate of oxygen consumption (V_{O_2}); l per ml O_2) of *Mytilus edulis* while 'regulating' (■) and while 'conforming' (●) their rates of oxygen consumption at reduced oxygen tension (P_{O_2}).

Marine mussels

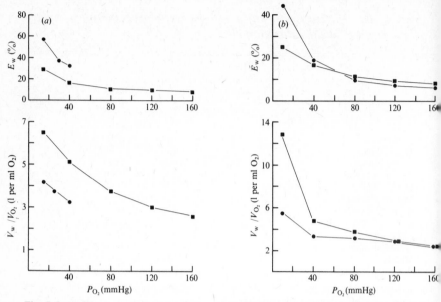

Fig. 7.5. (a) The 'convection requirement' (V_w/V_{O_2}) and the oxygen utilisation efficiency (E_w) of *Mytilus edulis* acclimated to 160 (■) and to 40 (●) mm Hg P_{O_2}. (b) The 'convection requirement' (V_w/V_{O_2}) and oxygen utilisation efficiency (E_w) of *Mytilus edulis* exposed to water at reduced oxygen tension (P_{O_2}) at 10 (●) and 22 (■) °C.

measured, the mechanisms that bring about increased extraction efficiency during the regulation of V_{O_2} cannot be thoroughly analysed (Mangum & Burnett, 1975). Bayne (1971*b*) suggested that an index of perfusion may be derived from the product of the heart beat frequency and apparent amplitude, the latter as estimated from recordings of the heart trace by impedance pneumography (p. 214). At best this is a very indirect indication of cardiac output, since its relationship with stroke volume is unknown. Nevertheless, when used in a ratio with V_w, this treatment suggested one mechanism by which utilisation efficiency may be increased during hypoxia (Fig. 7.6) and during increases in temperature (Widdows, 1973*a*). During hypoxia, individuals that regulated their rate of oxygen consumption showed a decline in this 'relative ventilation:perfusion ratio', with associated increase in E_w, whereas individuals with a more P_{O_2}-dependent oxygen uptake showed less change in the ratio (Fig. 7.6).

Mussels do not have a respiratory (or oxygen-transporting) pigment in the blood, and the carrying capacity of their blood for oxygen is therefore low, viz. 0.3–0.5 volumes per cent (Maloeuf, 1937; Bayne, 1971*b*). In these circumstances an increased cardiac output during hypoxia may be necessary in order to maintain a maximum diffusion gradient for oxygen across the gill (Taylor, 1974). Indeed, such a response may be typical of

266

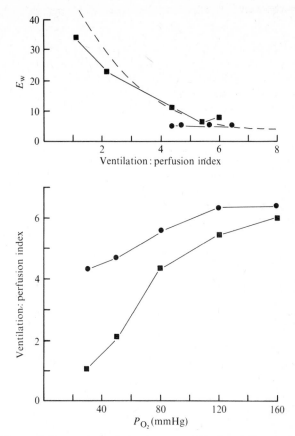

Fig. 7.6. The ventilation : perfusion index (see text) of *Mytilus edulis* related to oxygen tension (P_{O_2}) and to oxygen utilisation efficiency (E_w) by animals 'regulating' (■) and 'conforming' (●) their rates of oxygen consumption. The dashed line represents the relationship published earlier by Bayne (1971*b*).

many gas-exchange mechanisms which operate in the absence of an oxygen-transporting pigment (Cameron & Wohlschag, 1969; Hemmingsen & Douglas, 1970; Holeton, 1972; Hughes, 1973). The general response of the heart in bivalves when subjected to environmental hypoxia is a gradual increase in frequency and amplitude as P_{O_2} declines, followed by brady-cardia at very low oxygen tensions (p. 217), and in both *M. edulis* (Bayne, 1971*b*) and *Arctica islandica* (Taylor, 1974) an index of perfusion (frequency×amplitude of heart beat) increases as the P_{O_2} of the water declines. Further quantitative description of gas exchange in these animals must await the development of new techniques of physiological monitoring.

The oxygen:nitrogen (O:N) ratio

The ratio, by atomic equivalents, of oxygen consumed to nitrogen excreted can provide an index of the balance in the animal's tissues between the rates of catabolism of protein, carbohydrate and lipid substrates. If the amino acids which result from protein catabolism are deaminated, and the resulting ammonia is excreted while the carbon skeletons of the amino acids are completely oxidised, the theoretical minimum for the O : N ratio is about 7, signifying uniquely protein catabolism. Higher values for O:N indicate increased catabolism of carbohydrate and/or lipid. However, if following the degradation of protein, either the ammonia is used in a biosynthetic pathway (Chapter 6) or the amino acid carbon is utilised in gluconeogenesis, this will result in departures from these theoretical expectations for the O:N ratio. This ratio can therefore provide a useful integration for understanding 'the level of activity of the oxidative and protein metabolism' (Mayzaud, 1973) of the animal, but careful consideration must be given to the fates of the nitrogen and carbon that result from the breakdown of the proteins.

Bayne (1973a) described the seasonal changes in the O:N ratio for *Mytilus edulis* from a population on the North Sea coast of England. For the greater part of the year the O : N ratio for animals weighing 1 g dry flesh was relatively constant at about 100, signifying a considerable predominance of carbohydrate and/or lipid catabolism over the utilisation of protein in energy metabolism. In June and July, following spawning, the ratio was elevated to higher values, indicative of a marked utilisation of lipids at a time when glycogen in the body was scarce (Gabbott & Bayne, 1973; Thompson, Ratcliffe & Bayne, 1974) and protein synthesis was active.

Bayne (1973a) also recorded changes in O:N ratio resulting from starvation and exposure to increased temperatures. Changes in the ratio correlated reasonably with gross biochemical changes in the tissues (Gabbott & Bayne, 1973; Bayne, 1973b). In the winter and spring, when protein substrates contributed between 72 and 83% of the total energy loss from the body during starvation, the O:N ratio was reduced to very low values. In the summer and autumn, when the contribution from carbohydrate and lipid to energy losses during starvation increased to between 70 and 100% of the total, the O:N ratio also increased markedly. In these experiments the changes in O:N suggested an alteration of seasonally-controlled steady-state values to levels more typical of animals subjected to environmental stress (p. 289). Bayne (1975b) illustrated short-term changes in O:N following an increase in temperature (Fig. 7.7); a transient increase in the ratio reflected a period of rapid utilisation of carbohydrate during the early stages of temperature acclimation (Widdows & Bayne,

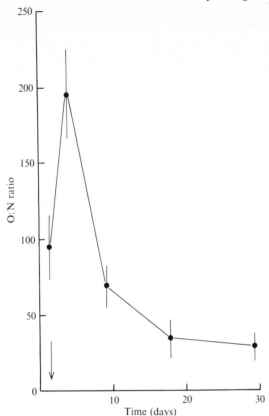

Fig. 7.7. The atomic ratio of oxygen consumed to ammonia-nitrogen excreted (O:N) in *Mytilus edulis* subjected to an increase in temperature (ambient+7 deg C) from the time indicated by the arrow; means±1 s.e. for five animals. (From Bayne, 1975*b*.)

1971; see also p. 172), and this was followed by a stabilisation of the response to a new, and lower, steady-state. Changes in the O:N ratio due to starvation were further documented for *M. edulis* by Bayne (1973*b*) and for *M. californianus* by Bayne *et al.* (1975*a*). In this latter species the O:N was naturally low (July–August), between 20 and 32, and following starvation in the laboratory these values were reduced to 9–14.

Bayne & Scullard (unpublished data) examined interactions between body size and ration as they affect the O:N ratio in *M. edulis* (Fig. 7.8). When the ration was in excess of the maintenance requirement (the concept of maintenance requirement is discussed later in this chapter), the O:N was higher in smaller than in larger individuals. During starvation, however, individuals of all sizes showed a reduction in O:N. This reduction was particularly marked in smaller individuals, which probably have less glycogen reserve to call upon during nutritive stress than do the larger

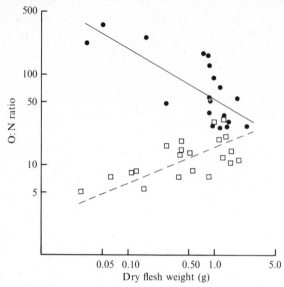

Fig. 7.8. The atomic ratio of oxygen consumed to ammonia-nitrogen excreted (O:N) in *Mytilus edulis* of different dry weights (log scales). ●, animals fed a ration in excess of maintenance; □, animals starved. Regression lines are shown for both sets of data.

animals. Working with *M. californianus* in the summer, Bayne *et al.* (1975a) did not observe a differential effect of body size on O:N.

Other studies on the O:N ratio in bivalve molluscs are scarce. Ansell & Sivadas (1973) calculated O:N ratios for the clam *Donax vittatus* at increased temperatures, and expressed some hope that the ratio may be useful in describing the physiological condition of the animal, a point made earlier by Bayne & Thompson (1970). These studies do suggest that integrations of oxygen consumption and nitrogen excretion can help in understanding interactions between oxidative and nitrogen metabolism, and therefore to provide a useful ecological 'tool'. Careful interpretation of O:N data is necessary, however, and it is important to realise that the ratio provides a relative, and not an absolute, indicator of substrate utilisation. In *M. edulis* there is evidence to suggest that during starvation under certain conditions the rate of protein 'turnover' is kept relatively constant, whereas carbohydrate and lipid catabolism is reduced, resulting in the decline in O:N (Bayne, 1975b); under these circumstances the ratio is probably a reliable index of metabolic balance. Under different conditions, however, not only may protein turnover be increased, but the degradation products of the proteins may be involved in metabolic pathways which render this use of the O:N ratio unreliable. Further research is required on the biochemical aspects of the ratio; in addition, more data on the normal range for O:N in natural populations are needed.

Integration of the effects of body size and temperature on the rate of oxygen consumption

The ecologist who is interested in production and the flow of energy through an ecosystem often requires of the physiologist information on the rates of oxygen consumption by component populations in order to estimate the metabolic heat loss, *R*, i.e. the proportion of the absorbed energy that is converted into heat by an organism or a population. Of all the terms in the energy-balance equation (see Crisp, 1971, for a discussion), the estimation of *R* is considered to be fraught with most potential for error. This is due to the large amount of variation that is normal in the respiration rate of any individual (Chapter 5), and to the difficulties of reproducing in the laboratory the critical conditions governing the respiration rate in the field. In addition, a lack of sensitivity in the statistical analysis of laboratory-based data has made it difficult to interpret (and therefore to extrapolate to field conditions with any confidence) the effects of particular environmental changes on respiration rate. The use of multiple regression techniques shows promise for clarifying the interpretation of laboratory determinations (Newell & Roy, 1973). The use of multivariable experimental procedures, coupled with complex multiple regression techniques, will be discussed later (p. 279). In this section we review some simpler attempts at the integration of respiration rate data, in which only two variables have been considered, in order to provide ecologically useful estimates of *R*.

These procedures involve the determination of the rates of oxygen consumption by individuals of different size at different temperatures, which cover the natural ecological range. The measurements should be repeated at least twice during the year in order to include animals undergoing gametogenesis and also animals in a neutral reproductive condition. The results are generalised by regression analysis and the analysis of covariance, to determine the statistical significance of variance in the experimental data, and equations established to describe the rate of oxygen consumption by animals of different size at all temperatures. These values may then be converted to their calorific equivalents by the use of oxycalorific coefficients, viz. 3.38 cal per mg O_2, or 4.83 cal per ml O_2 (Ivlev, 1934).

Pamatmat (1969) carried through such an analysis for the clam *Transennella tantilla*. He measured their respiration rates in each of the four seasons of the year at different temperatures and applied logarithmic transformation of the data. A multiple regression analysis yielded estimating equations of the form (for November–December):

$$\log_{10} \text{respiration} = -1.611 + 0.780 \log_{10} \text{weight} + 0.351 \log_{10} \text{temperature},$$

where respiration was in μl O_2 consumed h^{-1}, weight was mg total dr
weight, and temperature was in °C. The multiple coefficient of determina
tion, R^2 (see Snedecor & Cochran, 1972) was 0.924, signifying that mor
than 90% of the total variation in respiration rate was due to size and t
temperature. Three such equations (the data for spring and summer wer
pooled) allowed the construction of a nomogram from which the respira
tion rate of any animal at any size at any season could be predicted wit
some confidence.

A similar approach was used by Hughes (1970) and Dame (1972) for th
clam *Scrobicularia plana* and the oyster *Crassostrea virginica* respec
tively, although seasonal effects were not considered. Hughes (1970) foun
that the relationship between oxygen consumption and flesh weight a
temperatures from 0.5 to 22.5 °C could be expressed using a common valu
for the weight exponent b; oxygen consumption $= a \cdot$ weight^b. Values for
increased with increase in temperature, and the estimating equatio
($a = 0.0362TC + 1.8505$, where TC is the temperature in °C) could be use
to calculate a at any time of the annual cycle.

Ansell (1973) and Ansell & Sivadas (1973) used a slightly differen
procedure to derive an estimating equation for oxygen consumption b
Donax vittatus. Having estimated the normal allometric fit for oxyge
consumption and size, they recorded the temperature relationship in term
of Q_{10}:
$$Q_{10} = -0.184T + 5.2,$$

where T is the mid-point of the temperature range (in °C) in the calculatio
of Q_{10}. The following equation could then be written:
$$R_2 = R_1(-0.184T + 5.2)^{(T_2 - T_1)/10},$$

where R_1 is the rate of oxygen uptake at 15 °C (T_1); R_2 is the rate at T_2; an
$T = (T_2 + T_1)/2$.

These procedures all have the same objective, to describe the rate o
oxygen consumption as it is affected by body size and temperature. Wher
a comprehensive experimental programme in support of ecological studie
is not possible, the procedures described probably represent a minimun
requirement for deriving reasonably accurate values for respiratory hea
loss.

More complex integrations

The energy budget

The energy equation of Winberg (1956) provides a method for relating
growth to other physiological functions. Growth, as change in the caloric
equivalent of body weight with time ($\Delta W/\Delta t$), is defined as the difference
between the assimilated ration (Ab; the energy equivalent of the proportio

272

of food consumed, C, that is not rejected as faeces) and the respiratory heat loss, R (also in calories):

$$\frac{\Delta W}{\Delta t} = Ab - R.$$

The use of this expression in studies of fish production has been explored in detail by Paloheimo & Dickie (1965, 1966a, b).

As applied to mussels, the use of the basic energy equation requires measurement of respiration rate, filtration (= feeding) rate and assimilation efficiency (Thompson & Bayne, 1974), with all values expressed in caloric equivalents. These measurements may then be used to solve the Winberg equation for $\Delta W/\Delta t$ which, following Warren & Davis (1967), is termed the 'scope for growth'. The scope for growth represents the energy balance of an animal under the specified conditions, and it may be positive, when surplus energy is available for growth (P, the proportion of absorbed energy that is incorporated into somatic biomass) and for reproduction (G, the proportion of absorbed energy subsequently released as gametes), or it may be negative, resulting in weight loss due to the utilisation of the animals' own energy reserves. The scope for growth, therefore, provides an index of energy balance without distinction between somatic growth and gamete production. Growth is regarded as the net change in energy content of the animal in unit time. This is a useful concept, because it represents an integration of physiological processes to provide an index of the response of the 'whole organism' to changes in the environment. The manner in which the scope for growth is altered may be of greater significance than the change in the rate of any single physiological function, such as respiration.

Widdows & Bayne (1971) used the scope for growth to describe the initial and adaptive responses of *Mytilus edulis* to changes in temperature. When subjected to a rise in temperature, the scope for growth was reduced; when the ration level was low the scope became negative. There followed a gradual acclimation of respiration and feeding rates (p. 172) resulting in the establishment of a new, and positive, scope for growth. Thermal acclimation was therefore interpreted as an adjustment of certain fundamental physiological processes in order to maximise the energy available for growth.

Following acclimation, the scope for growth remains relatively independent of temperature, from 10 to 20 °C, although it is markedly dependent on ration (Fig. 7.9). Above 20 °C, however, there is a decline in the scope for growth which reflects a breakdown in the mechanisms of metabolic compensation at a temperature (25 °C) which approaches the lethal temperature for the species, as recorded by Read & Cumming (1967), and its southerly thermal limit as suggested by Wells & Gray (1960).

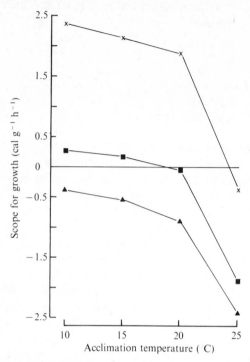

Fig. 7.9. The scope for growth (cal g^{-1} h^{-1}) in *Mytilus edulis* held for 21 days at 10, 15, 20 and 25 °C at three levels of ration; 0.91% (▲), 1.52% (■), and 3.04% (×) of the body weight per day. (From Bayne *et al.*, 1973.)

Although physiological compensation for changes in temperature can result in a stable scope for growth over a wide thermal range, the scope for growth is rather more sensitive to the level of ration. In Fig. 7.10 the scope for growth (here calculated as a percentage of the total body calories) is related to the daily ingested ration, which is expressed as a percentage of the dry flesh weight. At any given ration, smaller mussels have a greater scope for growth per unit of body size than larger ones, and their optimum scope occurs at a higher ingested ration relative to body size. These relationships essentially reflect the differences between the weight exponents for filtration rate (p. 135) and respiration rate (p. 159), because, as the animals grow in size, the increase in feeding rate (and hence in the ration obtained) is disproportionate to the increase in metabolic rate.

In an ecological context, growth relationships of this sort are best described as efficiencies. Gross growth efficiency (K_1) is the efficiency with which the animal utilises the ingested ration for growth and gamete production; net growth efficiency (K_2) is the efficiency with which the assimilated ration is utilised. Paloheimo & Dickie (1965, 1966*a, b*) analysed

274

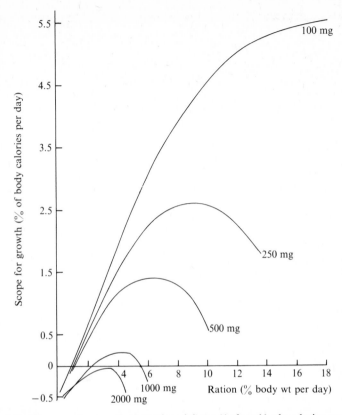

ig. 7.10. The scope for growth of *Mytilus edulis* (as % of total body calories per day) plotted against ingested ration (as % of body weight per day) for individuals of different size. (From hompson & Bayne, 1974.)

he growth of fish in terms of the relationship between K_1 and ingested ation, C. In general, the relationship was described by:

$$\frac{\Delta W}{\Delta t} = C \cdot e^{-a-bC},$$

$$\log K_1 = \log\left[\frac{\Delta W}{C\Delta t}\right] = -a-bC,$$

where ΔW is the change of body weight, C is the ingested ration during time Δt, and a and b are fitted parameters. For any specified food type, $\log K_1$ decreases linearly with increasing ration. Subsequent authors (Rafail, 1968; Brett, Shelbourn & Shoop, 1969; Kerr, 1971 *a*, *b*) have recognised an ascending phase at low ration in a plot of $\log K_1$ against C; that is, growth efficiency is an increasing function of ration until an inflection to negative

275

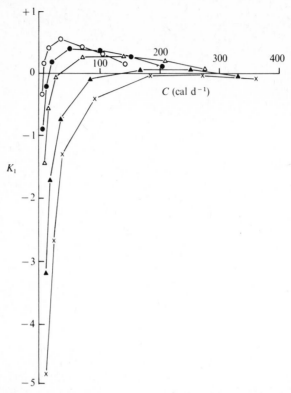

Fig. 7.11. The gross growth efficiency (K_1) for *Mytilus edulis* plotted against ingested ration (*C*) for individuals of different size; ○, 100 mg; ●, 250 mg; △, 500 mg; ▲, 1000 mg; ×, 2000 mg. (From Thompson & Bayne, 1974.)

slope is reached. In *Mytilus edulis*, K_1 is negative at very low ration levels (Fig. 7.11), but small increases in *C* result in greatly improved growth efficiencies, When the energy ingested is equal to the total energy metabolised, K_1 is zero and *C* is a measure of the maintenance ration. Further increase in *C* leads to greater values of K_1 until an inflection to negative slope occurs. Recently Winter (1974) has described an increase in K_1 with increasing ration for small *M. edulis* fed with *Dunaliella marina* and *Isochrysis galbana*.

Growth efficiency, K_1, is also dependent on the size of the mussel (Fig. 7.11); the smaller the animal the greater the maximum value of K_1 and the smaller the maintenance ration. The optimum ration, at which growth is most efficient, is an increasing function of weight, reflecting the greater energy input required to offset the total metabolism of a larger animal. At low absolute ration levels, the small animal is most efficient in converting food into body tissue, although the ration is large in relation to body weight.

276

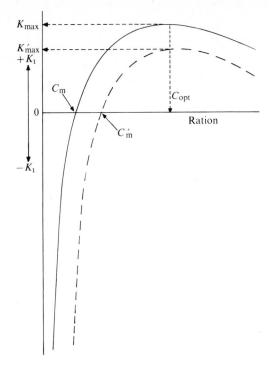

Fig. 7.12. Generalised curves for growth efficiency (K_1) of *Mytilus edulis* as related to ration. See text for details. (From Thompson & Bayne, 1974.)

Conversely, large animals grow more efficiently at higher absolute ration levels, which are nevertheless relatively low in comparison to body weight.

Fig. 7.12 represents a generalised plot of K_1 against ration (a 'K-line') for a mussel fed a single type of food (Thompson & Bayne, 1974). The curve has a negative and a positive component; at zero growth efficiency $C = C_m$, the maintenance ration. Values for particulate organic matter in coastal waters (Jørgensen, 1966) suggest that both these regions of the K-line may be relevant ecologically. The experiments on which this relationship is based employed a 'naked' flagellate (*Tetraselmis suecica*) at 15 °C as food; this ration probably represents optimum nutritional value. Variations in such factors as season, temperature and nutritional value may be expected to alter the position of the K-line relative to both body size and to ration level, C. For example, an increase in oxygen consumption, such as occurs during gametogenesis, will displace the K-line downwards (dashed line in Fig. 7.12). Maintenance ration then increases from C_m to C'_m, maximum growth efficiency is reduced, although optimum ration, C_{opt} remains unchanged. The energy equivalent of the additional oxygen

277

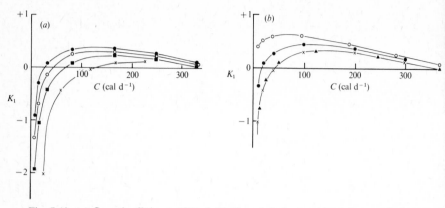

Fig. 7.13. (*a*) Growth efficiency (K_1) of *Mytilus edulis* from a North Sea population, as a function of ingested ration (C) and seasonal (ambient) temperature; ■, 5 °C; ○, 10 °C; ●, 15 °C; ×, 25 °C. (*b*) Growth efficiency (K_1) of *Mytilus edulis* from a population in Nova Scotia, as a function of ingested ration and seasonal (ambient) temperature; ○, 1 °C; ●, 5 °C; ×, 12 °C; ▲, 15 °C.

demand is represented by $C'_m - C_m$. Changes in the nutritional value of the food may be expected also to vary the numerical value of C_{opt}.

Ambient temperature differences, and in particular seasonal differences in the phasing of the reproductive cycle, will effect changes in the K-line. A population of *M. edulis* from the North Sea (Fig. 7.13*a*) has a maximum growth efficiency in the summer, when the seawater temperature is 15 °C; at temperatures on either side of this optimum, growth efficiency is reduced. In a population from Nova Scotia, Canada, however (Fig. 7.13*b*) maximum growth efficiency at constant ration (though not maximum growth, which depends on seasonally variable food levels) occurs in the winter, at temperatures as low as 1 °C. These differences may be explained, at least in part, by population differences in the reproductive cycle. During gametogenesis metabolism is increased (Chapter 5) and scope for growth is, in general, reduced. In Nova Scotia, gametogenesis occurs in the spring, whereas in the winter gametogenesis is inhibited (R. J. Thompson, unpublished data). In the North Sea, however, gametogenesis occurs in the winter and scope for growth is reduced.

So far in this chapter we have mainly considered the integration of physiological data from experiments in which environmental variables (such as ration) are changed singly, whilst other variables (such as temperature) are held constant. Of course, in the natural habitat, mussels are subjected to a wide variety of environmental changes acting together, and it is therefore of considerable interest to consider possible interactions between the effects of environmental change on the components of energy balance. This is discussed in the following section.

Multivariate experiments and response surface analysis

The response by an organism to its environment, when determined over different levels of several factors, forms a multidimensional surface or plane. Such a surface is often non-linear, but can be described accurately by means of multiple regression equations which are obtained by the method of orthogonal polynomials. The response surface may then be visualised by plotting isopleths of the response, or the dependent variable, against two independent variables (for a review see Alderdice, 1972). Investigation of the configuration and dynamic properties of a response surface is more instructive than the description of a two-dimensional transect of the surface. The plotting of such response surfaces following multifactorial experiments enables changes in tolerance, capacity and plasticity of response, and the interactions between independent variables, to be described.

The multifactorial, response surface approach has been applied in a number of recent studies of the combined effects of temperature and salinity on the development and survival of bivalve larvae (see Chapter 4). Some recent multifactorial experiments by Widdows (unpublished data) have determined the adaptive responses of *Mytilus edulis* adults of different body size to combinations of the factors ration, temperature and season. The following equation describes the estimated scope for growth, as calories per day, for individuals of an estuarine population from south-west England, over an annual cycle:

Scope for growth $=$
$$-1.87188 - (0.209459\,T) - (0.0534796\,T^2) + (0.0118615\,T^3)$$
$$-(0.74507E{-}3\,T^4) + (0.172659E{-}4\,T^5) - (0.132585E{-}6\,T^6) + (93.4622R)$$
$$-(112.585R^2) + (36.466R^3) - (27.3694\,W) + (7.3887\,W^2) - (1.26195\,W^3)$$
$$-(0.347338\,WT) + (82.3995\,WR) + (1.10224\,TR) + (0.604698E{-}2\,WT^2)$$
$$-(5.6765\,WR^2) - (11.1803\,W^2R) - (0.0230567\,T^2R),$$

where W is dry body weight from 0.1 to 3.0 g, C is available ration from 0 to 1.5 mg organic matter per litre, and T is time from 0 to 52 weeks. The equation explains 95% of the total variance in the experimental data, and therefore adequately represents the true response. Although the equation appears unwieldy, it is in fact quite simple to solve.

A response surface which is derived from this expression (Fig. 7.14) illustrates the effects of body size and available ration on the scope for growth of *M. edulis* in June when adapted to 15 °C. The maintenance ration is represented by the zero isopleth, and this shifts to higher ration levels with increasing body size. The maintenance ration also increases with rise in ambient temperature during spring and summer when, in this population, the gonads are ripe.

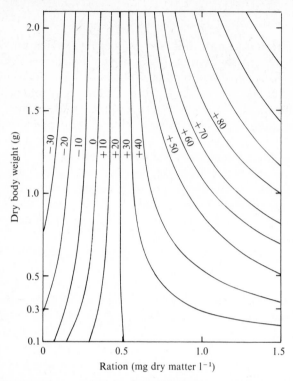

Fig. 7.14. The scope for growth of *Mytilus edulis* (cal d^{-1}) as a function of available ration and dry body weight; data for animals acclimated to 15 °C in June.

The response surface for gross growth efficiency (K_1) of mussels adapted to 15 °C in June (Fig. 7.15) is plotted from the following equation, which adequately describes 98% of the total variance in the experimental data:

$$K_1 = 0.231171 - (0.167763\,W) + (0.211136\,W^2) - (0.0515543\,W^3)$$
$$- (0.76773 \cdot \log R) + (0.245494 \cdot \log R^2) + (1.53133 \cdot \log R^3)$$
$$+ (0.79856\,W \cdot \log R) - (0.130526\,W^2 \cdot \log R),$$

where W is dry body weight from 0.1 to 3.0 g and C is the logarithm (to base 10) of available ration from 0 to 1.5 mg organic matter per litre. This diagram clearly shows the increase in optimum ration (at which growth efficiency is maximum) with increasing body size.

The multiple regression equation described above for scope for growth, also provides a simple, empirically based growth model for *M. edulis*. The energy that is available for growth can be predicted for any combination, within the stated limits, of size and available ration throughout the year. As indicated earlier, ration level is the major factor in growth, as reflected in the multifactorial experiments, where 64% of the total variance is due to the amount of available food. Such a model can be used in an analytical

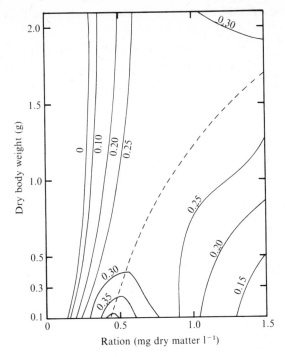

Fig. 7.15. The growth efficiency (K_1) of *Mytilus edulis* as a function of available ration and dry body weight; data for animals acclimated to 15 °C in June. The dashed line represents the 'ridge' of the response surface.

way to examine specific questions of ecological relevance, such as the capacity of particular organic components of the suspended particulate matter to support growth.

A rather different approach to the modelling of growth, which is also based on the energy-balance equation of Winberg (1956), aims to bring together physiological data into a simulation of the growth process as it is affected by the environment. We discuss this approach in the following section.

A simulation model of the growth of Mytilus edulis

A simulation of the growth of *Mytilus*, designed to incorporate the effects of changing seasons, temperature, ration level, salinity and tidal height, has recently been developed by Bayne & Radford (unpublished data). The simulation is based on the Continuous System Modelling Programme, or CSMP; the use of CSMP in modelling biological systems is discussed by Radford (1972) and by Brennan, De Wit, Williams & Quattrin (1970). The system flow diagram of the model is shown in Fig. 7.16. Change in 'dry

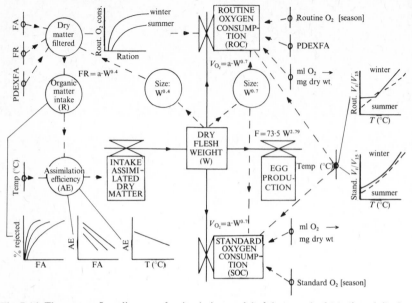

Fig. 7.16. The system flow diagram of a simulation model of the growth of *Mytilus edulis*. See text for details.

flesh weight' results from the difference between the 'intake of assimilated dry matter' and the sum of 'routine oxygen consumption', 'standard oxygen consumption' and 'egg production'.

Intake of assimilated dry matter

The rate of filtration (FR) is written as:

$$\text{Filtration rate (1 d}^{-1}) = 3.36W^{0.40},$$

where W is the dry flesh weight in milligrams. The amount of particulate material available in the water (FA) is a variable input, written in units of milligrams per litre, and quoted over a weekly or monthly time base. The amount of 'dry matter filtered' per day is then a function of dry matter available and filtration rate. The rate of filtration is specified as not being affected by gradual temperature change (Widdows, 1973a, b) or season (Bayne & Widdows, unpublished data) or the concentration of suspended particulates (Winter, 1969; Thompson & Bayne, 1974; Foster-Smith, 1975a) within the ranges used in the model to date. The effect of distribution within the tidal zone on the amount of material filtered per day is incorporated as a fractional multiplier (PDEXFA) which is the proportion of each day that the animal is exposed to the available food.

A proportion of filtered material is rejected by the animal as pseudofaeces (Winter, 1970, 1973; Davids, 1964; Tenore & Dunstan, 1973; Foster-Smith, 1975a). The production of pseudofaeces is a function of both body size and particle concentration, and there is a threshold concentration, which is variable with body size, below which there is no rejection of filtered material. There is also a maximum value to the proportion of particle matter rejected; estimates of this are scarce, but Bayne & Radford accepted 0.8, or 80% rejection, as an asymptote. The proportion of filtered particle matter that is rejected as pseudofaeces is specified as;

$$\text{Proportion rejected} = 0.8(1 - e^{K(I - FA)}),$$

where FA is the concentration of dry particulate matter (mg l^{-1}), I is the threshold concentration, and K is a fitted parameter. The numerical values of I and K are considered to bear the following relationships to body weight:

$$K = 0.25W^{-0.40},$$
$$I = W^{0.40}.$$

This treatment results in a family of curves relating the proportion of rejecta to the concentration of particulate matter. The difference between the dry matter filtered and the dry matter rejected represents dry matter ingested. This is converted to 'organic matter intake' per day by means of a conversion constant for the proportion of organic matter in the dry suspended material.

A certain proportion only of the organic material ingested is assimilated into the body; this 'assimilation efficiency' (AE) is influenced by body size, temperature (which is a variable input, or forcing function, to the model, quoted in °C over a weekly or monthly time base), and the weight of organic matter ingested (see Chapter 5). These effects are incorporated into two equations and one table;

$$AE_{(t=0)} = 0.916 - b \cdot R,$$

where $AE_{(t=0)}$ is the assimilation efficiency at 0 °C, R is the organic matter intake per day and b is a parameter that varies with body size according to the following table:

Body size (mg dry wt)	b
100	0.0288
250	0.0197
500	0.0144
1000	0.0120
2000	0.0110

Temperature effects on assimilation efficiency are incorporated as:

$$AE_{(t)} = AE_{(t=0)} - 0.007 \cdot t,$$

where t is temperature in °C. A minimum assimilation efficiency of 0.02 is specified, and the model computes an 'operating assimilation efficiency', which is the maximum of either 0.02 or the solution of:

$$0.916 - b \times \text{organic matter intake} - 0.007 \cdot t.$$

In use of the model to date, the operating assimilation efficiency has always been between 0.80 and 0.35.

The product of the operating assimilation efficiency and the organic matter intake per day yields the 'intake of assimilated organic matter' per day.

Rates of oxygen consumption

The standard rate of oxygen consumption (SOC) is specified as being independent of ration, but influenced by temperature, season and by body size. In the winter;

$$\text{SOC (ml } O_2 \text{ d}^{-1}) = 0.055W^{0.70},$$

where W is the dry body weight in milligrams. In the summer;

$$\text{SOC} = 0.041W^{0.70}.$$

The effects of temperature on the standard rates of oxygen consumption (Widdows, 1973a, b; Bayne et $al.$, 1973) are incorporated in the form of a table, listing the ratio of oxygen consumption at 15 °C (V_{15}) and oxygen consumption at temperature t °C;

	$V_t : V_{15}$	
t °C	Winter	Summer
0	0.617	0.466
5	0.689	0.626
10	0.803	0.803
15	1.000	1.000
20	1.284	1.162
25	1.708	1.512

The routine rate of oxygen consumption (ROC)* is specified in a similar way to SOC as affected by season, temperature and body size. But the routine oxygen consumption is also affected by ration level (Thompson & Bayne, 1974). As the weight of dry matter filtered per day increases, ROC increases to an asymptote; the form of the curve is different in winter and summer. Bayne & Radford specified these effects of ration in the form of a table listing ROC (as ml O_2 d^{-1}) per unit of 'metabolic body size' ($W^{0.70}$) as a function of milligrams dry matter filtered per day (Fig. 7.17). This use of

* The term 'routine rate of oxygen consumption' as defined in Chapter 5 (p. 163) is equivalent to SOC+ROC as used in this simulation.

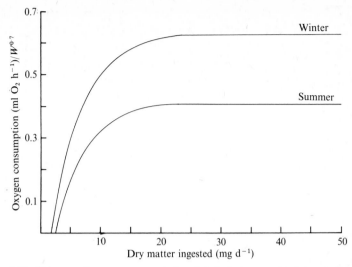

Fig. 7.17. The rate of oxygen consumption (ml O_2 h^{-1} per unit of metabolic body size; W = weight in milligrams) of *Mytilus edulis* related to ingested ration, in winter and summer.

tables (which are 'read' by the computer with linear interpolation) is a major strength of CSMP, allowing 'raw' data to be used in situations where a suitable mathematical expression describing the relationship between two variables is not available.

During periods of air exposure, the routine rate of oxygen consumption declines to zero, but standard rate is unaffected.

These various expressions for oxygen consumption rate are then brought together; for example:

$$SOC = 0.055 \cdot CSATW \cdot (W^{0.70}) \cdot (1.0 - SEASIN)$$
$$+ 0.041 \cdot CSATS \cdot (W^{0.70}) \cdot SEASIN,$$

where CSATW and CSATS refer to the effects of temperature, in winter and summer respectively, for which the model scans and interpolates in the relevant table. The seasonal effect is specified in the form of a switch (SEASIN) which either has the value 1.0 (in summer) or 0 (in winter). The numerical values for SOC and ROC are then summed and converted to an equivalent weight of tissue by the coefficient 0.96. This coefficient is an acceptable average, based on measured losses of carbohydrate, lipid and protein during starvation (Gabbott & Bayne, 1973), together with standard oxycalorific conversions (Winberg, 1956).

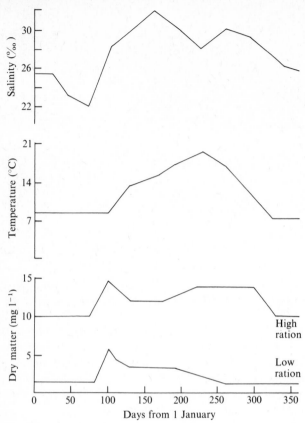

Fig. 7.18. 'Food available', temperature and salinity as specified in the simulation of growth of *Mytilus edulis*. See text for details.

Egg production

Estimates of the weight-related fecundity of *Mytilus edulis* are surprisingly scarce (see p. 96). In the simulation model an equation for fecundity (F) is used, which is based on measurements of the weight of the gonad before and after the release of gametes in a natural population of mussels;

$$\text{Fecundity (mg dry wt)} = 73.5W^{2.79},$$

where W is the dry flesh weight in grams.

Nitrogen balance

This simulation model also includes a sub-routine (not illustrated in Fig. 7.16) on nitrogen balance. The total body nitrogen is the difference between the nitrogen gained from the diet and nitrogen lost as excreted nitrogen and

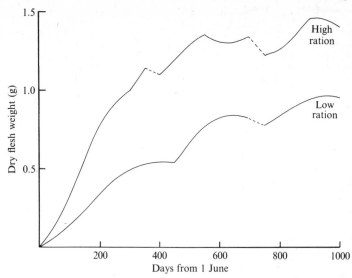

Fig. 7.19. The growth of *Mytilus edulis* over 1000 days at 'high' and at 'low' ration, as determined by a simulation model. The dashed portions of the curves signify weight loss due to spawning.

nitrogen content of the eggs. The rate of nitrogen excretion is considered to be dependent on ration, temperature, salinity, body size and season according to data presented by Bayne & Scullard (unpublished data) and as discussed in Chapter 6.

Outputs from the simulation model

This model of the growth of *Mytilus* has been used by Bayne & Radford (unpublished data) to examine general trends in growth and production of two hypothetical populations under similar conditions of temperature and at two different levels of food availability (Fig. 7.18). The temperature was set to an annual cycle typical of a small estuary in the south-west of England. Two feeding regimes were used; in the first (high ration), food levels were set similar to values recorded in a local estuary (high concentrations of particulate matter, little seasonal change, a low proportion of organic matter), whereas in the second (low ration), the food levels were set similar to offshore conditions (lower concentrations of particulate matter, marked seasonal changes and a high proportion of organic matter). Salinity conditions simulated those in a local estuary.

The model was started with a mussel weighing 1 mg dry flesh on 1 June. The resulting growth (Fig. 7.19) resembled recorded annual patterns of growth in temperate populations of *M. edulis* (see Chapter 2). Growth was rapid in the first summer and autumn and declined during the first winter;

287

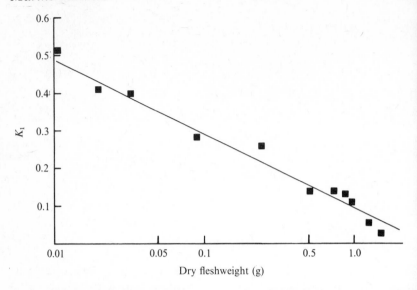

Fig. 7.20. The gross growth efficiency (K_1) of *Mytilus edulis* related to dry body weight, as determined in a simulation model.

spawning occurred at the end of the first spring period. In subsequent years growth occurred in the summer; there was a slight reduction in weight in the winter, and a marked loss in weight during the late spring and early summer due to spawning. Animals at 'high ration' grew very much better than animals at 'low ration' although the seasonal pattern of growth was similar in both cases.

Growth efficiency varied considerably during the year, as expected, with seasonally high values in the summer and negative values in the winter. The gross growth efficiency (K_1) for animals of different size at approximately equivalent times of the year (Fig. 7.20) declined exponentially, from a maximum of 0.52 for an animal of 10 mg dry weight, to 0.02 for an animal weighing 1.6 g dry flesh. The ratio of production:respiration ($P:R$) also varied seasonally. However, the annual average for $P:R$ during the second, third and fourth years (Table 7.2) was quite stable and showed little variation with ration.

The simulated oxygen:nitrogen ($O:N$) ratios are plotted in Fig. 7.21. The $O:N$ was relatively stable during the summer, autumn and winter, and showed little variation with animal size. In the spring, just prior to spawning, the $O:N$ values increased markedly. The overall level of the $O:N$ ratio was higher in animals fed at the 'high ration' than in animals at 'low ration'.

These results of the simulation of growth in *Mytilus* illustrate the potential of this technique for integrating experimental physiological data.

Table 7.2. *The mean ratios of production : respiration (both in units of mg dry flesh d^{-1}) for two hypothetical populations of mussels over four years, as estimated from growth simulations*

	$P:R$ for the year			
Ration level	1	2	3	4
High	1.04	0.10	0.08	0.06
Low	0.81	0.10	0.07	0.03

From Bayne & Radford (unpublished data).

The technique is also useful in identifying areas of ignorance and in suggesting problems for further research. In the present simulation it quickly became apparent that data for the amount of food available to animals in their natural habitats were very limited. Similarly, quantitative descriptions of the production of pseudofaeces and of weight-related fecundity were needed, but were not available in the literature. Finally, such simulations may be used analytically, to examine particular questions that may have an ecological relevance, such as the ways in which the $P:R$ or the $O:N$ ratios may vary with changes in food availability, temperature, salinity and distribution on the shore.

Stress

We suggested in the introduction to this chapter that physiological integrations could prove useful in describing the 'condition' of individual animals at any point in time. Such descriptions of condition might then be used to assess the 'health' of animals in their natural habitats and thereby to help in ecological understanding and in resource management.

Animals in poor condition are considered to be under stress. Bayne (1975b) offered the following definition of stress: 'Stress is a measurable alteration of a physiological (or behavioural, biochemical, or cytological) steady-state which is induced by an environmental change and which renders the individual (or the population or the community) more vulnerable to further environmental change.' The physiological conditions which have proved useful in describing a condition of stress in *Mytilus edulis* have been integrations such as the scope for growth and the $O:N$ ratio. The degree of disadvantage which accrues from a condition of stress has been assessed in terms of fecundity and the success of larval development (Bayne, 1975b).

An index of the energy available for growth and the production of

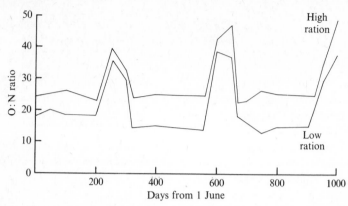

Fig. 7.21. The atomic ratio of oxygen consumed to ammonia-nitrogen excreted (O:N) in two hypothetical populations of *Mytilus edulis* (at 'high' and at 'low' ration) over 1000 days, as estimated in a simulation model.

gametes, or growth efficiency, which is a derivative of this, is a fundamental measure of physiological condition and the degree of stress. Much of our earlier discussion of the scope for growth and growth efficiency (p. 272) is pertinent here. For example, *M. edulis* acclimates to change in temperature by altering its feeding and respiration rates in such a way as to maintain scope for growth relatively stable over a wide range of temperature. At temperatures above 20 °C, however, there is an apparent failing of the adaptive response, leading to reduced scope for growth (Fig. 7.9). The stress that results from high temperature is measurable as a decline in the calories available for growth and gametogenesis. The O:N ratio has also been proposed as an index of stress (Bayne & Thompson, 1970; Ansell & Sivadas, 1973; Bayne, 1975*b*) and has been discussed in relation to ration (Bayne, 1973*c*), salinity (Bayne, 1975*b*) and temperature (Bayne & Scullard, unpublished data).

Other indices have been suggested for use in describing physiological condition. The proportion of the internal shell volume which is occupied by the body tissues (Baird, 1966) is often used in shellfish research. Gabbott & Stephenson (1974) recorded a significant correlation between glycogen level in the tissues of *Ostrea edulis* and a dry weight condition index (flesh dry weight as a proportion of internal shell volume). Ansell & Sivadas (1973) proposed the use of a carbon:nitrogen ratio for the tissues of *Donax* as an index of condition. Jeffries (1972), who studied the clam *Mercenaria mercenaria*, found that the molar ratio of taurine:free glycine could be used as an index of stress. When the ratio was less than 3, Jeffries considered the population to be normal; when between 3 and 5 a chronic stress condition was indicated, and if greater than 5, stress was acute. Jeffries suggested that 'a set of readily measured norms' could, collecively, indicate the 'status of homeostatic regulation' within a species. A departure from these norms would represent a symptom of stress, and a number of symptoms, taken together, might constitute a syndrome of physiological condition. Bayne (1975b) pointed out that the use of such indices may not disclose any of the causative factors involved, nor the detailed physiological compensations that participate in the response to a stressor, but, taken together, they may serve to quantify the degree of stress in a population.

In order for a condition of stress to be accepted as proven, some disadvantage must be shown to accrue from the physiological condition of the animal. In some cases the index of stress will itself indicate the degree of disadvantage; a reduced growth efficiency not only signifies a condition of stress, but also measures the disadvantage resulting to both the individual and the population. In other cases, the stress index will not indicate the degree of disadvantage; the oxygen:nitrogen, or taurine:glycine ratios require to be correlated with other physiological events (perhaps, in the case of the taurine:glycine ratio, with the capacity for isosmotic regulation) before 'disadvantage' can be quantified. The aspect of 'disadvantage' that has received most study in bivalves is the relationship between adult condition, fecundity and larval survival, as discussed elsewhere in this volume (Chapters 4 and 8).

The need for quantitative estimates of physiological condition and of 'fitness' is apparent in many areas of research on mussels. Ecological comparisons between populations would often benefit from the use of condition indices. Studies of population genetics may need to be related to measures of physiological fitness (Chapter 9). And the aquaculture and resource-management of mussels, not to mention pollution studies, would often benefit also from the deployment of quantitative estimates of the degree of stress in a population. The derivation, understanding and use of such condition indices represents a rich area for future research.

8. Energy metabolism

P. A. GABBOTT

Studies on energy metabolism are concerned with the ways in which the major carbohydrate, lipid and protein fuels are used by an organism for energy production. In this chapter the discussion is concerned not only with the physiology and biochemistry of the animal as a whole but with the inter-relationship between the different body tissues, their energy requirements and the metabolic transformation of the fuel reserves. The emphasis in the last section, on the control or regulation of metabolism, reflects one of the major developments that have taken place in the last decade in biochemical research. It is fortunate that the recent literature on the European mussel, *Mytilus edulis*, provides a perspective which includes metabolic regulation at all levels of organisation from the whole animal to individual enzymes. Where possible, comparisons have been made with other species of the genus *Mytilus*. In some instances it has been necessary to use indirect arguments by reference to studies on other bivalves, in particular the European, American and Pacific oysters, *Ostrea edulis*, *Crassostrea virginica* and *Crassostrea gigas*, or to work with mammalian systems.

   The first major review in this field was by Giese (1966) on the role of lipids in the economy of marine invertebrates; in the same year there were chapters by Goddard and by Martin, on carbohydrate metabolism, and by Florkin, on nitrogen metabolism, in *Physiology of Mollusca*, vol. 2 edited by Wilbur & Yonge (1966). Other more recent reviews are those by Giese (1969) on the biochemical composition of adult molluscs; Hammen (1969) on the metabolism of *C. virginica*; Campbell & Bishop (1970) on nitrogen metabolism in molluscs; and Gilles (1970) on intermediary metabolism and energy production. Also, there are chapters by Goudsmit on carbohydrate metabolism, by Voogt on lipids and sterols, by Florkin & Bricteux-Grégoire on nitrogen metabolism and by Gilles (1972a) on biochemical ecology in *Chemical zoology*, vol. 7, *Mollusca* edited by Florkin & Scheer (1972).

Biochemical composition

Most determinations of the biochemical composition of marine bivalves have been concerned with the gross changes in protein, lipid and carbohydrate content, and with changes in various subfractions such as protein and non-protein nitrogen; phospholipids and neutral lipids; and glycogen and free sugars. Methods of biochemical analysis for marine invertebrates have been reviewed by Giese (1967). More recently Holland & Gabbott (1971) and Holland & Hannant (1973) have described a microanalytical scheme for the sequential determination of protein, lipid and carbohydrate components from milligram quantities of bivalve larvae. Most procedures involve two distinct steps; (1) an initial fractionation of the material into the different biochemical constituents, and (2) the determination of the separated components. In some instances it is possible to use specific enzymatic procedures directly on the initial homogenate or unfractionated extract (see Bergmeyer, 1974). One recent example is the development of a rapid enzymatic method for the determination of oyster glycogen using a commercially available glucose oxidase test-kit (Gabbott & Hannant, unpublished data). Blood metabolites can also be determined by clinical methods using standard reagent kits (see Bayne, 1973a).

Changes in biochemical composition are usually reported as differences in the level of a given constituent (% dry weight) or as changes in biochemical content (weight per animal) (Giese, 1967). The disadvantage of expressing the results as a percentage is that changes in one biochemical component are reflected by reciprocal changes in all the other components. On the other hand, the biochemical content is dependent on size and on growth. When considering seasonal changes in biochemical composition, apart from growth, it is useful to express the results in terms of the composition of a 'standard animal' of a given size. The general procedure is as follows: each time a collection of animals is made for biochemical analysis, a calibration curve is set up relating body size to dry weight and the biochemical content of the standard animal is calculated from the results of a bulk analysis expressed in terms of dry weight (Barnes, Barnes & Finlayson, 1963).

Seasonal changes in adult mussels

Much of the early work on the seasonal changes in biochemical composition of *Mytilus edulis* and *M. galloprovincialis* has been reviewed by Giese (1969). More recent studies are those by Lubet & Le Feron de Longcamp (1969), Williams (1969b), De Zwaan & Zandee (1972a), Dare (1973a) and Gabbott & Bayne (1973) on *M. edulis*; by Chapat, Sany, Arnavielhebony & Gravange (1967) and Pavlović, Kekic & Mladenović (1970) on *M. galloprovincialis*; and by Alvarez (1968) on *Perna perna*. Surprisingly, the author has been unable to find any reference to recent work on the Californian mussel, *Mytilus californianus*.

All of the studies show that the changes in body weight are mainly due to changes in carbohydrate, or glycogen content. The seasonal cycles for storage and utilisation of glycogen reserves reflect the complex interactions between food supply and temperature, and between growth and the annual reproductive cycle. The form of the reproductive cycle varies considerably between species and with geographical locality; some species have definite annual cycles whereas others breed more or less continuously. In the south of England, the normal cycle of gonad development for *M. edulis* begins in the autumn and extends through the winter (Stages I–III); spawning takes place in the spring and then there is a short non-reproductive period (Stage 0) in the middle of the summer (Chipperfield, 1953; Seed, 1975). In the same locality, the seasonal cycle for *M. galloprovincialis* differs in several respects. First, the gonad index never falls to a very low value and there is a general absence of a well-marked spent phase; secondly, the spawning period is greatly extended (Seed, 1975). On the Atlantic coast of France (Basin d'Archachon), however, both species have distinct annual cycles with a definite non-reproductive period (Lubet, 1959). According to the biochemical data, glycogen accumulates mainly during the non-reproductive period in the summer. The same conclusion has been drawn from histochemical studies on the gonad tissue in *M. galloprovincialis* (Costanzo, 1966) and *Perna perna* (Lunetta, 1969).

In *M. edulis* the seasonal cycle of storage and utilisation of glycogen reserves is closely linked to the annual reproductive cycle. Fig. 8.1 shows the changes in glycogen content of the body tissues of mussels collected from the Waddenzee, Netherlands (De Zwaan & Zandee, 1972*a*). In the summer the mussels are spent (Stage 0) and the metabolic energy demand is low (Widdows & Bayne, 1971; Bayne, 1973*a*). Abundant food is available in the plankton and there is a marked increase in glycogen content with the highest accumulation in the mantle. Protein and lipid reserves are also built up but mainly in the non-mantle tissues (Gabbott & Bayne, 1973). During the autumn and winter the metabolic demand is high, due to gametogenesis, and the glycogen reserves fall to a minimum value in mid-winter (January-March). The loss of glycogen, in female mussels, is synchronous with Stage II – oogenesis and vitellogenesis – of the gametogenic cycle. The same synchrony between glycogen breakdown and vitellogenesis was noted previously by Lubet (1959) for *M. edulis* from the Basin d'Archachon.

In the European oyster, *O. edulis*, the glycogen storage cycle and the annual reproductive cycle are not clearly separated. During the winter there is a long period of sexual rest and gonad development does not begin until the spring, culminating in spawning in the summer and autumn. During the reproductive period abundant food is available so that growth and gametogenesis can take place at the same time. However, in all, or nearly all, the examples cited in the literature, including data for *C. virginica* and *C. gigas*, the accumulation of glycogen in the spring precedes gonad development (see data in review by Walne, 1970*b*).

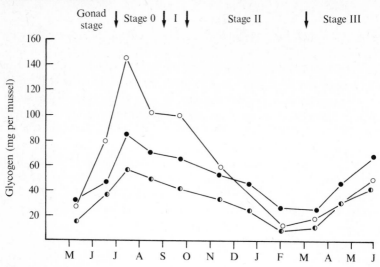

Fig. 8.1. Seasonal changes in glycogen content of *Mytilus edulis*. Data for mussels of length 5.5 cm (redrawn from De Zwaan & Zandee, 1972*a*). ○, mantle; ●, muscle and foot; ◐ digestive gland. Stages of gonad development: Stage 0, spent; Stage I, initiation; Stage II vitellogenesis; Stage III, maturation and spawning (after Chipperfield, 1953, modified b Lubet, 1959).

The seasonal changes in lipid content of *M. edulis* show an inverse correlation with the changes in glycogen content (Lubet & Le Feron de Longcamp, 1969; Williams, 1969*b*). There is an increase in the level of triglycerides and phospholipids during the winter, reaching a maximum during Stage III of gametogenesis; the lipid level is generally higher in the females than in the males, presumably due to the fatty reserves in the eggs (Lubet & Le Feron de Longcamp, 1969). In Mediterranean waters, *M. edulis* spawns repeatedly during the spring and early summer and between each successive period the gonad is reconstituted. The lipid level falls rapidly after spawning, and then increases again as the gametes mature (Lubet & Le Feron de Longcamp, 1969). In the middle of the summer, during the non-reproductive period, the levels of triglycerides and phospholipids remain low. It should be noted that the results presented by Lubet & Le Feron de Longcamp are expressed as a percentage by weight, so that the difference between the low levels of lipid in the summer and the high levels in the winter could be due to the seasonal changes in glycogen content (see Fig. 8.1). If the amount of lipid per mussel increases at the same time as the glycogen content declines the effect is to exaggerate the changes in percentage composition. The observation that the lipid level fell and then increased again during each successive spawning period supports the view that the fat content is higher during the winter due to the build-up of lipid reserves in the developing eggs.

Table 8.1. Mytilus edulis: *biochemical composition (µg per mg dry wt) of unfertilised eggs and 7-day-old veliger larvae*

	Lipid				Lipid:protein ratios	
	Neutral lipid	Phospho-lipid	Carbo-hydrate	Protein	Neutral lipid	Phospho-lipid
Eggs	89.3	66.3	21.8	296.4	0.30	0.22
Larvae	9.7	7.5	7.1	88.3	0.11	0.08

Data from Bayne, Gabbott & Widdows (1975).

Changes in composition during larval development

In most marine bivalves the egg reserves form a fatty yolk, consisting mainly of neutral lipids (Raven, 1966); this has been shown both histologically by examination of the gonad tissue in *M. californianus* (Worley, 1944) and by direct biochemical analysis of the eggs of *M. edulis* (Bayne, Gabbott & Widdows, 1975). Table 8.1 shows the biochemical data for unfertilised eggs and 7-day-old veliger larvae of *M. edulis*. It is apparent that there is a much greater fall in the lipid levels than in the other constituents during development from egg to larva. As a percentage of the total protein (lipid:protein ratio×100) the amount of neutral lipid fell from 30% in the eggs to 11% in the larvae and the amount of phospholipid from 22% in the eggs to 8% in the larvae. In contrast the carbohydrate:protein ratio remained the same in both eggs and larvae (0.07 and 0.08, respectively). The results suggest that phospholipids as well as neutral lipids (mainly triglycerides) make a significant contribution to energy metabolism during the early stages of larval development (Bayne, Gabbott & Widdows, 1975). In energetic terms both the triglyceride and phospholipid components can be considered as storage forms for fatty acid reserves (see later).

Triglycerides are also the main energy reserve in newly released oyster, *O. edulis*, larvae (Millar & Scott, 1967b; Holland & Gabbott, 1971; Holland & Spencer, 1973). *O. edulis* is larviparous and the early stages of embryonic development take place in the mantle cavity; it may be assumed however, that the eggs also have a fatty yolk. In marked contrast to the adult oyster, carbohydrate is the least important, quantitatively, of the larval reserves (Collyer, 1957). Fig. 8.2 is a composite diagram showing the changes in the percentage of neutral lipid and glycogen in *O. edulis* from newly released larvae to young adults (Holland & Spencer, 1973; Holland & Hannant, 1974). The reader will appreciate that the percentage data are sometimes difficult to interpret (see previous discussion); but, in this case, the same is true for the changes in biochemical content, because as growth

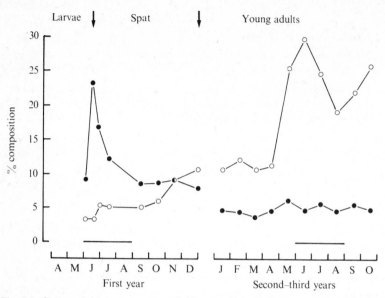

Fig. 8.2. Changes in the percentage of lipid and carbohydrate during development of *Ostrea edulis* (redrawn from Holland & Hannant, 1974). ●, neutral lipid; ○, glycogen. Horizontal bars indicate spawning season.

takes place all constituents must increase in weight. It is the percentage change that indicates the relative importance of the different reserves. The level of neutral lipid increased during larval development and then fell rapidly immediately after settlement. This supports the view that the newly settled spat are unable to feed during the early stages of metamorphosis (Cole, 1938; Hickman & Gruffydd, 1971). There was no loss of protein during metamorphosis, suggesting that the larval tissues are simply re-organised with little or no change in α-amino-nitrogen content (Holland & Spencer, 1973). The loss of neutral lipid is presumably associated with the utilisation of triglyceride reserves to meet the energy demand during metamorphic development. The level of glycogen was low in the larvae but increased in the spat; in the young adults the glycogen content rose dramatically in the spring prior to gonad development and then followed the established seasonal cycle for mature oysters (see Walne, 1970*b*).

Holland & Spencer (1973) also carried out a series of short-term starvation experiments with *O. edulis* larvae at different stages of development. Both lipid and protein reserves were lost during starvation but there was little change in carbohydrate content. The loss of reserves is related to the energy consumption of the larvae and the calculated values for metabolic rate agree well with determinations of oxygen consumption by direct respirometry (see review by Crisp, 1975). It should be noted, however, that the calculated values are for starved larvae and might be depressed in comparison with the rates for actively feeding larvae. Using the biochemical approach Holland and co-workers have shown that neutral lipids are the main energy reserves not only in bivalve larvae but also in cirripedes (Holland & Walker, 1975) and gastropod larvae (Holland, Tantarasiriwong & Hannant, unpublished data).

Gabbott & Holland (1973) have used the biochemical data to calculate an energy budget for *O. edulis* in terms of growth and metabolism. This gives an independent means of calculating the net growth efficiency, K_2, which is defined as the percentage of the assimilated ration converted into growth:

assimilated ration = growth (P)+respiration (R),

$$K_2 = P/(P+R).$$

Both components of K_2 can be estimated from the changes in composition during larval development (P) and from the starvation experiments (R). Calculated values for K_2 ranged from 78% for newly released larvae on day 0, to 55% on day 8. The results confirm the view that bivalve larvae have high growth efficiencies such that 50–80% of the assimilated ration is converted into growth (Jørgenson, 1952; Walne, 1965; see Chapter 4).

Whereas fully planktotrophic larvae can survive for indefinite periods in the sea, the non-feeding stages can remain alive only as long as stored reserves are available. This is true for bivalve larvae during early embryonic development and again during metamorphosis. Crisp (1975) has calculated the maximum time of survival when different proportions of the body weight are used as energy reserves, assuming a value of 5 ml O_2 h^{-1} per g dry wt as the lower limit for oxygen consumption of continuously active larvae. The data in Table 8.2 show the great advantage of storing lipid rather than protein or carbohydrate, because of the higher caloric content for fats. Furthermore, the survival times make it clear that, without feeding, the length of time the larvae can spend in the plankton is only a matter of a few days; in terms of energy reserves there may be little in excess of the requirement for settlement and metamorphosis (Crisp, 1975). In the case of *O. edulis* larvae the calculated energy loss during starvation is equivalent to an oxygen consumption rate of 5–6 ml O_2 h^{-1} per g dry wt for larvae kept at 20–22 °C (Holland & Spencer, 1973; Gabbott & Holland, 1973). The highest level of lipid reserves (approximately 25% by weight) is reached immediately prior to metamorphosis. This corresponds to a survival time of 4–5 days which is in good agreement with the time taken for the development of the adult feeding mechanisms (Cole, 1938; Hickman & Gruffyd, 1971).

There is a negative correlation between the time for which the larvae of *M. edulis* can delay metamorphosis and temperature, from approximately 40 days at 10 °C to only 2 days at 20–22 °C (Bayne, 1965). During the delay

299

Marine mussels

Table 8.2. *Maximum survival time in days for non-feeding larvae*

Form of energy reserve	Proportion of tissue weight as metabolisable energy reserve			
	5%	10%	25%	50%
Lipid	0.8	1.7	4.2	8.3
Protein	0.5	1.0	2.5	5.0
Carbohydrate	0.3	0.7	1.7	3.3

The table is based on an assumed oxygen consumption rate of 5 ml h^{-1} per g dry wt. Data from Crisp (1975).

feeding is reduced and Bayne observed a reduction in the number of stored oil-droplets around the alimentary tract. Although there are no data on the relationship between temperature and metabolic rate for bivalve larvae, it is interesting to note that at 20–22 °C the delay of metamorphosis is reduced almost to zero, as predicted by the data in Table 8.2; all of the stored energy reserves are required for metamorphosis.

Effects of temperature and nutritive stress

A long and detailed series of experiments on the physiological and biochemical effects of temperature and nutritive stress on *M. edulis* has been undertaken by Dr B. L. Bayne and his collaborators (Bayne & Thompson, 1970; Widdows & Bayne, 1971; Thompson & Bayne, 1972; Bayne, 1973a, b; Bayne, Thompson & Widdows, 1973; Gabbott & Bayne, 1973; Thompson, Ratcliffe & Bayne, 1974; Bayne, 1975a; Bayne, Gabbott & Widdows, 1975). A general aim in these studies has been to extend the observations on the seasonal changes in biochemical composition of *M. edulis* into the laboratory, where the experimental conditions are controlled and the physiological state of the mussels can be measured. In this way it has been possible to gain a better insight into the relationships between food level and temperature and the energy requirements for growth and gametogenesis. The physiological aspects of this work are discussed in Chapters 5 and 7 and we shall be concerned mainly with the biochemical changes. Related studies have been carried out on *O. edulis* by Gabbott & Walker (1971) and Gabbott & Stephenson (1974); on *C. virginica* by Quick (1971); and on *Donax vittatus* by Ansell (1972) and Ansell & Sivadas (1973).

High temperatures and low food levels result in a decline in the body condition (dry weight) of adult *M. edulis* (Bayne & Thompson, 1970; Gabbott & Bayne, 1973). In most of the experiments carried out by Bayne's group the temperature has been held constant at 15 °C, close to summer ambient for the British Isles. Under laboratory conditions, stress

300

Table 8.3. Mytilus edulis: *loss of reserves during stress experiments*

Time of year	Duration (days)	Percentage of total energy loss			Calories lost per day	Reference
		Protein	Carbo-hydrate	Lipid		
Summer	41	0	100	0	8.6	Gabbott & Bayne (1973)
Autumn	60	30	19	51	9.4	Bayne (1973b)
Winter	60	83	10	7	15.8	Bayne & Thompson (1970)
Winter	81	75	10	15	29.2	Gabbott & Bayne (1973)
Winter	154	72	8	20	35.4	Thompson (1972)

can be quantified in terms of the energy balance or 'scope for growth' which is the difference between the assimilated ration and the energy lost in respiration $(A - R)$. According to Bayne and co-workers *M. edulis* can fully acclimate its routine rate of oxygen consumption between 5 °C and 20 °C; above this range there is an increase in oxygen consumption with temperature (Widdows & Bayne, 1971; Widdows, 1973a; Bayne *et al.*, 1973; see Chapter 5). Up to 20 °C the index of energy balance $(A - R)$ is relatively independent of temperature but markedly dependent upon ration (see Bayne *et al.*, 1973 and Chapter 7). In *M. edulis*, temperature is only important as a stress factor during the initial 14-day acclimation period (Widdows & Bayne, 1971), or when the temperature exceeds 20 °C. This is not true, however, for other bivalves such as *D. vittatus* which do not acclimate; in which case temperature stress can be an important aspect of environmental change (Ansell, 1973; Ansell & Sivadas, 1973).

The maintenance ration is defined as the ration level which results in a zero value for the index of energy balance (Chapter 7). In the winter, this corresponds to a food level equal to approximately 1.5% of the body weight per mussel per day (Bayne *et al.*, 1973) and is equivalent, in terms of the assimilated ration, to a metabolic rate of 0.5 ml O_2 h^{-1} per g dry wt (Widdows & Bayne, 1971). When the food level is below the maintenance ration the mussels are out of energy balance and must utilise their body reserves to meet the metabolic energy demand. Table 8.3 shows the loss of reserves during stress experiments, taken from a table drawn up by Bayne (1973b). The metabolic rate is low in summer and high in the winter, due to the increased energy demand for gametogenesis (Widdows & Bayne, 1971; Bayne, 1973a; Gabbott & Bayne, 1973). This difference accounts largely for the variation in daily calorific loss (Table 8.3). In the summer, all of the energy loss is accounted for by the breakdown of carbohydrate, but in the autumn, during more prolonged starvation, there is a marked increase in the utilisation of lipid reserves. Then, in the winter, there is a change-over to protein as the main respiratory substrate. The time course for utilisation

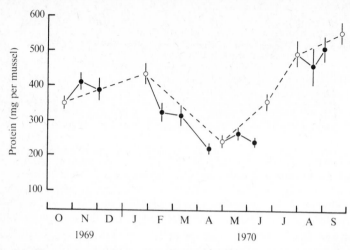

Fig. 8.3. Changes in protein content of the non-mantle tissues of *Mytilus edulis*. Data for mussels of length 5.2 cm (redrawn from Gabbott & Bayne, 1973). ○, seasonal values; ●, stress experiments. Vertical lines represent the standard error of the mean.

of body reserves in the summer is first carbohydrate, then lipid and protein. In the winter, however, it is almost entirely protein that is lost during starvation. This is because the glycogen reserves are at a seasonal minimum and the high lipid levels in the winter are largely associated with the accumulation of egg reserves. The idea that the winter months represent a period of physiological stress for *M. edulis* is borne out by observations in the field as well as the laboratory. Food levels are generally low compared with those in the summer and this is the time when the body weight and condition index decline and protein is lost from the non-mantle tissues (Dare, 1973*a*; Gabbott & Bayne, 1973) (Fig. 8.3). Increased protein catabolism is indicated by a high level of ammonia excretion and a decline in the O:N ratio (Bayne, 1973*a*).

In the European oyster, *O. edulis*, the level of glycogen remains high during the winter months (Walne, 1970*b*). Walne also points out that in oysters there is a positive correlation between the carbohydrate (or glycogen) level and the percentage of body fat (data for *C. virginica*, *C. gryphoides*, *C. gigas*, *C. angulata*, *O. edulis* and *O. lutaria*, taken from the literature; Walne, 1970*b*). This suggests that when conditions are good glycogen and lipid accumulate together and when conditions are poor both are lost to a similar degree; in no case was there a negative correlation between glycogen and fat content (Walne, 1970*b*). This is quite different to the situation in *M. edulis* where there is an inverse relationship between the levels of glycogen and lipid (Lubet & Le Feron de Longchamp, 1969; Williams, 1969*b*) and where protein is the main energy reserve in the

302

winter. The difference between mussels and oysters may be related to the form of the reproductive cycle. In the European oyster, *O. edulis*, gonad development takes place in the spring and summer when food levels are high. Growth and gametogenesis occur at the same time and there is an increase in glycogen and lipid content. During the winter there is a long period of sexual rest (when, presumably, the metabolic demand is low) and both carbohydrate and lipid are available as energy reserves. When oysters are kept in the laboratory at low ration, during the spring and summer, gonad development must take place at the expense of stored reserves. In an unfed group of oysters, Gabbott & Stephenson (1974) showed that the fall in glycogen condition index was proportionately greater than in the dry weight index; egg development took place largely at the expense of glycogen reserves. This is similar to the situation in *M. edulis*, during the autumn and early winter.

Low food levels result in a decline in the body condition of mussels kept in the laboratory; but, in spite of the loss of reserves, *M. edulis* is able to continue gonad development at an increased rate when the temperature is above ambient (Gabbott & Bayne, 1973; Bayne, 1975a). Although gametogenesis appears to be normal under these conditions, there is evidence that stress in the adult affects subsequent larval development in *M. edulis* (Bayne, 1972; Bayne, Gabbott & Widdows, 1975) and in the oyster, *O. edulis* (Helm, Holland & Stephenson, 1973). In mussels, stress resulted in an increase in abnormal embryonic development during cleavage, gastrulation and development to the first shell stage (Bayne, 1972). In oysters, the viability of the newly released larvae, expressed in terms of growth rate and per cent spat yield was less in larvae from adults kept at low ration than from adults kept at high ration (Helm *et al.*, 1973). Irrespective of the feeding regime, the viability of the larvae declined as the length of the conditioning period increased and this could be correlated with a fall in body condition of the parent stock. The growth of the larva during the first 96 h after liberation was positively correlated with the lipid level in the newly released larvae; larvae liberated late in the experiment had the lowest lipid levels and grew least well. This general point, that larvae derived from adults kept under the greatest degree of stress have the lowest growth rate, has also been confirmed for *M. edulis* by Bayne, Gabbott & Widdows (1975).

In bivalves, the early stages of larval development represent a period of intense morphogenetic activity when there is complete dependence on the stored energy reserves acquired from the adult. This is followed by a period of rapid growth as the shell is deposited and the larvae increase in size (Chapter 4). During the second stage of growth the larvae are feeding and there is less reliance on stored energy reserves. Recognition of these stages suggests that stress in the adult will have a maximum effect on early embryonic development and a minimal effect during the main larval growth phase (Bayne, 1972; Bayne, Gabbott & Widdows, 1975). It is possible that the effects of stress on the growth of newly released oyster larvae described by Helm *et al.* (1973) were partially due to effects on earlier embryonic development when the larvae were still in the mantle cavity of the adults.

Williams (1969*b*) has looked at the effect of the copepod parasite, *Myticola intestinalis*, on the biochemical composition of *Mytilus edulis*. There were only small differences, however, between the infected and non-infected mussels; in the summer the level of carbohydrate was slightly lower in the parasitised mussels. A similar decline in the level of carbohydrate has been reported for infected clams, *Mercenaria mercenaria*, by Jeffries (1972). Hard clams were collected from polluted and clean habitats in Narragansett Bay, Rhode Island. Jeffries (1972) recognised a stress syndrome (see Chapter 7) in response to environmental pollution and infection of the shells by the polychaete *Polydora*; the levels of carbohydrate and free amino acids fell and the molar ratio of free taurine to glycine increased. The stress syndrome had many other facets but could be recognised simply by observing the amount of taurine and glycine; when the molar ratio was less than 3, the population was normal but with a ratio above 3 the clams were stressed. A similar pattern in the free amino acids in the haemolymph of *C. virginica* has been described by Feng, Khairallah & Canzonier (1970) for oysters infected with *Bucephalus* sp. and *Minchinina nelsoni*. Jeffries (1972) calculated the molar ratios of taurine to glycine from the data given by Feng *et al.* (1970) and showed that the ratio ranged from 0.5 to 2.0 (mean 1.2) for normal oysters and from 1.3 to 9.1 (mean 4.0) for the infected oysters. Feng *et al.* (1970) have proposed a compensatory mechanism in the host, in which keto acids are converted into amino acids that have been lost to the parasite; at the same time the taurine level is increased to maintain osmotic balance.

Relationships between carbohydrate, lipid and protein metabolism

Storage cycle in Mytilus edulis

In *M. edulis* the distribution of assimilated food to the body tissues is controlled by the digestive gland (Thompson, 1972; Chapter 5). Thompson showed that, in mussels fed with ^{14}C-labelled *Tetraselmis*, the rate of transfer of the ^{14}C-label from the digestive gland varied with season. In the summer (June) and autumn (September) transfer of the assimilated food took place rapidly and was essentially complete within 7–10 days. At this time of year ration had little or no effect. In January, however, the rate of transfer was much slower in starved mussels and continued for at least 35 days; at high ration the transfer was again complete within 7–10 days (see

Table 8.4. Mytilus edulis: *distribution of radioactivity* (*c.p.m. per mg eggs*) *in unfertilised eggs*

	Radioactivity		
Condition of adults	Total	Lipid	Neutral lipid
Experiment 1, 1972			
High ration	340	86	54
Low ration	1380	312	177
Experiment 2, 1973			
High ration	—	51	36
Low ration	—	204	134

Adult mussels were fed ^{14}C-labelled *Tetraselmis* for 6 h and then spawned 3–4 weeks later. Both experiments were conducted at 16 °C.
Data from Bayne, Gabbott & Widdows (1975).

Chapter 5). In a series of experiments carried out in the winter of 1972 and 1973, Bayne and co-workers fed ^{14}C-labelled *Tetraselmis* to two separate groups of mussels which had been kept in the laboratory at 'high ration' or 'low ration' (Bayne, Gabbott & Widdows, 1975). Both groups were fed the same amount of algae for a period of 6 h; then 3–4 weeks later the mussels were spawned and the amount of radioactivity in the eggs determined (Table 8.4). The results showed there was more ^{14}C-label in the eggs from mussels kept at low ration than in the eggs from mussels kept at high ration. In each case the distribution of the label between the lipid and non-lipid components was similar, with 22–28% of the activity in the lipid fraction and 58–65% of the lipid activity in the neutral lipid components. Thin-layer chromatography examination of the neutral lipids showed that most of the ^{14}C was in the triglycerides (see Bayne, 1975a). This data has been interpreted as follows. When the mussels were kept at high ration the ^{14}C-label was lost rapidly from the digestive gland. Some would be transferred to the developing eggs and be retained in the yolk but most of the assimilated food would be used for maintenance metabolism. At low ration, the transfer would be slower and take place over a longer period so that, ultimately, more of the ^{14}C-label accumulated in the eggs than in mussels kept at high ration (Bayne, Gabbott & Widdows, 1975).

These tracer experiments can be related to the seasonal changes in glycogen content of adult mussels. In the autumn and early winter (October–January) the loss of glycogen reserves is synchronous with Stage II – oogenesis and vitellogenesis of the reproductive cycle. At this time of year the food level is above the maintenance ration because there is no overall loss in body condition (Dare, 1973a; Gabbott & Bayne, 1973); the assimilated food is transferred rapidly from the digestive gland and is

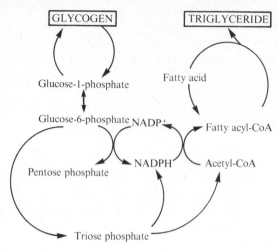

Fig. 8.4. Metabolic pathways for the conversion of carbohydrate into lipid.

available for respiratory metabolism. This is analogous to the 'high ration' condition in Table 8.4. It is difficult to see why there should be a loss of glycogen unless this is directly associated with vitellogenesis and it is assumed that there is a net conversion of pre-stored glycogen into the lipid reserves of the developing eggs (Bayne, Gabbott & Widdows, 1975; Gabbott, 1975). In contrast, from January to March the glycogen reserves are at a minimum and the food levels are low. The metabolic rate remains high, however, and the increased energy demand is met mainly from protein reserves in the non-mantle tissues (Gabbott & Bayne, 1973). During this time vitellogenesis may still be taking place (but at a reduced rate) and the transfer of nutrients from the digestive gland to the mantle tissue may be important for the continuation of gonad development (see Thompson *et al.*, 1974).

Fig. 8.4 shows the metabolic pathways for the conversion of glycogen into fatty acid or triglyceride reserves. These involve glycolysis in which glucose-6-phosphate is converted into acetyl-CoA; lipogenesis in which acetyl-CoA is converted into long-chain fatty acyl-CoA; and, finally, formation of the triglycerides by esterification of glycerol phosphate. All three precursors for triglyceride synthesis, namely acyl-CoA, NADPH and triose phosphate, are provided by the breakdown of glycogen. Glucose-1-phosphate is readily converted into glucose-6-phosphate by phosphoglucomutase. This enzyme, together with phosphorylase, has been detected in *M. californianus* by Bennett & Nakada (1968). Glucose-6-phosphate is at the branchpoint of several metabolic pathways. It can be dephosphorylated to glucose and transported around the body in the haemolymph; it can be converted into fructose-6-phosphate and enter the

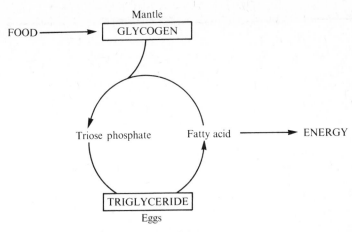

Fig. 8.5. Glycogen storage cycle in *Mytilus edulis*.

glycolytic sequence; or it may be oxidised to ribulose-5-phosphate via the pentose phosphate pathway resulting in the production of NADPH for fatty acid and sterol synthesis. The pentose phosphate pathway also leads to the production of pentose sugars and in particular ribose-5-phosphate which is a basic constituent of the nucleic acids. In this way glycogen provides five-carbon sugars for nucleic acid synthesis as well as the necessary intermediates for lipogenesis. Both synthetic pathways are of major importance during gonad development; vitellogenesis takes place in the eggs and DNA and RNA synthesis in both the eggs and spermatozoa.

In adult mussels, the conversion of pre-stored glycogen into lipid reserves in the developing eggs can be considered as a 'storage' cycle, analogous to the glucose–fatty acid cycle in vertebrates (Fig. 8.5). The reader should note, however, that the evidence for a carbohydrate-lipid storage cycle in marine bivalves is largely circumstantial; nonetheless the cumulative evidence is sufficient for the cycle to be put forward as a working hypothesis (see Gabbott, 1975). There are two important features of such a cycle. First, it can be distinguished from the normal 'metabolic' cycles by the fact that the different steps are essentially separated in time (Krebs, 1972). In *M. edulis* glycogen is synthetised in the mantle tissues during the summer and converted into triglyceride reserves in the autumn and early winter. The fatty acids are then oxidised and used as an energy source during early embryonic development. Secondly, there is the metabolic cost of converting carbohydrate into lipid; in fact, this is quite small (Krebs, 1972). For each two-carbon fragment from glucose which is converted into a two-carbon fragment of fatty acid, 2 moles of ATP are consumed. One three-carbon fragment of glucose yields 18 moles of ATP on oxidation while the two-carbon fatty acid fragment yields 17 moles of

ATP. The net energy loss is 6 moles of ATP per mole of glucose. Using a slightly more sophisticated calculation Krebs (1972) has shown that less than 7% of the energy to be stored is lost during the conversion of carbohydrate to lipid. In adult bivalves the level of glycogen can be as high as 40–50% of the dry tissue weight, but in the eggs and larvae there are clear advantages to be gained from storing fat. First, it is a more concentrated energy form (see Table 8.2) and secondly there is an increase in buoyancy due to its lower density than carbohydrate or protein. The loss of utilisable energy in converting pre-stored glycogen into lipid reserves in the eggs is the price paid for this measure of adaptability.

There are two possible levels of control for the glycogen–lipid storage cycle in *M. edulis*. One is physiological and concerns the distribution of assimilated food to the body tissues, and is regulated by the digestive gland (see previous discussion). The second is biochemical and concerns the regulation of glycogen metabolism (see Gabbott, 1975). In mammals glycogen is stored in the liver, where it can be converted back to glucose and be distributed throughout the rest of the body, and in the muscles where it is available for immediate use. In bivalves, the general reduction in hepatic function has resulted in the proliferation of Leydig tissue for the local storage of glycogen, together with the enzymes associated with its metabolism (Eble, 1969). The second assumption we must make is that the control of glycogen metabolism in marine bivalves is essentially the same as that in mammalian liver and muscle. In the liver the main regulatory mechanisms are hormonal, mediated by the action of cyclic AMP as a second messenger (Hers, De Wulf & Stalmans, 1970). In addition there are a number of other controls that operate at the metabolic level, such as the control of muscle phosphorylase by the adenosine nucleotides and the control of liver phosphorylase by blood glucose (see reviews by Hers *et al.*, 1970; Hochachka, 1973; and Newsholme & Start, 1973).

Cyclic AMP acts antagonistically on the glycogen synthetase and phosphorylase systems so that when glycogen synthesis is activated, glycogen breakdown is inhibited and vice versa. This is important because it means that glycogen synthesis and breakdown do not normally occur at the same time and the control is mediated by an on–off switch mechanism. It is interesting to note that Bayne, Gabbott & Widdows (1975) failed to detect any ^{14}C-label in the Leydig cells of the mantle tissue of *M. edulis* during the late stages of gametogenesis; at this time the glycogen content is low and one might have expected some synthesis of glycogen, at least under high ration conditions. Thompson *et al.* (1974) have shown that in the early summer there is an accumulation of carbohydrate in the digestive gland, most of which is lost during July at the time when glycogen reserves are built up in the mantle. This may be taken as further evidence that the synthesis of glycogen does not take place in the mantle tissues until gonad development is complete. Lubet (1959) also noted that the glycogen content of *M. edulis* increased in the spring (April and May) but there was no storage in the connective tissue. The main difference between glycogen synthesis and utilisation in mammalian liver and the glycogen storage cycle in marine bivalves is the time scale of the metabolic changes. In the liver glycogen has a half-life of 24–36 h; synthesis is rapid and the hormonal control mechanisms represent a true modulation of enzyme activity with response times in the order of seconds or minutes. In contrast, glycogen synthesis and breakdown in bivalves takes place on a seasonal basis and the metabolic responses are essentially long-term.

Two assumptions have been made; first, that vitellogenesis takes place at the expense of stored glycogen reserves and secondly, that glycogen metabolism is regulated by hormonal mechanisms. If these assumptions are true then glycogen metabolism and gametogenesis may both be controlled by the same regulators (Gabbott, 1975). According to Lubet (1959; see review, 1973) gonad development in *Mytilus* is controlled by internal neuro-endocrine factors and the external factors such as temperature and food act as synchronisers, which extend or shorten the different stages (I–III) of gametogenesis. More recently Houtteville & Lubet (1974) have distinguished between (*a*) the effect of the cerebral ganglia on gonad development and on the utilisation of glycogen and protein reserves in the connective tissues, and (*b*) the action of the visceral ganglia which seems to control the accumulation of reserves during the period of sexual rest. This is clearly an area for further investigation; studies on the adenyl cyclase–cyclic AMP reaction and on glycogen synthetase and phosphorylase may provide an in-vivo system on which to test the action of the neurosecretory products as hormonal messengers.

Lipid and sterol metabolism

Studies on the lipid and sterol composition of the Mollusca have been reviewed by Voogt (1972). Most attention has been focused on the fatty acid components and the structure of the molluscan sterols. Many of the early investigations have become outdated with the introduction of modern methods of analysis based on thin-layer (TLC) and gas–liquid (GLC) chromatography, and infrared and mass spectrometry.

Table 8.5 summarises the available data on the fatty acid composition of *Mytilus* species (Rodegker & Nevenzel, 1964; Bannatyne & Thomas, 1969; Gardner & Riley, 1972). The most complete analysis is that given by Gardner & Riley (1972), and includes separate values for the different lipid classes separated by preparative TLC. All of the authors have commented on the high proportion of long-chain polyunsaturated fatty acids of the C_{20} and C_{22} series. This distribution of penta- and hexa-enoic acids (20:5 and

Table 8.5. *Percentage distribution of fatty acids in* Mytilus *species*

Fatty acid	M. edulis*			M. californianus[†]			Perna canaliculus[‡]
	TG	PL	STE	Male gonad	Female gonad	Body tissue	
12:0	0.3	1.1	7.0	Not determined			
14:0	3.0	1.9	1.5	1.3	1.0	3.5	5.0
16:0	13.6	10.7	11.5	35.5	25.1	24.5	17.3
16:1	8.3	4.8	6.7	8.2	3.5	1.1	11.2
16:2	2.0	1.6	Trace	2.6	2.6	3.0	0.9
16:3	0.4	0.6	2.5	Not determined			
16:4	1.0	1.0	1.2	Not determined			
18:0	4.7	10.0	5.9	2.4	3.2	1.7	4.8
18:1	7.4	6.8	9.2	4.5	4.1	3.2	4.7
18:2	2.7	1.3	2.0	2.7	2.2	3.2	0.9
18:3	0.3	2.8	5.1	—	—	—	0.9
18:4	6.1	4.8	4.4	1.4	1.3	1.6	—
20:1	8.8	6.2	6.1	2.7	4.4	2.6	6.5
20:2	1.0	1.3	Trace	4.5	2.8	7.5	—
20:3	0.5	1.0	Trace	Not determined			
20:4	4.1	4.8	4.1	0.6	0.9	1.0	2.2
20:5	12.9	9.6	7.9	12.3	15.2	14.0	25.2
22:1	3.4	7.0	1.6	—	—	—	Trace
22:2	1.8	1.2	Trace	1.3	1.5	2.2	—
22:3	1.6	0.4	Trace	—	—	—	1.7
22:4	—	—	Trace	Not determined			
22:5	1.5	2.5	Trace	1.1	1.2	1.1	2.0
22:6	5.9	3.2	3.2	18.2	29.7	27.7	15.6

TG, triglycerides; PL, phospholipids; STE. sterol esters.
Data from * Gardner & Riley (1972); [†] Rodegker & Nevenzel (1964); [‡] Bannatyne & Thomas (1969).

20:6) is a general feature of all marine bivalves so far investigated (Gardner & Riley, 1972; Voogt, 1972). In contrast, the freshwater bivalve *Anodonta* sp. had proportionately less total polyunsaturated acids (32.8%) compared to the marine species (43–49%); this is analogous to the difference between lipids of marine and freshwater fish (Gardner & Riley, 1972). Gardner & Riley (1972) also showed that marine bivalves, including *M. edulis*, contain small quantities of odd-numbered C_{13}–C_{21} saturated and unsaturated fatty acids and C_{13}–C_{19} and C_{21} branched-chain acids. This may be important because, in contrast to the even-numbered fatty acids which are completely oxidised to carbon dioxide and water, the odd-numbered acids result in the production of propionyl-CoA which can then be converted into succinyl-CoA (see later).

A number of studies have been carried out on the relationship between component fatty acids of the (algal) diet and the fatty acid composition of marine zooplankton (see, for example, Jezyk & Penicnak, 1966; Culkin &

Morris, 1970; Hinchcliffe & Riley, 1972). Recently Watanabe & Ackman (1974) have examined the effect of feeding a meal of *Dicrateria inornata* or *Isochrysis galbana* on the fatty acid composition of the oysters *C. virginica* and *O. edulis*. The two algal species were chosen because they have a somewhat different fatty acid content. Thus, *D. inornata* lacks long-chain C_{18}–C_{22} polyunsaturated fatty acids whereas *I. galbana* contains high quantities of 18:4 and 20:5 acids (Watanabe & Ackman, 1974). The results, however, were essentially negative and showed that the fatty acid composition of the oysters was species-orientated rather than dependent on the diet; although in only one combination, *O. edulis* and *D. inornata*, was enough fatty acid assimilated to modify the oyster composition in an easily detectable way. The most interesting difference between the composition of the oysters and that of the diet was the absence of a 16:3 acid in *O. edulis* (the deposition of unchanged 16:3 from *D. inornata* would have resulted in approximately 1% of the acid in *O. edulis*, and this would have been easily measured). Watanabe & Ackman (1974) suggest that the short-chain unsaturated fatty acids, commonly found in phytoplankton species (see, for example, Chuecas & Riley, 1969) are either degraded by β-oxidation or converted into the common higher homologues of the C_{20}–C_{22} series. The virtual absence of C_{16} polyunsaturated fatty acids extends to other invertebrates in the marine food chain (see, for example, Culkin & Morris, 1970; Gardner & Riley, 1972). The conclusion that the fatty acid composition of oysters is species-orientated rather than diet-orientated is the same as that drawn by Hinchcliffe & Riley (1972) for *Artemia* fed on a range of uni-algal diets.

In comparison with the work on the distribution of fatty acids in bivalves, we know very little about their metabolism (see Voogt, 1972). There are two outstanding problems:

(1) Does the synthesis of polyunsaturated C_{20}–C_{22} fatty acids take place *de novo* or are acetate units added to unsaturated C_{16} and C_{18} acids from the diet? One implication of the observation made by Watanabe & Ackman (1974), that *O. edulis* does not contain any 16:3 fatty acid when fed *D. inornata* is that C_{16} acids are converted into higher homologues by chain elongation. This question is not easy to answer. The main difficulty concerns the use of specifically labelled [^{14}C]fatty acids; most labelled fatty acids have the ^{14}C in the carboxyl position e.g. [1-^{14}C]palmitic acid. In the absence of data defining the position of the ^{14}C-label incorporated into the higher polyunsaturated fatty acids it is not possible to distinguish between de-novo synthesis by direct elongation and desaturation reactions and the addition of two-carbon acetate units (derived by partial oxidation of the [^{14}C]fatty acid precursor) to C_{16} and C_{18} polyunsaturated acids of dietary origin (see discussion in Morris & Sargent, 1973). An alternative approach is suggested by recent work on the development of microencapsulated

311

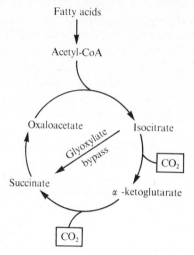

Fig. 8.6. Metabolism of acetate via the Krebs cycle and the glyoxylate cycle. For explanation see text.

diets for suspension-feeders (Jones, Munford & Gabbott, 1974; Jones & Gabbott, 1975). It might be possible to rear bivalve (or crustacean) larvae on a chemically defined diet and, in this way, determine whether or not there is an essential fatty acid requirement. Furthermore, if the larvae can be grown on a fat-free diet, then by definition there can be no dietary source of fatty acids so that the interpretation of ^{14}C-tracer experiments would be unequivocal.

(2) In higher animals, fatty acids cannot be used as a sole source of energy unless there is some other means of replenishing the Krebs cycle intermediates. This is because the cycle is used not only for oxidative metabolism but also to generate α-keto acids for biosynthesis. The reason why fatty acids cannot be used for this purpose is that in the first part of the Krebs cycle two carbon atoms are lost as carbon dioxide (Fig. 8.6). Normally the intermediates of the cycle are replenished from glucose via pyruvate (or phosphoenolpyruvate) and from the glucogenic amino acids. A problem arises during the metamorphosis of bivalve larvae; immediately after settlement the spat cannot feed and there is considerable use of lipid reserves to meet the metabolic energy demand (Holland & Spencer, 1973). The carbohydrate levels are low and there is no loss of body protein (it would not be a good adaptive strategy to lose protein at a time of maximum morphogenetic activity). How does the oxidation of acetyl-CoA continue without carbohydrate or protein to replenish the intermediates of the Krebs cycle? There are a number of possibilities. First, there may be sufficient glycogen or other stored carbohydrate for this purpose or

alternatively, glycerol (from breakdown of the triglycerides) could be used to generate pyruvate or phosphoenolpyruvate. Secondly the larvae may contain odd-numbered or esterified volatile fatty acids which can be converted via propionyl-CoA into succinate. Finally the larvae might be able to bypass the two carboxylation steps by means of the glyoxylate cycle (see Fig. 8.6). The glyoxylate cycle is found in a wide variety of micro-organisms and fatty seeds (see review by Kornberg & Elseden, 1961) and in several parasitic metazoans (Prichard & Schofield, 1968; Barrett, Ward & Fairbairn, 1970).

The key enzymes of the cycle are isocitrate lyase and malate synthase:

$$\text{Isocitrate} \xrightarrow[\text{lyase}]{\text{isocitrate}} \text{succinate} + \text{glyoxylate.}$$

$$\text{Glyoxylate} + \text{acetyl-CoA} \xrightarrow[\text{synthase}]{\text{malate}} \text{malate.}$$

Malate is converted into oxaloacetate via the normal Krebs cycle, and then combines with acetyl-CoA to give citrate. The net result is the conversion of 2 moles of acetyl-CoA into succinate. Recently Leonard (1975) has shown that isocitrate lyase activity can be detected spectrophotometrically (by the method of Dixon & Kornberg, 1959) in crude extracts of oyster larvae. In *O. edulis* activity increased from 9.7 nmoles glyoxylate per mg protein per hour in newly released larvae to 56.6 nmoles glyoxylate per mg protein per hour in eyed larvae, immediately prior to settlement. Some activity persisted in 1-day-old and 1-week-old spat, but the activity was completely absent in 5- and 8-week-old oysters. Furthermore, there is no isocitrate lyase in the adult body tissues of *M. edulis* (De Zwaan & Van Marrewijk, 1973b; Leonard, 1975). The changes in isocitrate lyase activity in oyster larvae are similar to the developmental changes in *Ascaris* eggs in which the glyoxylate cycle operates only for a short period during the conversion of triglyceride reserves into trehalose and glycogen (Barrett *et al.*, 1970). Although the circumstantial evidence is persuasive, further studies are required before it can be firmly established that bivalve larvae have a functional glyoxylate cycle.

Studies on the structure and composition of molluscan sterols have been reviewed by Idler & Wiseman (1972); results prior to the application of GLC and mass spectrometry are discussed in relation to more recent studies. Many of the older papers contain references to supposedly new sterols, but in most cases proper elucidation of the structures is lacking or has proved to be incorrect because the preparations were not homogeneous (Idler & Wiseman, 1972). Nowadays pure sterols can be isolated by preparative GLC and TLC. Lamellibranch bivalves contain the most varied and complex sterol mixtures found in the Mollusca; with few exceptions cholesterol is the principal Δ^5-sterol, but as many as thirteen different components may be present in one species (Idler & Wiseman, 1972). The quantity of provitamin D $\Delta^{5,7}$-sterols varies, but generally they are only minor components.

M. edulis contains nine different sterols with cholesterol (58.7%), brassicasterol (9.6%), 22-dehydrocholesterol (9.5%), desmosterol (8.1%), poriferasterol or 24-methylenecholesterol (6.4%) and 22-*trans*-24-norcholesta-5, 22-dien-3β-0l (5.9%) as the major components (Idler & Wiseman, 1971). Essentially the same compounds were found in a later study by Teshima & Kanazawa (1974; see below). So far, however, there is no information on seasonal changes in sterol composition in *Mytilus*, or on their physiological function apart from an unspecified role in gonad development (see Longcamp, Lubet & Drosdowsky, 1974).

Synthesis takes place from acetate according to the following general scheme:

Acetate ⟶ mevalonate ⟶ [isoprenoid unit]

⟶ squalene

lanosterol

Lanosterol is a precursor of cholesterol and probably all the other Δ^5-sterols. Reports on the conversion of [^{14}C]acetate or [^{14}C]mevalonate into sterols of marine bivalves are confusing. Fagerlund & Idler (1960) were the first to show the conversion of [^{14}C]acetate into the sterol components of *M. californianus* and the clam, *Saxidomus giganteus*. Clams also converted [^{14}C]squalene into sterols, suggesting a role for squalene in the biosynthetic pathway. On the other hand Salaque, Barbier & Lederer (1966) found that there was no incorporation of radioactivity from [^{14}C]mevalonate into any sterols, after incubation of *Ostrea gryphea* for 70 h. Similarly Voogt (1972) has reported that, apart from one experiment with *O. edulis*, injection of oysters and mussels (*M. edulis*) with [^{14}C]acetate does not result in radioactive sterols even after 120 h incubation. It has been suggested by Idler & Wiseman (1972) that possibly the duration of the experiments was too short, particularly if synthesis of sterols in bivalves is slow. The claim made by Voogt (1972) that, when *Cardium edule* was injected with [2-^{14}C]acetate the sterols became labelled, whereas with [1-^{14}C]acetate they were not, is difficult to understand in relation to the supposed biosynthetic pathway from mevalonate. More recently Teshima & Kanazawa (1974) have shown that after incubation for 12 h and 8 d, respectively, following an injection of [^{14}C]mevalonate both the abalone, *Haliotis gurneri*, and the mussel, *M. edulis*, contained ^{14}C-labelled squalene and Δ^5-sterols. Teshima & Kanazawa (1974) identified cholesterol, 22-dehydrocholesterol, desmosterol and 24-methylenecholesterol as ^{14}C-labelled metabolites in *M. edulis* but it was not clear whether mevalonate had been incorporated into any of the other sterol components.

Finally, Longcamp *et al.* (1974) have used ^{3}H- and ^{14}C-labelled precursors to follow the interconversion of a number of sterol compounds in the male and female gonad of *M. edulis*. There was very little conversion of [^{3}H]acetate into any of the sterols and none into cholesterol, but quite possibly this was due to the short incubation times of 2 and 6 h (see previous discussion). Four types of enzyme reactions were demonstrated: a 3β-hydroxysteroid dehydrogenase-$\Delta^{5.4}$-isomerase; a C_{17-20}-lyase; a 17β-hydroxysteroid dehydrogenase; and a 5α-reductase (Longcamp *et al.*, 1974). Seasonal changes in 17β-hydroxysteroid dehydrogenase activity in *C. gigas* have been studied histochemically; activity increased as maturation of the gonad took place and then declined after spawning (Mori, Tamate & Imai, 1966). According to Mori *et al.* (1966) there is a close relationship between 17β-hydroxysteroid dehydrogenase activity (oestrone$\rightleftharpoons$17β-oestradiol) and between glycogen metabolism and sexual development in *C. gigas*; breakdown of glycogen via the pentose phosphate pathway provides NADPH for sterol synthesis.

Amino acid metabolism

The comparative aspects of nitrogen metabolism in molluscs have been comprehensively reviewed by Campbell & Bishop (1970). Other recent reviews are those by Florkin & Bricteux-Grégoire (1972); Allen & Garrett (1971*a*) on taurine; by Lange (1972) and Schoffeniels & Gilles (1972) on osmoregulation; and by Campbell (1973) on nitrogen excretion.

Studies on the levels of free amino acids in different species and tissues of lamellibranchs have been summarised by Campbell & Bishop (1970) and by Florkin & Bricteux-Grégoire (1972). The quantitatively most important amino acids are (in alphabetical order) alanine, arginine, aspartic acid, glutamic acid, glycine and proline. All are 'non-essential' in the sense that they can be formed from carbohydrate (the reader should note, however, that the concept of 'essential' and 'non-essential' amino acids is based on experimental nutrition and the fact that synthesis can take place from glucose does not necessarily mean that the rate of synthesis is adequate

315

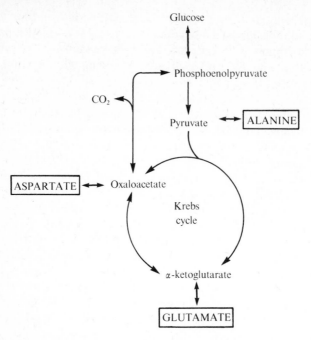

Fig. 8.7. Metabolic pathways for the interconversion of alanine, aspartate and glutamate.

for the physiological needs of the animal; see comment on arginine metabolism). Conversely, the six major free amino acids are all glucogenic. Alanine, aspartic acid and glutamic acid are directly interconvertible with the corresponding keto acid intermediates of the Krebs cycle (Fig. 8.7). These reactions are catalysed by the aminotransferases glutamate–pyruvate transaminase (GPT) and glutamate–oxaloacetate transaminase (GOT). Read (1962b) has studied the specificity of the aminotransferases in the digestive gland of *M. edulis*. Transamination of amino acids other than alanine or aspartate with α-ketoglutarate and amino acids other than glutamate with pyruvate or oxaloacetate were negligible under the conditions employed. Both the GPT and GOT reactions were freely reversible;'GOT activity was found in all the body tissues of *M. edulis* except for the haemolymph.

In mammals glycine and serine are interconvertible via the action of serine hydroxymethylase:

$$\text{L-serine} + \text{tetrahydrofolate} \rightleftharpoons \text{glycine}$$
$$+ N^5, N^{10}\text{-methylenetetrahydrofolate}.$$

Serine can be formed by oxidation of glycerate and transamination with glutamate or alanine.

Phosphorylated pathway:

$$NAD^+ \quad NADH \qquad\qquad glutamate \quad \alpha\text{-ketoglutarate}$$

3-phosphoglycerate $\rightarrow$ phosphohydroxypyruvate $\rightarrow$ phosphoserine $\longrightarrow$ serine.

Non-phosphorylated pathway:

$$NAD^+ \quad NADH \qquad alanine \quad pyruvate$$

D-glycerate $\rightarrow$ hydroxypyruvate $\rightarrow$ serine.

The incorporation of radioactivity from [^{14}C]bicarbonate and [U-^{14}C]-glucose into serine and glycine indicates that these pathways are operative in molluscs (data for land snails; Campbell & Bishop, 1970).
Proline is formed from glutamate in a pathway involving two reductions by NADH:

$$NADH \quad NAD^+ \qquad\qquad NADH \quad NAD^+$$

Glutamate $\rightarrow$ pyrroline-5-carboxylate $\rightarrow$ proline.

Labelling of proline from [^{14}C]bicarbonate has been taken as evidence for the glutamate $\rightarrow$ proline pathway in molluscs (Campbell & Bishop, 1970). So far, however, none of the biosynthetic pathways for serine, glycine and proline have been studied in marine bivalves apart from the demonstration of serine hydroxymethylase activity in the scallop, *Pecten caurinus* (Whiteley, 1960).

Arginine is also related to glutamate. But, in most mammals, arginine is considered to be an essential amino acid because the rate of formation is too slow to satisfy the requirement for protein synthesis. Arginine metabolism in molluscs has been reviewed by Campbell & Bishop (1970) and by Florkin & Bricteux-Grégoire (1972) (see Chapter 6).

Formation of alanine from pyruvate requires a source of glutamate or ammonia (reductive amination). Glutamic acid is formed from α-keto-glutarate in a reaction catalysed by glutamic dehydrogenase (GDH):

$$NADH/NADPH + \alpha\text{-ketoglutarate} + NH_4^+ \rightleftharpoons NAD^+/NADP^+$$
$$+ glutamate + H_2O.$$

Lange (1972) has pointed out that formation of amino acids via transamination with glutamate is an overall oxidation–reduction process, resulting in the oxidation of NADH formed during glycolysis. Parenthetically, the reductive amination of pyruvate is an essential feature of the scheme proposed by De Zwaan and co-workers for anaerobic metabolism in

317

Marine mussels

M. edulis (see later). This argument can be extended to aspartic acid, glycine and proline so that formation of all the quantitatively important amino acids is associated with the reoxidation of NADH. The possible role of proline in the maintenance of redox balance under anaerobic conditions has been discussed by Hochachka, Fields & Mustafa (1973). In a sense these particular amino acids can be considered as end-products of energy metabolism and are the ones most likely to be lost by excretion (Lange, 1972).

Glutamate dehydrogenase is important for another reason. None of the interconversions of amino acids and keto acids can result in the net synthesis of α-amino-nitrogen without, at some stage, fixation of ammonia. This takes place via the GDH reaction. At the same time GDH is generally considered to be the main enzyme responsible for deamination *in vivo*. Apart from glutamate, other amino acids can be oxidised by a combination of transamination reactions (to alanine and then to glutamate) followed by deamination to α-ketoglutarate (see review by Campbell, 1973). However, in molluscs there is no convincing experimental evidence that amino acids can be catabolised by transdeamination and GDH appears to operate mainly in the direction of ammonia fixation (Lange, 1972; Campbell, 1973). This point is discussed again on p. 333 in relation to the reductive amination of pyruvate (see De Zwaan & Van Marrewijk, 1973a; De Zwaan, Van Marrewijk & Holwerda, 1973). Clearly it is a very important area for further investigation, not only in relation to the direction of the GDH reaction but also because of the possibility of enzyme–enzyme complex formation with the aminotransferases (Fahien & Smith, 1974). The alternative possibility, that ammonia is formed directly by the activity of L-amino acid oxidases, would seem to be unlikely since, in the molluscs, oxidation is essentially limited to the basic amino acids and is inhibited by those amino acids, such as alanine, glycine and proline, which do not act as substrates (see reviews by Campbell & Bishop, 1970; Lange, 1972; Campbell, 1973; also Chapter 6).

Intermediary carbohydrate metabolism

Glycogen metabolism

In most marine bivalves glycogen is the major carbohydrate storage reserve. In *M. edulis* the blood sugar levels are low and the 'index of blood sugar reserve' (time for which the total blood sugar would meet the metabolic energy demand), varies from approximately 15 min in the mid-winter to 200 min in the late summer (Bayne, 1973a). In many insects, trehalose rather than glucose is the main blood sugar reserve, but in molluscs the role of trehalose is still obscure. Badman (1967) has studied the seasonal changes in trehalose content of *C. virginica*; compared to

318

glycogen the levels of trehalose were very low except for the period after spawning when the glycogen content was at a seasonal minimum. Badman concluded that trehalose was not a storage reserve but suggested a possible role in maintaining low blood glucose levels and establishing a steep gradient for absorption of glucose from the alimentary tract. Recently L-Fando, García-Fernández & R-Candela (1972) have demonstrated the conversion of [U-^{14}C]trehalose into glycogen in homogenates of *O. edulis*. So far, however, there is no information on the occurrence or distribution in bivalves of the enzymes responsible for the interconversion of glucose and trehalose.

Surprisingly, little is known about the interconversion of glucose and glycogen in marine bivalves. UDPG–glycogen transglucosylase (glycogen synthase) has been found in cell-free extracts of *M. californianus* by Wang & Scheer (1963) and in *M. edulis* by Lamana (1973). Glycogen phosphorylase activity has been demonstrated histochemically in *C. virginica* by Eble (1969). Phosphorylase was found in Leydig tissue close to active metabolic sites such as the developing gonad, the digestive gland and alimentary tract, the adductor and heart muscles, and the mantle and gills (Eble, 1969). Phosphorylase has also been detected in cell-free extracts of the digestive gland of *M. californianus* (Bennett & Nakada, 1968). Using tissue slices of the digestive gland Bennett & Nakada (1968) have shown that [^{14}C]glucose is converted into $^{14}CO_2$ (glycolysis) and into labelled glycogen by *M. californianus*. L-Fando *et al.* (1972) have investigated some of the factors affecting glycogen synthesis in *O. edulis*. Synthesis from [^{14}C]glucose was inhibited under anaerobic conditions and there was an inverse correlation between the incorporation of glucose into glycogen and the level of glycogen in the gill tissues. In oysters glycogen may act as a negative feedback inhibitor of its own synthesis.

Glycolysis and gluconeogenesis

It is generally agreed that the classical Embden–Meyerhof pathway for glycolysis operates in bivalves as far as the level of phosphoenolpyruvate (PEP). All of the enzymes of the glycolytic pathway have been demonstrated in *M. californianus* and *Haliotis rufescens* by Bennett & Nakada (1968). More recent reports by Engel & Neat (1970) on *Mercenaria mercenaria* and by O'Doherty & Feltham (1971) on *Plactopen magellanicus* have been concerned with the relative activities of the key enzymes which regulate the flow of glycolytic intermediates in the direction of glycolysis or gluconeogenesis (see below).

Fig. 8.8 shows the metabolic pathways for glycolysis and gluconeogenesis. In the glycolytic sequence three of the steps are non-equilibrium reactions (those catalysed by hexokinase, phosphofructokin-

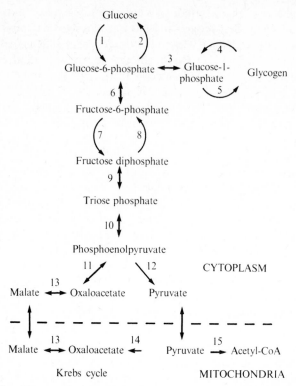

Fig. 8.8. Metabolic pathways for glycolysis and gluconeogenesis. 1, HK, hexokinase; 2, glucose-6-phosphatase; 3, PGM, phosphoglucomutase; 4, glycogen phosphorylase; 5, glycogen synthetase; 6, PGI, phosphogluco-isomerase; 7, PFK, phosphofructokinase; 8, FDPase, fructose diphosphatase; 9, aldolase; 10, triose phosphate isomerase, glyceraldehyde-3-phosphate dehydrogenase, phosphoglycerate kinase, phosphoglycerate mutase and enolase; 11, PEPCK, phosphoenolpyruvate carboxykinase; 12, PK, pyruvate kinase; 13, MDH, malate dehydrogenase; 14, PC, pyruvate carboxylase; 15, pyruvate dehydrogenase.

ase and pyruvate kinase). These 'energy barriers' are bypassed in the gluconeogenic pathway by glucose-6-phosphatase, fructose diphosphatase and by the combined activity of pyruvate carboxylase and phosphoenolpyruvate carboxykinase. In mammalian tissues the main gluconeogenic precursors are lactate, glycerol and amino acids. In bivalves lactic acid is not the main end-product of anaerobic metabolism (see later) and the low levels of lactate dehydrogenase preclude lactate as a major source of glucose. Glycerol is derived from breakdown of triglycerides and can be activated by glycerokinase to enter the Embden–Meyerhof pathway at the level of triose phosphate. So far, however, glycerokinase has not been identified in marine bivalves. All of the amino acids, except leucine, can be converted into gluconeogenic precursors which enter the Krebs

Table 8.6. *Relative activities of glycolytic enzymes based on values of 100 (or 1) for hexokinase*

Enzyme	*Mytilus californianus**	*Haliotis rufescens**	Skeletal muscle†	Vertebrate liver†
Hexokinase	100	100	1	1
Phosphogluco-isomerase	720	673	117	560
Phosphofructokinase	115	115	37	10
Aldolase	27	68	52	12
Triose phosphate isomerase	56	56	1768	—
Glyceraldehyde-3-phosphate dehydrogenase	95	80	295	300
Phosphoglycerate kinase	75	70	113	260
Phosphoglycerate mutase	85	93	67	—
Enolase	41	41	105	106
Pyruvate kinase	180	116	258	74

Data from * Bennett & Nakada (1968); † Newsholme & Start (1973).

cycle to form oxaloacetate or are converted into pyruvate by transamination (Fig. 8.7).

In the adductor muscle glucose is degraded to PEP by the normal glycolytic reactions. The fate of the PEP then depends on whether the pathway continues aerobically (conversion to pyruvate and oxidation by the Krebs cycle) or anaerobically in which case PEP is converted into alanine (via pyruvate) and into oxaloacetate and then to succinate by the reverse of the Krebs cycle (see Fig. 8.10). In the mantle tissue the situation is more complicated. During gluconeogenesis, oxaloacetate is converted into PEP by the action of phosphoenolpyruvate carboxykinase (PEPCK), so that PEPCK must be able to function in both directions. At the same time, under conditions favouring decarboxylation of oxaloacetate, pyruvate kinase activity must be reduced or 'shut-off' to prevent the futile cycling of carbon at the expense of high-energy phosphate. Another problem is that PEPCK is almost entirely located in the cytoplasm (see later) whereas oxaloacetate is formed in the mitochondria and does not easily pass through the mitochondrial membrane (Scrutton & Utter, 1968). One way in which this problem can be overcome is by conversion of oxaloacetate into malate as shown in Fig. 8.8.

Table 8.6 shows the activities (relative to hexokinase) of the glycolytic enzymes in the soluble or cytoplasmic fraction of the digestive gland of *M. californianus* and *H. rufescens* (Bennett & Nakada, 1968). Comparisons have been made with data for skeletal muscle and vertebrate liver, taken from the compilation of Newsholme & Start (1973). Maximum catalytic rates can be used to predict non-equilibrium or control reactions which normally have low activities (see Newsholme & Start, 1973). It is important to note that the assays should be carried out under 'optimum' conditions so that accurate estimates are obtained for the in-vivo rate. Although there is no direct evidence that the values given by Bennett & Nakada (1968) are maximum rates, nonetheless the data are worth comparing with those for vertebrate tissues. In muscle and liver, hexokinase (HK), phosphofructokinase (PFK) and aldolase have low activities. However, taken in conjunction with the mass-action ratios (see later) only HK and PFK can be definitely described as non-equilibrium enzymes (Newsholme & Start, 1973). Pyruvate kinase (PK) presents a problem; according to the catalytic data it belongs in the high-activity group but the mass-action ratio and kinetic data on allosteric control in liver strongly suggest that it catalyses a non-equilibrium reaction (Newsholme & Start, 1973). In *M. californianus* and *H. rufescens* the enzymes HK, PFK and PK all have high activities relative to those catalysing interconversions at the triose phosphate level (aldolase, triose phosphate isomerase and enolase); the properties of aldolase might be worth further investigation particularly in relation to the breakdown of triglycerides and the entry of glycerol into the Embden–Meyerhof pathway. Although PFK is not rate-limiting, Bennett & Nakada (1968) showed that the enzyme was subject to allosteric control by ATP and it is likely that PFK plays an important role in the regulation of glycolysis. More recent studies on the kinetics and allosteric regulation of pyruvate kinase from *M. edulis* (De Zwaan, 1972; De Zwaan & Holwerda, 1972; Holwerda & De Zwaan, 1973; Holwerda, De Zwaan & Van Marrewijk, 1973; Livingstone & Bayne, 1974; Livingstone, 1975) and *C. gigas* (Mustafa & Hochachka, 1971) have shown that PK plays a key role in the control of the PEP branchpoint during glycolysis and gluconeogenesis.

Van Marrewijk, Holwerda & De Zwaan (1973) and De Zwaan & Van Marrewijk (1973b) have made a comparative study of the carbon-dioxide-fixing enzymes pyruvate carboxylase (PC), PEPCK and 'malic enzyme' (ME) in *M. edulis* (Table 8.7). The activity of PEPCK was high in all three tissues (adductor muscle, mantle and digestive gland) and the enzyme was located almost entirely in the cytoplasm. A similar distribution of PEPCK in the cytoplasm has been observed for *Rangia cuneata* (mantle) by Chen & Awapara (1969), for *P. magellanicus* (mantle and adductor) by O'Doherty & Feltham (1971) and for *C. gigas* (adductor; negligible activity was found in the mantle and gills) by Mustafa & Hochachka (1973a, b). PC activity was low in *M. edulis* and the enzyme was entirely located in the mitochondrion; carbon dioxide fixation of pyruvate is clearly less important than that of PEP. Because of its higher activity, the small fraction of PEPCK in the mitochondrion was as active as the total PC (De Zwaan & Van Marrewijk, 1973b).

Malic enzyme catalyses the following reaction:

$$Malate + NADP^+ \overset{Mn^{2+}}{\rightleftharpoons} NADPH + pyruvate + CO_2.$$

Table 8.7. *Mytilus edulis:* *tissue and subcellular distribution of pyruvate carboxylase(PC), PEP-carboxykinase(PEPCK) and 'malic enzyme' (ME)*

Fraction	Adductor muscle			Mantle tissue			Digestive gland		
	ME	PEPCK	PC	ME	PEPCK	PC	ME	PEPCK	PC
	Total activity*								
(500 g supernatant)	3	3874	63	2	2187	28	1	1872	44
	% activity recovered in subcellular fractions								
Cytoplasm	58	98	1	94	98.5	2	77	97	0.2
Mitochondria	42	2	99	6	1.5	98	23	3	99.8

Activity as nmoles $H^{14}CO_3$-incorporated per minute per gram fresh tissue.
Data from De Zwaan & Van Marrewijk (1973b).

Kinetic studies on the adductor enzyme from *C. gigas* have shown that *in vivo* 'malic enzyme' functions mainly in the direction of decarboxylation of malate to pyruvate (Hochachka & Mustafa, 1973). Its activity is regulated by pH and the redox balance of the cell. The optimum pH for the forward reaction is around pH 8.0 and the optimum for the carboxylation (reverse) reaction is about pH 5.2–5.5 (Hochachka & Mustafa, 1973). In *M. edulis* the activity of 'malic enzyme' was extremely low (Table 8.7), but since this was measured at pH 7.6 in the direction of carbon dioxide fixation it is difficult to assess the importance of the enzyme *in vivo*; clearly further studies are required. Previously Hammen (1966) had reported 'malic enzyme' activity in the mantle tissue of *C. virginica*, assayed in the direction of $NADP^+$ reduction; only slight reoxidation of NADPH occurred on addition of pyruvate and bicarbonate. Malic enzyme has also been reported in the mantle tissue and adductor muscle of *P. magellanicus* but again in the direction of $NADP^+$ reduction (O'Doherty & Feltham, 1971).

Research on the key enzymes of the gluconeogenic pathway between PEP and glucose has been limited to the demonstration of FDPase activity in the foot, mantle and gill tissue of *Mercenaria mercenaria* (Engel & Neat, 1970) and in the adductor muscle and mantle tissue of *P. magellanicus* (O'Doherty and Feltham, 1971). In *M. mercenaria* there was a significant increase in the specific activity of FDPase in the foot following prolonged starvation. This suggested an activation of gluconeogenesis once the normal glycogen reserves had been depleted; however, the expected inhibition of PFK under such conditions was not observed (Engle & Neat, 1970).

Marine mussels

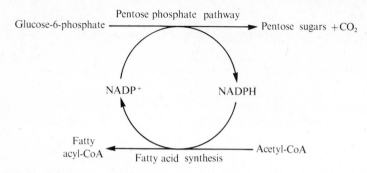

Fig. 8.9. Role of the pentose phosphate pathway.

Pentose phosphate pathway

It is generally accepted that the main function of the pentose phosphate pathway is the provision of NADPH for the synthesis of fatty acids and sterols and the supply of pentoses for nucleic acid synthesis (Fig. 8.9). Glucose-6-phosphate (G6P) is oxidised to 6-phosphogluconate by G6P dehydrogenase and then oxidatively decarboxylated to form the pentose sugar ribulose-5-phosphate; both these reactions result in the reduction of NADP$^+$ to NADPH. The activity of the pentose phosphate pathway is probably regulated by the availability of NADP$^+$ which is controlled by the rate of fatty acid synthesis; this is an example of metabolic regulation by co-factor availability (see Newsholme & Start, 1973). The ribulose-5-phosphate can be converted into other pentose sugars and used for nucleic acid synthesis or be converted back to glucose-6-phosphate in a series of transaldolase and transketolase reactions; the details are given in most biochemistry textbooks. The net result is the complete oxidation of the C-1 carbon atom of glucose to carbon dioxide and the reduction of 2 moles of NADP$^+$ to NADPH.

A second possible source of NADPH for fatty acid synthesis is 'malic enzyme' which catalyses the conversion of malate into pyruvate during the transfer of acetyl-CoA (and oxaloacetate) across the mitochondrial membrane into the cytoplasm. One turn of the pyruvate–malate shuttle transfers 1 mole of acetyl-CoA into the cytoplasm and provides 1 mole, or half of the NADPH required for the subsequent reduction of acetyl-CoA during fatty acid synthesis (Newsholme & Start, 1973). Pyruvate carboxylase which catalyses the conversion of pyruvate to oxaloacetate, prior to condensation with acetyl-CoA, is also involved in the pyruvate–malate shuttle. One consequence of the low activity of 'malic enzyme' (but see previous discussion) and PC in *M. edulis* is that the pentose phosphate pathway is probably the major source of NADPH for lipogenesis. Acetyl-CoA must then be transported out of the mitochondria in some other way; for a discussion of alternative mechanisms see Newsholme & Start (1973).

The evidence for operation of the pentose phosphate pathway in marine bivalves and the association of NADP$^+$ reduction with gametogenic activity has been reviewed by Gabbott (1975). The presence of G6P dehydrogenase, gluconate-6-phosphate dehydrogenase and the trans-aldolase–transketolase system has been demonstrated in *M. californianus* by Bennett & Nakada (1968). When slices of whole mussels, or slices of the digestive gland and gill tissues were incubated with [1-^{14}C]glucose or [6-^{14}C]glucose the ^{14}CO$_2$ ratios (C-1/C-6) confirmed the operation of the pentose phosphate pathway (Bennett & Nakada, 1968).

Krebs cycle and oxidative phosphorylation

The enzymes of the Krebs cycle and the components of the electron transport chain are located in the mitochondrial compartment. Evidence for a functional Krebs cycle in bivalves comes from (*a*) demonstration of the carboxylic acid intermediates and enzymes involved in the cycle and (*b*) studies on the flow of ^{14}C from glucose through the cycle into ^{14}CO$_2$. Much of the early work has been reviewed by Goddard (1966) and by Hammen (1969); most authors have concluded that a normal Krebs cycle operates under aerobic conditions. Production of ^{14}CO$_2$ from labelled glucose has been shown for *M. californianus* by Bennett & Nakada (1968) and for *M. edulis* by De Zwaan, De Bont & Kluytmans (1975).

In addition to the Krebs cycle, terminal oxidation of pyruvate requires the operation of the electron transport chain to regenerate oxidised coenzymes and produce ATP. Several authors have studied the properties of 'respiratory particles' isolated from bivalves by differential centrifugation, in the same way as vertebrate mitochondria. The distribution of respiratory pigments and enzymes is qualitatively similar to that found in mammalian cells (Kawai, 1959; Tappel, 1960; Ryan & King, 1962; Mattisson & Beechey, 1966). Cytochromes *b*, *c* and ($a+a_3$) have been detected spectroscopically in *Mytilus crassitesta* (Kawai, 1959) and in *M. edulis* (Ryan & King, 1962). The properties of succinic oxidase and the cytochrome oxidase (a_3) system in respiratory particles from the adductor muscle of *M. edulis* and *M. californianus* have been described by Ryan & King (1962). But the most recent and comprehensive work, including electron microscopy studies, is that of Mattisson & Beechey (1966) on subcellular particles sedimented at 10 000 *g*, designated P$_{10}$ particles, from the adductor muscle of *Pecten maximus*. Direct electron microscopic examination of the striated and non-striated muscle showed very few mitochondria with poorly developed cristae. Examination of the P$_{10}$ pellet from striated muscle showed the material to be vesicular; some of the vesicles contained cristae and others had electron-dense areas of membranous appearance. The authors concluded that this fraction was of mitochondrial origin but that the particles had been drastically altered during the isolation procedure. It is important to note that even with the best techniques available, the procedures for isolation of mitochondria from vertebrate tissues do not give 'intact' preparations when applied to marine bivalves (see also Ryan & King, 1962).

The reduced-minus-oxidised difference spectra showed the presence of cytochromes b and $(a+a_3)$ and probably cytochrome c, in the P_{10} particles; cytochrome c was also isolated from the intact striated muscle and shown to have properties similar to mammalian cytochrome c. Treatment with carbon monoxide confirmed the presence of cytochrome oxidase (a_3). Other respiratory enzymes present in the P_{10} particles were NADH-cytochrome c reductase, succinate–cytochrome c reductase, NADH oxidase and succinate oxidase (Mattisson & Beechey, 1966).

Anaerobic metabolism

Succinate and alanine, not lactate, are the major end-products of anaerobic metabolism in marine bivalves (Stokes & Awapara, 1968; Chen & Awapara, 1969; Hammen, 1969; Malanga & Aiello, 1972; De Zwaan & Zandee, 1972b; De Zwaan & Van Marrewijk, 1973a; Hammen, 1975). Although lactic dehydrogenase (LDH) activity is very low in the adductor muscle, it is not entirely absent. LDH has been found in *M. edulis* by Gäde & Zebe (1973) and by Van Marrewijk, Holwerda & De Zwaan (1973). Presumably some lactate is formed under anaerobic conditions but this is of minor importance compared to the production of succinate and alanine; the subsequent fate of the lactic acid is not known. In the following discussion it is assumed that intertidal bivalves such as *Mytilus* are facultative anaerobes whose mode of existence results in regular periods of aerobic (submerged) and anaerobic (exposed) metabolism.

Metabolic pathways

Metabolic pathways for the conversion of glucose into succinate and alanine have been discussed by Stokes & Awapara (1968), Chen & Awapara (1969), Hochachka & Mustafa (1972), Hochachka, Fields & Mustafa (1973), De Zwaan, Van Marrewijk & Holwerda (1973) and De Zwaan, De Bont & Kluytmans (1975). Glucose is first converted into phosphoenolpyruvate (PEP) by the classical Embden–Meyerhof pathway and then PEP (not pyruvate) acts as the branchpoint for anaerobic metabolism (Fig. 8.10). Starting from glucose, three carbons of succinate and all three carbons of alanine are derived from PEP; the fourth carbon in

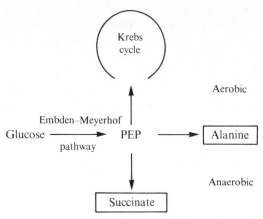

Fig. 8.10. Metabolic fate of phosphoenolpyruvate (PEP) under aerobic and anaerobic conditions.

succinate is derived from carbon dioxide. In *M. edulis* the main carbon dioxide fixation step is the conversion of PEP into oxaloacetate (Van Marrewijk *et al.*, 1973; De Zwaan & Van Marrewijk, 1973*b*; see Table 8.7). The activity of PEPCK is high in the adductor muscle and the mantle tissue; both tissues also contain an active pyruvate kinase (PK) (Van Marrewijk *et al.*, 1973; Livingstone & Bayne, 1974) and these two enzymes PEPCK and PK compete for PEP under anaerobic conditions (see later).

Tables 8.8 and 8.9 show the distribution of radioactivity under anaerobic versus aerobic conditions when mussels, *M. edulis*, were injected with [^{14}C]sodium carbonate (Table 8.8; data from Loxton & Chaplin, 1974) and with [U-^{14}C]glucose and [U-^{14}C]glutamate (Table 8.9; data from De Zwaan *et al.*, 1975). In Table 8.8, the Group I and II mussels were kept in seawater under aerobic conditions for 72 h prior to the injection of [^{14}C]carbonate; Group III mussels were left dry in air (anaerobic conditions). After treatment, the Group I mussels were returned to aerated seawater and the Group II and III mussels left dry for 3 h. Under aerobic conditions (Group I) the ^{14}C-label was found mainly in aspartate and glutamate with only low activities in the Krebs cycle carboxylic acids. This observation is in agreement with the enzyme data which show high activities for PEPCK in *M. edulis*; oxaloacetate and α-ketoglutarate can be readily transaminated by GOT (glutamate–oxaloacetate transaminase) and GPT (glutamate–pyruvate transaminase) (Read, 1962*b*; see Fig. 8.7). Under anaerobic conditions (Groups II and III) a high proportion of the ^{14}C-label was found in the organic acids, malate and succinate; presumably succinate is formed from oxaloacetate by reversal of the Krebs cycle reactions (see below). The low activity in alanine under both aerobic and anaerobic conditions shows that the PEPCK reaction operates mainly in

Table 8.8. Mytilus edulis: distribution of radioactivity in ethanol-soluble intermediates, 3 h after injection of [14C]sodium carbonate. Results are expressed as % of total activity in the soluble fraction.

Compound	Group I (Aerobic)				Group II (Anaerobic; exposed 3 h)				Group III (Anaerobic; exposed 75 h)			
	GL	DG	MAN	MUS	GL	DG	MAN	MUS	GL	DG	MAN	MUS
Alanine	3.9	6.6	—	2.9	6.2	5.6	6.3	5.7	7.1	6.4	8.0	7.6
Aspartate	79.4	61.1	75.7	74.1	55.7	24.6	44.9	39.4	50.7	31.2	37.1	41.0
Glutamate	12.5	24.9	21.9	16.6	13.7	7.0	6.6	5.5	7.8	2.7	4.3	4.6
Fumarate	0.6	0.4	—	0.2	1.8	2.6	2.0	1.9	3.5	1.7	1.6	1.8
Malate	3.6	4.2	2.2	5.5	10.5	18.5	8.7	13.4	16.5	14.1	11.9	13.3
Succinate	—	2.7	—	1.1	8.6	50.6	29.0	31.4	11.3	42.5	35.5	30.0

GL, gills; DG, digestive gland; MAN, mantle; MUS, muscle.
Data from Loxton & Chaplin (1974).

Table 8.9. Mytilus edulis: *distribution of radioactivity 2.5 h after injection of [U-¹⁴C]glucose and [U-¹⁴C]glutamate. Results expressed as % of total activity recovered*

Component	Aerobic conditions		Anaerobic conditions	
	[U-¹⁴C]-glucose	[U-¹⁴C]-glutamate	[U-¹⁴C]-glucose	[U-¹⁴C]-glutamate
Metabolic carbon dioxide	22.9	35.2	2.0	6.1
Proteins	11.2	6.6	1.1	3.5
Lipids	0.1	< 0.1	< 0.1	< 0.1
Glycogen + glucose	5.6	0.4	1.3	0.4
Organic acids	11.1	10.4	56.6	38.0
Amino acids	45.3	50.6*	38.6	50.6
Malate	4.3	0.9	1.3	2.8
Fumarate	0.1	0.7	1.6	2.8
Succinate	4.3	5.6	36.3	18.7
Alanine	24.5	2.6	18.3	4.7
Aspartate	11.4	22.6	14.3	20.5
Glutamate	5.6	—	3.6	—

* Calculated from the total radioactivity in the amino acid fraction minus the radioactivity in glutamate.
Data from De Zwaan *et al.* (1975).

the direction PEP → oxaloacetate. The changeover from aspartate to succinate as the main end-product of the PEPCK reaction takes place rapidly (after 3 h under anaerobic conditions). Loxton & Chaplin (1974) also made another very interesting comment; it might be expected that during exposure tissues nearest the respiratory surfaces such as the gills would show smaller changes in metabolism than tissues deeper inside the animal. This is certainly borne out by the data in Table 8.8 where there is less change in the gills in the Group II and III mussels than in the other tissues.

Table 8.9 shows the results of the experiments carried out by De Zwaan *et al.* (1975) on mussels which had been kept aerated in seawater (aerobic conditions), or dry under a stream of nitrogen (anaerobic conditions) for 15 h and were then injected with [U-¹⁴C]glucose or [U-¹⁴C]glutamate. The distribution of radioactivity was determined 2.5 h after treatment. Under aerobic conditions the highest activities were found in the exhaled carbon dioxide and in the amino acid and protein fractions. Unexpectedly there was a marked incorporation of ¹⁴C-label from glucose into alanine, as well as into aspartate and glutamate (De Zwaan *et al.*, 1975). The main product from [U-¹⁴C]glutamate was aspartate. This pattern of results is consistent with the operation of the normal oxidative pathway (glycolytic sequence plus Krebs cycle) with transamination of pyruvate to alanine, oxaloacetate to aspartate and α-ketoglutarate to glutamate (Fig. 8.7). Under anaerobic conditions there was a marked increase in the amount of ^{14}C-label in the organic acid fraction, particularly in succinate. In spite of the fact that the activity of the Krebs cycle is considerably reduced (low yield of $^{14}CO_2$) there was still a marked incorporation of ^{14}C-label into aspartate (Table 8.9). In the case of [U-^{14}C]glucose this can be explained by a reduction in pyruvate kinase activity and conversion of PEP into oxaloacetate (De Zwaan *et al.*, 1975). But, in the case of [U-^{14}C]glutamate the ^{14}C-label can only enter into aspartate via α-ketoglutarate and the forward operation of the Krebs cycle to oxaloacetate; this results in the loss of 1 mole of carbon dioxide per mole of α-ketoglutarate (De Zwaan *et al.*, 1975).

The physiological condition of the experimental animals and the time of year when the experiments are carried out have often been forgotten in the interpretation of ^{14}C tracer experiments. For example, in *M. edulis* the level of glycogen is low during the winter (from January onwards; Fig. 8.1) and protein, not carbohydrate, is used as the main energy reserve during stress (Gabbott & Bayne, 1973). What is not known is whether the level of glycogen in the winter is sufficient to maintain anaerobic metabolism during exposure, or whether anaerobic as well as aerobic metabolism takes place at the expense of protein reserves. The experiments described by De Zwaan *et al.* (1975) were carried out in January, on mussels that had been kept in the laboratory without food for approximately 3 weeks prior to treatment. Under these conditions there will be an increase in protein catabolism and this must be reflected in the distribution of the ^{14}C-label. It might, for example, explain the high level of radioactivity found in alanine under aerobic conditions by De Zwaan *et al.* (1975) (see Table 8.9; data for [^{14}C]glucose). During stress there is a marked increase in the loss of α-amino-nitrogen (Bayne, 1973*a*) and, in several species of bivalves, Hammen (1968) has shown that amino acid excretion is closely correlated with the activity of aminotransferases such as GPT.

Fig. 8.11 shows the overall scheme, proposed by De Zwaan *et al.* (1973), for anaerobic carbohydrate metabolism in *M. edulis*. This differs from that originally suggested by Stokes & Awapara (1968) in one important respect; it is recognised that NAD^+ cannot pass through the mitochondrial membrane so that redox balance must be maintained on both sides of the barrier. According to De Zwaan and co-workers both malate and alanine are formed in the cytoplasm by a reduction step involving the reoxidation of NADH to NAD^+ (see below). In the mitochondria, redox balance is maintained by conversion of malate partly into succinate by the reverse of the Krebs cycle and partly into α-ketoglutarate in the forward direction of the cycle. In this way one reduction is balanced by three oxidation steps so that succinate and glutamate are formed in the approximate ratio of three to

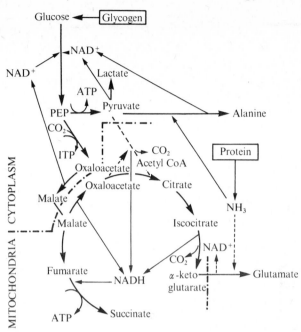

Fig. 8.11. General scheme for anaerobic carbohydrate metabolism in *Mytilus edulis* (redrawn from De Zwaan *et al.*, 1973). For explanation see text. Redox balance is maintained by reduction of pyruvate and oxaloacetate in the cytoplasm and by operation of the Krebs cycle in both forward and reverse directions. Energy yield: 2 moles of ATP per mole of glucose in the cytoplasm and between 1 and 2 moles of ATP in the mitochondria.

one (De Zwaan & Van Marrewijk, 1973*a*; De Zwaan *et al.*, 1973). The formation of glutamate from α-ketoglutarate results in the oxidation of NADH to NAD$^+$ in the cytoplasm (see De Zwaan & Van Marrewijk, 1973*a*) which balances the requirement for NAD$^+$ during the production of pyruvate. In a later paper De Zwaan *et al.* (1975) have shown that, under anaerobic conditions, the Krebs cycle does not necessarily stop at α-ketoglutarate but can carry on to oxaloacetate. However, the major pathway for the production of succinate from [^{14}C]glucose cannot be via the forward direction of the Krebs cycle because of the low yield of ^{14}CO$_2$ (see Table 8.9). The main difficulty with this scheme, as discussed by De Zwaan *et al.* (1975), is that the Krebs cycle between oxaloacetate and succinate operates in both directions at the same time.

An alternative scheme for anaerobic metabolism, in which redox balance is maintained by the simultaneous utilisation of both carbohydrate (glucose) and protein (aspartate and glutamate) has been proposed by Hochachka and co-workers (Hochachka & Mustafa, 1972; Hochachka, Fields & Mustafa, 1973) (Fig. 8.12). The link between glucose and amino acid metabolism is achieved primarily by the redox couple formed between

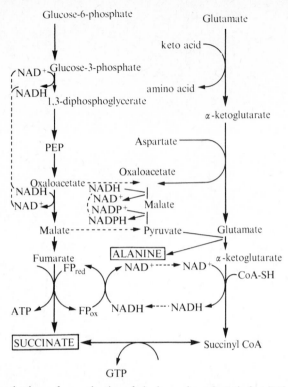

Fig. 8.12. General scheme for production of alanine and succinate in facultative anaerobes (redrawn from Hochachka *et al.*, 1973). For explanation see text. Redox balance is maintained by the simultaneous utilisation of 2 moles of aspartate and 2 moles of glutamate per mole of glucose in the cytoplasm, and by the redox couple between fumarate reductase and α-ketoglutarate dehydrogenase. Energy yield: 2 moles of ATP per mole of glucose in the cytoplasm and 4 moles of ATP in the mitochondria (from 1 mole glucose plus 2 moles of glutamate). FP, flavoprotein.

fumarate reductase and α-ketoglutarate dehydrogenase. For every mole of glucose degraded, 2 moles of aspartate and 2 moles of glutamate are used up to maintain redox balance. Aspartate is converted into alanine whereas glucose and glutamate are converted into succinate. According to Hochachka and co-workers, pyruvate kinase is largely 'shut-off' during anaerobiosis and PEP cannot be the source of alanine; instead pyruvate is formed from malate in a redox couple with oxaloacetate (Fig. 8.12).

There are two major difficulties in accepting this scheme, at least as far as *M. edulis* is concerned. First, there is no evidence that the loss of glycogen under anaerobic conditions (or under stress) is accompanied by an equal loss of protein; the general pattern for *M. edulis* seems to be that the carbohydrate reserves are used up first and then protein and lipid. Secondly, if aspartate is converted into alanine in a redox couple, then conversion of [^{14}C]glutamate into [^{14}C]aspartate should also result in labelled alanine as well as an increased production of $^{14}CO_2$ (De Zwaan *et al.*, 1975). But this did not happen when *M. edulis* was injected with [U-^{14}C]glutamate, under anaerobic conditions; only aspartate and not alanine became heavily labelled (Table 8.9; De Zwaan *et al.*, 1975). It seems to the author that the relationship between carbohydrate and protein catabolism postulated by Hochachka and co-workers might be related to time and not to an obligatory metabolic link. For example, there are clear differences in the balance of carbohydrate and protein reserves in *M. edulis*, between summer and winter animals (Bayne, 1973*b*; Gabbott & Bayne, 1973). During the winter, anaerobic metabolism may take place largely at the expense of protein reserves and this, in turn, will involve the conversion of amino acids into succinate.

Redox balance

In both reaction schemes redox balance is maintained by substrate-level oxidation and reduction (Figs. 8.11 and 8.12). According to De Zwaan and co-workers (De Zwaan & Van Marrewijk, 1973*a*; De Zwaan *et al.*, 1973) alanine is the preliminary end-product of anaerobic metabolism and must be formed in an oxidation–reduction step. This could take place by the action of an alanine dehydrogenase (reaction (3) below) or by an initial transamination reaction followed by reduction of α-ketoglutarate by glutamate dehydrogenase (reactions (1) and (2) below) (De Zwaan & Van Marrewijk, 1973*a*).

(1) Pyruvate + glutamate $\rightleftharpoons$ alanine + α-ketoglutarate.

(2) α-ketoglutarate + NH_3 + NADH + H$^+$ $\longrightarrow$ glutamate + NAD$^+$ + H_2O.

(3) Pyruvate + NH_3 + NADH + H$^+$ $\longrightarrow$ alanine + NAD$^+$ + H_2O.

The hypothetical alanine dehydrogenase reaction ((1) plus (2)) has a possible structural counterpart in mammalian liver as an enzyme–enzyme complex between the transaminases (GOT or GPT) and glutamate dehydrogenase (GDH) (Fahien & Smith, 1974). In the reverse direction the complex catalyses the oxidative deamination of amino acids which do not normally react with GDH but which react with the transaminases. Since GDH is likely to be a source of ammonia production in bivalves, there must be provision for the GDH reaction to operate in both directions under anaerobic conditions, possibly in different cellular compartments or as a complex with aminotransferases. It is surprising that virtually nothing is known about the properties and function of GDH in molluscs (see Campbell & Bishop, 1970).

Production of alanine from pyruvate requires a source of glutamate

12 BMM

(transamination) or ammonia (reduction), both of which are derived from protein catabolism. According to De Zwaan & Van Marrewijk (1973*a*) the reductive amination of pyruvate to alanine may be necessary to prevent a build-up of ammonia under anaerobic conditions, i.e. when the mussels are exposed on the shore (see Chapter 6). However, it has already been pointed out that the breakdown of glycogen during anaerobiosis is not necessarily accompanied by a loss of protein; and one of the limiting factors for alanine production from pyruvate may be the supply of ammonia. This could be important when considering the end-products and time course of anaerobic metabolism in different body tissues (such as the adductor muscle and mantle tissue of *M. edulis*), or when looking at animals at different times of the year under temperature and nutritive stress. The subsequent fate of alanine is not known, only that alanine accumulates under anaerobic conditions. In vertebrate muscle, lactic acid is removed by the blood and then converted back into glucose in the liver. In *M. edulis*, the possibility of alanine being converted back to pyruvate and then into glucose via oxaloacetate and PEP seems to be limited by the low activity of pyruvate carboxylase (Table 8.7; De Zwaan & Van Marrewijk, 1973*b*). Alanine may simply be excreted when the mussels are re-immersed in seawater or be converted to pyruvate and oxidised, under aerobic conditions, via the normal Krebs cycle.

There are two other substrate-level reactions for the reoxidation of NADH; these are octopine formation and the operation of the Büchler shuttle. Thoai & Robin (1959) have shown that octopine is formed in the muscles of *P. maximus* and *Cardium edule* by a reductive condensation of arginine and pyruvate, resulting in the reoxidation of NADH:

$$\text{arginine} + \text{pyruvate} + \text{NADH} \rightleftharpoons \text{octopine} + \text{NAD}^+.$$

Octopine dehydrogenase (ODH) has been reported in the adductor muscle of *M. edulis* by Gäde & Zebe (1973) but was surprisingly absent in the scallop *P. magellanicus* (O'Doherty & Feltham, 1971). According to Robin & Thoai (1961) the synthesis of octopine is analogous to the formation of lactic acid in other species and serves to regenerate NAD^+ for glycolysis; nothing is known, however, as to the source of arginine necessary to sustain anaerobic metabolism in bivalves.

Secondly there is the Büchler shuttle, or α-glycerol phosphate cycle, which serves to reoxidise cytoplasmic NADH in insect flight muscle; during the initial stages of flight pyruvate accumulates at a faster rate than it can be oxidised by the Krebs cycle. Under these conditions dihydroxyacetone phosphate (DHAP) is converted into α-glycerol phosphate by cytoplasmic α-glycerol phosphate dehydrogenase and NADH is oxidised to NAD^+. α-Glycerol phosphate enters the mitochondria where it is converted back to DHAP; the DHAP returns to the cytoplasm and the cycle is complete. Mitochondrial α-glycerol phosphate dehydrogenase is a flavoprotein which is reoxidised by the electron transport chain. Both cytoplasmic and mitochondrial forms of α-glycerol phosphate dehydrogenase have been reported in the adductor muscle of *P. magellanicus* by O'Doherty & Feltham (1971). This is the first demonstration of the Büchler shunt enzymes in a marine invertebrate; one possible role of the shunt in scallops might be to initiate a rapid swimming action during the escape response. In *M. edulis*, no α-glycerol phosphate dehydrogenase activity was detected in the posterior adductor muscle by De Zwaan (1971).

One of the problems of redox balance in facultative anaerobes is to maintain the different $NAD^+/NADH$ ratios in the cytoplasmic and mitochondrial compartments. In vertebrate cells the redox state in the cytoplasm is kept at a more oxidised level ($NAD^+/NADH$ ratio is about 10^3 in liver) than in the mitochondria. This ensures that the flux through the glyceraldehyde-3-phosphate dehydrogense reaction is in favour of glycolysis (see Newsholme & Start, 1973). However, in the mitochondria the $NAD^+/NADH$ ratio has to be much smaller (about 10 in liver) to provide the 'driving force' for the electron transport chain. This difference explains why the mitochondrial membrane has to be impermeable to NAD^+ and NADH and it is difficult not to accept that this must be true also for bivalves. According to the original scheme proposed by Stokes & Awapara (1968) and by Chen & Awapara (1969), redox balance is maintained *between* the two compartments, because in the mantle of *Rangia* the mitochondria are permeable to NADH (Chen & Awapara, 1969). This assertion is based on the fact that fumarate reduction by isolated mitochondria only takes place in the presence of NADH and according to Chen & Awapara (1969) it does not matter whether the mitochondria are 'broken' or 'intact'. But we have already seen that it is very difficult to prepare intact mitochondria from marine bivalves (Ryan & King, 1962; Mattisson & Beechey, 1966). The alternative explanation for the NADH requirement is simply that both mitochondrial fractions were 'broken' and that the pyridine nucleotides had been lost during the preparation step. Both of the alternative schemes, proposed by De Zwaan and co-workers and by Hochachka's group, assume that redox balance must be maintained separately in the cytoplasmic and mitochondrial compartments. It would be very interesting to determine the $NAD^+/NADH$ ratios in the different tissues and cell compartments under aerobic and anaerobic conditions – the necessary experimental procedures are discussed by Newsholme & Start (1973).

Finally we must briefly consider the possibility of reoxidation of NADH by specific, non-substrate electron acceptors. According to Zs-Nagy and co-workers (see review by Zs-Nagy, 1974) the neural tissues of *M. galloprovincialis* contain cytosome particles which can support the oxidation of NADH and act as electron acceptors during the production of ATP (Zs-Nagy & Ermini, 1973*a*, *b*). This process has been termed 'anoxic endogenous oxidation' by Zs-Nagy. The cytosomes contain a lipochrome pigment which functions at a redox potential between cytochrome *a* and molecular oxygen; in this way approximately 55% of the normal production of ATP can be realised under anaerobic conditions (Zs-Nagy & Ermini, 1973*b*). There are two problems. First, it is not known how important the cytosome granules are in tissues other than the nerve cells and secondly anoxic endogenous oxidation is not effected by immediate anoxia but requires some time to activate the cytosomal mechanism (Zs-Nagy & Ermini, 1973*b*). This is brought about by the gradual decrease in oxygen tension in the body tissues. Both problems require further experimental analysis in relation to short-term anaerobic excursions between tides and to longer-term exposure of species found high on the shore.

Succinate formation and ATP yield

The yield of ATP in the normal glycolytic sequence is 2 moles of ATP per mole of glucose converted into lactate. In marine bivalves and other facultative anaerobes this yield can be increased to between 3 and 4 moles of ATP per mole of glucose, due to production of succinate (Fig. 8.11) and up to 6 moles of ATP from the simultaneous utilisation of 1 mole of glucose and 2 moles of glutamate (Fig. 8.12). In both schemes succinate is formed in the reverse direction of the Krebs cycle, by reduction of fumarate. Oxaloacetate is reduced to malate by cytoplasmic malate dehydrogenase and is then converted into fumarate. Malate (or fumarate) passes into the mitochondria and becomes reduced to succinate by fumarate reductase in a reaction coupled to the formation of ATP. According to O'Doherty & Feltham (1971) fumarate reductase is located in the mitochondria in both the mantle and adductor muscle of *P. magellanicus*. Hammen (1975) has recently looked at the relationship between substrate concentration and reaction velocity for the forward (succinate oxidation) and reverse (fumarate reduction) reactions of the succinate–fumarate step in the adductor muscle of *Mercenaria mercenaria*, *C. virginica* and *M. edulis*. Among micro-organisms the ratio K_m (SUC oxid.)$/K_m$ (FUM redn.) was 0.58 for an aerobe, 17.6 for an anaerobe and 3.14 for a facultative anaerobe (Singer, 1971, quoted by Hammen, 1975). The values for *M. mercenaria* (1.3), *C. virginica* (4.8) and *M. edulis* (1.1) suggest that all three bivalves are facultative anaerobes (Hammen, 1975). In the case of *M. edulis* the reduction of fumarate by NADH was stimulated by the addition of ADP, and the oxidation of succinate by NAD^+ had an absolute requirement for ATP; this provides experimental evidence for the coupling of phosphorylation and fumarate reduction (Hammen, 1975).

Formation of succinate from fumarate results in the oxidation of NADH to NAD^+; according to De Zwaan and co-workers and to Hochachka's group, redox balance is maintained in the mitochondria by coupling the forward and reverse reactions of the Krebs cycle. There is, however, another possibility suggested by some early work on *Ascaris* sarcosome preparations by Seidman & Entner (1961). In this scheme 1 mole of malate acts as a hydrogen donor and another, in the form of fumarate, as the final hydrogen acceptor. Formation of ATP is coupled to the oxidation and reduction of $NADH/NAD^+$ in a dismutation reaction which results in the formation of equimolar amounts of succinate and pyruvate (see Gilles, 1970, 1972*a*). Pyruvate can be transaminated to alanine; however, the aminotransferase reaction requires a source of glutamate or ammonia and this could become a limiting factor to the production of mitochondrial ATP (see previous discussion). The production of succinate and alanine, rather than lactate, as end-products of glycolysis may be related to the absence of a hepatic function in marine bivalves (see review by Van Weel, 1974) and the consequent inability to reoxidise lactic acid from the muscles. The advantages of succinate formation are twofold; first, the reduction of fumarate acts as an additional step for the production of ATP from glucose and secondly, succinate can be readily oxidised on return to aerobic conditions by the Krebs cycle.

Control of carbohydrate metabolism

It is a common observation that metabolic processes operate most of the time at considerably less than their maximum rate (Atkinson, 1971). This generalisation is evident in mussels when one considers the differences between the routine and active rates of metabolism for *Mytilus edulis* and the effects of temperature on oxygen consumption (Chapter 7). Major advances have been made in the past ten to fifteen years in our understanding of metabolic regulation so that it is possible, now, to present a generally acceptable account for the control of carbohydrate metabolism. This has been discussed in detail by Newsholme & Start (1973) in their book on *Metabolic regulation*. It is this control function that links metabolic pathways to particular biochemical conditions within the cell and eventually to the needs of the organism as a whole; such controls are important sites for interaction with the external environment leading to evolutionary changes and to differences between species (see Hochachka, 1973).

In this section on metabolic regulation it is possible only to summarise the main features of the control theories, so that the later sections in this chapter are intelligible to the reader. A more general review, dealing with the comparative aspects of metabolic regulation is that by Hochachka (1973).

Metabolic regulation

Metabolic regulation is concerned mainly with the control of enzyme activities and there are two ways in which this can be achieved; (*a*) the amount of the enzyme can be controlled and (*b*) the activity – or catalytic potential – of the enzyme can be varied. The first method is essentially slow, taking place over a period of hours or days and represents a 'coarse' level of control. The second is rapid, taking place almost instantaneously and provides the 'fine' control or modulation of enzyme activity. In any sequence of biochemical reactions such as the glycolytic pathway, only certain enzymes act as metabolic controls. These catalyse non-equilibrium reactions and can be recognised experimentally by comparing the equilibrium constants for the reaction with the mass-action ratios, or by measurement of the maximal enzyme activities in the pathway (see Newsholme & Start, 1973); in general, enzymes that have a low activity are considered to catalyse non-equilibrium reactions. Ideally the two methods should give a similar classification but this is not always the case. For example, pyruvate kinase is often placed in the high-activity category but the mass-action ratio and kinetic data on allosteric control strongly suggest that it catalyses a non-equilibrium reaction (Newsholme & Start, 1973). Both procedures have been used extensively in the study of metabolic pathways in mammalian systems but not, so far, in any marine invertebrates.

There are two other approaches that can be used. First, regulatory enzymes are usually strategically positioned, either at the beginning of a pathway or at a metabolic branchpoint where two or more pathways diverge. The last criterion has been used by Hochachka & Mustafa (1972) to identify pyruvate kinase (PK) and phosphoenolpyruvate carboxykinase (PEPCK) as regulatory enzymes controlling the phosphoenolpyruvate (PEP) branchpoint. This, together with the kinetic data on the regulation of PK and PEPCK, suggests that the branchpoint plays an important role *in vivo* in the regulation of anaerobic metabolism in marine bivalves (see later). Secondly, there is the 'crossover theorem'. If the substrate concentration of a non-equilibrium enzyme changes in the opposite direction to the flux, i.e. if the flux is decreased and the substrate concentration increases, or vice versa, this shows that the enzyme is regulatory. The rationale is simple: a change in substrate concentration in the opposite direction to the flux is inconsistent with the idea of control by substrate concentration alone. Glycolytic flux can be inhibited by the use of alternative substrates to glucose (such as fatty acids) or increased by the use of uncoupling agents for oxidative phosphorylation and by anoxic conditions – see, for example, the experiments on locust flight muscle by Ford & Candy (1972). This leads to positive and negative crossover points when the substrate concentrations are determined for control and treated tissues. An important point to note, however, is that it is not possible to conclude that an enzyme is *not* regulatory (see Newsholme & Start, 1973) simply because the flux and substrate concentration change in the same direction. This second approach, involving measurements of substrate concentration in response to added glucose, has been used by Hochachka, Freed, Somero & Prosser (1971) to identify the control sites for glycolysis in the leg muscle of the arctic king crab, *Paralithodes camtschatica*.

Newsholme & Start (1973) have suggested an operational definition of a regulatory enzyme as 'an enzyme which catalyses a non-equilibrium reaction and whose activity is controlled by factors other than substrate concentration'. One of the most important of these factors is feedback regulation. In the sequence:

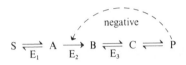

it is assumed that the reaction A → B is non-equilibrium so that the activity of enzyme E_2 limits the flux from S to P. If the activity of E_2 is inhibited by P, this provides a feedback mechanism for control of the flux through the pathway. The same concept can be extended to feed-forward activation of the reaction by S and, in its widest sense to control by any specific metabolic modulator. In most cases the molecules that act as modulators bear no structural relationship to the substrates or products of the enzymes which they regulate. Thus, the regulatory enzymes have distinct regions for binding substrates and other sites for binding modulators; the modulator-binding sites have been termed 'allosteric' (meaning sterically different) to emphasise their separation from the catalytic sites. There are two important kinetic properties of allosteric or regulatory enzymes. First, the change in velocity or rate of the enzyme reaction with respect to substrate concentration is sigmoidal, and not hyperbolic (Fig. 8.13a). The apparent Michaelis constant, K_m, is replaced by the value $K_{0.5S}$ which is the substrate concentration for half-maximum velocity. In the case of sigmoid kinetics the Lineweaver–Burk double reciprocal plots of $1/V$ against $1/S$ are not linear but curve upwards showing a positive co-operative effect (Fig. 8.13b). The various models to explain sigmoidal behaviour are discussed by Newsholme & Start (1973). Values for $K_{0.5S}$ are obtained from a Hill plot of $\log_{10} (V/V_{max} - V)$ against $\log_{10} [S]$; $K_{0.5S}$ is the concentration of S where $\log_{10} (V/V_{max} - V) = 0$. Secondly, there is the fact that physiological substrate concentrations are normally of the same order of magnitude, or lower than, the values of K_m or $K_{0.5S}$. The effect of positive

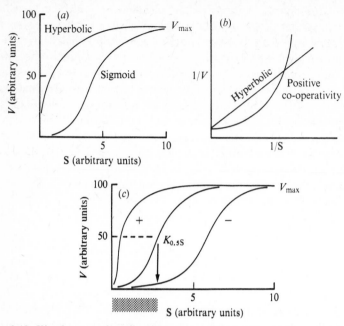

Fig. 8.13. Kinetic properties of regulatory enzymes. (*a*) Hyperbolic and sigmoid plots of reaction velocity *(V)* against substrate concentration (S); (*b*) Lineweaver–Burk double reciprocal plots for hyperbolic and sigmoidal kinetics showing positive co-operativity; (*c*) Effect of positive (+) and negative (−) modulators. The stippled area represents the substrate concentration below $K_{0.5S}$.

(+) and negative (−) modulators is to decrease or increase the values of $K_{0.5S}$ (Fig. 8.13*c*). In the case of a positive modulator the enzyme is able to function at a rate nearer to its V_{max} potential and is more sensitive to small changes in substrate concentration. The opposite is true for a negative modulator and the catalytic rate is reduced. In most cases, only $K_{0.5S}$ and not V_{max} is affected by enzyme modulation.

Most metabolic regulators perform two functions; they act as inter-mediates in the metabolic pathway and as specific feedback or feed-forward regulators for key enzymes in another part of the pathway. This dual role is part of the reason for the sigmoidal type of reaction curve, because the small changes in substrate concentration normally produced in complex metabolic pathways have to be 'amplified' in order to have a regulatory effect. If the interaction between an enzyme and its regulator is hyperbolic (Michaelis–Menten kinetics) then large changes (up to eighty-fold) are required to alter significantly the enzymic activity. If the response curve is sigmoid, however, much smaller changes (fourfold for an exaggerated sigmoid curve) will result in the same relative effect; see discussion in Newsholme & Start (1973).

Mechanisms of temperature adaptation

Molecular mechanisms of temperature adaptation for aquatic poikilotherms have been reviewed by Hochachka & Somero (1971, 1973) and Hazel & Prosser (1974). The body temperature of such animals corresponds closely to that of the environment; despite this fact, many marine invertebrates (and fish) are able to survive and function normally in habitats characterised by marked variations in temperature. They do so because, as the temperature changes, their physiology and biochemistry change in a compensatory manner. It is now generally recognised that such temperature compensation may occur over at least three distinct time scales: (*a*) immediate or short-term changes which result in thermal independence; (*b*) longer-term changes after a period of acclimation; and (*c*) evolutionary changes involved in the adaptation of new species to their environment (see Hochachka & Somero, 1973).

In *M. edulis* we are concerned mainly with two phenomena (Chapter 5). First, inactive or starved mussels have a low standard rate of oxygen consumption which is largely temperature-independent and does not acclimate (Newell & Pye, 1970*a*, *b*; Bayne *et al.*, 1973; Widdows, 1973*b*). Secondly, the routine rate of oxygen consumption for fed mussels is markedly temperature-dependent but acclimates completely, within a period of 14 days, between 5 and 20 °C (Widdows & Bayne, 1971; Bayne *et al.*, 1973; Widdows, 1973*b*). Factors affecting thermal independence and temperature acclimation in *M. edulis* and in the winkle, *Littorina littorea*, have been reviewed by Newell (1973), by Newell & Bayne (1973) and by Pye & Newell (1973).

Mechanisms for immediate temperature compensation involve changes in the flow of carbon through metabolic branchpoints resulting in the operation of alternative pathways or temperature modulation of enzyme–substrate and enzyme–regulator interactions (Hazel & Prosser, 1974). The flux of carbon through the Krebs cycle can be controlled by the relative channelling of acetyl-CoA into the cycle or into fatty acid synthesis, and by the operation of the pentose phosphate pathway as an alternative to glycolytic breakdown of glucose-6-phosphate; both changes are related to each other because of the requirement for NADPH during lipogenesis (Fig. 8.4). A second possibility is that increased glycogen synthesis may take place at high temperature. Either mechanism could control the flow of carbon away from the Krebs cycle in the presence of high endogenous substrate and maintain the metabolic rate independent of temperature (see Introduction in Newell & Pye, 1971*a* and review by Hochachka & Somero, 1971). Newell & Pye (1971*a*) have investigated these two possibilities in the case of *L. littorea*. There was no evidence, however, of a net change in glycogen or triglyceride content even though

341

whole-animal homogenates showed a marked metabolic homeostasis over the temperature range 10–35 °C, with respect to oxygen uptake. Whilst glycolysis may have been dependent on temperature, the end-products evidently did not enter the Krebs cycle, neither were they channelled into lipid. The authors concluded that the rate-limiting step controlling the flow of carbon through the Krebs cycle and maintaining metabolic homeostasis is located within the mitochondria (Newell & Pye, 1971*a*).

The effects of temperature on enzyme–substrate affinities has been discussed extensively by Hochachka & Somero (1971, 1973) and reviewed by Hazel & Prosser (1974). Many enzymes from poikilothermic animals are characterised by a direct relationship between the assay temperature and K_m for the substrate such that as the assay temperature is reduced the value for K_m becomes smaller. Consequently, when catalytic rates are measured at physiological substrate concentrations (below K_m) the effect of the reduced kinetic energy at low temperature is compensated for by the enhanced affinity of the enzyme for its substrate. This phenomenon has been termed 'positive thermal modulation' to emphasise the similarity to the effect of positive modulators on enzyme activity (Hochachka & Somero, 1973). In the case where the enzyme exhibits sigmoidal kinetics, the quantity K_m is replaced by $K_{0.5S}$; regardless of the mechanistic implications of the two different constants, both represent the concentration of substrate for half-maximum velocity and are indicators of the sensitivity of the reaction to changes in substrate concentration. To determine whether positive thermal modulation is the underlying mechanism for immediate temperature compensation requires the measurement of the kinetic properties (with respect to temperature) of all the regulatory enzymes in a given pathway, as well as the determination of substrate concentrations *in vivo*, but to date no such rigorous analysis has been accomplished (Hochachka & Somero, 1973). For *L. littorea*, however, Newell & Pye have presented circumstantial evidence that this type of modulation may be a part of the compensatory mechanism. First, the low Q_{10} values characteristic of the standard metabolic rate for intact winkles have also been observed for tissue homogenates and for isolated mitochondria (Newell & Pye, 1971*a*, *b*). Furthermore the rate of oxygen consumption of the mitochondria was markedly dependent on temperature at high substrate concentration (2 mM pyruvate) but virtually temperature-independent at low concentrations of pyruvate between 0.01 and 1.0 mM (Newell & Pye, 1971*b*). The temperature responses were similar to the differences between active and standard metabolism in the intact animal (Newell & Pye, 1970*b*). Experiments were also carried out to examine the effects of other Krebs cycle intermediates on temperature compensation. Despite the fact that the substrate concentration (1 mM) was close to the upper limit at which homeostasis might be expected to occur, it was found

342

that oxygen uptake was essentially temperature-independent between 5 and 17–18 °C when succinate, malate, fumarate and oxaloacetate were used as substrates (Newell & Pye, 1971*b*). The results suggest that variations in enzyme–substrate affinity of one or more of the intramitochondrial enzymes might be responsible for the control of metabolic homeostasis in *L. littorea*; according to this hypothesis the differences between active (or routine) and standard metabolism with respect to temperature changes may be due, in part, to the level of available substrates (Newell & Bayne, 1973; Pye & Newell, 1973). One wonders, however, about the validity of the experiments with isolated mitochondria, particularly since it is very difficult to get an 'intact' preparation (see previous section) and the determinations of oxygen uptake were carried out in seawater with no further additions other than substrate. How, for example, did oxaloacetate enter the mitochondria? More recently, Pye (1973) has shown that ADP-limited state IV respiration of mitochondria from potatoes, mung bean seedlings and frog skeletal muscle is relatively temperature-independent even at high substrate concentrations. It must be stressed that in these instances the lack of ADP was known to be a limiting factor for respiratory metabolism (Pye & Newell, 1973). In a different sense Newell (1973) has recognised this problem when he states that with isolated mitochondria many of the possibilities for metabolic control have been removed and that the strictly 'reductionist' approach has a limited value because it is unlikely that there will be one, single controlling factor.

Compared to immediate temperature compensation we know even less about the molecular mechanisms of temperature acclimation in marine molluscs. In *M. edulis* the routine metabolic rate is highly temperature-dependent but acclimates fully within 14 days. According to Widdows (1973*b*) acclimation is largely due to changes in ventilation rate and a concomitant reduction (or increase) in the metabolic cost of pumping water through the gills. In contrast the standard metabolic rate for starved mussels is temperature-independent and does not acclimate (see previous discussion). Standard metabolism represents the minimum energy requirement to maintain all the basic functions at a cellular level and routine metabolism includes, in addition, the energy demands of spontaneous activity (see Chapter 5). But, there is a difference between the standard metabolic rate for starved mussels (in which temperature-independence is maintained by low substrate concentrations) and the 'basal' rate in fed mussels when the substrate concentrations are higher. In this case the standard or basal metabolic rate, as a component of the total energy demand, must acclimate or be temperature-independent at high substrate concentrations.

The various mechanisms of temperature acclimation have been reviewed by Hazel & Prosser (1974). In many cases the adaptations in

343

whole-animal respiration are reflected in comparable changes in tissue metabolism; however, not all tissues show the same acclimation pattern nor is acclimation always achieved to the same extent. At the cellular level, temperature acclimation may affect the flow of carbon through a given metabolic pathway and determine the relative contribution of alternative pathways. These differences between tissues and in the reorganisation of metabolism make it very difficult to discuss acclimation in mechanistic terms for the organism as a whole. Even when the rate of oxygen consumption is taken as an indicator of oxidative phosphorylation and the problem is reduced to the mitochondrial level, we do not remove this ambiguity, because the products of respiratory metabolism (ATP and NAD^+) may be utilised by different pathways at different temperatures (Hochachka & Somero, 1973). The changes in cellular metabolism can be explained, in a general sense, by temperature-induced changes in enzyme activity. However, the basic strategies of biochemical adaptation are different in the case of temperature acclimation from those described previously for immediate temperature compensation. One of the most important features of the acclimatory response is that it takes place over a relatively long time period (usually days or weeks) and this has led to the idea that the process of adaptation involves a feedback to the genetic material of the organism resulting in changes in protein synthesis. These changes can be quantitative, resulting in differences in specific activity, or qualitative, resulting in new isoenzymes with different temperature-dependent kinetic properties (see Hazel & Prosser, 1974).

Glycolysis and gluconeogenesis in Mytilus edulis

Mechanisms for the control of glycolysis in vertebrate muscle and for the control of glycolysis and gluconeogenesis in liver have been discussed by Newsholme & Start (1973). Implicit in their discussion is the fact that muscle is glycolytic whereas the liver is both glycolytic and gluconeogenic; this provides a point of control at the tissue level. In the same way, the physiological significance of the control mechanism for glycolysis and gluconeogenesis in *M. edulis* depends on a knowledge of the basic functions of the body tissues.

Adductor muscle

The low concentrations of mitochondria and cytochromes in the adductor muscle of marine bivalves suggest that the muscle is primarily glycolytic (Mattisson & Beechey, 1966). There may be differences, however, between the striated or 'fast' portion of the muscle and the smooth or 'slow' part. According to Suryanarayanan & Alexander (1971) the slow muscle of *Lamellidens corrianus* has a lower glycogen and higher lipid content than the fast muscle and may utilise lipids as the main fuel for oxidative metabolism. O'Doherty & Feltham (1971) have shown that the adductor muscle of *Plactopen magellanicus* contains fructose diphosphatase (FDPase; see Fig. 8.8) and from this they imply that glycogen synthesis takes place from gluconeogenic precursors. It would be generally accepted that the glucose → glycogen step takes place in the adductor muscle, but the presence of FDPase does not necessarily mean the muscle is gluconeogenic. Newsholme & Start (1973) point out that FDPase is often present in vertebrate muscle, under conditions where there is no evidence of quantitatively significant rates of gluconeogenesis. Instead they suggest that FDPase is concerned in the substrate cycling of fructose-6-phosphate and fructose diphosphate and that this 'amplifies' the control of phosphofructkinase (PFK) by the adenosine nucleotides during glycolysis. Neither can the presence of phosphoenolpyruvate carboxykinase (PEPCK) in the adductor muscle of marine bivalves be taken as evidence for gluconeogenesis, because the end-products of glycolysis are different to those in vertebrates and involve the conversion of phosphoenolpyruvate (PEP) into oxaloacetate (see previous section).

Mantle

In *M. edulis* the mantle is both glycolytic and gluconeogenic. Indirect evidence for this view is based on the utilisation of glycogen reserves during gametogenesis (glycolysis) and on the seasonal changes in $K_{0.5s}$ for mantle pyruvate kinase which favour glycolysis in the winter and gluconeogenesis in the summer (Livingstone & Bayne, 1974; Livingstone, 1975). As yet, there is no direct evidence for the operation of the Krebs cycle and oxidative phosphorylation but, presumably, the cycle is active under aerobic conditions. The activity of mantle PEPCK is high in *M. edulis* (Table 8.7; De Zwaan & Van Marrewijk, 1973*b*) and the enzyme must catalyse the interconversion of PEP and oxaloacetate in both directions; *in vivo* the direction of carbon flow is probably determined by changes in the apparent K_m for PEP and oxaloacetate (see discussion in Mustafa & Hochachka, 1973*a*). In the Pacific oyster, *C. gigas*, the activity of PEPCK is very low in the mantle tissues (Mustafa & Hochachka, 1973*a*) and this can be explained by differences in physiological function in oysters and mussels. In *M. edulis* the mantle plays an active role in gametogenesis; it is the main site for the storage of glycogen reserves and the gonad develops within the mantle tissue. In contrast, oysters have a separate gonad which lies between the digestive gland and the mantle. Gonad development takes place in the spring and summer and most probably the digestive gland, and the gonad itself, are the main storage organs for the glycogen reserves.

Digestive gland

The physiology of the digestive gland in molluscs (and Crustacea) has been discussed recently by Van Weel (1974) in relation to the so-called 'hepar' (= liver) and 'pancreas' (= digestive) functions. Van Weel concluded that there was prima facie evidence for a role in digestion but that the liver function was strictly limited, at least in vertebrate terms. In a more general sense, however, the digestive gland has a storage and distribution function and is concerned with the transfer of assimilated food to the body tissues (Sastry & Blake, 1971; Thompson, 1972; Vassallo, 1973; Thompson *et al.*, 1974). In *M. edulis* carbohydrate is accumulated in the early summer and then lost in July–August when glycogen reserves are built-up in the mantle; lipid is synthesised and stored in the digestive gland during the mid-summer when the carbohydrate content is low (Thompson *et al.*, 1974). All of the enzymes of the glycolytic pathway have been demonstrated in the digestive gland of *M. californianus* (Table 8.6; Bennett & Nakada, 1968), but there has been no work on the key gluconeogenic enzymes (although one suspects that gluconeogenesis must take place). One of the main functions of the Krebs cycle in the digestive gland may be to provide intermediates for biosynthesis (see Goddard, 1966).

Blood

The relationship between the digestive gland and the blood sugar level is not fully understood. In marine molluscs the level of blood sugar is controlled at the upper limit by conversion into tissue glycogen (Barry & Munday, 1959). But a concentration minimum, below which the digestive gland would add glucose to the blood, does not seem to exist, possibly because of the necessity for a steep diffusion gradient for the rapid transfer of glucose from the digestive tubules into the blood (see Badman, 1967). The loss of a true hepatic function has led to the proliferation of Leydig tissue for the local storage of glycogen (Eble, 1969). This removes the requirement for a high blood glucose level because the body tissues have their own supply of carbohydrate.

The key enzymes regulating the flow of glycolytic intermediates in the direction of glycolysis or gluconeogenesis are hexokinase and glucose-6-phosphatase; phosphofructokinase (PFK) and FDPase; and pyruvate kinase—PEPCK—pyruvate carboxylase. In the subsequent discussion reference has been made to Newsholme & Start (1973) and Hochachka

(1973) and statements on the control of PFK activity are made as a matter of generally accepted fact without further justification.

$$\text{Fructose-6-phosphate} + \text{ATP} \xrightarrow{\text{PFK}} \text{fructose diphosphate} + \text{ADP}.$$

PFK is inhibited by ATP (above a certain optimum concentration) and this inhibition is relieved by AMP, phosphate, fructose diphosphate (FDP) and fructose-6-phosphate. Fructose-6-phosphate acts as both a substrate and positive modulator for PFK. The simplest theory of glycolytic control is that PFK is regulated by feedback inhibition by ATP in a closed loop between glycolysis and the Krebs cycle. However, large changes in the concentration of ATP do not occur *in vivo* and instead the effects of ATP are amplified by AMP (see Newsholme & Start, 1973). In the cell the concentration of AMP is related to that of ATP by the equilibrium reaction of adenylate kinase:

$$2\text{ADP} \rightleftharpoons \text{ATP} + \text{AMP}.$$

Since the level of ATP is always very much higher than that of AMP, small changes in ATP result in much larger fractional changes in AMP concentration. Under glycolytic conditions (when there is a demand for ATP) the ratio ATP/AMP declines and the increased AMP level stimulates PFK activity. The efficiency of the system is further improved by the action of FDP and ADP as product activators; the process is autocatalytic and leads to a large increase in PFK activity initiated by AMP. This point is discussed again on p. 355 in relation to the effect of FDP on pyruvate kinase activity and the control of the PEP branchpoint by pH changes.

In *M. californianus* PFK is inhibited by ATP and this inhibition is relieved by cyclic AMP (Bennett & Nakada, 1968); *in vivo* AMP is probably the activator of PFK. This suggests that PFK may play an important role in the regulation of glycolysis in marine bivalves. Most attention, however, has been centred on the control of the PEP branchpoint and on the kinetic properties of pyruvate kinase and PEPCK in *M. edulis* and *C. gigas*, and each enzyme will now be considered in turn.

Properties of pyruvate kinase (PK)

The kinetic and regulatory properties of PK from *M. edulis* have been described by De Zwaan (1972), De Zwaan & Holwerda (1972), Holwerda & De Zwaan (1973), and Holwerda *et al.* (1973) for the adductor muscle enzyme and by Livingstone & Bayne (1974), and Livingstone (1975) for the mantle enzyme (Table 8.10).

$$\text{Phosphoenolpyruvate} + \text{ADP} \xrightarrow{\text{PK}} \text{pyruvate} + \text{ATP}.$$

In *M. edulis*, PK has properties which are very similar to the L-type enzyme from rat liver. Both the adductor and mantle enzymes are inhibited

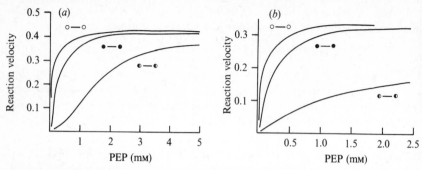

Fig. 8.14. Effects of fructose diphosphate (FDP) and alanine on the activity of pyruvate kinase from *Mytilus edulis*. (*a*) Adductor muscle. ●, control; ○, plus 0.1 mM FDP; ◐, plus 2.0 mM alanine: reaction velocity expressed as μmole NADH reduced min⁻¹ per mg protein. (Redrawn from De Zwaan, 1972.) (*b*) Mantle. ●, control; ○, plus 0.1 mM FDP; ◐, plus 1.0 mM alanine: reaction velocity expressed as ΔA_{334} min⁻¹ i.e. change in absorbance at 334 nm. (Redrawn from Livingstone & Bayne, 1974.) For other reaction conditions see references.

by alanine and ATP, and activated by FDP. There are some differences, however, between the enzymes in *Mytilus* and rat liver with respect to the interactions between substrate and pH, and between pH ·and FDP activation (see De Zwaan & Holwerda, 1972; Holwerda & De Zwaan, 1973; Livingstone & Bayne, 1974). Fig. 8.14 shows the effects of FDP and alanine on PK activity in *M. edulis*. Addition of 0.1 mM FDP changes the kinetics from sigmoid to hyperbolic, resulting in an activation of PK (positive modulation); in the mantle this activation is pronounced even at concentrations as low as 0.01 mM FDP (Livingstone & Bayne, 1974). Alanine is a strong inhibitor of PK resulting in an increase in sigmoidicity and in $K_{0.5S}$ for PEP (negative modulation). In the presence of 0.1 mM FDP, however, the inhibition by alanine is reversed resulting in a hyperbolic reaction curve and a reduction in $K_{0.5S}$ (Holwerda & De Zwaan, 1973; Livingstone & Bayne, 1974); within the physiological range of substrate and modulator concentrations the activity of PK would be negligible without the feed-forward activation by FDP (see De Zwaan, 1972). For the mantle enzyme the interaction between alanine and FDP is best seen in

FDP (0.01 mM)	Alanine (1 mM)	$K_{0.5S}$ PEP (μM)
−	−	212
+	−	67
−	+	1080
+	+	311

+, present in the assay; −, absent from the assay.

Table 8.10. Mytilus edulis: *kinetic properties of pyruvate kinase*

	Adductor muscle		Mantle
	De Zwaan (1972)	De Zwaan & Holwerda (1972)	Livingstone & Bayne (1974)
$K_{0.5S}$ (μM)	650 (pH 7.4, 25 °C) Sigmoidal	630 (pH 7.6; 25 °C) Sigmoidal	212–240 (pH 7.5–8.0; 16 °C) Sigmoidal
pH optimum	—	7.0–7.5	7.0–8.0
Alanine inhibition	$K_{0.5S} = 1580$ (2 mM alanine) V_{max} slightly reduced	—	$K_{0.5S} = 1080$ (1 mM alanine) V_{max} reduced
FDP activation (0.1 mM FDP)	$K_{0.5S} = 250$ Hyperbolic V_{max} unaffected	$K_{0.5S} = 150$ Hyperbolic V_{max} unaffected	$K_{0.5S} = 40$ Hyperbolic V_{max} unaffected
FDP+alanine	Inhibition slight	—	Inhibition slight

terms of the changes in $K_{0.5S}$ for PEP (Livingstone & Bayne, 1974): PK is also inhibited by ATP; in the adductor muscle the inhibition is overcome by FDP (Holwerda & De Zwaan, 1973). FDP activation and inhibition by alanine and ATP have been reported for the mantle and adductor enzymes from *C. gigas* but in both tissues PK showed non-allosteric kinetics with linear double reciprocal plots (Mustafa & Hochachka, 1971).

Fig. 8.15 shows the effect of pH on PK activity in *M. edulis*. In the absence of FDP (and alanine) both forms of the enzyme show a marked dependence of reaction velocity on pH with optima between pH 7.0–7.5 for adductor PK and between 7.5–8.0 for mantle PK. It is difficult to compare the results at low pH because Livingstone & Bayne (1974) did not work below pH 7.0; however, from the data available it seems that below pH 7.0 the activity of PK is markedly inhibited (De Zwaan & Holwerda, 1972). When 0.1 mM FDP is added to the reaction mixture the effect is to increase the activity of PK at low pH and to 'flatten' the pH profile so that *in vivo* the inhibition by increased [H$^+$] is counteracted by an increase in FDP concentration. Inhibition of mantle PK by alanine is increased at low pH (below 8.0) but this is again counteracted by FDP (Livingstone & Bayne, 1974). Essentially the same effects of FDP on the pH profiles of the mantle and adductor enzymes from *C. gigas* have been reported by Mustafa & Hochachka (1971) (see later).

Seasonal changes in $K_{0.5S}$ PEP for mantle PK in *M. edulis* have been described by Livingstone & Bayne (1974) and by Livingstone (1975). The values of $K_{0.5S}$ varied between different populations of mussels from Heacham and Plymouth (Livingstone & Bayne, 1974) and from the River

Marine mussels

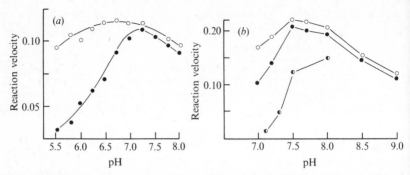

Fig. 8.15. Effect of pH on pyruvate kinase (PK) activity in *Mytilus edulis*. (*a*) Adductor muscle. ●, control; ○, plus 0.1 mM FDP: reaction velocity expressed as ΔA_{340} min^{-1}. (Redrawn from De Zwaan & Holwerda, 1972.) (*b*) Mantle. ●, control; ○, plus 0.1 mM FDP; ◑, plus 0.5 mM alanine: reaction velocity expressed as ΔA_{334} min^{-1}. (Redrawn from Livingstone & Bayne, 1974.) For other reaction conditions see references.

Lyner (Livingstone, 1975); in all cases, however, $K_{0.5S}$ was very much higher in the summer (100–400 μM) than in the winter (20–100 μM). This can be related to the alternate gluconeogenic and glycolytic function of the mantle tissues and is similar to the difference between $K_{0.5S}$ PEP for the mantle and adductor enzymes from *C. gigas* (Mustafa & Hochachka, 1971). The differences in the population means for $K_{0.5S}$ PEP for mussels from different localities are related to the seasonal gametogenic cycle and to the ambient temperature range (for discussion see Livingstone, 1975).

Fig. 8.16 shows the seasonal changes in $K_{0.5S}$ PEP for mantle PK from the Lyner population; in both male and female mussels there is a short period of 3–4 months in the summer when the value of $K_{0.5S}$ is markedly increased compared with the winter period. This corresponds to the time of maximum accumulation of glycogen reserves (Gabbott & Bayne, 1973). When gluconeogenesis is taking place and PEP is produced from oxaloacetate, any simultaneous activity of PK would result in the futile cycling of carbon at the expense of ATP (see Atkinson, 1971). In the mantle tissues this is avoided because of the high value of $K_{0.5S}$ for PEP and the generally low concentration of FDP during gluconeogenesis. It is interesting to compare the summer and winter values of $K_{0.5S}$ PEP with those for type-L pyruvate kinase from liver, and muscle PK. Newsholme & Start (1973) quote values for $K_{0.5S}$ of 840 and 100 μM for the liver and muscle enzymes, respectively. This is further evidence that the kinetic properties of mantle PK in *M. edulis* are related to the seasonal changes in gluconeogenic and glycolytic function. In contrast, $K_{0.5S}$ PEP for the adductor enzyme does not vary seasonally and has a mean value of approximately 100 μM for the Lyner population (Livingstone, 1975); low values of $K_{0.5S}$ PEP are indicative of a glycolytic function in the adductor muscle.

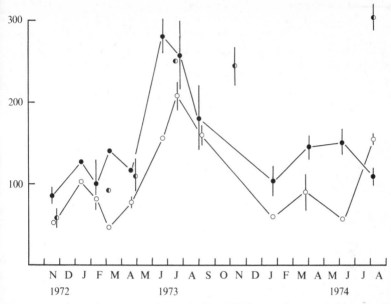

Fig. 8.16. *Mytilus edulis*: seasonal changes in $K_{0.5S}$ PEP for mantle pyruvate kinase (redrawn from Livingstone, 1975). Lyner population. ●, females; ○, males; ◑, sex indeterminate. Vertical lines represent the standard error of mean.

Properties of phosphoenolpyruvate carboxykinase (PEPCK)

The kinetic properties of adductor PEPCK have been described by Mustafa & Hochachka (1973*a*, *b*) for *C. gigas* and by De Zwaan & De Bont (1975) for *M. edulis*. In mammalian tissues interest has centred on the role of PEPCK in gluconeogenesis and on the properties of PEPCK assayed in the direction of decarboxylation of oxaloacetate; in marine bivalves, however, we are concerned primarily with the PEPCK-catalysed carboxylation of PEP under anaerobic conditions.

$$\text{Phosphoenolpyruvate} + \text{IDP} + \text{CO}_2 \overset{\text{PEPCK}}{\rightleftharpoons} \text{oxaloacetate} + \text{ITP}.$$

PEPCK has a requirement for divalent cation and Mn^{2+}, Mg^{2+} and Zn^{2+} all support activity in *M. edulis*, but in the case of *C. gigas* there was no activity in the presence of Mg^{2+}; the specific differences between the enzymes from *C. gigas* and *M. edulis* and the kinetic properties with respect to their cation requirements are discussed by Mustafa & Hochachka (1973*a*), and by De Zwaan & De Bont (1975). Double reciprocal plots with PEP as substrate are linear in the presence of all three cations for *M. edulis* and in the presence of Zn^{2+} for *C. gigas*. In *C. gigas* the PEP saturation curves with Mn^{2+} as the cation were aberrant and this led

351

Mustafa & Hochachka (1973*a*) to consider that *in vivo* Zn^{2+} was the most likely cofactor. This is an important point because it is directly related to the non-overlapping nature of the pH profiles for PEPCK and PK (see later). ITP acts as a competitive inhibitor of PEPCK with respect to PEP as substrate; in the case of *C. gigas*, but not *M. edulis*, alanine counteracts the inhibition by ITP. None of the other metabolites tested, including FDP and AMP had any effect on the activity of PEPCK (Mustafa & Hochachka, 1973*b*; De Zwaan & De Bont, 1975).

Fig. 8.17 shows the pH profiles for PEPCK from *M. edulis* in the presence of Zn^{2+}, Mn^{2+} or Mg^{2+} as cofactors. The shape of the curves depends on the cation used; Zn^{2+} and Mg^{2+} give a marked pH optimum at pH 6.0 and 6.8, respectively, but in the presence of Mn^{2+} the curve is essentially flat between pH 6.0–8.0 (De Zwaan & De Bont, 1975). These results, particularly with Zn^{2+} as the cation, suggest that *in vivo* PEPCK is activated by an increase in $[H^+]$ whereas the opposite is true for PK. In the absence of FDP activation, the activity of PK is inhibited by low pH (Fig. 8.15). This is clearly shown by comparing the activity of the two enzymes from *M. edulis* at pH 6.5 and 7.4 (De Zwaan & De Bont, 1975). Changing the pH from 7.4 to 6.5 has the opposite effect on the two enzymes; PEPCK is activated and PK is inhibited.

	PEPCK		PK	
	pH 6.5 (Zn^{2+})	ph 7.4 (Mn^{2+})	pH 6.5	pH 7.4
Adductor muscle	5.16	2.20	1.81	16.33
Mantle	0.80	0.32	Nil	7.97
Digestive gland	1.17	0.27	0.98	14.66
Gills	0.18	0.09	0.93	10.88

Activities as ΔA_{340} min^{-1} per g fresh tissue with 0.1 mM PEP as substrate.

Control of the PEP branchpoint

In marine bivalves PEP can be further metabolised by conversion to pyruvate, catalysed by PK, or to oxaloacetate, catalysed by PEPCK. Both reactions generate a group-transfer compound; ATP in the case of PK and ITP (or GTP) in the case of PEPCK, and both enzymes are subject to product inhibition by ATP or ITP. In this sense they behave in accordance with the energy charge concept (Atkinson, 1971) and, undoubtedly, the activity of both enzymes is dependent on the level of ATP in the cells. At the same time, the 'fine' control of the carbon flux through pyruvate and oxaloacetate is determined by the regulatory properties of PK and PEPCK with respect to their interaction with FDP, alanine and pH (Fig. 8.18).

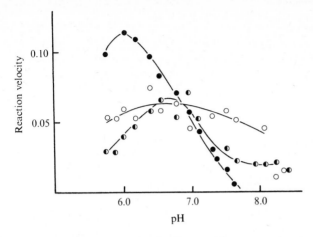

Fig. 8.17. *Mytilus edulis*: effect of pH on adductor muscle phosphoenolpyruvate-carboxykinase (PEPCK) activity (redrawn from De Zwaan & De Bont, 1975). ●, 0.5 mM Zn^{2+}; ○, 1 mM Mn^{2+}; ◐, 1 mM Mg^{2+}: reaction velocity expressed as ΔA_{340} min^{-1}. For other reaction conditions see reference.

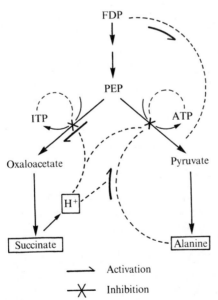

Fig. 8.18. Regulation of pyruvate kinase (PK; catalysing PEP → pyruvate) and phosphoenol-pyruvate (PEPCK; catalysing PEP → oxaloacetate) activity at the PEP branchpoint.

353

According to Hochachka & Mustafa (1972) the two enzymes operate on a reciprocal or 'on–off' basis, and both cannot be active at the same time. Under aerobic conditions PEP is converted into pyruvate, but under anaerobic conditions PEPCK activity is favoured by the inhibition of PK by alanine and low pH so that PEP is converted into oxaloacetate and ultimately into succinate. This, in turn, has led to the postulate that alanine is not formed from PEP but from aspartate in a redox couple between oxaloacetate and malate (Fig. 8.12; Mustafa & Hochachka, 1973*b*). There are a number of objections to this theory (Livingstone & Bayne, 1974).

(1) The argument depends largely on the fact that the pH curves for PEPCK and PK are non-overlapping (see Hochachka & Somero, 1973). But, this is only true if the activity of PK is determined in the absence of FDP; at a concentration of 0.1 mM, FDP results in a marked change in the pH profile such that the activity of PK is nearly constant between pH 6.0 and 8.0 (Fig. 8.15; Mustafa & Hochachka, 1971). Furthermore, in *M. edulis* the activity of PEPCK is relatively independent of pH when Mn^{2+} is used as the cofactor; it is only with Zn^{2+} that the profiles for PK and PEPCK are partially non-overlapping (Figs. 8.15 and 8.17).

(2) In *M. edulis*, FDP overrides the inhibition of PK by alanine (Holwerda & De Zwaan, 1973; Livingstone & Bayne, 1974). The concentrations of FDP are physiologically realistic, viz. 0.01–0.1 mM; De Zwaan (1972) quotes a concentration of 0.18 mM FDP in whole-mussel extracts decreasing to 0.11 mM during anaerobiosis. The role of FDP as a positive modulator of PK was noted previously by Mustafa & Hochachka (1971) for *C. gigas* but this has been largely ignored in their subsequent discussions. Indeed, in their 1971 paper Mustafa & Hochachka state that 'the physiological function of fructose-1,6-P_2 activation may be to allow P-enolpyruvate conversion to pyruvate during extended periods of anaerobiosis, when intracellular pH might be reduced'.

(3) Alanine has no effect on PEPCK activity in *M. edulis* (De Zwaan & De Bont, 1975).

(4) There is a fundamental paradox when one considers the effect of pH on the activity of PEPCK (see Livingstone & Bayne, 1974). According to Hochachka and co-workers there must be an increase in $[H^+]$ before PEPCK is activated and takes precedence over PK; however, such a reduction in pH cannot be due to the acid end-products of anaerobic metabolism until PEPCK is activated. As yet no other means of increasing the $[H^+]$ has been found, apart from the possible contribution of a small amount of lactic acid.

There may be some specific differences in the control mechanisms in *C. gigas* and *M. edulis* and between different tissues in the same animal. But,

354

in a general sense, it is difficult to accept that there is a strict aerobic–anaerobic change in the activity of PK and PEPCK as proposed by Mustafa & Hochachka (1973a, b).

The alternative view put forward by Livingstone & Bayne (1974) and supported by De Zwaan & De Bont (1975) is that both pathways operate together and that the flow of carbon through the PEP branchpoint is determined by the degree of tissue hypoxia. This can be related to the known effects of anaerobiosis on the regulation of glycolysis in liver and kidney cortex (see Newsholme & Start, 1973). Under anaerobic conditions the content of ATP is reduced (with a concomitant increase in the level of AMP) resulting in the activation of PFK; this may increase the level of FDP and activate PK. Pyruvate is then converted into alanine as the first end-product of anaerobic metabolism (De Zwaan & Van Marrewijk, 1973a; De Zwaan et al., 1973). The feedback inhibition of PK by alanine is overridden by FDP, but eventually the activity of PK will be reduced; at the same time carboxylation of PEP to oxaloacetate, catalysed by PEPCK, will result in the accumulation of succinate and a fall in pH. This will potentiate the inhibition of PK by alanine and increase the activity of PEPCK. The balance between the flux of carbon through pyruvate and oxaloacetate is determined by the changes in alanine concentration and pH, and by the effect of FDP on PK (during glycolysis the activities of PFK and PK are closely integrated). The extent to which one pathway will predominate over the other will depend on the particular conditions within the body tissues; where the level of hypoxia is reduced, succinate will be removed by oxidative metabolism and the pH will increase resulting in a reversal of the regulatory control. It may be that PK and PEPCK operate on a reciprocal basis and that there are oscillations in the net production of alanine and succinate.

9.   Population genetics of mussels

J. S. LEVINTON & R. K. KOEHN

Mussels, like most marine epifaunal invertebrates, have a pelagic larval stage followed by sedentary adulthood. Difficulties of rearing the larval stages hamper genetic and evolutionary studies of mussel populations. Crossing experiments that are essential to establish the heritability of morphological traits have not been performed. Therefore, our understanding of the genetics of mussel populations is still in its infancy. We know nothing, for example, of the genetic basis of morphological polymorphisms, such as shell colour, and only a limited amount of data exists on the cytogenetics of mussel species (Ahmed & Sparks, 1967, 1970; Menzel, 1968).

The recent discovery and routine application of electrophoretic separation of isoenzymes has led to extensive surveys of genetic variation in populations of a variety of marine species. Electrophoretic separation of isoenzymes permits inference of Mendelian variation in populations that are not easily amenable to crossing experiments, so that genetic variation can now be surveyed in most marine invertebrate species.

Much of the early descriptive work was done on fishes (see de Ligny, 1969, and Gooch, 1975, for review). A diversity of marine invertebrates has been studied, including bivalves (O'Gower & Nicol, 1968; Wilkins & Mathers, 1973), copepods (Manwell, Baker, Ashton & Corner, 1967), ectoprocts (Gooch & Schopf, 1970, 1971) and horseshoe crabs (Selander, Yang, Lewontin & Johnson, 1970). Fortunately, mussels have received considerable attention (Milkman & Beaty, 1970; Koehn & Mitton, 1972; Levinton, 1973; Mitton, Koehn & Prout, 1973; Koehn, 1975; Koehn, Milkman & Mitton, 1975; Levinton & Fundiller, 1975) and these and other results will be discussed in detail in this chapter.

These genetic data are not only of importance to our understanding of adaptation and evolution in sedentary mussel populations, but provide a means to examine the role of genetic variation in population dynamics. The life cycle of mussels is complex and knowledge of genetic variation of populations, or their genetical structure, cannot be gained without consideration of population parameters of dispersal, settlement, growth, reproductive patterns and potential, and mortality. *Mytilus edulis* is a very fecund species, females producing as many as 2.5×10^7 gametes per season (Field, 1922). Gametes are liberated into the water upon spawning, and fertilisation subsequently takes place there (Chipperfield, 1953). A planktotrophic larval stage feeds in the plankton for a period of normally as much as 22 days, with subsequent settling on filamentous algae (Bayne,

1964 *b*; Chapter 4). Secondary movement to favourable sites for adult attachment occurs (Bayne, 1964*b*). Newly metamorphosed juveniles are highly vagile, moving about with an active foot on the surface. Even after the final slowing of movement the adult blue mussel is highly mobile and tends to move upward and outside of a clump of mussels (Harger, 1968; 1972*b*). This is in contrast to the slight movement of the California sea mussel (*Mytilus californianus*; Harger, 1972*b*), although movement into adjacent areas where open space is created is very common (R. T. Paine, verbal communication).

The high dispersal ability of mussel larvae, vagility in the sedentary adult stage, plus their dominance in the intertidal zone, are characteristics which raise several lines of research on the genetics of mussel populations.

(1) How does the great dispersal potential interact with selection to yield observed macrogeographic patterns of genetic variation?

(2) How is the patchy larval settlement and early movement of juveniles important in the distribution of genotypes?

(3) If larval settling is not genotype-specific, then do we observe repetitive patterns of adjustment, through selective mortality, during the history of the cohort?

(4) Does the environmental heterogeneity imparted by differences at different levels in the intertidal zone (exposure time, associated flora and fauna) exert an effect on genetic variation? Is there a difference in genotype-specific reproductive effort at different levels in the intertidal zone?

Work on genetic polymorphisms of mussels (principally of the blue mussel, *Mytilus edulis*) done in several laboratories throughout the world provides partial answers to points (1), (3) and (4). Surveys of genetic variation, geographical changes, microgeographical variation, and changes in experimental populations have been examined. Though we still know relatively little, these data suggest that the study of isoenzyme polymorphisms will help us to understand the selective forces operating on populations of mussels, and will serve as a model system for the study of other sedentary invertebrate species. Hopefully work will be initiated in the future to assess the importance of juvenile and adult vagility on genetic heterogeneity within the intertidal zone. It is our intent to illustrate the point that an understanding of the evolutionary biology of mussels should become an integral part of studies of their population biology.

Electrophoresis and population genetics

The development of methods for the visualisation of individual proteins led to a considerable increase in studies on molecular, cellular and population

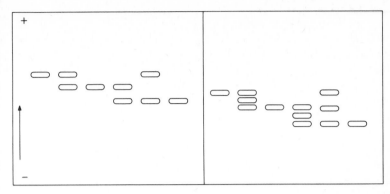

Fig. 9.1. Expected electrophoretic phenotypes for three allele polymorphisms when the enzyme is a functional monomer (left) or functional dimer (right). Leucine aminopeptidase in *Mytilus edulis* appears like the left-hand patterns, while most other enzymes discussed in text appear like the patterns on the right.

aspects of protein variation. The majority of individual enzymes were found to exist in a variety of forms, called *isozymes* (Hunter & Markert, 1957), though the term *isoenzyme* is currently preferred. Multiple enzyme forms can be due to a variety of causes such as multiple gene loci, protein polymerisation, developmental changes and genetic variation at individual loci. It is this last basis for multiple enzyme forms that concerns us here, because the frequencies of allozymes (allelic isoenzymes) can be used to estimate the genetic compositions of mussel populations.

Visualisation of allozymes depends on their separation in a supportive medium (e.g. starch, acrylamide gels) on the basis of net molecular charge, molecular weight and configuration. When an electric potential is applied to a mixture of proteins in a suitable medium, electrophoresis results in differential migration of proteins which differ in the above characteristics. Subsequent to electrophoretic separation, individual proteins are visualised by histochemical techniques, whereby an insoluble dye is precipitated on the gel where a specific enzyme is located. Detailed explanations of electrophoresis may be found in Brewer (1970) and Gordon (1969). Since more than fifty different enzymes can be studied by this technique, an even greater number of loci can be examined for allelic variants. Though there are a few exceptions, most allozymes are codominant and allele and genotype frequencies can be determined directly by counting the number of each allozyme type in a sample. A single electrophoretic band does not necessarily represent a unique gene product because the functional structure (i.e. monomer, dimer, etc.) will influence the electrophoretic phenotype of certain genotypes (Fig. 9.1).

The genetic characterisation of natural populations involves the identification of allozyme types at a single locus and computation of the frequency of individual allozymes (allele frequency) as well as of the various genotypic combinations (genotypic frequency) that may occur among them. Ideally, we would hope to characterise all gene loci of the genome of a particular species population, but realistically we must rely on sampling the entire genome by examining only a few gene loci. For example, when two alleles A and B are present at a single locus, the observed genotypic frequencies of AA, AB and BB are expressed as their proportion of the total number of genotypes observed. From the distribution of genotypic classes, we can further characterise our sample by expressing the frequency of the A and B alleles independent of the particular genotypic combinations in which they are found. Thus, the frequency of allele A (designated as p) will be equal to the total number of A alleles observed divided by the total number of alleles of all types observed. The frequency of allele B (designated q) is computed similarly. Because we are estimating the true population frequency and the confidence limits of our estimates are directly related to the size of our sample, it is preferable to express the allele frequency as $p\pm$ standard error, or $p\pm\sqrt{[p(1-p)/2N]}$. By using the familiar Hardy–Weinberg formulation, we can compute the expected zygotic frequency distribution based on our estimates of p and q, if we assume no selection, migration, deviations from random mating, and an infinite size population. Comparing observed frequencies of zygotic classes with those expected with Hardy–Weinberg equilibrium is one way of inferring that the observed variation is Mendelian. However, the test of comparison is notoriously weak and can not easily reject the null hypothesis of conformance to a Hardy–Weinberg distribution. Moreover, systematic deviations from Hardy–Weinberg distributions have been observed in mussel populations that may be explained by (a) selection against unfavourable genotypes or (b) mixture of populations with differing allele frequencies. The latter effect may be very common in mussels, as a population of settling larvae might consist of individuals from various parental populations each with differing allele frequencies. At a locus of two alleles, settlement of larvae with initial differing allele frequencies will result in a 'population' of mussels with allele frequencies intermediate to the parental populations. When sampled, this population will exhibit a net deficiency of heterozygotes relative to Hardy–Weinberg expectations (Wahlund effect).

Empirical population genetic studies of mussels (and other species) depend on detecting apparent violations of the Hardy–Weinberg equilibrium. That is, we are not only interested in whether the genetic compositions of populations differ from place to place, but what the forces may be that bring about these differences. Various population effects which bring about a departure of the zygotic frequency distribution from that expected at Hardy–Weinberg equilibrium are of great significance because they may

Table 9.1. *Levels of genetic polymorphism in* Mytilus edulis

Number of loci	Average heterozygosity/ individual	% loci polymorphic
22	0.13	31

Data furnished by J. Beardmore; they do not include esterases.

reflect the action of natural selection, migration, non-random mating, or small population sizes. We can estimate departures from expectations by examining the heterozygote classes and determining to what degree and in what direction these departures occur. A deviation of heterozygote numbers from Hardy–Weinberg expectations can be formulated with an index, D (Selander, 1970; it should be parenthetically noted that there are many indices that measure this deviation and we use D here because it is most common in the literature dealing with mussels). The deviation is equal to

$$D = (H_0 - H_e)/H_e,$$

where H_0 is the observed number of heterozygotes (either a specific heterozygote class or the sum of all heterozygote classes) and H_e is the number of heterozygotes expected from the estimated allele frequencies. D will be positive or negative if there is an excess or deficiency of heterozygotes, respectively. If N is the number of genomes sampled, the standard error of D is:

$$\{(1-D)[2pq(1+D)+D(2-D)(1-2p)^2]\}/2Npq.$$

Geographic differentiation

Mytilus edulis

A very large number of loci have been studied. J. Beardmore & R. K. Koehn (unpublished data) surveyed twenty-two loci (Table 9.1) and found levels of heterozygosity similar to many other species (Table 9.1). Most studies, however, have been concerned with a few polyallelic loci that exhibit varying magnitudes of geographic variation.

Leucine aminopeptidase (*LAP*)

Five alleles have been detected at the locus for leucine aminopeptidase (L-leucylpeptide-hydrolase; E.C. 3.4.1.1; *LAP*). These have been designated LAP^{100}, LAP^{98}, LAP^{96}, LAP^{94}, and LAP^{92}, in order of decreasing

Table 9.2. *Comparison of allele frequencies at three loci of* Mytilus edulis *from various areas of the world*

Locality	LAP^{100}	LAP^{98}	LAP^{96}	LAP^{94}	LAP^{92}	*GPI^{100}	GPI^{83}	GPI^{89}	AP^{105}	AP^{100}	AP^{97}	AP^{92}	AP^{89}
Western North America (San Juan Is., Wash.)	0.006	0.087	0.462	0.431	0.012	0.119	0.606	0.274	0.000	0.000	0.137	0.787	0.075
Melbourne, Australia	0.000	0.070	0.289	0.641	0.000	0.099	0.823	0.078	0.140	0.521	0.148	0.000	0.000
Eastern North America South Cape Cod	0.009	0.270	0.210	0.550	0.000	0.290	0.250	0.190	0.000	0.010	0.440	0.360	0.200
North Cape Cod	0.050	0.530	0.390	0.090	0.001	0.380	0.250	0.370	0.001	0.005	0.439	0.353	0.198
Iceland	0.217	0.519	0.217	0.023	0.023	0.537	0.381	0.081	0.000	0.001	0.366	0.561	0.073
Western Ireland	0.130	0.600	0.210	0.050	0.010	—	—	—	—	—	—	—	—

* Frequencies of the three *GPI* alleles are composites of several less frequent alleles.

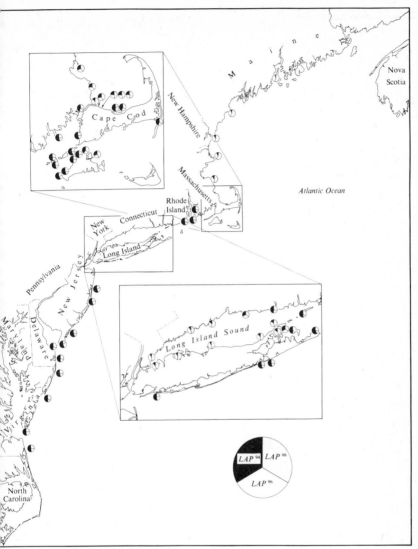

Fig. 9.2. Sampled localities for *M. edulis* on the east coast of the United States. Pie diagrams illustrate relative frequencies of leucine aminopeptidase loci *LAP⁹⁸*, *LAP⁹⁶*, and *LAP⁹⁴* alleles. (Modified from Koehn *et al.*, 1975.)

electrophoretic mobility (Koehn *et al.*, 1975). Though all five alleles occur on both coasts of North America and on those of northern Europe, only three are common in any one area (Table 9.2). Moreover, there is considerable variation in the frequency of the alleles within each major geographic region. On the east coast of North America the frequency of *LAP⁹⁴* is rather invariant from Virginia to Cape Cod, but declines abruptly from 0.58 to 0.10 and remains less common throughout the Gulf of Maine (Figs. 9.2 and 9.3). The frequencies of the LAP^{98} and LAP^{96} alleles vary in a complimentary pattern to LAP^{94} in this region. In Nova Scotia the frequency of LAP^{94} rapidly increases to 0.35 and remains constant in all samples from Nova Scotia and Newfoundland. In Iceland, LAP^{94} declines to 0.09 whereas LAP^{98} is 0.53, the same frequency as is found in the Gulf of Maine. LAP^{100}, which occurs with no greater frequency than 0.05 throughout the east coast of North America, increases to 0.21 in Iceland. In western Ireland (Galway; J. Murdock, unpublished data) LAP allele frequencies are very similar to Iceland: LAP^{96}, LAP^{94}, and LAP^{9} frequencies are 0.21, 0.05 and 0.01, respectively, in Ireland and Iceland. In Ireland, LAP^{98} is slightly more frequent and LAP^{100} less frequent than in Iceland. There is significant variation in the frequencies of LAP^{98}, LAP^{9} and LAP^{94} throughout Northern Ireland and the Irish Republic. Reductions in frequencies of LAP^{96} and LAP^{94} with a concomitant increase in LAP^{98} occur in areas of more severe exposure to wave action (J. Murdock, unpublished data).

The three alleles common in the eastern North Atlantic (LAP^{98}, LAP^{96} and LAP^{94}) were also the most common in a sample of seventy-three individuals from Melbourne, Australia (Koehn, unpublished data). In fact, at the LAP locus, there is a greater identity between samples from Australia and the North American coast south of Cape Cod than between the latter region and the Gulf of Maine!

Aminopeptidase (*AP*)

Seven alleles have been observed at the locus for aminopeptidase (aminoacyl-dipeptide-hydrolase; E.C. 3.4.1.3; AP). Of these, only three are common in the eastern North Atlantic (AP^{97}, AP^{92} and AP^{89}). Unlike the LAP locus, alleles of AP are of constant frequency over most of the studied areas of the eastern Atlantic, but in Nova Scotia and Newfoundland AP^{92} increases from 0.32 to 0.52 with a concomitant decrease of AP^{97} (Fig. 9.3). Frequencies of all alleles in Iceland are very similar to Newfoundland.

Of the seven alleles in the eastern North Atlantic, four are found in Australia. The two fastest-migrating alleles (AP^{100} and AP^{105}) are extremely rare in eastern North America, but have frequencies of 0.52 and 0.14, respectively in Melbourne, Australia (Koehn, unpublished data).

Only three alleles have been observed on the west coast of North America, though extensive sampling has not been done (Levinton, unpublished data). Frequencies of AP^{97}, AP^{92} and AP^{89} were quite different from other studied geographic regions (Table 9.2). No information on geographic variation at this locus is available from Europe.

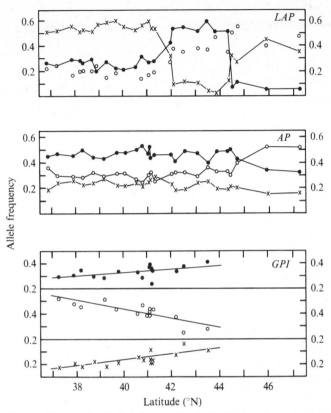

Fig. 9.3. Geographic variation in allele frequencies in *M. edulis* on the east coast of the United States at the leucine aminopeptidase (*LAP*), aminopeptidase (*AP*), and glucose phosphate isomerase (*GPI*) loci. Different alleles at each locus are represented by different types of line. (Modified from Koehn *et al.*, 1975.)

Glucose phosphate isomerase (GPI)

More alleles have been observed for this enzyme (D-glucose-6-phosphate ketolisomerase; E.C. $5 \cdot 3 \cdot 1 \cdot 9$; GPI) than for any other. Nine allozymes occur in populations throughout eastern North America, with up to seven in an individual sample. Similar to *LAP* and *AP*, only three are relatively common (*GPI^{100}*, *GPI^{89}* and *GPI^{83}*) and each show a significant linear correlation with the latitude of samples (Fig. 9.3). The largest change in frequency with latitude is in the *GPI^{83}* allele, which changed in frequency from 0.54 at 37° N to 0.30 at 44° N.

Samples from Long Island Sound were homogeneous at a frequency expected by the regression on latitude of coastal samples. Nearly all samples corresponded with Hardy–Weinberg expectations (Koehn *et al.*, 1975).

Table 9.3. *Protein systems examined in* Mytilus californianus

Protein	Loci	Number of alleles observed	N
Esterase	EST-1	4	40
	EST-2	3	40
	EST-3	4	40
Aspartate aminotransferase	AAT	2	80
Aminopeptidase	AP	3	40
Sorbitol dehydrogenase	SDH	Monomorphic	20
Tetrazolium oxidase	TO-1	Monomorphic (very rare second allele)	L*
	TO-2	Monomorphic	L
Leucine aminopeptidase	LAP-I	5	L
	LAP-II	3	L
Glucose phosphate isomerase	GPI	7	L

* L represents $N > 1000$.

J. Murdock (unpublished data) reports the *GPI* locus to be monomorphic in Ireland, but several alleles have been observed in Wales (J. Beardmore, unpublished data). No co-electrophoresis has been done with North American samples and thus specific identity of *GPI* alleles in Europe is unknown.

Of the nine alleles in eastern North America, seven occur in western North America (Levinton, unpublished data) and four in Australia (Koehn, unpublished data). Frequencies of the most common alleles vary considerably among various regions of the world (Table 9.2).

Other loci

Finally, some data are available for four additional loci. At the *malate dehydrogenase* locus (L-malate: NAD oxidoreductase; E.C. 1.1.1.37; MDH) a common allele (*MDH^{100}*) and a single alternate allele (*MDH^{82}*) never exceeding 0.04 are found from Virginia to Iceland. The *MDH^{100}* allele is fixed in the sample from Australia. No information is available from other regions.

Four alleles were observed at the *6-phosphogluconic dehydrogenase* locus (6-phospho-D-gluconate: NAD(P) oxidoreductase; E.C. 1.1.1.43; 6-PGD) in the western North Atlantic (Koehn *et al.*, 1975). Three alleles (*6-PGD^{77}*, *6-PGD^{91}* and *6-PGD^{100}*) never exceeded 0.05 while *6-PGD^{83}* varied only between 0.95 and 0.99. *6-PGD^{83}* was significantly less common in Newfoundland and Iceland.

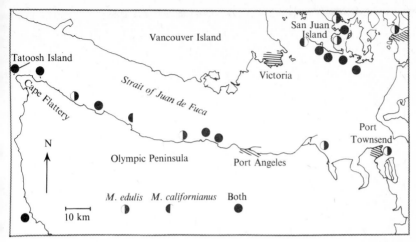

Fig. 9.4. Sampled localities for *M. californianus* in the Strait of Juan de Fuca, Washington, United States.

At the *IDH* locus (threo-DS Isocitrate: NADP oxidoreductase; E.C. 1.1.1.42; IDH), two rare (*IDH¹¹⁰* and *IDH⁹⁰*) and one common (*IDH¹⁰⁰*) alleles have been observed in the western North Atlantic. *IDH¹⁰⁰* varied from 0.95 at latitude 37° N to 0.99 at latitude 47° N.

Two loci synthesise *aspartate aminotransferase* (L-aspartate: 2-oxo-glutarate aminotransferase; E.C. 2.6.1.1; AAT) and the least anodal of these exhibits six alleles in samples from the west coast of the United States (Johnson & Utter, 1973). Although no frequency data are available, this locus is also polymorphic in the eastern North Atlantic (F. M. Utter, unpublished data).

Mytilus californianus

Fewer enzymes have been surveyed than for *Mytilus edulis*, but we have investigated eleven loci, eight of which are polymorphic (Table 9.3). Of these loci, we have geographic variation data for the *LAP-I* and *GPI* loci. Samples were collected between Santa Barbara, California, and Torch Bay, Alaska, but primarily from Cape Flattery, Washington, to San Juan Island, within the Strait of Juan de Fuca (Fig. 9.4); and around Tatoosh Island, off Cape Flattery, Washington, (discussed below). Although the coverage on the west coast of North America is not nearly as extensive as for *Mytilus edulis* on the east coast, we can at least obtain a picture of geographic variation over a comparable range of latitude.

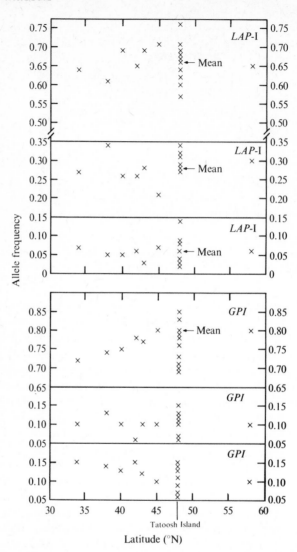

Fig. 9.5. Geographic variation in allele frequencies in *M. californianus* on the west coast of the United states at the leucine aminopeptidase-I (*LAP-I*) and glucose phosphate isomerase (*GPI*) loci.

Leucine aminopeptidase (*LAP*)

Levinton (unpublished data) observed and defined two zones of polymorphic LAP activity as the *LAP-I* and *LAP-II* loci, in order of decreasing electrophoretic mobility towards the anode. The *LAP-II* zone of activity was usually too faint for consistent scoring, so we will discuss only the

368

Table 9.4. *Heterogeneity of juvenile samples versus homogeneity of adult
samples, in allele frequencies at the GPI locus of* Mytilus californianus,
along the west coast of North America.

Contrast	Degrees of freedom	χ^2 s	p
GPI^m			
All adult samples	8	5.1	N.S.†
All juvenile samples	7	28.4	< 0.005***
All juvenile/all adults grouped	1	16.5	< 0.005***
GPI^s			
All adult samples	8	5.3	N.S.
All juvenile samples	7	23.0	< 0.005***
All juvenile/all adults grouped	1	0.3	N.S.
GPI^f			
All adult samples	8	7.1	N.S.
All juvenile samples	7	21.7	< 0.005***
All juvenile/all adults grouped	1	0.9	N.S.

† Not significant.
*** Highly significant.

LAP-I locus. *LAP-I* variation is very similar to the *LAP* locus of *Mytilus
edulis*, with three common (*LAP^s*, *LAP^m*, and *LAP^f*) and two rare (*LAP^{ss}*
and *LAP^{ff}*) alleles. Large variations in activity among individuals include
some individuals with insufficient activity for scoring *LAP* phenotypes.
The reason for this remains unclear, but individual variation in regulation,
genetic variation involving activity, etc., may explain it.

Geographic variation of the three common alleles along the west coast of
North America is minimal and shows no pattern. The range of variation
observed at Tatoosh Island is as great as that over 24 degrees of latitude
(Fig. 9.5). Clearly, microgeographic variation and its control is of great
significance. This small range of variation differs greatly from that for
Mytilus edulis on the east coast, as discussed above. Preliminary data of
D. L. Alford show larger ranges of variation for *M. edulis* along the
west coast of North America than for *M. californianus* (D. L. Alford,
unpublished data).

Glucose phosphate isomerase (GPI)

Similar to the results for *Mytilus edulis*, one zone of polymorphic activity,
corresponding to one locus, was observed. However, it is much less
polymorphic. Samples of adults usually contained only four or five alleles,
though seven alleles have thus far been recognised. Three alleles are most
common: *GPI^s*, *GPI^m*, and *GPI^f*. In samples of adults, the frequencies of

369

Table 9.5. *Mean interlocality genetic difference ($\overline{X}$; arc sine of Hedrick's coefficient) at four different magnitudes for geography for LAP-I and GPI loci of* Mytilus californianus.

Scale	Sample sites	$\overline{X}$	S.E.	No. of two-way comparisons
	GPI			
m	False Bay	80.7	0.78	15
10^2 m	Tatoosh Island	79.4	0.42	55
10^4 m	Strait of Juan de Fuca	79.3	0.45	105
10^5 m	West Coast, N. America	78.0	0.70	36
	LAP			
m	False Bay	82.9	1.08	15
10^2 m	Tatoosh Island	74.0	1.01	55
10^4 m	Strait of Juan de Fuca	73.9	0.78	105
10^5 m	West Coast, N. America	78.8	0.78	36

Larger values of $\overline{X}$ mean less differentiation.

these three alleles change little from Santa Barbara to within the Strait of Juan de Fuca. In contrast, juveniles show significant between-locality heterogeneity (Table 9.4) and tend to have more alleles for a given sample size than adults collected at the same site. There is very little geographic differentiation between Santa Barbara and Torch Bay, Alaska (Fig. 9.5), and samples taken at Tatoosh Island show as much variation as the entire west coast of North America.

Heterogeneity at different scales of geography

Because we have samples taken from average distances of metres, hundreds of metres, kilometres, and hundreds of kilometres, it is possible to test hypotheses relating the contribution of migration to genetic differentiation. If interpopulation gene flow were important in maintaining genetic similarity, we would predict that interlocality differences over the largest scale of geography would be the greatest. Hedrick's (1971) coefficient of genetic similarity was calculated for *M. californianus* for data at the *GPI* and *LAP-I* loci (Table 9.5). For the *GPI* locus we found no difference in average interlocality difference over the four scales of geography. Some significant differences were found at the *LAP-I* locus, but they are not consistent with increasing average interlocality distance. Further, within the outer coast samples, product-moment correlations between Hedrick's coefficient and distance for two-way locality contrasts are not significant for the *LAP-I* or *GPI* loci.

Table 9.6. *Comparison of mean interlocality genetic similarity, measured by arc sine transformation of Hedrick's coefficient ($\bar{X}$), for samples of* Mytilus edulis *and* M. californianus *taken from localities in the Strait of Juan de Fuca*

Species	$\bar{X}$	S.E.	No. of two-way comparisons
	GPI		
M. edulis	69.23	0.581	136
M. californianus	79.30	0.448	105
	LAP		
M. edulis	68.03	0.795	136
M. californianus	73.94	0.781	105

Larger values of $\bar{X}$ mean less differentiation.

The differences in general ecology of *Mytilus edulis* and *Mytilus californianus* would make a study comparing genetic heterogeneity for these two species over the same geographic range of considerable interest. *Mytilus edulis* lives in habitats ranging from estuaries (sometimes in freshwater at low tide) to outer coast habitats such as Tatoosh Island. Levinton (unpublished data) collected samples of both species from localities along the Strait of Juan de Fuca (Fig. 9.4) and computed Hedrick's coefficient of genetic similarity at the *LAP-I* and *GPI* loci for all pairwise locality contrasts for each species. Table 9.6 shows that *M. edulis* shows more average interlocality geographic heterogeneity than *M. californianus*. Because both species have apparently similar modes of dispersal and times of pelagic larval life, the greater variety of habitats occupied by *M. edulis* may contribute to this greater interlocality heterogeneity (see below).

Microgeographic variation

In this section we will discuss studies of mussels which have demonstrated genetic heterogeneity over small distances, primarily over tidal flats, in estuaries, and among different levels in the intertidal zone. Size of individuals has also been taken into account, as this parameter will indicate age-dependent changes and perhaps differences in growth rates.

We will show that there is an impressive amount of genetic differentiation among samples less than one metre apart, indeed among individuals of different sizes in a single sample. This variation is superimposed upon the macrogeographic differentiation described in earlier sections.

Microgeographic variation is of particular interest in mussels because as

371

groups of larvae settle in various heterogeneous environments, differential selective mortality can occur. Since larvae are not likely to settle at the exact site of the parents, there seems to be a cyclicity initiated by larvae settling on a patchy environment, involving differential spatial mortality with eventual production of larvae that will be mixed and settle in new areas. This condition has been discussed by Koehn *et al.* (1975) who conclude that it closely corresponds to Levene's (1953) model of selection involving the random distribution of mixed genotypes having different respective fitnesses in different environmental patches.

Variation within the intertidal zone

An obvious source of environmental heterogeneity that may result in microgeographic variation is that different levels within the intertidal zone are dramatically different in exposure time, temperature, heat transfer, water retention, characteristic fauna and flora and other features (see Newell, 1970, for an excellent discussion). Many workers have found that species that live in higher levels of the intertidal show greater physiological tolerance than those lower in the tide zone (Ushakov, 1965; Kinne, 1970*c*; Newell, 1970). In fact, temperature tolerance has been shown to determine behavioural patterns important in tidal zonation (Newell, 1970). These findings suggest that selection may operate differently at different levels in the tide zone, producing a systematic microgeographic variation. If such variation exists, then two questions arise. Is larval settlement in the intertidal zone independent of genotype, with subsequent divergence through selective mortality? Is larval settling and early movement of mussels in the intertidal zone correlated with genotype? At present, we have only circumstantial data supporting the hypothesis that genetic differentiation occurs as a function of intertidal height.

At initial study assessing the extent of microgeographic variation in *Mytilus edulis* examined variation along two transects on a sand flat outside the Nissequogue River estuary on the north shore of Long Island, New York (Balegot, 1971). A 50-metre transect (2) with eleven sample stations and a vertical drop of 22 cm was established, along with a 20-metre transect (1) with ten stations and a vertical drop of 19 cm. All samples were analysed for variation at the *LAP* and *AP* loci. Allele frequencies at both these loci showed noticeable inter-sample variation. Along transect 1, the frequency of the *LAP*[98] allele varied from 0.43 to 0.56. Along transect 2, variation of this allele was from 0.40 to 0.55. The *AP*[97] allele varied from 0.41 to 0.57 in transect 2. Neither locus exhibited significant among-sample variation, but a significant positive correlation was observed between the deficiency of one *LAP* heterozygote class and distance along transect 2 from high to low (Fig. 9.6). A positive correlation was also observed

372

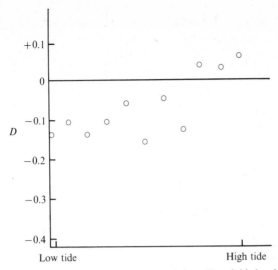

Fig. 9.6. Relationship between heterozygote deviation (*D*) and tide level at the *LAP* locus in *M. edulis* from Nissequogue River estuary, New York. (From Balagot, 1971.)

between the frequency of the LAP^{94} allele and mean length in a sample. These observations are part of a common theme in most studies of mussels.

Pooled samples of each transect showed significant deficiencies of heterozygotes relative to Hardy–Weinberg expectations. Inbreeding cannot be a significant cause for this deficiency because of large population size and extended disperal. Using the range of allelic variation from Long Island to Woods Hole, Massachusetts (from Milkman & Beaty, 1970), Balegot suggested the observed deficiencies were consistent with the Wahlund effect. However, we now know (Koehn *et al.*, 1975) that deficiencies are too large to be adequately explained by the Wahlund effect except in populations between other populations of considerable genetic differences. We will return to this point later.

The work of Balegot (1971) cited above prompted Koehn and others (Koehn, Turano & Mitton, 1973) to examine populations of the ribbed marsh mussel, *Modiolus demissus*, on an eroding channel of *Spartina* marsh peat in the Nissequogue estuary. Two samples were collected, one 0.66 m higher than the other. The tetrazolium oxidase (*TO*) locus was selected for study because of its relatively simple two-allele, three-zygotic class pattern. Size-frequency analysis of the two samples, after the method of Bhattacharya (1967), revealed three modes which could be interpreted as year classes. There was no apparent difference in mean rate of growth at the two tide levels. It was assumed that trends among classes approximated the changes that occur to a given year class over time. In both samples, an enrichment of heterozygotes occurred with increasing size (as a measure of

373

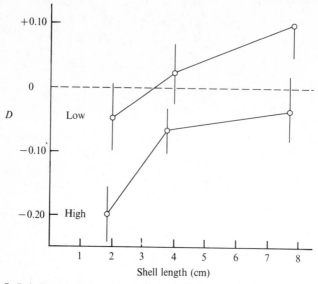

Fig. 9.7. Relationship between heterozygote deviation (*D*) and size in *Modiolus demissus* from Nissequogue River estuary, New York at high and low regions of the intertidal. (From Koehn *et al.*, 1973.)

age) (Fig. 9.7). This is most logically explained as being the result of selective mortality, favouring the heterozygote class. Of further interest was the greater deficiency of heterozygotes found in the higher intertidal sample.

In considering the probable selective mortality occurring at this locus, it is interesting that juveniles displayed a significant deficiency of heterozygotes. Koehn and others argued that the variance in allele frequencies in the Long Island Sound region was too small to account for the deficiency by the Wahlund effect. Since inbreeding can also safely be excluded, selection probably occurs against the heterozygote class. While increased proportions of heterozygotes might be explained by directional selection for one allele, the magnitude of change in allelic frequencies in this study was not sufficient to support this hypothesis. It was therefore concluded that overdominance was the most likely explanation for the observed data.

This pattern, namely increased heterozygosity with size, was also observed in a study of size class differences at the *LAP* locus in *M. edulis*. Milkman (see Koehn *et al.*, 1975) examined over 2000 individuals in the Cape Cod Canal, over a size range of length from 4 to 45 mm. In the smallest size class, the frequency of the LAP^{94} allele was intermediate between populations north (0.10) and south (0.55) of Cape Cod. In the Cape Cod Canal, the frequency of this allele increased with length. In addition,

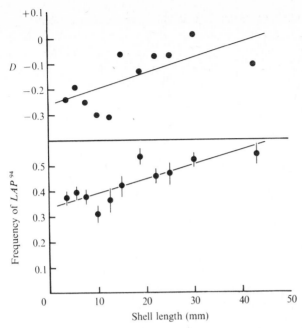

Fig. 9.8. Size-dependent variation in the frequency of LAP^{94} and net deviation of observed heterozygotes from expected frequencies in *M. edulis* from Cape Cod. (Data of R. Milkman; from Koehn *et al.*, 1975.)

net heterozygosity (D for all heterozygote classes) also showed a regular increase with length (Fig. 9.8).

Levinton (Levinton & Fundiller, 1975) examined variation at the *LAP-I* locus in *Mytilus californianus*. Eleven samples of adult mussels were collected at Tatoosh Island, off Cape Flattery, Washington. Mean individual length was calculated for each sample. Levinton found that mean length was correlated with both the frequency of the LAP^m allele and deviation of the LAP^{sm} heterozygote class from Hardy–Weinberg expectation (Fig. 9.9). This is strikingly similar to the results of Milkman for *Mytilus edulis*. This correlation was examined further with information on intertidal height. Samples were taken from high and low areas of the intertidal from two localities. In each case, the lower sample exhibited larger mean individual length, a more positive value of D, and greater frequency of the most common allele, LAP^m (Table 9.7). For most of these stations, algal data was also recorded. Rigg & Miller (1949) described the algal zonation at a nearby locality, Postelsia Point, near Neah Bay, Washington. Relative intertidal height of algal dominants was quantified on a simple ranking basis, and plotted against several parameters. These analyses revealed a significant negative correlation between mean length of

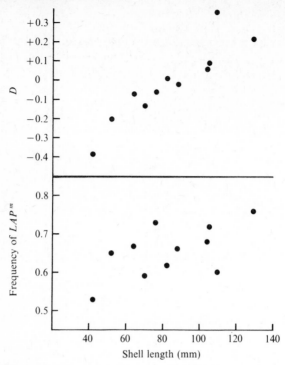

Fig. 9.9. Size-dependent variation in frequency of *LAP*^{m} and deviation of the *LAP*^{sm} heterozygote class in *M. californianus*. (From Levinton & Fundiller, 1975.)

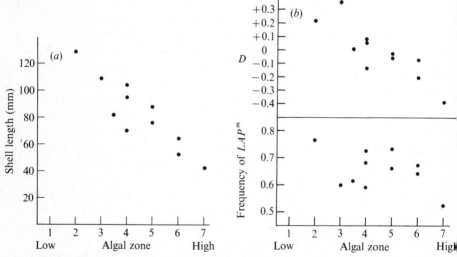

Fig. 9.10. Relationship between mean shell length (*a*), heterozygosity and frequency of *LAP*^{m} (*b*) and height above low water in *M. californianus*. (From Levinton & Fundiller, 1975.)

Table 9.7. *Comparison of mean length, frequency of the LAP^m allele, frequency of the LAP^sm heterozygote class relative (D) to Hardy–Weinberg expectation, and mean gonad weight of females at two sites on Tatoosh Island, off Cape Flattery, Washington*

Site	Sample*	Mean length (mm)	Mean gonad weight (g)	$LAP^m \pm$ s.e.	D
I (1 m)	High	51.9	0.46	0.645±0.0430	−0.200
	Low	104.5	3.87	0.746±0.0408	+0.092
II (2 m)	High	42.3	0.43	0.545±0.0471	−0.372
	Low	108.5	6.27	0.677±0.0477	+0.204

* At each site one sample was taken at the top of the mussel bed, the other towards the bottom. (Height difference is indicated in the site column.)

individual mussels and height above low water (Fig. 9.10a); a negative correlation between heterozygosity and intertidal height; and a weak negative correlation between the frequency of the most common (LAP^m) allele and intertidal height (Fig. 9.10b). These findings differ from the TO locus of *Modiolus demissus* (Koehn *et al.*, 1973) in that not only zygotic classes, but allele frequency changes with intertidal height.

The negative correlation of size with intertidal height also suggests that size is not only a reflection of age but of growth rate. To explain the size differences by age alone would require individuals settling in the high intertidal and continually migrating towards the low intertidal as they age. While this process can happen on certain high-angle slopes (R. T. Paine, personal communication) it seems to be an unlikely explanation for these samples collected from various localities around an island, including many horizontal benches. Mean length of individuals varied by as much as a factor of three in our collections. However, growth experiments performed at high and low tide by Harger (1970b) show differences in growth rate that would in itself explain the differences in length observed in the samples. Clearly the observed differences are some combination, though unknown, of age and growth rate. Perhaps size, *per se*, is an important correlative parameter in our genetic data. This proposition is reasonable, since selective mortality in the intertidal zone is likely to be correlated with size. A given physiological shock of, say, high temperature is likely to be lethal to smaller individuals (Kinne, 1970c; but see Chapter 5). Levinton & Fundiller (1975) have temperature-shocked juveniles of the soft shell clam, *Mya arenaria*, with a resultant preferential mortality in the smaller size classes, when preferential mortality was observed at all. When mortality was correlated with size, changes in the genetic composition of the survivors were also noted. This process may require that larger and faster

Table 9.8. *Comparison between three successive spatfalls, and resident large adults on Tatoosh Island. D is calculated with respect to the LAP^sm heterozygote class*

	Juveniles			Adults 1972
	1972	1973	1974	
LAP^ss	0.000	0.018	0.000	0.000
LAP^s	0.200	0.227	0.302	0.188
LAP^m	0.672	0.618	0.613	0.758
LAP^f	0.082	0.118	0.085	0.054
LAP^ff	0.036	0.027	0.000	0.000
D	−0.121	−0.222	−0.082	+0.190 (+0.101)*
N	55	55	53	56

* This value of *D* calculated on the assumption that the mean value of *D* for the juvenile populations is the true 'starting' allele frequency of the population before selective mortality.

growing individuals be of different genetic properties than smaller, slower growing individuals. This seems to be the case in *Mya arenaria*. In a large sample of *Mytilus californianus* the mean length of *LAP^ss* mussels was found to be significantly less than that of *LAP^sm* and *LAP^mm* (Levinton & Fundiller, 1975), suggesting different mean growth rates for different genotypes. Physiological shock and predation are often size-selective and could thus shift allele frequencies in mussel populations.

One means of avoiding growth rate problems in estimating the magnitude of selective mortality is to contrast newly settled juveniles with adults at the same site. At Tatoosh Island, juveniles (as judged by small size, lack of erosion of shell and no gonadal development) that probably had settled within 6 months of sampling, were taken at one locality in the summers of 1972, 1973 and 1974, and were contrasted with adults (mean length 130 mm) at the same locality. All juvenile samples showed greater deficiency of the *LAP^sm* heterozygote class and lower frequency of the *LAP^m* allele than the co-occurring adults (Table 9.8). Thus, as in the study of Milkman described above in *Mytilus edulis*, selective mortality tends to increase the frequency of heterozygotes and changes the frequency of an allele, at a locus coding for leucine aminopeptidase. Change in allele frequency over time complicates the basis for calculating *D* (see Table 9.8).

A study of microgeographic variation explored the interactions that might occur between two loci. Mitton *et al.* (1973) examined size-dependent changes in frequencies of *LAP* and *AP* dilocus zygotic classes in *M. edulis*. Individuals were scored for both *LAP* and *AP* phenotypes. A 6×6 contingency table, consisting of the number of individuals of each

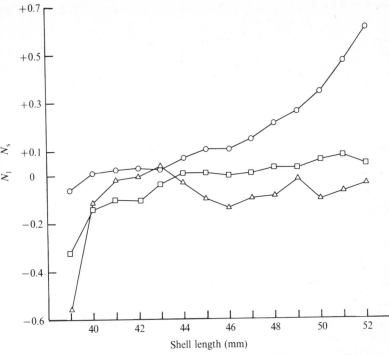

Fig. 9.11. Relationships of differences between coefficients of dependence with increments of shell size for samples of *M. edulis* from an intertidal zone transect. ○ , high tide; □, mid-tide; and △, low tide. χ^2_l/N_l is the coefficient of dependence for the 'large' group and χ^2_s/N_s that for the 'small' group at each increment. (Modified from Mitton *et al.*, 1973.)

dilocus phenotype, was tested for randomness. Disequilibrium occurred between the *LAP* and *AP* loci. The magnitude of this interaction was enhanced with increasing size in high intertidal samples, but not in samples taken from mid tide or low tide (Fig. 9.11). The change in magnitude of interlocus dependency with increasing size in the high intertidal sample was interpreted as epistatically-determined differences in fitnesses, that changed the proportion of dilocus genotypes during the history of the cohort. Details of this argument will be found in Mitton *et al.* (1973).

Thus, investigations to date reveal that systematic variation on the microgeographic scale may be related to intertidal height, size and age, and is most plausibly explained through selective mortality. Interactions between loci are also apparent and should be explored further in future studies.

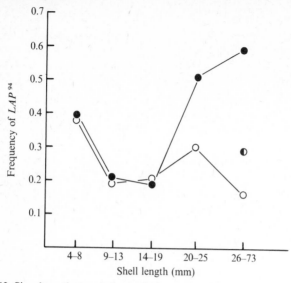

Fig. 9.12. Size-dependent variation in the frequency of *LAP*⁹⁴ in estuarine (○) and marine (●) samples of *M. edulis* from Mill Creek, Barnstable, Mass. ◑ represents an intermediate station. (Data of R. Milkman; from Koehn *et al.*, 1975.)

Estuarine variation

The hypothesis that environmental heterogeneity influences spatial variation in allele frequencies bears further investigation. Estuaries are useful vehicles for such work because they provide an obvious source of spatial heterogeneity which may be contrasted and compared with genetic variation. Milkman (in Koehn *et al.*, 1975) investigated populations of *M. edulis* within and outside an estuarine environment at Barnstable, Massachusetts, on the north shore of Cape Cod. Although small individuals (4–19 mm long) had essentially identical frequencies of the *LAP*⁹⁴ allele, significant divergence occurred in the large size classes. Within the estuary, larger individuals showed a lower frequency of this allele than outside the estuary. A single sample of large mussels from an intermediate station had an intermediate frequency of this allele (Fig. 9.12). The obvious inference is that newly established juveniles have initially the same genetic characteristics, with subsequent divergence in the older (larger) populations along the gradient in the estuary. Other estuaries have been studied by Milkman with similar results. Reductions of salinity are always correlated with a reduction in frequency of *LAP*⁹⁴, though the converse need not be true.

Similarly, the degree of heterogeneity within an estuary is greater when contrasted with environments outside the estuary. Replicate samples of

380

M. edulis from within the Nissequogue estuary discussed above, were compared with replicates from outside the estuary. At the *LAP*, *AP* and *GPI* loci, allelic frequencies in the two areas did not differ in observed or expected heterozygosity, average allele frequency or effective number of alleles. However the among-sample variance was higher within the estuary. Between-habitat variances were compared (Koehn *et al.*, 1975) and significant differences were found for the *AP^{92}*, *AP^{82}* and *GPI^{100}* alleles. The among-sample variance was greater in the estuary for *LAP* alleles, but not significantly so. These data support the hypothesis that the greater environmental heterogeneity of estuaries selects for greater variance in allelic frequencies.

Variation in allele frequency, specifically a reduction in *LAP^{94}*, has been repeatedly observed by R. Milkman and R. K. Koehn. Several estuaries studied exhibited this pattern; no exceptions have so far been observed. It appears that great spatial differences in *LAP* allele frequencies occur whenever either of two conditions are met: the immigration of larvae from areas of differing allele frequency and/or a reduction from normal salinity. A single massive example of this variation occurs in Long Island Sound. The frequency of *LAP^{94}* declined rapidly along the Connecticut shore from 0.55 in the Atlantic Ocean to 0.10 near New York City (Koehn *et al.*, 1975). The frequency on the Long Island shore did not parallel the Connecticut shore, but remained low (Fig. 9.2). These differences between the northern and southern shores of the Sound are consistent with the net current patterns and surface salinities (Hardy, 1970).

Relative to the Cape Cod Canal study site, size-dependent changes of the *LAP^{94}* allele and heterozygosity were reversed at Orient Point, the eastern tip of the north fork of Long Island (R. Milkman & R. K. Koehn, unpublished data). Mussels living at this locality experience reduced salinity but larvae apparently originate from areas of normal salinity outside the Sound. Thus the higher heterozygosity and *LAP^{94}* allele frequency of newly settled larvae are subsequently reduced by selective mortality in the estuary. These data strongly suggest selective mortality at the *LAP* locus which may partially involve selection for or against heterozygotes in areas of normal and reduced salinity, respectively.

These studies have demonstrated genetic response to both different environments and to different levels of environmental heterogeneity. This latter response is illustrated by a study on correlated patterns of variation between *M. edulis* and *Modiolus demissus* which co-occurred broadly within the Nissequogue estuary (Koehn & Mitton, 1972). This estuary is dominated by *Spartina* salt marsh. If selection, particularly some physical factors, is important in regulating variation at this locus, there should be concomitant variation in allelic frequencies and zygotic classes of the species from one locality to another. For example, if one allele is rare in

381

Table 9.9. *Homogeneity* χ^2 *values from pairwise comparisons of zygotic distributions at the* LAP *locus in samples of* Mytilus *and* Modiolus

	Mytilus			
Modiolus	I	II	III	IV
	Comparisons between pairs of localities			
I	—	6.11*	12.48**	3.77
II	0.99	—	9.38**	4.29
III	9.98*	6.80*	—	19.88*
IV	3.14	3.96	3.08	—
	Comparisons between the two species at each of the four localities			
Mytilus vs. *Modiolus*	0.747	3.14	5.54	5.37

NOTE: two degrees of freedom in all tests.
* $P < 0.05$.
Modified from Koehn & Mitton (1972).

one species at a site, the analogous allele should be rare in the second species, i.e. spatial change in the frequency of this allele in one species should be paralleled in the other species. Four sample sites were established within the estuary and collections made of *Mytilus* and *Modiolus* at each site. Both species showed a similar pattern of three common and two rare alleles. With the exception of one pair of alleles of *Modiolus* at one of the localities, the rank order of relative abundances of alleles of a particular electrophoretic mobility, as well as their phenotypes, were nearly identical between species at all localities (Table 9.9). Koehn & Mitton concluded that this parallel variation might represent an optimal response to particular patterns of environmental heterogeneity in the estuary.

Discussion

We have attempted to describe the qualitative and quantitative patterns of genetic differentiation that have been observed in species of mussels, primarily *Mytilus edulis*. These studies, in addition to providing information on differentiation, raise many important questions that must be considered in future investigations. A fundamental question that arises from consideration of the geographic variation data along the east and west coasts of North America is the relative contributions of migration and/or selection. The answers to this question are basic to our understanding of how population differentiation and, ultimately, speciation occurs in the sea. While it is not surprising to observe genetic differences between populations of mussels as widely separated as those in North America, Iceland and Ireland, the abrupt frequency changes in North America, variations with exposure in Ireland, and intertidal zone differentiation, all suggest that patterns of geographic variation mainly reflect adaptations to different environmental conditions rather than geographic isolation. In view of the extended planktotrophic larval dispersal stage of these species, this is a somewhat unexpected finding. For example, if we estimate the potential dispersal of *Mytilus edulis* by assuming a larval life span of from 20 to 45 days, and unidirectional longshore currents of 10 cm s^{-1}, dispersal of a minimum of 173 km is possible. Koehn *et al.*, (1975) attempted to assess the relative contributions of migration facilitated during larval dispersal and/or genotype-specific mortality to the pattern of variation of *LAP* allele frequencies in eastern North America. They suggested several predictions which should hold if migration were important to the observed pattern. For example, at any locus there should be correlations between magnitudes of spatial variance of allele frequency and heterozygote deficiencies, due to the Wahlund effect. Secondly, loci which exhibit significant spatial heterogeneity in allele frequency variation should, on the average, exhibit greater heterozygote deficiency than loci at which allele frequencies are geographically more homogeneous. Although the largest deficiencies were associated with abrupt spatial changes in allele frequency, no overall patterns of correlation could be demonstrated. Also, the magnitudes of heterozygote deficiencies at the *LAP* locus, the most spatially heterogeneous locus, were not significantly larger than other, more geographically homogeneous loci. These authors concluded that migration plays only a minor role in affecting observed patterns of geographic variation. Allele frequencies in various areas more likely reflect the action of natural selection.

Discordant patterns of variation between the *AP* and *LAP* loci provide some further evidence relating the roles of migration and selection. There are great differences in allele frequency at the *LAP* locus, but not at the *AP* locus, between the open coast at 41° N latitude and within Long Island Sound. The differing patterns of geographic change at different loci suggest different sources and magnitudes of natural selection affecting each of the three best-studied loci discussed above. For example, allele frequencies at the *AP* locus are invariant over a distance of eight degrees of latitude from Virginia to southern Nova Scotia. This homogeneity could be maintained by very large inter-regional migration that obscures potential differentiation due to selection, i.e. selection for different phenotypes in different areas is not sufficiently strong to counteract the effects of migration. However, once such a hypothesis has been considered, it is impossible to explain the patterns of geographic variation at the *LAP* and *GPI* loci without postulating selective coefficients that differ from region to region

383

and are of sufficient magnitude to obscure totally inter-regional migration. Even between the *LAP* and *GPI* loci the effects of selection must be markedly different. The *GPI* allele frequencies change in a smooth linear pattern. Frequencies of *LAP* alleles are uniform within major coastal regions, but change in abrupt step-clines between regions. Considering patterns of geographic variation *in toto*, the evidence that selection is an overriding force maintaining the patterns of variation is inescapable, though it is not yet clear if selection is acting on these specific loci.

The difference in magnitude of variation for *Mytilus edulis* on the east coast of North America, and for *M. californianus* on the west coast, also bears on this question. The great geographic homogeneity of *M. californianus* at the *LAP* and *GPI* loci, relative to the significant changes in *M. edulis* over approximately the same range of latitude, suggests that the sharper latitudinal temperature gradient on the east coast influences geographic variation. The lack of increased genetic differentiation with interlocality difference also argues against geographic separation as a major factor in differentiation.

Geographic data provide a static indication that selection is paramount in maintaining enzyme polymorphisms. But microgeographic variation further demonstrates that these polymorphisms are maintained dynamically and can be observed to change over time, and within the framework of the tidal zone. It is this framework that causes us to conclude that the extensive present knowledge of the population biology of mussels, plus the newly available data on genic polymorphisms, provide the potential for an integrated approach to the study of population dynamics and micro-evolutionary events. The vast array of tools available will allow us to discover the evolutionary importance of ecological parameters and their role in adaptation in both mussels and other marine invertebrate species. Without such knowledge, we cannot hope to know the interaction of genetic differences with ecological variables, and the evolutionary potential of ecologically correlated changes in populations over time and space.

10. Cultivation

J. MASON

The total world harvest of mussels in 1971 was 370000 tonnes. The common European mussel, *Mytilus edulis*, and the closely related *M. galloprovincialis*, on the Atlantic coast of Europe accounted for 76% of these and *M. galloprovincialis* on the Mediterranean coast of Europe for a further 8% (FAO, 1972). Despite its image as the oyster's poor relation, the mussel is itself a luxury commodity, and the demand for mussels of good quality exceeds the supply. Market promotion and suitable presentation could further stimulate the demand.

Unlike many of the fin-fish species exploited by man, mussels are low in the food web, feeding directly on the primary producers, phytoplankton and bacteria, and non-living organic material. For each stage an animal is removed from this basic production of carbon, there is an energy loss of some 80–90% for herbivores and 90% for carnivores in converting food into flesh (Havinga, 1964; Webber, 1968; see also Ryther & Bardach, 1968). Therefore, while the predatory fish caught by man are relatively wasteful of the basic production, mussels are more efficient converters and so lend themselves well to cultivation, producing greater yields per unit area than species at higher trophic levels (Ryther & Bardach, 1968).

The mussel's sessile habit also makes it suited to cultivation. Currents transport food from a great volume of water, and small areas of sea can be made to yield large quantities of flesh (Havinga, 1964; Ryther & Bardach, 1968; Ryther, 1969). In the summer and autumn 20–30% of filtered organic matter may be fixed as meat weight by *Mytilus edulis* (Boje, 1965). Furthermore, sessile animals are easy to harvest and frequently require less capital investment for equipment, buildings and space than motile forms (Andreu, 1968a; Kinne, 1970a).

While man's use of the production of the sea has been largely confined to the exploitation of the natural production, much publicity has recently been given to sea farming, or cultivation, as a possible supplement to this exploitation (see e.g. Waugh, 1966; Richardson, 1967; Cole, 1968; Kinne, 1970a). Bivalve molluscs, however, especially oysters and mussels, have been cultivated for centuries.

Iversen (1968) defined sea farming as a 'means to promote or improve growth, and hence production, of marine and brackish-water plants and animals for commercial use by protection and nurture on areas leased or owned'. As Ryther & Bardach (1968) pointed out, farming includes one or more manipulations that alter or interfere with the natural life cycle of an animal or plant.

Kinne (1970*b*) distinguished four ascending classes of cultivation, each of which includes those preceding it.

(1) Maintenance: keeping alive of organisms, without significant growth, for scientific or commercial purposes.

(2) Raising: bringing up (fattening) of young adults.

(3) Rearing: bringing up of early ontogenetic stages (e.g. fertilised eggs, larvae).

(4) Breeding: production and bringing up of offspring.

The basic principles of shellfish farming have been outlined recently by a number of authors (e.g. Havinga, 1956*a*, *b*; Walne, 1963; Cronin, 1967; Iversen, 1968; Ryther & Bardach, 1968; Korringa, 1970; Milne, 1972; Bardach, Ryther & McLarney, 1972). The switch from free fishing to well-organised cultivation is radical, involving the lease or ownership of an area to which the operator has the sole rights. Private rights are essential to farming. Settlement of larvae and young post-larvae (spat) is promoted by presenting a suitable substratum; these young are transferred to areas where food is abundant and growth consequently fast, and possibly later to other areas for fattening prior to marketing. Growth is regulated by controlling density according to the food supply. Where necessary and practicable, the cultivated species is protected from enemies. Intensive and thorough harvesting takes place at the appropriate age. Cultivation results in a greater stability of supply and price, which increases economic efficiency and attracts capital investment.

Rearing larvae is often the most difficult part of cultivation, and it is costly (Ryther & Bardach, 1968; Davis, 1969). Thus, while hatchery techniques have been developed for producing oyster larvae and spat (Davis, 1969), the cost of rearing mussel larvae could not be supported owing to the lower price fetched by mussels than oysters. For this reason mussel cultivation is based on the collection and raising of naturally settled spat, and falls in Kinne's (1970*b*) second class.

Methods of cultivation in Europe

Much of this section is based on the descriptions given in the recent review by Mason (1972). The account will therefore be brief. The methods currently used in cultivating mussels in Europe fall into three categories: (1) on poles projecting from the substratum between tide marks; (2) on the sea-bed itself; (3) suspended, either from fixed frames or from freely-floating structures.

Fig. 10.1 Collector bouchots. (Crown copyright, by courtesy of Dr H. A. Cole.)

Cultivation on poles

This is the oldest method currently employed, having been used on the west coast of France since the thirteenth century and remaining to this day the principal method used in France (Audouin, 1954). Its origin and development have been described in great detail in a number of papers of which, in addition to Audouin (1954), the most recent include Field (1922) and Lambert (1939).

In the traditional bouchot method the mussel spat settle on 'collector bouchots' (Fig. 10.1), rows of pine stakes 35 cm apart, which are placed furthest from the land and exposed only at low water of spring tides (Field, 1922). Settlement occurs in the spring and is enhanced by the presence of the hydroid *Tubularia mytiliflora* (Lambert, 1939). By the end of the first summer these seed mussels reach a length of up to 31 mm, with a mode of 20 mm. From this stage the largest mussels are taken off at intervals and transferred to the 'rearing bouchots' closer to the land (Audouin, 1954). The posts are 75 cm apart, are set in rows at right-angles to the shore, and have branches of willow or chestnut interwoven horizontally between them and reaching to within 30 cm of the surface of the mud (Fig. 10.2) (Lambert, 1939; Audouin, 1954). The seed mussels are attached to the rearing bouchots in bags of fine mesh netting which rot and fall away after the mussels have attached themselves by the byssus (Field, 1922; Andreu, 1968a). As the mussels grow they are thinned out in order to reduce competition for food, and the thinnings are transferred to other rearing bouchots. Successive thinnings are transplanted higher on the shore and the animals gradually become accustomed to remaining closed during exposure to the air by the tide, so travelling better to market and remaining fresh longer (Calderwood, 1895; Anonymous, 1937). Some residual mussels are ultimately left on the collector bouchots to complete their growth there (Audouin, 1954). By the end of their second summer, or when perhaps 2-years-old, the mussels are marketed, having reached a size of 40–50 mm (Field, 1922; Lambert, 1939; Andreu, 1968a).

As the bouchots often contain mussels arising from more than one spatfall, many of which are not of marketable size, it is necessary to sort the catch by means of a flat grille or a rotary sorter. The mussels are then washed, packed in sacks or in wicker baskets containing 25 or 50 kg, and transported to market, mainly by road. The principal markets are in central and south-west France, and the 'Midi' and even Algeria and Tunisia (Audouin, 1954).

While the bouchot method has been practised for over 700 years on the west coast of France, a modification of it has become established on the north coast of Brittany during the past hundred years and has expanded rapidly since the Second World War. Its development has been described by Ryther (1968). The main difference from the traditional method is that, while conditions in northern Brittany are good for growth and fattening of mussels, there is little natural settlement of mussels there. Seed has, therefore, to be brought in from elsewhere, mainly from the region of La Rochelle on the west coast. Loosely woven ropes are suspended in the intertidal region near natural mussel beds at La Rochelle and these obtain a settlement of spat in late spring or summer. The ropes are then transported to Brittany and wrapped spirally round oak poles driven into the intertidal flats. The mussels quickly become established on the poles. The procedure is then similar to the traditional method, the mussels, as they grow, being thinned out and wrapped round other poles or hung between them in netting.

A modification of the traditional bouchot method has now been adopted

Fig. 10.2. Rearing bouchots. (Crown copyright, by courtesy of Dr H. A. Cole.)

on Jersey, again with transplanted seed from the west of France (Anonymous, 1974). Furthermore, this modification is now replacing the traditional method on the west coast of France, and has already (August, 1973)
done so in the Baie de l'Aiguillon (G. Davies, personal communication).
The chief reasons for this are the laborious nature of transferring the
mussels to the fencing 'rearing bouchots' and of harvesting them later, and
the fact that the fencing tends to cause silting up. The collector bouchots,
furthest from the shore, are of two types; plain poles as before on which
spat settles directly, and double rows of poles with crossbars joining pairs
of adjacent poles across which lengths of coir rope are draped to collect
spat. Rearing bouchots now consist of rows of plain poles, which are
scraped clean of barnacles and old mussels prior to the transfer of seed in
order to provide a firm surface to which to cling.

389

Fig. 10.3. Modified bouchot method used in Jersey. (*a*) Newly seeded poles with mussels wound spirally round poles; (*b*) crop ready for harvesting. (By courtesy of La Rocque Fisheries Ltd and Miss Helen Sayers.)

In the summer, seed from the collector posts is placed in clumps in tubular net bags. The tubes are then wound spirally round the rearing posts and the mussels escape and become attached to the posts (Fig. 10.3). Similarly, coir ropes on which mussels have settled in the spring are wound in 2.5 m lengths round the bare rearing posts, to which the mussels become attached.

The main advantage of the bouchot method is that the mussels are less exposed to the bottom-living predatory crabs and star-fish. Predation is further reduced by the fairly recent adoption of plastic sheaths round the bases of the poles (Ryther, 1968; Milne, 1972). The main drawback, however, is that the mussels are exposed at low tide, especially on spring tides, and so less time is available for feeding, and the growth and fattening are not so good as at sites where the mussels are immersed all the time. Also, the bouchots are vulnerable to storms.

There appears to be little prospect of a great increase in the yield from this method in France, since almost all the available area is now in use, especially in the Baie de l'Aiguillon (Audouin, 1954; Ryther, 1968). The French demand for mussels is so great, however, that it seems certain that large quantities will continue to be imported, as they are at present, chiefly from Spain and Holland.

Cultivation on the sea-bed

While it is practised in a number of European countries, including Denmark and West Germany (Korringa, 1970), bottom cultivation of mussels has reached its greatest development in Holland, where it is used exclusively (Lambert, 1951; Havinga, 1956*b*, 1964; Korringa, 1970). The principle of bottom cultivation is the transfer of young, or seed, mussels from areas of great abundance where growth is often poor owing to overcrowding, or from temporary patches which form from time to time, to areas of good growth and fattening potential (Lambert, 1951; Havinga, 1956*a*, 1964; Korringa, 1970).

In the Dutch Wadden Sea the spat of *M. edulis* settles chiefly in the spring, and in many years dense mussel beds are formed, covering many hectares on the flats. Less extensive beds are also formed below low-water mark along the slopes of channels (Havinga, 1956*b*). In the deeper places where currents bring rich food supplies, the mussels can grow and fatten well and are harvested, by means of toothless dredges, when they have attained marketable size. In contrast, the mussels on the flats, which dry out at low tide and so have a restricted time for feeding, grow poorly and even after several years do not exceed 3–4 cm long (Havinga, 1956*b*).

Each mussel farmer is allocated one or more plots, or 'parks', of deeper ground, varying in area from 5 to 10 ha, for which he pays a rent to the Government, having, like the French mussel farmer, exclusive use of his parks. Seed mussels which have settled in the spring are dredged from public grounds the following autumn or spring at a size of 13–25 mm and spread thinly and evenly on the parks. By the end of their third summer, when they are some 2½ years old, they attain a size of 63 mm and marketing commences (Havinga, 1964); but on the deepest grounds they reach 55 mm in two growing periods (Havinga, 1956*b*) and are ready for marketing in the early winter.

As the mussels grow they are thinned out to enable the good growth to continue. The seed may be laid first on shallow grounds, where they will grow, and they are later transferred to deeper grounds to fatten (Havinga, 1956*b*). While sowing too thickly inhibits growth, a mussel farmer will occasionally deliberately sow a thick layer in order to be sure of a reserve of half-grown mussels. When needed they are thinned out or moved to other parks (Havinga, 1964; Iversen, 1968).

When ready for marketing, the mussels are dredged up and dumped in a thick layer in an area of little tidal movement free from drifting sand, where they are left for 48 h to rid themselves of silt (Lambert, 1951; Havinga,

Fig. 10.4. Dutch mussel boat and dredges. (Crown copyright, by courtesy of Mr A. C. Simpson.)

1964). They may even be stored in these areas until required for marketing. They are then dredged up again and delivered to highly mechanised factories for separation, cleaning of the shell, grading and packing.

An artificial alternative to these cleansing and storage beds of the East Scheldt is being sought. A pilot plant has been set up adjacent to the Experimental Mussel Research Station on the island of Texel in the Wadden Sea (Drinkwaard, 1972a, b; Anonymous, 1973). The pilot plant is designed to handle 3–5% of the Wadden Sea mussel production, but the proposed capacity is twenty times larger. Water from the Marsdiep is pumped into a pressure basin, where sand settles out. A controlled flow is fed thence through concrete sedimentation basins to six mussel holding basins. When drained, the water from these either runs to waste or is stored in reservoirs for recirculation if required. It is envisaged that each operator will ultimately manage his own 8-ha unit, based on the Texel prototype, built adjacent to the culture plots in the Wadden Sea.

Bottom cultivation of mussels in Holland has allowed a high degree of mechanisation, which increases the yield and consequently lowers the price. The boats used are large, 15–20 m or more long (Fig. 10.4), with a shallow draught. Hauling and shooting as many as four 2-m dredges mechanically, they have a capacity of 40 t or more of mussels per hour (Walne, 1963). According to Havinga (1964) and Iversen (1968) 200 or more such boats are involved in the Dutch mussel industry.

Predators, especially starfish, present a problem and must be controlled. When present in large numbers in deeper water they can substantially reduce mussel density and even eliminate entire beds (Havinga, 1956*b*, 1964). Before transplanting, the parks are prepared by removing as many predators as possible (Iversen, 1968). Dredging especially for starfish is carried out by using either an ordinary mussel dredge short-hitched (Lambert, 1951) or a modified 'roller dredge', which is said to have a 30–50% efficiency (Edwards, 1968).

Bottom cultivation is the only method so far used on an appreciable scale in Great Britain. It is practised in the Menai Straits, where dredges are involved and seed mussels are transferred some distance, and on a smaller scale in the Wash (Anonymous, 1966; Edwards, 1968; Davies, 1970). The method has also been adopted on a small scale in Ireland (Meaney, 1970*c*).

The chief advantage of bottom cultivation is that the mussels are submerged for most or all of the time and therefore feed longer than those exposed by the tide as in the bouchot method. Furthermore, a high degree of mechanisation is possible. The principal drawbacks are exposure to bottom-living predators and the need to cleanse the mussels of silt.

Cultivation on suspended ropes

This is the method which has shown the greatest development in recent years and which appears to offer the best prospects for future expansion. There are two basic types of suspended culture; fixed and floating (Andreu, 1968*a*).

Fixed suspended culture

This method is usually used in shallow areas with a small tidal range and gently sloping bottom. It has for many years been practised with *Mytilus galloprovincialis* on the European coasts of the Mediterranean and Adriatic, including a number of places in southern France (Lambert, 1939), Spain (Korringa, 1970), Yugoslavia (Lubet, 1961; Nikolić & Stojnić, 1963) and Italy (Field, 1922; Favretto, 1968). Usually seed mussels are collected from the natural settlement in the tidal zone and then attached to ropes in netting which rots away when the mussels have become fixed by the byssus (Andreu, 1968*a*); seed may also be allowed to settle on coarse ropes stretched horizontally in the surface layers (Korringa, 1970). Ropes of mussels are then hung in the water from fixed wooden or metal frames. Often the seed is transported some distance from areas of settlement to areas of growth.

Growth is generally fast. At Toulon, France, a size of 60–70 mm is attained in 15 months to 2 years (Lambert, 1939), and in the Bay of Pula, Yugoslavia, 70–80 mm in 15–18 months (Lubet, 1961).

Marine mussels

Mussel cultivation on fixed ropes is a flourishing industry in a number of shallow lakes near Naples, Italy, which are connected to the sea by narrow canals (Korringa & Postma, 1957; Renzoni & Sacchi, 1961; Sacchi & Renzoni, 1962). A similar system of culture is practised in the Etang de Thau, a land-locked salt-water lake in southern France with canals connecting it to the Mediterranean (Lubet, 1973). In addition, this method has been adopted at Salses-Leucate and at three places on the east coast of Corsica (Raimbault & Tournier, 1973), so that by 1971 more than one-fifth of the total French mussel production of 39800 t came from the Mediterranean (FAO, 1972).

An experiment recently undertaken in Guernsey is unusual in that it involves seed of *Mytilus edulis* brought in from other areas and suspended from a fixed structure in an area of large tidal range and strong currents. They are suspended from wires stretched between beams about 2 m above low water of spring tides below a jetty. The tops of the ropes are accessible at low tide but the mussels remain submerged throughout. Results so far are encouraging.

Floating suspended culture

The most impressive developments in mussel cultivation have, however, occurred in floating culture, especially in Spain, where it has formed the basis of an entirely new industry (Waugh, 1966; Andreu, 1958, 1968a, b; Paz-Andrade & Waugh, 1968; Wiborg & Bøhle, 1968). This method is generally used in water more than 3 m deep at low water of equinoctial spring tides (Andreu, 1968a), so that the nature of the bottom is not important. Since rafts are used to suspend the mussel ropes, a good degree of shelter from rough weather is desirable (Andreu, 1968a). Floating culture in Spain was first tried on the Mediterranean coast, at Barcelona among other places (Andreu, 1958; Wiborg & Bøhle, 1968). It was not until the late 1940s, however, that the possibilities of the Galician coast of north-west Spain were considered. Experiments there were so successful (Andreu, 1958) that suspended mussel culture based on the floating system has developed rapidly and Spain is now the world's leading producer of mussels. The current annual production of Spanish mussels, almost all from the Galician rias, has been estimated at around 150000 t (see Andreu, 1968a; Joyner & Spinelli, 1969; Anonymous, 1970), though the most recent figures issued by FAO, those for 1971 (FAO, 1972), quote a figure of only 109000 t. The species cultivated in Galicia is, according to Andreu (1968b) and Lubet (1973), *Mytilus galloprovincialis*, which is abundant on all the coasts of Spain and Portugal.

The Galician rias are river valleys which have become flooded by the sea. They are up to 25 km long, 3–12 km wide, and have a maximum depth of

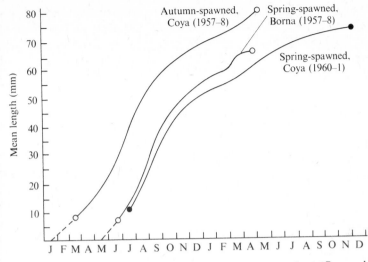

Fig. 10.5. Growth in length of mussels cultivated on rafts at two places (Coya and Borna) in the Ria de Vigo. (After Andreu, 1968*b*.)

Fig. 10.6. Original Galician raft constructed from an old hulk. (Crown copyright.)

Fig. 10.7. Modern Galician raft. (Crown copyright.)

60 m. Conditions are excellent for both growth and fattening of mussels. The water is practically oceanic, with a salinity of about 35‰. Plankton and suspended organic matter provide abundant food. Primary production (Andreu, 1968b,) ranges from 0.07 to 7.6 g C m^{-2} d^{-1}, with an annual average of 0.9 g C m^{-2} d^{-1}. Photosynthesis ranges from 1.1 to 61.6 μg C l^{-1} h^{-1}, with an annual average of 10.5 μg C l^{-1} h^{-1}. Primary production is usually three to four times higher in the inner waters of the rias than near the mouths (Vives & Fraga, 1961). The average tidal range is 4 m. Temperatures are high, ranging on the surface from 9 °C in the winter to 21 °C in the summer (Andreu, 1968b). Consequently mussel growth is especially rapid in the summer (June–September) and even continues, though at a reduced rate, in the winter (Andreu, 1958, 1968b; see Fig. 10.5).

Recent accounts of the development of suspended mussel culture in the Galician rias and of the equipment used have been given by Andreu (1968a, b), Paz-Andrade & Waugh (1968), Ryther (1968) and Wiborg & Bøhle (1968).

During the early years of cultivation in the rias, old hulks were used to provide flotation (Fig. 10.6), being equipped with a wooden framework from which ropes were suspended. As the industry developed, lighter, more durable and more stable structures were required, and so specialised rafts have been developed (Fig. 10.7). Modern rafts most commonly have

Fig. 10.8. Modern Galician raft; close-up showing frame and suspended ropes. (Crown copyright.)

four or more wooden floats, covered with cement (or more recently fibre-glass: Andreu, 1968*b*; Ryther, 1968) to protect the wood against marine boring organisms. On the top of the floats is built a framework of eucalyptus beams, about 60 cm apart, from which the ropes are hung (Fig. 10.8). The framework is supported by steel stays from the ends of the beams to the tops of eucalyptus masts. The raft is usually fitted with a working deck and a shelter for the operators (Andreu, 1968*a*). Andreu (1960) had suggested that shelters should be fitted to the raft owing to the detrimental effect of bright sunlight on the growth of the mussels. A typical raft is about 20 m square and carries an average of 500–600 ropes (Andreu, 1968*a*).

In 1968 there were almost 2800 rafts in the Galician rias. Most (1800), were in the Ria de Arosa, 483 in the Ria de Vigo, 210 in the Ria de Pontevedra and small numbers in other nearby rias (Andreu, 1968*a*). As with the French bouchot method, raft cultivation of mussels in Spain is usually a family business, each family owning on average two or three rafts and involving two or three people in the operations. Larger operators may have up to twenty-five rafts and employ ten people. The culture areas are owned by the Spanish Government and leased to the operators for limited periods (Andreu, 1968*b*; Ryther, 1968; Wiborg & Bøhle, 1968).

In the early days the ropes were about 3 cm in diameter, made of loosely woven local esparto grass and treated with tar, periodic treatment prolonging the use of ropes for up to ten years (Andreu, 1968*a*). Now thinner (1.5 cm) nylon ropes are being increasingly used owing to their greater durability (Andreu, 1968*b*), though Mason (1968) found in experiments in Scotland that smoother synthetic ropes acquired a poorer settlement of spat than ropes of more hairy fibres such as coir. The ropes are from 3 to 12 m long according to the depth of water and they are hung 50–70 cm apart from the beams of the frame (Andreu, 1968*b*; Paz-Andrade & Waugh, 1968). Small wooden pegs are inserted through the lay of the rope every 40 cm to prevent mussels sliding down the rope and being lost (Andreu, 1968*a*, *b*; Wiborg & Bøhle, 1968).

The mussel in the Galician rias has two major spawning periods, the main one in the spring and a smaller one in the autumn (Andreu, 1958, 1968*a*, *b*). The heaviest spatfall is in May, so that empty ropes are hung from the rafts in April. They soon become covered with young mussels attached by the byssus (Fig. 10.9). The mussels grow rapidly and generally reach a size of 30–40 mm by October the same year and 75–90 mm by their second summer or autumn, when they are marketed at an age of 14–18 months (Andreu, 1968*a*; Paz-Andrade & Waugh, 1968); smaller mussels, as small as 60 mm, are sometimes harvested (Wiborg & Bøhle, 1968).

During the period from August to December the mussel flesh is in excellent condition, reaching at its peak perhaps 50% of the total wet weight, though 35–40% is considered good (Andreu, 1968*a*, *b*). During cooking, the wet flesh weight is reduced by some 45–50% (Andreu, 1968*a*), so that a cooked meat yield of 18–22% of the total mussel weight would be acceptable; according to Wiborg & Bøhle (1968) 20–25% is considered good.

Rafts are moored by either one or two iron chains. Those moored at one point swing with the tide so that mussels nearest to the point of attachment grow fastest because they are always presented with the food first (Paz-Andrade & Waugh, 1968). Andreu (1968*a*) suggested that if the rafts were arranged in rows along the current and not staggered across it, the productivity of the ropes would be less.

As the mussels grow it is necessary to thin them out in order to reduce competition for food and promote better growth and fattening. Thinning commences in October when the mussels are 30–40 mm long (Paz-Andrade & Waugh, 1968). The mussels are stripped off the ropes into bins and are then bound onto new ropes with netting; at first cotton netting was used, but now it is usually synthetic fibre. One 'settlement' or 'collector' rope yields enough mussels for twelve to fifteen 'growing-on' ropes.

Larvae from the autumn spawning settle largely on rocks, cliffs and breakwaters on the shore and not on the suspended ropes (Andreu, 1968*a*,

Fig. 10.9. Young mussels attached to rope. (Crown copyright.)

b). Seed mussels from the autumn settlement are collected from the rocky shores, especially on islands at the mouths of the rias. They are tied onto ropes and suspended from the rafts, conveniently filling gaps left on the rafts in the winter after the sale of marketable mussels (Andreu, 1968*b*). Autumn-spawned mussels also grow rapidly, those attached to ropes in March at a size of 9 mm in the Ria de Vigo taking advantage of the excellent conditions from the early spring and reaching marketable size one year

14-2

Fig. 10.10. Work-boat with crane and basket for harvesting mussels, north-west Spain. (Crown copyright.)

later, presumably some 14–15 months after their initial settlement (Andreu, 1958, 1968*b*) (Fig. 10.5).

The method of harvesting is very simple. The ropes are lifted onto the deck of the raft in a basket by means of a crane which is either on the raft or in the bows of an open-decked work-boat (Wiborg & Bøhle, 1968) (Fig.

10.10). A vigorous shake detaches the mussels. Those of 80 mm or more are washed and packed into bags to be marketed alive or sent to canning factories. The rest are sorted according to size and bound onto other ropes to allow further growth (Andreu, 1968a). The annual yield of a typical 500–600-rope raft is some 50–60 t of mussels (Andreu, 1968a; Ryther, 1968).

Experimental raft culture has been carried out on the Atlantic coast of France (Brienne, 1960; Marteil, 1961), but conditions there are not as favourable as in the Spanish rias and no commercial project has resulted.

Cultivation on ropes has a number of advantages, some of which arise from the mussels being raised above the sea-bed and so are shared with cultivation on poles (Waugh, 1966; Andreu, 1968a; Paz-Andrade & Waugh, 1968). The main advantage is that it is three-dimensional, giving more efficient use of the available space and using the food at all depths of the rope. The mussels are removed from the bottom-living predators, except a few which happen to settle on the ropes, but they are subject to attack by predatory fish in Spanish waters including the golden mackerel, *Sparus aurata* (Andreu, 1968a), and in more northerly waters they are taken by diving birds such as the eider (Mason, 1969; Dunthorn, 1971). In floating culture the mussels are covered at all states of the tide, with all the resultant advantages. These advantages outweigh the disadvantages, the greatest of which are perhaps the amount, and therefore expense, of labour and the specialised equipment involved in suspended culture (Andreu, 1968a).

In the Spanish rias the more serious competitors of the mussels for space and food, especially the ascidian *Ciona* (Andreu, 1968a), are picked from the ropes and destroyed when the mussels are thinned out (Waugh, 1966). It is impracticable to destroy the predators by exposing the ropes to the air, owing to the large numbers of ropes and the lack of space on the rafts (B. Andreu, personal communication). Furthermore, Andreu thinks that a prolonged exposure would be necessary there and that this would affect the vitality of mussels not accustomed to tidal exposure.

However, in the fixed cultivation in the Gulf of Trieste (Favretto, 1968), all the fouling organisms except barnacles are killed by leaving the ropes out of water in the shade for not more than a day, a treatment which the mussels are able to survive. The ropes are then replaced in the water. Similarly in the Neapolitan lakes (Korringa & Postma, 1957) the ropes are hung exposed to the air over the topmost bars of the frames. Exposure of the ropes is also practised on the French Mediterranean coast (Lambert, 1939) and was recommended by Lubet (1961) for use in Yugoslavia, though Hrs-Brenko & Igić (1968) recommended immersion in fresh water for not longer than a day.

The cultivation of other species

There are many different species of mussels throughout the world, some highly prized commercially and others largely ignored and unexploited (Davies, 1970; Mason, 1971). Many of these are much larger than *Mytilus edulis* and *M. galloprovincialis* and grow rapidly. But, while there is a long history of exploitation and cultivation of mussels in Europe, recognition of their importance and the possibility of cultivation have been slow to develop elsewhere. Very recently, however, experimental and commercial cultivation has commenced in the Indo-Pacific region, including Australasia, in South Africa and in South America.

In the Indo-Pacific region the green mussel *Mytilus smaragdinus* is widespread, grows well in most of the tropical waters and has a high potential for cultivation (Ling, 1973). It is already cultivated intensively in Thailand, moderately in the Philippines and on an experimental scale in Singapore. It is seen variously as a means of increasing production of animal protein to meet the needs of fast-growing populations; as a means of producing high-priced commodities for export and so earning foreign exchange; as creating employment opportunities; and as a way of using large areas which are at present unproductive. In countries with such areas and cheap labour but poor financial resources, simple techniques are used, even though the yield might be low. More complicated methods with higher yields but high costs of production are more suitable for more prosperous countries where the earnings and purchasing power of the people are high (Ling, 1973).

In Thailand the annual production of *M. smaragdinus* is about 45 000 t. The biology appears to be similar to that of the European species (see Obusan & Urbano, 1968; Tham, Yang & Tan, 1973). Bamboo poles are planted as spat collectors in shallow water on mudflats in December–March. Settlement, which reaches a peak in February–March, is usually so dense that the young mussels have to be thinned out at least twice, those removed being used as animal feed. Those remaining are allowed to grow to a marketable size, which they attain in 8–12 months (Sribhibhadh, 1973).

In the Philippines, where it is known as 'tahong' (Obusan & Urbano, 1968), *Mytilus smaragdinus* was at first considered a fouling organism by oyster farmers, but is now increasingly being cultivated (Bardach *et al.*, 1972). Slender bamboo stakes 4–6 m long are placed in the sea-bed in shallow areas, some 3 m deep at low water, and acquire a settlement of spat which is allowed to grow. A variety of seed collectors, including coconut shells and oyster shells strung on wire or rubber sheets or strips from old tyres, are hung from bamboo frames. Horizontal logs are also used as collectors in calmer, silt-free areas (Obusan & Urbano, 1968; Blanco,

1973). A thick growth of barnacles is conducive to the mussel's secondary settlement (Tham *et al.*, 1973) and so the collectors are put out early to acquire a settlement of barnacles (Obusan & Urbano, 1968). Seed from collectors or from natural sources may be placed in trays on the frames where it may be allowed to grow to a commercial size. Alternatively, bamboo shoots may be placed in the trays to become coated with young mussels, and are then driven into the sea-bed. There the mussels are allowed to grow to commercial size (Bardach *et al.*, 1972). They are known to exceed 30 cm in length, but they are marketed at 3–8 cm, which they attain in 4–10 months; they are in greatest demand at 5 cm, after 6 months. Those on poles are stripped by means of hired divers and those in trays are harvested at low tide. Transplanting of mussels is done by cutting up seeded stakes and taking them to good growing areas. This is also done to seed new areas (Obusan & Urbano, 1968).

In 1966, only 8 ha of culture area was in use in the Phillipines and produced about 2000 t of mussels (Bardach *et al.*, 1972). In 1970, 300 ha were used for oyster and mussel cultivation in Bacoor Bay alone. The potential area in bays, coves and estuaries in Bacoor Bay, Manila Bay and elsewhere totals 100000 ha (Blanco, 1973), so that the prospects appear most promising, for both home consumption and export.

In Singapore, *Mytilus smaragdinus* is abundant on firm substrata in the Johore Straits, where it is becoming increasingly exploited. Because the natural stocks are becoming depleted, the possibility of culture by means of poles and suspended ropes and trays is being investigated. Here the mussel has been known to reach a commercial size of 5–7 cm in only 4–6 months, a growth rate which compares favourably with those in Thailand and the Philippines (Tham *et al.*, 1973). The prospects appear promising. Jones & Alagarswami (1968) and Jones (1973) consider that *Mytilus smaragdinus* and a faster-growing brown species, *Mytilus* sp., both of which form the basis of a subsistence fishery of considerable local importance in various parts of India, could be cultivated. There is a regular and abundant supply of seed and cheap labour is available.

Several species of mussel are found on the coasts of New Zealand, but only two are economically important. The fast-growing green mussel, *Perna canaliculus*, occurs subtidally and is the basis of a dredge fishery for human consumption, and the slower-growing blue mussel occurs littorally and is commonly hand-gathered for food or bait (Watkinson & Smith, 1972). Landings of green mussels have fluctuated, and beds have been depleted, but the fishery has been maintained by exploiting other stocks. Landings in 1971 were 1195 t.

In the 1960s experimental cultivation, particularly of green mussels, was commenced, involving hanging seed mussels in cages or netting or bound on ropes. Seed was also collected on twigs and ropes. Growth was rapid. In

Hauraki Gulf, seed which settled in October reached a size of 10 cm in 1 year, though a further year was required to grow another 2 cm (New Zealand Fishing Industry Board, 1971). As a result of this success, a fishermen's association in 1969 and 1970 set out commercial rafts with ropes in the Marlborough Sounds. The area of sheltered water suitable for mussel cultivation on rafts is much greater in New Zealand than in Spain, and with a faster-growing species, much less pollution and unsatisfied internal demand and good export potential, the prospects appear promising. However, insufficient is yet known about the productivity of the waters and their ability to support high-density stocks (New Zealand Fishing Industry Board, 1971; Watkinson & Smith, 1972).

The mussel has little appeal in Australia, though in Victoria *Mytilus edulis planulatus* is dredged along with scallops, and at Sydney it is harvested by diving (Maclean, 1972). Experimental raft culture is being undertaken in Sydney Harbour, but the prospects at present are limited by the price offered by processors. Market promotion is needed for domestic consumption, though export possibilities exist.

In South Africa few mussels are eaten, though they are used for bait. However, Mostert (1972) suggested that *Choromytilus meridionalis* would be a suitable subject for cultivation. In the Langebaan lagoon naturally-settled spat suspended in 'Netlon' bags grew to a size of 62–70 mm within 15–18 months. Gitay (1972) has suggested that the same species, together with *Aulocomya magellanica*, might be cultivated on the west coast.

In South America experimental mussel cultivation has been tried in Chile (Hancock, 1969*b*; Lozada, Rolleri & Yañez, 1971) and in Venezuela (Iversen, 1968).

In Chile, three species, *Choromytilus chorus* ('choro'), *Aulocomya ater* ('cholga') and *Mytilus edulis chilensis* ('chorito'), are exploited commercially, all being collected by hand, usually from submerged banks by divers (Hancock, 1969*b*). Choro command a high price and are sold fresh while most cholga and chorito are canned, some for export. All are large and grow rapidly, reaching maximum lengths of 260 mm, 200 mm or more, and 120 mm, respectively. Heavy landings of choro were made in the 1930s and the stocks declined, and now cholga and chorito are being heavily exploited and cholga beds are showing signs of diminution. Legislation has been instituted to protect them. Various close seasons have been introduced, together with minimum landing sizes of 120 mm for choro, 70 mm for cholga and 50 mm for chorito. Since 1961 landings of choro had been prohibited in various areas (Hancock, 1969*b*). Hancock stated that the Chilean mussel fishery would survive only if urgent steps were taken to introduce cultivation.

Rafts have been built and settlement of choro and chorito obtained on suspended branches put out in September, just before spawning. The seed has been taken to various areas and attached to ropes by means of netting. Growth on the suspended ropes has been more rapid than on the bottom (Lozada *et al.*, 1971; Hancock, 1969*b*). The environment was found to be favourable at Putemún in the Castro estuary, Mejillones Bay and other areas.

Cultivation of *Perna perna* has been tried in the Gulf of Cariaco, Venezuela, under the guidance of Spanish scientists (Iversen, 1968). Rafts 7 m square, floated on styrofoam, were put out, each with 100 bamboo pole collectors suspended from them. The experiments at first looked promising. Mussels settled on the poles, grew well and produced good flesh. Unfortunately the bamboo poles were lost owing to boring animals. The results were, however, sufficiently good to encourage further experiments (Milne, 1972).

Public health aspects

With the increase in mussel cultivation has come an increasing awareness of public health hazards, which are of two main kinds; poisoning due to phytoplankton blooms and bacterial pollution.

Dinoflagellates of the genus *Gonyaulax* are chiefly associated with the paralytic toxin found in filter-feeding molluscs (see Halstead, 1965; Robinson, 1968), but *Prorocentrum*, *Exuviaella* and *Gymnodinium* spp. have also caused poisoning (Ingham, Mason & Wood, 1968). Normally these dinoflagellates would be harmless, but occasionally blooms occur, the so-called 'red tides', and vast numbers are then taken in by filter-feeders which accumulate the toxin in their flesh. These blooms have been associated with higher-than-average water temperature, a high nutrient content, calm conditions, proximity to land and usually a lowered salinity (Robinson, 1968).

If affected shellfish are eaten by humans the outcome can be serious, resulting in neurotoxic symptoms and occasionally death (Halstead, 1965). Most recorded outbreaks of paralytic shellfish poisoning have been in North America, where shellfish and plankton are now monitored and warnings issued when necessary (Halstead, 1965; Quayle, 1969; Prakash, Medcof & Tennant, 1971). Sporadic outbursts of paralytic shellfish poisoning in *Choromytilus meridionalis* in South Africa have contributed to the prejudice against this mussel, giving rise to the belief that it is unfit for human consumption in certain months (Mostert, 1972). Occurrences of poisoning are rare in Europe, but an outbreak on the east coast of Britain in the spring of 1968 was found to be due to the toxin produced by *Gonyaulax tamarensis* (Ingham *et al.*, 1968). Other recent European outbreaks of mussel poisoning have been recorded in Holland (Korringa & Roskam, 1961; Korringa, 1968) and Oslofjord, Norway (Bøhle, 1965). Bøhle pointed out that outbreaks seldom occur in the autumn and winter, when the mussel flesh is at its best. When plankton blooms occur off the Galician coast, cultivators are advised not to sell mussels until it is once again safe to do so; no cases of paralysis have been recorded there (Wiborg & Bøhle, 1968). Similar precautions are applied in Oslofjord (Bøhle, 1965).

Bacterial pollution is dealt with in a recent pamphlet by Wood (1969). Many coastal areas of the sea are polluted, especially near towns, and mussels and other filter-feeders pick up pathogenic bacteria originating from sewage. Cooking the mussels by immersing them in boiling water and leaving them in for 2 minutes after the resumption of boiling is sufficient to sterilise the flesh (Sherwood, 1957). If polluted mussels are to be marketed alive, however, they must first be purified. This is done by allowing the mussels to rid themselves of bacteria by storing them in tanks of sterile sea water for 2 days (Wood, 1969). The bacteria are thrown out within 48 hours in mucous threads of faeces which prevent re-pollution of the water.

In many countries mussels sold alive for human consumption require a certificate of cleanness. This has necessitated the installation of purification plants, which has allowed cultivation in areas which would otherwise be too dirty (Lubet, 1973). However, the provision of these facilities adds to the costs of production. The original large-scale installation at Conwy, North Wales (Dodgson, 1928) used water which was purified by chlorination and subsequently dechlorinated. In Galicia purification plants are run by the larger operators and provide a service for smaller operators (Wiborg & Bøhle, 1968; Anonymous, 1970). The water is sterilised either by chlorination or by ultraviolet light from the sun. The process of purification is expensive and is estimated to account for 75% of the cost of Spanish cultivated mussels (New Zealand Fishing Industry Board, 1971). In France, either chlorination or ozone is used (Lubet, 1973).

General considerations

Sessile, bottom-dwelling filter-feeders, such as the mussel, by virtue of their ability to filter food from water brought to them by currents from a much greater region than that actually occupied by the animals, often attain great productivity in areas of intensive cultivation (Ryther, 1969). Bottom-living bivalves, without cultivation, might yield 150 kg (wet weight) of flesh ha^{-1} yr^{-1} (Ryther, 1969). In comparison, mussel culture on poles in France yields about 5 t of flesh ha^{-1} yr^{-1} (Ryther, 1968) and bottom culture yields 12–25 t ha^{-1} annually (Waugh, 1966). Suspended cultivation on ropes vastly increases the production. Ryther (1968) estimated that intensive cultivation in the Galician rias with four 20×20 m rafts per acre (roughly ten per hectare) and an average annual yield of 60 t of mussels (30 t of wet meat, assuming the maximum yield of 50%) per raft, yields 120 t of meat per acre, or approximately 300 t of wet meat ha^{-1} yr^{-1}. Using the figure of 50 t of mussels per raft per year (25 t of flesh) quoted by Andreu (1968b), the annual yield would be 250 t ha^{-1}. This wet meat yield from intensive mussel cultivation in the Galician rias is roughly 500–600 times greater than yields from any other form of husbandry or culture in which animals are grown naturally with no supplemental or artificial feeding. The yield is put in perspective when compared with the increase of only 300 lb live weight of beef cattle per acre (340 kg ha^{-1}) per year quoted by Hickling (1968) for good permanent grassland in England or 0.25 ton of cattle and sheep per acre (635 kg ha^{-1}) produced by a 100-acre (40-ha) Welsh farm (Davies, 1970).

As Ryther (1969) rightly emphasised, his figures were based on the maximum meat yield of 50% quoted by Andreu (1968a, b). Mussels are marketed with meat yields down to 35–40% (Andreu, 1968a, b, c) so that the average yield will in fact be slightly less than 50%. The yield per unit area is still very impressive, however.

Prospects

Mytilus edulis and *M. galloprovincialis* are successfully cultivated on the coasts of central and southern Europe. One of the major factors against economic cultivation on ropes in more northerly countries, such as Great Britain and Norway, is the lower sea temperature, with the consequence of reduced feeding and growing seasons. The slower growth rate results in a longer time required to reach a marketable size and so adds to the cost of production (Waugh, 1966). Thus, while cultivation on the sea bottom in Great Britain and Holland can produce mussels economically with at least 2½ years to reach marketable size, suspended culture, which is more labour-intensive, might not be economically viable with a similar growth rate, unless the yield could be made much greater by mechanisation or other means (Edwards, 1968; Mason, 1971) or the costs reduced.

However, mussels growing on floating objects and in experimental floating culture in northern Europe have shown much more rapid growth. Fraser (1938) found submerged mussels on buoys and a lightship in Liverpool Bay, England, which had grown up to 60 mm in 1 year and 90 mm in 2½ years.

Experimental cultivation of *Mytilus edulis* on ropes hung from rafts has recently been carried out in Linne Mhuirich, an inlet on the west coast of Scotland in which the temperature of the water follows that of the air rather than that of the adjacent open sea (Mason, 1968, 1969). For much of the year temperatures are appreciably higher than in the open sea, reaching 20 °C in the summer. As a result, mussels of high quality have been grown quickly. From a settlement in the early summer the mussels reached

407

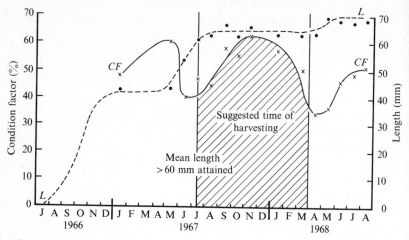

Fig. 10.11. Growth in length (*L*) and condition factor (*CF*; wet flesh volume as a percentage of shell space) of mussels settled on ropes in Linne Mhuirich, Scotland, in summer 1966. (From Mason, 1969; Crown copyright.)

commercial size, averaging 67 mm, late in their second summer, only 14 months after settlement (Fig. 10.11). With this rapid growth, and provided that the costs of labour and floating installations can be minimised, suspended culture in this and other Scottish inlets with similar conditions could be profitable (Mason, 1968, 1969). Indeed pilot commercial operations are already under way. Harvesting should occur in the second autumn and early winter when flesh condition is best (Fig. 10.11).

Similarly, some Norwegian inlets, including Oslofjord, have relatively high summer temperatures and experiments there have also yielded results which augur well for suspended culture (Bøhle & Wiborg, 1967). Experiments are also being carried out in Ireland (Edwards, 1968; Murray, 1971). Mussels have been grown in suspension in the Flensburger Förde, Germany (Meixner, 1971), but the process there has not yet been shown to have economic possibilities.

Interest has been shown in cultivating *Mytilus galloprovincialis* in other Mediterranean countries, including Greece (Rigopoulos, 1972) and Tunisia (Ricci, 1957). Little use is made of mussels in the Soviet Union, where landings of *M. galloprovincialis* from the Black Sea are small. However, experiments by Ivanov (1971) led him to conclude that the prospects for raising mussels in the Azov–Black Sea basin are very promising owing to the high productivity of certain areas.

Although *Mytilus edulis* is abundant in North America there is little demand for it. There are, however, signs that the American market is increasing, and mussels which were formerly considered a nuisance on the cultivated oyster and clam beds of Long Island Sound are now being marketed on the east coast (Coffey, 1972). Similarly, while Japan produces large quantities of oysters and other molluscs, mussels are not cultivated (Iversen, 1968).

The potential for mussel cultivation is now beginning to be appreciated in the Far East and the southern hemisphere also. In south-east Asia cultivation of the green mussel, *Mytilus smaragdinus*, is the subject of a growing industry in Thailand and the Philippines. Experiments on other species in New Zealand, Australia, South Africa and South America have yielded encouraging results.

The prospects of a continued expansion of mussel culture appear, therefore, to be good. The method of cultivation will depend on the physical conditions in any particular area. If conditions are suitable, suspended culture is the most efficient existing method, owing to the enhanced growth rate and consequent quick turnover and best use of equipment. It is, however, more labour-intensive than other methods, so that for it to be economic, capital expenses have to be kept to a minimum and a high level of efficiency maintained. Cheaper forms of flotation for suspension might be sought, together with improved methods of harvesting and sorting. Costs might be reduced by processing near the place of production, thus avoiding the expense of transporting shell.

Bøhle (1970) has successfully used cylinders of plastic netting for thinning out *Mytilus edulis* seed 10–20 mm long or transferring it to new areas. The young crawl out through the meshes and attach themselves to the outside of the netting so that the netting then forms a rope-like core. This procedure is now being used commercially in southern France (M. N. Mistakidis; personal communication) and is now being adopted commercially in Scotland. Dare (1971) has shown also that instead of transporting seed mussels (20–30 mm long) from areas of abundant spatfall to areas of poor spatfall but good growth and fattening, much smaller spat can be transported on collectors over considerable distances for cultivation in good areas. This is less laborious and enables much greater numbers to be handled.

Cultivation of mussels of various species is at present more likely to provide a business opportunity in the advanced countries of the world than to provide a means of helping to overcome the protein deficiency caused by rapid population increase in the less-developed countries (Ryther & Bardach, 1968; Webber, 1968; Korringa, 1971). In the long term, however, as production increases and costs are reduced, the price of mussels will fall as it has in the Spanish mussel industry. Then, especially in areas with cheap labour and high sea temperatures giving rapid growth, cultivation of mussels might help to improve the diets of under-developed peoples and contribute towards a solution of the world's food shortage (Ryther &

Marine mussels

Bardach, 1968; Davies, 1970; Kinne, 1970*b*). With this in view, experiments are being carried out aimed at producing from mussels a cheap protein concentrate suitable for human consumption (Joyner & Spinelli, 1969). One such preparation has been produced which is rich in protein and has a pleasing taste and smell. The present limitation is the inadequate supply of raw material, which cultivation might help to overcome.

References

Abd-el-Wahab, A. & Pantelouris, E. M. (1957). Synthetic processes in nucleated and non-nucleated parts of *Mytilus* eggs. *Experimental Cell Research*, **13**, 78–82.

Ahmed, M. & Sparks, A. K. (1967). Chromosomes of oysters, clams and mussels. *Proceedings of the National Shell-fisheries Association*, **58**, 10.

Ahmed, M. & Sparks, A. K. (1970). Chromosome number, structure and autosomal polymorphism in the marine mussels *Mytilus edulis* and *M. californianus*. *Biological Bulletin. Marine Biological Laboratory, Woods Hole, Mass.*, **138**, 1–13.

Aiello, E. L. (1957). The influence of the branchial nerve and 5-hydroxytryptamine on the ciliary activity of *Mytilus* gill. *Biological Bulletin. Marine Biological Laboratory, Woods Hole, Mass.*, **113**, 325.

Aiello, E. L. (1960). Factors affecting ciliary activity on the gill of the mussel, *Mytilus edulis*. *Physiological Zoology*, **33**, 120–35.

Aiello, E. L. (1962). Identification of the cilioexcitatory substance present in the gill of the mussel *Mytilus edulis*. *Journal of Cellular and Comparative Physiology*, **60**, 17–21.

Aiello, E. L. (1965). Pharmacology of cilio-excitation in the mussel. *Federation Proceedings. Federation of American Societies for Experimental Biology*, **24**, 758.

Aiello, E. L. (1970). Nervous and chemical stimulation of the gill cilia in bivalve molluscs. *Physiological Zoology*, **43**, 60–70.

Aiello, E. L. (1974). Control of ciliary activity in Metazoa. In: *Cilia and flagella* (ed. M. A. Sleigh), pp. 353–76. Academic Press, New York and London.

Aiello, E. L. & Guideri, G. (1965). Distribution and function of the branchial nerve in the mussel. *Biological Bulletin. Marine Biological Laboratory, Woods Hole, Mass.*, **129**, 431–8.

Aiello, E. L. & Guideri, G. (1966). Relationship between 5-hydroxytryptamine and nerve stimulation of ciliary activity. *Journal of Pharmacology and Experimental Therapeutics*, **154**, 517–23.

Aiello, E. L. & Paparo, A. (1968). A cholinergic mechanism for cilio-excitation and inhibition. *Federation Proceedings. Federation of American Societies for Experimental Biology*, **27**, 756.

Aiello, E. L. & Sleigh, M. A. (1972). The metachronal wave of lateral cilia of *Mytilus edulis*. *Journal of Cell Biology*, **54**, 493–506.

Alder, J. & Hancock, A. (1851). On the branchial currents in *Pholas* and *Mya*. *Annual Magazine of Natural History*, **8**, 370.

Alderdice, D. F. (1972). Factor combinations. In: *Marine ecology* (ed. O. Kinne), vol. 1, part 3, pp. 1659–1722. Wiley-Interscience, New York.

References

Ali, R. M. (1970). The influence of suspension density and temperature on the filtration rate of *Hiatella arctica*. *Marine Biology*, **6**, 291–302.

Allen, F. E. (1955). Identity of breeding temperature in southern and northern hemisphere species of *Mytilus* (Lamellibranchia). *Pacific Science*, **9**, 107–9.

Allen, J. A. (1970). Experiments on the uptake of radioactive phosphorus by bivalves and its subsequent distribution within the body. *Comparative Biochemistry and Physiology*, **36**, 131–41.

Allen, J. A. & Garrett, M. R. (1971*a*). Taurine in marine invertebrates. *Advances in Marine Biology*, **9**, 205–53.

Allen, J. A. & Garrett, M. R. (1971*b*). The excretion of ammonia and urea by *Mya arenaria* L. (Mollusca: Bivalvia). *Comparative Biochemistry and Physiology*, **39A**, 633–42.

Allen, K. (1961). The effect of salinity on the amino-acid concentration in *Rangia cuneata* (Pelecypoda). *Biological Bulletin. Marine Biological Laboratory, Woods Hole, Mass.*, **121**, 419–24.

Allen, K. & Awapara, J. (1960). Metabolism of sulfur amino acids in *Mytilus edulis* and *Rangia cuneata*. *Biological Bulletin. Marine Biological Laboratory, Woods Hole, Mass.*, **118**, 173–82.

Allen, E. J. & Todd, R. A. (1902). The fauna of the Exe Estuary. *Journal of the Marine Biological Association of the United Kingdom*, **6**, 295–335.

Alvarez, J. B. (1965). Some observations on the organic nitrogen of sea water and the chemical composition of the mussel *Mytilus edulis* in Ria de Vigo, Spain. *Boletin del Instituto Oceanografico de la Universidad de Oriente, Venezuela*, **4**, 172–83.

Alvarez, J. B. (1968). Variacion mensual de la composicion quimica del mejillon, *Perna perna* (L.). *Boletin del Instituto Oceanografico de la Universidad de Oriente, Venezuela*, **7**, 137–47.

Alyakrinskaya, I. O. (1966). On the behaviour and filtrational ability of black sea mussels (*Mytilus galloprovincialis*) in oil polluted water. *Zoologicheskii Zhurnal*, **45**, 998–1002. (In Russian.)

Alyakrinskaya, I. O. (1967). Distribution of mussels and some data on their chemical composition in connection with the pollution of the Bay of Novorossiisk. *Trydȳ Institua Okeanologii. Akademiya Nauk SSSR, Moscow*, **85**, 66–76.

Anderson, J. W. (1975). The uptake and incorporation of glycine by the gills of *Rangia cuneata* (Mollusca: Bivalvia) in response to variations in salinity and sodium. In: *Physiological ecology of estuarine organisms* (ed. F. J. Vernberg), pp. 239–58. University of South Carolina Press, Columbia, South Carolina.

Anderson, J. W. & Bedford, W. B. (1973). The physiological responses of the estuarine clam, *Rangia cuneata* (Gray), to salinity. II. Uptake of glycine. *Biological Bulletin. Marine Biological Laboratory, Woods Hole, Mass.*, **144**, 229–47.

Andreu, B. (1958). Sobre el cultivo del mejillón en Galicia. Biologia, crecimiento y producción. *Industrias Pesqueras*, **745/6**, 44–7.

Andreu, B. (1960). Ensayos sobre el efecto de la luz en el ritmo de crecimiento del mejillón (*Mytilus edulis*) en la Ria de Bigo. *Boletín de la Real Sociedad Española de Historia Natural Biologica*, **58**, 217–36.

Andreu, B. (1963). Propagacion del copépodo parasito *Mytilicola intestinalis* en el mejillón cultivado de las rias gallegos (N.W. de Espana). *Investigación Pesquera*, **24**, 3–20.

Andreu, B. (1968a). The importance and possibilities of mussel culture. Working Paper 5, Symposium on possibilities and problems of fisheries development in south-east Asia. German foundation for developing countries, Berlin (Tegel), 10–30 September, 1968.

Andreu, B. (1968b). Pesqueria y cultivo de mejillónes y ostras en España. *Publicaciones Tecnicas de la Junta des Estudios de Pesca, Madrid*, **7**, 303–20.

Andreu, B. (1968c). Fishery and culture of mussels and oysters in Spain. *Proceedings of the Symposium on Mollusca*, **3**, 835–46.

Andrew, W. (1965). *Comparative hematology*. Grune & Stratton, New York. 188 pp.

Andrews, T. R. & Reid, R. G. B. (1972). Ornithine cycle and uricolytic enzymes in four bivalve molluscs. *Comparative Biochemistry and Physiology*, **42B**, 475–91.

Anonymous (1937). Mussel cultivation. *Fisheries Notice, London*, No. 13. 6 pp.

Anonymous (1952). *Marine fouling and its prevention*. US Naval Institute, Annapolis. 388 pp.

Anonymous (1966). Making mussel farming pay. *World Fishing*, **15**(3), 51–3.

Anonymous (1970). Cleansing 200 tons of mussels a day. *World Fishing*, **19**(10), 30–3.

Anonymous (1973). Pilot plant for Dutch mussel culture. *Fish Farming International*, **1**, 62–7.

Anonymous (1974). A successful first year for Jersey's mussel farm. *Jersey Magnet Magazine*, 23 January, 23.

Ansell, A. D. (1961). Reproduction, growth and mortality of *Venus striatula* (Da Costa) in the Kames Bay, Millport. *Journal of the Marine Biological Association of the United Kingdom*, **41**, 191–215.

Ansell, A. D. (1967). Burrowing in *Lyonsia norvegica* (Gmelin). *Proceedings of the Malacological Society of London*, **37**, 387–93.

Ansell, A. D. (1972). Distribution, growth and seasonal changes in biochemical composition for the bivalve *Donax vittatus* (da Costa) from Kames Bay, Millport. *Journal of Experimental Marine Biology and Ecology*, **10**, 137–50.

Ansell, A. D. (1973). Oxygen consumption by the bivalve *Donax vittatus* (da Costa). *Journal of Experimental Marine Biology and Ecology*, **11**, 311–28.

Ansell, A. D. & Sivadas, P. (1973). Some effects of temperature and starvation on the bivalve *Donax vittatus* (da Costa) in experimental laboratory populations. *Journal of Experimental Marine Biology and Ecology*, **13**, 229–62.

References

Antheunisse, L. J. (1963). Neurosecretory phenomena in the zebra mussel *Dreissena polymorpha* Pallas. *Archives Néerlandaises de Zoologie*, **15**, 237–314.

Armstrong, D. A. & Millemann, R. E. (1974). Effects of the insecticide serin and its first hydrolytic product, 1-napthol, on some early developmental stages of the bay mussel, *Mytilus edulis. Marine Biology*, **28**, 11–16.

Atkins, D. (1937*a*). On the ciliary mechanisms and interrelationships of lamellibranchs. II. Sorting devices on the gills. *Quarterly Journal of Microscopical Science*, **79**, 339–73.

Atkins, D. (1937*b*). On the ciliary mechanisms and interrelationships of lamellibranchs. III. Types of lamellibranch gills and their food currents. *Quarterly Journal of Microscopical Science*, **79**, 375–421.

Atkinson, D. E. (1971). Adenine nucleotides as stoichiometric coupling agents in metabolism and as regulatory modifiers: the adenylate energy charge. In: *Metabolic pathways* (ed. H. J. Vogel), vol. V, pp. 1–21. Academic Press, New York and London.

Audouin, J. (1954). La Mytiliculture en Baie de l'Aiguillon. *Science et Pêche*, **1**(16), 7–10.

Ayers, J. C. (1956). Population dynamics of the marine clam, *Mya arenaria. Limnology and Oceanography*, **1**, 26–34.

Badman, D. G. (1967). Quantitative studies on trehalose in the oyster, *Crassostrea virginica* Gmelin. *Comparative Biochemistry and Physiology*, **23**, 621–9.

Bagge, P. & Salo, A. (1967). *Biological detectors of radioactive contamination in the Baltic*. Report SFL-A9, Institute of Radiation Physics, Helsinki.

Baggerman, B. (1953). Spatfall and transport of *Cardium edule* L. *Archives Néerlandaises Zoologie*, **10**, 315–42.

Bailey, D. F. & Benjamin, P. R. (1968). Anatomical and electrophysiological studies on the gastropod osphradium. *Symposium of the Zoological Society of London*, **23**, 263–8.

Bailey, D. F. & Laverack, M. S. (1966). Aspects of the neurophysiology of *Buccinum undatum* L. (Gastropoda). I. Central responses to stimulation of the osphradium. *Journal of Experimental Biology*, **44**, 131–48.

Baird, R. H. (1966). Factors affecting the growth and condition of mussels (*Mytilus edulis*). *Fishery Investigations. Ministry of Agriculture, Fisheries and Food, London*, Ser. II, **25**, 1–33.

Baird, R. H. & Drinnan, R. E. (1956). The ratio of shell to meat in *Mytilus* as a function of tidal exposure to air. *Journal du Conseil. Conseil permanent International pour l'Exploration de la Mer*, **22**, 329–36.

Balagot, B. P. (1971). Microgeographic variation at two biochemical loci in the Blue Mussel, *Mytilus edulis*, pp. 1–55. MA thesis, State University of New York, Stony Brook.

Ballantine, R. (1940). Analysis of the changes in respiratory activity accompanying the fertilisation of marine eggs. *Journal of Cellular and Comparative Physiology*, **15**, 217–32.

Bannatyne, W. R. & Thomas, J. (1969). Fatty acid composition of New Zealand shellfish lipids. *New Zealand Journal of Science*, **12**, 207–12.

Barber, V. C. (1968). The structure of mollusc statocysts, with particular reference to cephalopods. *Symposium of the Zoological Society of London*, **23**, 37–62.

Bardach, J. E., Ryther, J. H. & McLarney, W. O. (1972). *Aquaculture.* Wiley, New York. 868 pp.

Baret, R., Mourgue, M., Broc, A. & Charmoit, J. (1965). Etude comparative de la désamidination de l'acide-guanidobutyrique et de l'arginine par l'hépatopancréas on le foie de divers Invertébrés. *Comptes Rendus des Séances de la Société de Biologie*, **159**, 2446–50.

Barnes, H. (1957). The northern limits of *Balanus balanoides* (L.). *Oikos*, **8**, 1–15.

Barnes, H., Barnes, M. & Finlayson, D. M. (1963). The seasonal changes in body weight, biochemical composition, and oxygen uptake of two common boreo-artic cirripedes, *Balanus balanoides* and *B. balanus*. *Journal of the Marine Biological Association of the United Kingdom*, **43**, 185–211.

Barrett, J., Ward, C. W. & Fairbairn, D. (1970). The glyoxylate cycle and the conversion of triglycerides to carbohydrates in developing eggs of *Ascaris lumbricoides*. *Comparative Biochemistry and Physiology*, **35**, 577–86.

Barry, R. J. C. & Munday, K. A. (1959). Carbohydrate levels in *Patella*. *Journal of the Marine Biological Association of the United Kingdom*, **38**, 81–95.

Barsotti, G. & Meluzzi, C. (1968). Osservazioni su *Mytilus edulis* L. e *M. galloprovincialis* Lamarck. *Conchiglie*, **4**, 50–8.

Bartels, H. & Moll, W. (1964). Passage of inert substances and oxygen in the human placenta. *Pflügers Archiv fur die gesamte Physiologie des Menschen und der Tiere*, **280**, 165–77.

Bartlett, M. S. (1949). Fitting a straight line when both variables are subject to error. *Biometrics*, **5**, 207–12.

Baskin, R. J. & Allen, K. (1963). Regulation of respiration in the molluscan heart. *Nature, London*, **198**, 448–50.

Battle, H. (1932). Rhythmical sexual maturity and spawning of certain bivalve mollusks. *Contributions to Canadian Biology and Fisheries*, N.S., **7**, 257–76.

Bayne, B. L. (1963). Responses of *Mytilus edulis* larvae to increases in hydrostatic pressure. *Nature, London*, **198**, 406–7.

Bayne, B. L. (1964a). The responses of the larvae of *Mytilus edulis* L. to light and to gravity. *Oikos*, **15**, 162–74.

Bayne, B. L. (1964b). Primary and secondary settlement in *Mytilus edulis* L. (Mollusca). *Journal of Animal Ecology*, **33**, 513–23.

Bayne, B. L. (1965). Growth and the delay of metamorphosis of the larvae of *Mytilus edulis* (L.). *Ophelia*, **2**, 1–47.

Bayne, B. L. (1967). The respiratory response of *Mytilus perna* L. (Mollusca: Lamellibranchia) to reduced environmental oxygen. *Physiological Zoology*, **40**, 307–13.

References

Bayne, B. L. (1971a). Oxygen consumption by three species of lamellibranch mollusc in declining ambient oxygen tension. *Comparative Biochemistry and Physiology*, **40A**, 955–70.

Bayne, B. L. (1971b). Ventilation, the heart beat and oxygen uptake by *Mytilus edulis* L. in declining oxygen tension. *Comparative Biochemistry and Physiology*, **40A**, 1065–85.

Bayne, B. L. (1971c). Some morphological changes that occur at the metamorphosis of the larvae of *Mytilus edulis*. In: *Fourth European marine biology symposium* (ed. D. J. Crisp), pp. 259–80. Cambridge University Press, London.

Bayne, B. L. (1972). Some effects of stress in the adult on the larval development of *Mytilus edulis*. *Nature, London*, **237**, 459.

Bayne, B. L. (1973a). Physiological changes in *Mytilus edulis* L. induced by temperature and nutritive stress. *Journal of the Marine Biological Association of the United Kingdom*, **53**, 39–58.

Bayne, B. L. (1973b). Aspects of the metabolism of *Mytilus edulis* during starvation. *Netherlands Journal of Sea Research*, **7**, 399–410.

Bayne, B. L. (1973c). The responses of three species of bivalve mollusc to declining oxygen tension at reduced salinity. *Comparative Biochemistry and Physiology*, **45A**, 793–806.

Bayne, B. L. (1975a). Reproduction in bivalve molluscs under environmental stress. In: *Physiological ecology of estuarine organisms* (ed. F. J. Vernberg), pp. 259–77. University of South Carolina Press, Columbia, South Carolina.

Bayne, B. L. (1975b). Aspects of physiological condition in *Mytilus edulis* (L.), with special reference to the effects of oxygen tension and salinity. In: *Proceedings of the ninth European marine biology symposium* (ed. H. Barnes), pp. 213–38. Aberdeen University Press, Aberdeen.

Bayne, B. L., Bayne, C. J., Carefoot, T. C. & Thompson, R. J. (1975a). The physiological ecology of *Mytilus californianus* Conrad. 1. Aspects of metabolism and energy balance. *Oecologia*, in press.

Bayne, B. L., Bayne, C. J., Carefoot, T. C. & Thompson, R. J. (1975b). The physiological ecology of *Mytilus californianus* Conrad. 2. Adaptations to exposure to air. *Oecologia*, in press.

Bayne, B. L., Gabbott, P. A. & Widdows, J. (1975). Some effects of stress in the adult on the eggs and larvae of *Mytilus edulis* L. *Journal of the Marine Biological Association of the United Kingdom*, **55**, 675–89.

Bayne, B. L. & Thompson, R. J. (1970). Some physiological consequences of keeping *Mytilus edulis* in the laboratory. *Helgoländer Wissenschaftliche Meeresuntersuchungen*, **20**, 526–52.

Bayne, B. L., Thompson, R. J. & Widdows, J. (1973). Some effects of temperature and food on the rate of oxygen consumption by *Mytilus edulis* L. In: *Effects of temperature on ectothermic organisms* (ed. W. Wieser), pp. 181–93. Springer-Verlag, Berlin.

Bedford, W. B. & Anderson, J. W. (1972). The physiological response of the estuarine clam, *Rangia cuneata* (Gray). I. Osmoregulation. *Physiological Zoology*, **45**, 255–60.

Bedford, J. J. (1971). Osmoregulation in *Melanopsis trifasciata*. III. The nitrogenous compounds. *Comparative Biochemistry and Physiology*, **40A**, 899–910.

Belopolskii, L. O. (1961). *Ecology of the sea colony birds of the Barents Sea.* H. A. Humphrey, London.

Bennett, R. & Nakada, H. I. (1968). Comparative carbohydrate metabolism of marine molluscs. I. The intermediary metabolism of *Mytilus californianus* and *Haliotus rufescens*. *Comparative Biochemistry and Physiology*, **24**, 787–97.

Berg, W. E. (1950). Lytic effects of sperm extracts on the eggs of *Mytilus edulis*. *Biological Bulletin. Marine Biological Laboratory, Woods Hole, Mass.*, **98**, 128–38.

Berg, W. E. & Kutsky, P. B. (1951). Physiological studies of differentiation in *Mytilus edulis*. I. The oxygen uptake of isolated blastomeres and polar lobes. *Biological Bulletin. Marine Biological Laboratory, Woods Hole, Mass.*, **101**, 47–61.

Berg, W. E. & Prescott, D. M. (1958). Physiological studies of differentiation in *Mytilus edulis*. II. Accumulation of phosphate in isolated blastomeres and polar lobes. *Experimental Cell Research*, **14**, 402–7.

Bergmeyer, H. U. (1974). *Method of enzymatic analysis.* (ed. H. U. Bergmeyer assisted by K Gawehn), vols. I–IV. Academic Press, New York and London.

Berner, L. (1935). La reproduction des moules comestibles (*Mytilus edulis* L. et *M. galloprovincialis* Lmk) et leur répartition géographique. *Bulletin de l'Institut Oceanographique, Monaco*, **680**, 1–8.

Beverton, R. J. H. & Holt, S. J. (1957). On the dynamics of exploited fish populations. *Fishery Investigations. Ministry of Agriculture, Fisheries and Food, London*, Ser. II, **19**, 1–533.

Bhattacharya, C. G. (1967). A simple method of resolution of a distribution into Gaussian components. *Biometrics*, **23**, 115–35.

Black, R. E. (1962a). Respiration, electron transport enzymes and Krebs cycle enzymes in early developmental stages of the oyster *Crassostrea virginica*. *Biological Bulletin. Marine Biological Laboratory, Woods Hole, Mass.*, **123**, 58–70.

Black, R. E. (1962b). The concentrations of some enzymes of the citric acid cycle and electron transport system in the large granule fraction of eggs and trochophores of the oyster *Crassostrea virginica*. *Biological Bulletin. Marine Biological Laboratory, Woods Hole, Mass.*, **123**, 71–9.

Blake, J. W. (1960). Oxygen consumption of bivalve prey and their attractiveness to the gastropod *Urosalpinx cinerea*. *Limnology and Oceanography*, **5**, 273–80.

References

Blanco, G. J. (1973). Status and problems of coastal aqua-culture in the Philippines. In: *Coastal aquaculture in the Indo-Pacific region* (ed. T. V. R. Pillay), pp. 60–7. Fishing News (Books) Ltd, London.

Blaschko, H. & Hope, D. B. (1956). The oxidation of L-amino acids by *Mytilus edulis. Biochemical Journal*, **62**, 335.

Blaschko, H. & Milton, S. (1960). Oxidation of 5HT and related compounds by *Mytilus edulis* gill plates. *British Journal of Pharmacology and Chemotherapy*, **15**, 42–6.

Blegvad, H. (1914). Food and conditions of nourishment among the communities of invertebrate animals found on or in the sea bottom in Danish waters. *Report of the Danish Biological Station*, **22**, 41–78.

Blegvad, H. (1929). Mortality among animals of the littoral region in ice winters. *Report of the Danish Biological Station*, **35**, 49–62.

Boëtius, I. (1962). Temperature and growth of *Mytilus edulis* (L.) from the Northern Harbour of Copenhagen (The Sound). *Meddelelser fra Danmarks Fiskeri-og Havundersøgelser, Kobenhavn*, N.S., **3**, 339–46.

Bøhle, B. (1965). Undersøkelser av blåskjell (*Mytilus edulis* L.) i Oslofjorden. *Fiskets Gang*, **51**, 388–94.

Bøhle, B. (1970). Forsøk med dyrking av blåskjell (*Mytilus edulis* L.) ved overføring av yngel til nettingstromper. *Fiskets Gang*, **56**, 267–71.

Bøhle, B. (1971). Settlement of mussel larvae (*Mytilus edulis*) on suspended collectors in Norwegian waters. In: *Fourth European marine biology symposium* (ed. D. J. Crisp), pp. 63–9. Cambridge University Press, London.

Bøhle, B. (1972). Effects of adaptation to reduced salinity on filtration activity and growth of mussels (*Mytilus edulis*). *Journal of Experimental Marine Biology and Ecology*, **10**, 41–9.

Bøhle, B. & Wiborg, K. F. (1967). Forsøk med dyrking av blåskjell. *Fiskets Gang*, **53**, 391–5.

Boje, R. (1965). Die bedeutung von nahrungsfaktoren für das wachstum von *Mytilus edulis* L. in der Kieler Förde und im Nord-Ostsee Kanal. *Kieler Meeresforschungen*, **21**, 81–100.

Bolster, G. C. (1954). The biology and dispersal of *Mytilicola intestinalis* Steuver, a copepod parasite of mussels. *Fishery Investigations. Ministry of Agriculture, Fisheries and Food, London*, Ser. II, **18**, 1–30.

Borisjak, A. (1909). Pelecypoda du plankton de la Mer Noire. *Bulletin Scientifique de la France et de la Belgique*, **42**, 149–81.

Bourcart, C., Lavallard, R. & Lubet, P. (1965). Ultrastructure du spermatozoide de la Moule (*Mytilus perna* von Ihering). *Compte Rendus, Acadamie des Sciences, Paris*, **12**, 5096–9.

Bourcart, C. & Lubet, P. (1965). Cycle sexuel et évolution des réserves chez *Mytilus galloprovincialis* Lmk (Moll. Bivalve). *Rapports et Procès-verbaux des Réunions. Conseil permanent International pour l'Exploration de la Mer*, **18**, 155–8.

Bourcart, C., Lubet, P. & Ranc, H. (1964). Métabolisme des lipides au cours du cycle sexuel chez *Mytilus galloprovincialis* Lmk (Moll. Lamellibr.). *Comptes Rendus des Séances de la Société de Biologie*, **158**, 1638–40.

Bouxin, H. (1931). Influence des variations rapides de la salinité sur la consommation d'oxygène chez *Mytilus edulis* var. *galloprovincialis* (Lmk.). *Bulletin de l'Institut Oçéanographique*, **569**, 1–11.

Bouxin, H. (1956). Observations sur le frai de *Mytilus edulis* var *galloprovincialis* (Lmk) dates precises de frai et facteurs provoquant l'emission de produits génitaux. *Rapport et Procès-verbaux des Réunions. Conseil permanent International pour l'Exploration de la Mer*, **140**, 43–6.

Boyce, R. & Herdman, W. A. (1897). On a green leucocytosis in oysters associated with the presence of copper in the leucocytes. *Proceedings of the Royal Society of London*, **62**, 30–8.

Boyden, C. R. (1972a). The behaviour, survival and respiration of the cockles *Cerastoderma edule* and *C. glaucum* in air. *Journal of the Marine Biological Association of the United Kingdom*, **52**, 661–80.

Boyden, C. R. (1972b). Aerial respiration of the cockle *Cerastoderma edule* in relation to temperature. *Comparative Biochemistry and Physiology*, **43A**, 697–712.

Brand, A. R. (1968). Some adaptations to the burrowing habit in the class Bivalvia. PhD thesis, Hull University.

Brand, A. R. (1972). The mechanism of blood circulation in *Anodonta anatina* (L.) (Bivalvia, Unionidae). *Journal of Experimental Biology*, **56**, 361–79.

Brand, A. R. & Roberts, D. (1973). The cardiac responses of the scallop *Pecten maximus* (L.) to respiratory stress. *Journal of Experimental Marine Biology and Ecology*, **13**, 29–43.

Breese, W. P., Millemann, R. E. & Dimick, R. E. (1963). Stimulation of spawning in the mussels, *Mytilus edulis* Linnaeus and *Mytilus californianus* Conrad, by Kraft Mill effluent. *Biological Bulletin. Marine Biological Laboratory, Woods Hole, Mass.*, **125**, 197–205.

Brennan, R. D., De Wit, C. T., Williams, W. A. & Quattrin, E. V. (1970). The utility of a digital simulation language for ecological modelling. *Oecologia*, **4**, 113–32.

Brett, J. R., Shelbourn, J. E. & Shoop, C. T. (1969). Growth rate and body composition of fingerling sockeye salmon, *Oncorhynchus nerka*, in relation to temperature and ration size. *Journal of the Fisheries Research Board of Canada*, **26**, 2363–93.

Brewer, G. J. (1970). *Introduction to isozyme techniques*. Academic Press, New York and London. 186 pp.

Bricteux-Gregoire, S., Duchâteau-Bosson, G., Jeuniaux, C. & Florkin, M. (1964). Constituants osmotiquement actifs des muscles adducteurs de *Mytilus edulis* adaptée à l'eau de mer ou à l'eau saumâtre. *Archives Internationales de Physiologie et de Biochemie*, **72**, 116–23.

Brienne, H. (1960). Essai de culture de moules sur cordes dans le pertuis Breton. *Science et Pêche*, **83/4**, 4 pp.

Brodtmann, N. V. (1970). Studies on the assimilation of 1,1,1-trichloro-2,2-bis(*p*-chlorophenyl)ethane(DDT) by *Crassostrea virginica* Gmelin. *Bulletin of Environmental Contamination and Toxicology*, **5**, 455–62.

References

Brooks, R. R. & Rumsby, M. G. (1965). The biogeochemistry of trace element uptake by some New Zealand bivalves. *Limnology and Oceanography*, **10**, 521–7.

Brown, B. & Newell, R. C. (1972). The effect of copper and zinc on the metabolism of *Mytilus edulis*. *Marine Biology*, **16**, 108–18.

Bruce, J. R. (1926). The respiratory exchange of the mussel (*Mytilus edulis* L.). *Biochemical Journal*, **20**, 829–46.

Bruce, J. R., Knight, M. & Parke, M. W. (1939). The rearing of oyster larvae on a algal diet. *Journal of the Marine Biological Association of the United kingdom*, **24**, 337–74.

Brunies, A. (1971). Taste of mineral oil in sea mussels. *Archiv für Lebensmittelhygiene*, **22**, 63–4.

Bruyne, C. de (1896). Contribution a l'étude de la phagocytose (I). *Archives de Biologie*, **14**, 161–87.

Bryan, G. W. (1971). The effects of heavy metals (other than mercury) on marine and estuarine organisms. *Proceedings of the Royal Society of London*, Ser. B, **177**, 389–410.

Bubel, A. (1973*a*). An electron-microscope study of periostracum formation in some marine bivalves. II. The cells lining the periostracal groove. *Marine Biology*, **20**, 222–34.

Bubel, A. (1973*b*). An electron-microscope study of periostracum repair in *Mytilus edulis*. *Marine Biology*, **20**, 235–44.

Bucquoy, E., Dautzenberg, P. & Dolfus, G. (1887–98). *Les mollusques marins du Roussillon*, vol. 2, *Pélécypodes*. (Parts 14–16 and separate atlas.) Paris.

Bullock, T. H. (1955). Compensation for temperature in the metabolism and activity of poikilotherms. *Biological Reviews*, **30**, 311–42.

Bullock, T. H. & Horridge, G. A. (1965). *Structure and function in the nervous systems of invertebrates*. W. H. Freeman, San Francisco. 1719 pp.

Butler, P. A. (1965). Reactions of some estuarine molluscs to environmental factors. In: *Biological problems in water pollution, third seminar 1962*, pp. 92–104. USPHS Publication No. 999–WP–25.

Butler, P. A. (1966). The problem of pesticides in estuaries. *American Fisheries Society, Special Publication*, **3**, 110–15.

Butler, P. A. (1971). Influence of pesticides on marine ecosystems. *Proceedings of the Royal Society of London*, Ser. B, **177**, 321–9.

Butler, P. A. (1973). Residues in fish, wildlife and estuaries. *Pesticides Monitoring Journal*, **6**, 238–362.

Cain, T. D. (1973). The combined effects of temperature and salinity on embryos and larvae of the clam *Rangia cuneata*. *Marine Biology*, **21**, 1–6.

Calabrese, A., Collier, R. S., Nelson, D. A. & MacInnes, J. R. (1973). The toxicity of heavy metals to embryos of the american oyster (*Crassostrea virginica*). *Marine Biology*, **18**, 162–66.

Calabrese, A. & Davis, H. C. (1966). The pH tolerance of embryos and larvae of *Mercenaria mercenaria* and *Crassostrea virginica*. *Biological*

Bulletin. Marine Biological Laboratory, Woods Hole, Mass., **131**, 427–36.

Calabrese, A. & Davis, H. C. (1967). The effects of 'soft' detergents on embryos and larvae of the american oyster (*Crassostrea virginica*). *Proceedings of the National Shellfisheries Association*, **57**, 11–16.

Calabrese, A. & Davis, H. C. (1970). Tolerances and requirements of embryos and larvae of bivalve molluscs. *Helgoländer Wissenschaftliche Meeresuntersuchungen*, **20**, 553–64.

Calabrese, A. & Nelson, D. A. (1974). Inhibition of embryonic development of the hard clam *Mercenaria mercenaria* by heavy metals. *Bulletin of Environmental Contamination and Toxicology*, **11**, 92–7.

Calderwood, W. L. (1895). *Mussel culture and bait supply*. Macmillan & Co., London, 121 pp.

Calow, P. & Fletcher, C. R. (1972). A new radiotracer technique involving ^{14}C and ^{51}Cr, for estimating the assimilation efficiencies of aquatic, primary producers. *Oecologia*, **9**, 155–70.

Cameron, J. N. & Wohlschag, D. E. (1969). Respiratory response to experimentally induced anaemia in the pinfish (*Lagodon rhomboides*). *Journal of Experimental Biology*, **50**, 307–17.

Campbell, J. W. (1973). Nitrogen excretion. In: *Comparative animal physiology*, 3rd edition (ed. C. L. Prosser), vol. I, pp. 279–316. W. B. Saunders, Philadelphia.

Campbell, J. W. & Bishop, S. H. (1970). Nitrogen metabolism in molluscs. In: *Comparative biochemistry of nitrogen metabolism* (ed. J. W. Campbell), vol. I, pp. 103–206. Academic Press, New York and London.

Campbell, S. A. (1969). Seasonal cycles in the carotenoid content in *Mytilus edulis*. *Marine Biology*, **4**, 227–32.

Campbell, S. A. (1970). The occurrence and effects of *Mytilicola intestinalis* in *Mytilus edulis*. *Marine Biology*, **5**, 89–95.

Carlson, A. J. (1905). Comparative physiology of the invertebrate heart. *Biological Bulletin, Marine Biological Laboratory, Woods Hole, Mass.*, **8**, 123–68.

Carriker, M. R. (1950). Notes on the killing and preservation of bivalve larvae in fluids. *Nautilus*, **64**, 14–17.

Carriker, M. R. (1961). Interrelation of functional morphology, behaviour, and autecology in early stages of the bivalve *Mercenaria mercenaria*. *Journal of the Elisha Mitchell Scientific Society*, **77**, 168–241.

Carter, G. S. (1924). On the structure and movements of the latero-frontal cilia of the gills of *Mytilus*. *Proceedings of the Royal Society of London*, Ser. B, **96**, 115–22.

Carvajal, R. J. (1969). Fluctuación mensual de las larvas y crecimiento del mejillón *Perna perna* (L.) y las condiciones ambientales de la ensenada de Guatapanare, Edo, Sucre, Venezuala. *Boletin del Instituto Oceanografico de la Universidad de Oriente, Venezuala*, **8**, 13–20.

Caspers, H. (1939). Uber Vorkommen und Metamorphose von *Mytilicola*

References

intestinalis Steuer (Copepoda parasitica) in der südlichen Nordsee. *Zoologischer Anzeiger*, **126**, 161–71.

Cassie, R. M. (1954). Uses of probability paper in analysis of size frequency distributions. *Australian Journal of Marine and Freshwater Research*, **5**, 513–22.

Castagna, M. & Chanley, P. (1973). Salinity tolerance limits of some species of pelecypods from Virginia. *Malacologia*, **12**, 47–96.

Castilla, J. C. (1972). Responses of *Asterias rubens* to bivalve prey in a Y maze. *Marine Biology*, **12**, 222–8.

CEGB (Central Electricity Generating Board) (1965). Marchwood power station marine fouling report, pp. 1–13.

Chanley, P. (1955). Possible causes of growth variations in clam larvae. *Proceedings of the National Shellfisheries Association*, **45**, 84–94.

Chanley, P. (1970). Larval development of the hooked mussel, *Brachidontes recurvus* Rafinesque (Bivalvia: Mytilidae) including a literature review of larval characteristics of the Mytilidae. *Proceedings of the National Shellfisheries Association*, **60**, 86–94.

Chanley, P. & Andrews, J. D. (1971). Aids for identification of bivalve larvae of Virginia. *Malacologia*, **10**, 45–120.

Chanley, P. & Normandin, R. F. (1966). Use of artificial foods for larvae of the hard clam, *Mercenaria mercenaria* (L.). *Proceedings of the National Shellfisheries Association*, **57**, 31–7.

Chanley, P. & van Engel, W. A. (1969). A three-dimensional representation of measurement data. *Veliger*, **12**, 78–83.

Chapat, M., Sany, C., Arnavielhebony, M. & Gravange, G. (1967). Variations des stérides et des triglycérides chez *Mytilus galloprovincialis* Lmk. au cours d'un cycle annuel. *Compte Rendus des Séances de la Société de Biologie*, **161**, 2571–4.

Chappuis, J. G. & Lubet, P. (1966). Etude du débit palléal et de la filtration de l'eau par une methode directe chez *Mytilus edulis* L. et *M. galloprovinciallis* Lmk. (Mollusques Lamellibranches). *Bulletin de la Société linnéenne de Normandie*, **10**, 210–16.

Chen, C. & Awapara, J. (1969). Intracellular distribution of enzymes catalyzing succinate production from glucose in *Rangia* mantle. *Comparative Biochemistry and Physiology*, **30**, 727–37.

Cheng, T. C. (1967). Marine molluscs as hosts for symbioses. In: *Advances in marine biology*, (ed. F. S. Russell), vol. 5. Academic Press, New York and London.

Cheng, T. C. & Rifkin, E. (1970). Cellular reactions in marine molluscs in response to helminth parasitism. In: *Diseases of fishes and shellfishes* (ed. S. F. Snieszko), pp. 443–95. American Fisheries Society Special Publication No. 5, Washington.

Cheng, T. C., Shuster, C. N. & Anderson, A. H. (1966). A comparative study of the susceptibility and response of eight species of marine pelecypods to the trematode *Himasthla quissetensis*. *Transactions of the American Microscopical Society*, **85**, 284–95.

Chew, K. K. & Eisler, R. E. (1958). A preliminary study of the feeding habits of the Japanese oyster drill *Ocenebra japonica. Journal of the Fisheries Research Board of Canada*, **15**, 529–35.

Chipman, W. A. (1972). Ionizing radiation: animals. In: *Marine ecology* (ed. O. Kinne), part 3, pp. 1621–43. Wiley-Interscience, New York.

Chipperfield, P. N. J. (1953). Observations on the breeding and settlement of *Mytilus edulis* (L) in British waters. *Journal of the Marine Biological Association of the United Kingdom*, **32**, 449–76.

Chuecas, L. & Riley, J. P. (1969). Component fatty acids of the total lipids of some marine phytoplankton. *Journal of the Marine Biological Association of the United Kingdom*, **49**, 97–116.

Clapp, W. F. (1950). Some biological fundamentals of marine fouling. *Transactions of the American Society of Mechanical Engineers*, **72**, 101–7.

Clarke, C. D. (1947). Poisoning and recovery in barnacles and mussels. *Biological Bulletin. Marine Biological Laboratory, Woods Hole, Mass.*, **92**, 73–91.

Clarke, R. B. (1965). Endocrinology and the reproductive biology of polychaetes. *Oceanography and Marine Biology Annual Review*, **3**, 211–55.

Clasing, T. (1923). Beitrag zur Kentnis der Nervensystems und der Sinnesorgane der Mytiliden. *Jenaische Zeitschrift für Medizin und Naturwissenschaft*, **59**, 261–310.

Cleland, K. W. (1950). Respiration and cell division in developing oyster eggs. *Proceedings of the Linnean Society of New South Wales*, **75**, 282–95.

Coe, W. R. (1932). Season of attachment and rate of growth of sedentary marine organisms at the pier of the Scripps Institute of Oceanography, La Jolla, California. *Bulletin of the Scripps Institute of Oceanography*, Technical Series **3**, 37–86.

Coe, W. R. (1945). Nutrition and growth of the Californian bay mussel (*Mytilus edulis diegensis*). *Journal of Experimental Zoology*, **99**, 1–14.

Coe, W. R. (1946). A resurgent population of the Californian bay mussel *Mytilus edulis diegensis. Journal of Morphology*, **78**, 85–103.

Coe, W. R. & Fox, D. L. (1942). Biology of the Californian sea mussel *Mytilus californianus*. I. Influence of temperature, food supply, sex and age on the rate of growth. *Journal of Experimental Zoology*, **90**, 1–30.

Coe, W. R. & Fox, D. L. (1944). Biology of the Californian sea mussel *Mytilus californianus*. III. Environmental conditions and rate of growth. *Biological Bulletin. Marine Biological Laboratory, Woods Hole, Mass.*, **87**, 58–72.

Coffey, B. T. (1972). Aquaculture sustains L.I. Sound shellfishery. *National Fisherman*, **52**(10), 12A.

Cole, H. A. (1937). Experiments on the breeding of oysters (*Ostrea edulis*) in tanks, with special reference to the food of the larvae and spat. *Fisheries Investigations. Ministry of Agriculture, Fisheries and Food, London.* Ser. II, 25(4).

References

Cole, H. A. (1938). The fate of larval organs in the metamorphosis of *Ostrea edulis. Journal of the Marine Biological Association of the United Kingdom*, **22**, 469–84.

Cole, H. A. (1956). Benthos and the shellfish of commerce. In: *Sea fisheries.* (ed. A. Graham), pp. 139–206.

Cole, H. A. (1968). The scientific cultivation of sea fish and shell-fish. *Fishing News International*, **7**(6), 20–8.

Cole, H. A. & Savage, R. E. (1951). The effect of the parasitic copepod *Mytilicola intestinalis* (Steuer) upon the condition of mussels. *Parasitology*, **41**, 156–61.

Coleman, N. (1973*a*). The oxygen consumption of *Mytilus edulis* in air. *Comparative Biochemistry and Physiology*, **45A**, 393–402.

Coleman, N. (1973*b*). Water loss from aerially exposed mussels. *Journal of Experimental Marine Biology and Ecology*, **12**, 145–55.

Coleman, N. (1974). The heart rate and activity of bivalve molluscs in their natural habitats. *Oceanography and Marine Biology Annual Review*, **12**, 301–13.

Coleman, N. & Trueman, E. R. (1971). The effect of aerial exposure on the activity of the mussels *Mytilus edulis* L. and *Modiolus modiolus* (L.) *Journal of Experimental Marine Biology and Ecology*, **7**, 295–304.

Collip, J. B. (1921). A further study of the respiratory processes in *Mya arenaria* and other marine mollusca. *Journal of Biological Chemistry*, **49**, 297–310.

Collyer, D. M. (1957). Viability and glycogen reserves in the newly liberated larvae of *Ostrea edulis* L. *Journal of the Marine Biological Association of the United Kingdom*, **36**, 335–7.

Comfort, A. (1957). The duration of life in molluscs. *Proceedings of the Malacological Society of London*, **52**, 219–41.

Connell, J. H. (1961*a*). The influence of interspecific competition and other factors on the distribution of the barnacle *Chthamalus stellatus. Ecology*, **42**, 710–23.

Connell, J. H. (1961*b*). Effects of competition, predation by *Thais lapillus* and other factors on natural populations of the barnacle *Balanus balanoides. Ecological Monographs*, **31**, 61–104.

Connell, J. H. (1970). A predator–prey system in the marine intertidal region. I. *Balanus glandula* and several predatory species of *Thais. Ecological Monographs*, **40**, 49–78.

Connell, J. H. (1972). Community interactions on marine rocky intertidal shores. *Annual Review of Ecology and Systematics*, **31**, 169–92.

Conover, R. J. (1966). Assimilation of organic matter by zooplankton. *Limnology and Oceanography*, **11**, 338–54.

Costanzo, G. (1966). Some histochemical aspects of the male and female gonads of *M. galloprovincialis* L. *Atti dell'Accademia gioenia di scienze naturali*, **18**, 93–8.

Costlow, J. D., Bookhout, C. G. & Monroe, R. (1960). The effect of salinity and temperature on larval development of *Sesarma cinereum* (Bosc) reared in the laboratory. *Biological Bulletin. Marine Biological Laboratory, Woods Hole, Mass.*, **118**, 183–202.

Costlow, J. D., Bookhout, C. G. & Monroe, R. (1962). Salinity-temperature effects on the larval development of the crab *Panopeus herbstii* Milne-Edwards, reared in the laboratory. *Physiological Zoology*, **35**, 79–93.

Coughlan, J. (1969). The estimation of filtering rate from the clearance of suspensions. *Marine Biology*, **2**, 356–8.

Coughlan, J. & Ansell, A. D. (1964). A direct method for determining the pumping rate of siphonate bivalves. *Journal du Conseil. Conseil permanent International pour l'Exploration de la Mer*, **29**, 205–13.

Coulthard, H. S. (1929). Growth of the sea mussel. *Contributions to Canadian Biology and Fisheries*, **4**, 123–36.

Courtright, R. C., Breese, W. P. & Krueger, H. (1971). Formulation of a synthetic seawater for bioassays with *Mytilus edulis* embryos. *Water Research*, **5**, 877–88.

Cowell, E. B., Baker, J. M. & Crapp, G. B. (1972). The biological effects of oil pollution and oil cleaning materials on littoral communities, including salt marshes. In: *Marine pollution and sea life*. Fishing News (Books) Ltd, London.

Cragg, S. M. & Gruffydd, L. D. (1975). The swimming behaviour and the pressure responses of *Ostrea edulis* L. veliconcha larvae. In: *Proceedings of the ninth European marine biology symposium* (ed. H. Barnes), pp. 43–57. Aberdeen University Press, Aberdeen.

Craig, G. Y. & Hallam, A. (1963). Size frequency and growth ring analyses of *Mytilus edulis* and *Cardium edule* and their palaeological significance. *Mémoires de la Société géologique de France. Paléontologie*, **6**, 731–50.

Cranfield, H. J. (1973*a*). A study of the morphology, ultrastructure and histochemistry of the foot of the pediveliger of *Ostrea edulis*. *Marine Biology*, **22**, 187–202.

Cranfield, H. J. (1973*b*). Observations on the behaviour of the pediveliger of *Ostrea edulis* during attachment and cementing. *Marine Biology*, **22**, 203–9.

Cranfield, H. J. (1973*c*). Observations on the functions of the glands of the foot of the pediveliger of *Ostrea edulis* during settlement. *Marine Biology*, **22**, 211–23.

Crapp, G. B. (1971*a*). Field experiments with oil and emulsifiers. In: *The ecological effects of oil pollution on littoral communities* (ed. E. B. Cowell), pp. 114–28. Elsevier, Amsterdam.

Crapp, G. B. (1971*b*). Laboratory experiments with emulsifiers. In: *The ecological effects of oil pollution on littoral communities* (ed. E. B. Cowell), pp. 129–49. Elsevier, Amsterdam.

Crapp, G. B. (1971*c*). The biological consequences of emulsifier cleaning. In: *The ecological effects of oil pollution on littoral communities* (ed. E. B. Cowell), pp. 150–68. Elsevier, Amsterdam.

Crisp, D. J. (1971). Energy flow measurements. In: *Methods for the study of marine benthos, IBP Handbook No. 16* (ed. N. A. Holme and A. D. McIntyre), pp. 197–279. Blackwell Scientific Publications, Oxford.

References

Crisp, D. J. (1974). Factors influencing the settlement of marine inverte-brate larvae. In: *Chemoreception in marine organisms* (ed. P. T. Grant and A. M. Mackie), pp. 177–265. Academic Press, New York and London.

Crisp, D. J. (1975). The role of the pelagic larva. In: *Perspectives in experimental biology* (ed. P. Spencer Davies), vol. I. Pergamon Press, Oxford. (In press.)

Crisp, M. (1973). Fine structure of some prosobranch osphradia. *Marine Biology*, **22**, 231–40.

Cronin, L. E. (1967). The role of man in estuarine processes. In: *Estuaries* (ed. G. H. Lauff), pp. 667–89. American Association for the Advance-ment of Science, Washington.

Crosby, N. D. & Reid, R. G. B. (1971). Relationships between food, phylogeny and cellulose digestion in the Bivalvia. *Canadian Journal of Zoology*, **49**, 617–22.

Crowley, M. (1970). *The edible mussel – Mytilus edulis.* Leaflet 15, Department of Agriculture and Fisheries, Fisheries Division, Dublin.

Culkin, F. & Morris, R. J. (1970). The fatty acid composition of two marine filter-feeders in relation to a phytoplankton diet. *Deep Sea Research*, **17**, 861–5.

Culliney, J. L. (1971). Laboratory rearing of the larvae of the mahogany date mussel, *Lithophaga bisulcata. Bulletin of Marine Science of the Gulf and Caribbean, Miami*, **21**, 591–602.

Dakin, W. J. (1910). The visceral ganglion of *Pecten*, with some notes on the physiology of the nervous system, and an enquiry into the innervation of the osphradium in the Lamellibranchiata. *Mitteilungen aus der Zoologischen Station zu Neapel*, **20**, 1–40.

Dame, R. F. (1972). The ecological energies of growth, respiration and assimilation in the inter-tidal American oyster, *Crassostrea virginica. Marine Biology*, **17**, 243–50.

Dan, J. C. (1962). The vitelline coat of the *Mytilus* egg. 1. Normal structure and effect of acrosomal lysin. *Biological Bulletin. Marine Biological Laboratory, Woods Hole, Mass.*, **123**, 531–41.

Dan, J. C., Kakizawa, Y., Kushida, H. & Fujita, K. (1972). Acrosomal triggers. *Experimental Cell Research*, **72**, 60–8.

Daniel, R. J. (1921). Seasonal changes in the chemical composition of the mussel (*Mytilus edulis*). *Report of the Lancashire Sea Fisheries laboratory*, **30**, 74–84, 205–21.

Daniel, R. J. (1922). Seasonal changes in the chemical composition of the mussel (*Mytilus edulis*) *Report of the Lancashire Sea Fisheries Laboratory*, **31**, 27–50.

Dare, P. J. (1966). The breeding and wintering populations of the oyster-catcher (*Haematopus ostralegus* L.) in the British Isles. *Fishery Investigations. Ministry of Agriculture, Fisheries and Food, London*, Ser. II, **25**, 1–69.

Dare, P. J. (1969). The settlement, growth and survival of mussels, *Mytilus edulis* L. in Morecambe Bay, England. International Council for the Exploration of the Sea, CM 1969/K: 18, pp. 1–8 (Mimeo.)

References

Dare, P. J. (1971). Preliminary studies on the utilisation of the resources of spat mussels, *Mytilus edulis* L. occurring in Morecambe Bay, England. International Council for the Exploration of the Sea, CM 1971/K: 11, pp. 1–6 (Mimeo.)

Dare, P. J. (1973*a*). Seasonal changes in meat condition of sublittoral mussels (*Mytilus edulis* L.) in the Conwy fishery, North Wales. International Council for Exploration of the Sea; Shellfish and Benthos Committee, C.M./K: 31, 6 pp. (Mimeo.)

Dare, P. J. (1973*b*). The stocks of young mussels in Morecambe Bay, Lancashire. *Ministry of Agriculture Fisheries and Food, Shellfish Information Leaflet*, **28**, 1–14.

Dare, P. J. (1975). Settlement, growth and production of the mussel *Mytilus edulis* L. in Morecambe Bay. *Fishery Investigations. Ministry of Agriculture, Fisheries and Food, London*, Ser. II, in press.

Davids, C. (1964). The influence of suspensions of micro-organisms of different concentrations on the pumping and retention of food by the mussel (*Mytilus edulis* L.). *Netherlands Journal of Sea Research*, **2**, 233–49.

Davies, C. C. (1972). The effects of pollutants in the reproduction of marine organisms. In: *Marine pollution and sea life*, pp. 305–11. Fishing News (Books) Ltd, London.

Davies, G. (1969). Observations on the growth of *Mytilus edulis* in the Menai Straits in the period 1962–68. International Council for the Exploration of the Sea, CM 1969/K: 39, pp. 1–5. (Mimeo.)

Davies, G. (1970). Mussels as a world food resource. In: *Proceedings of the Symposium on Mollusca, Marine Biological Association of India*, pp. 873–84.

Davies, G. (1974). A method for monitoring the spatfall of mussels (*Mytilus edulis* L.). *Journal du Conseil. Conseil International pour Exploration de la Mer*, **36**, 27–34.

Davies, P. S. (1966). Physiological ecology of *Patella*. 1. The effect of body size and temperature on metabolic rate. *Journal of the Marine Biological Association of the United Kingdom*, **46**, 647–58.

Davies, P. S. (1969). Physiological ecology of *Patella*. III. Desiccation effects. *Journal of the Marine Biological Association of the United Kingdom*, **49**, 291–304.

Davis, D. S. & White, W. R. (1966). Molluscs from a power station culvert. *Journal of Conchology*, **26**, 33–8.

Davis, H. C. (1953). On food and feeding of larvae of the American oyster, *Crassostrea virginica. Biological Bulletin. Marine Biological Laboratory, Woods Hole, Mass.*, **104**, 334–50.

Davis, H. C. (1958). Survival and growth of clam and oyster larvae at different salinities. *Biological Bulletin. Marine Biological Laboratory, Woods Hole, Mass.*, **114**, 296–307.

Davis, H. C. (1961). Effects of some pesticides on eggs and larvae of oysters (*Crassostrea virginica*) and clams (*Venus mercenaria*). *Commercial Fisheries Review*, **23**, 8–23.

427

References

Davis, H. C. (1969). Shellfish hatcheries – present and future. *Transactions of the American Fisheries Society*, **98**, 743–50.

Davis, H. C. & Guillard, R. R. (1958). Relative value of ten genera of micro-organisms as foods for oyster and clam larvae. *Fishery Bulletin of the United States Fish and Wildlife Service*, **58**, 293–304.

Davis, H. C. & Hidu, H. (1969). Effects of pesticides on embryonic development of clams and oysters and on survival and growth of the larvae. *Fishery Bulletin of the United States Fish and Wildlife Service*, **67**, 393–404.

Dayton, P. K. (1971). Competition, disturbance and community organisation: the provision and subsequent utilisation of space in a rocky intertidal community. *Ecological Monographs*, **41**, 351–89.

De Block, J. W. & Geelen, H. J. (1958). The substratum required for the settling of mussels (*Mytilus edulis* L.) *Archives Néerlandaises*, Jubilee Volume, 446–60.

De Ligny, W. (1969). Serological and biochemical studies on fish populations. *Oceanography and Marine Biology Annual Review*, **7**, 411–513.

De Zwaan, A. (1971). PhD thesis, University of Utrecht.

De Zwaan, A. (1972). Pyruvate kinase in muscle extracts of the sea mussel *Mytilus edulis* L. *Comparative Biochemistry and Physiology*, **42B**, 7–14.

De Zwaan, A. & De Bont, A. M. Th. (1975). Phosphoenolpyruvate carboxykinase from adductor muscle tissue of the sea mussel *Mytilus edulis* L. *Journal of Comparative Physiology*, **96**, 85–94.

De Zwaan, A., De Bont, A. M. Th. & Kluytmans, J. H. F. M. (1975). Metabolic adaptations on the aerobic–anaerobic transition in the sea mussel *Mytilus edulis* L. In: *Proceedings of the ninth European marine biology symposium* (ed. H. Barnes), pp. 121–38. Aberdeen University Press, Aberdeen.

De Zwaan, A. & Holwerda, D. A. (1972). The effect of phosphoenolpyruvate, fructose 1,6-diphosphate and pH on allosteric pyruvate kinase in muscle tissue of the bivalve *Mytilus edulis* L. *Biochimica et Biophysica Acta*, **276**, 430–3.

De Zwaan, A. & Van Marrewijk, W. J. A. (1973a). Anaerobic glucose degradation in the sea mussel *Mytilus edulis* L. *Comparative Biochemistry and Physiology*, **44B**, 429–39.

De Zwaan, A. & Van Marrewijk, W. J. A. (1973b). Intracellular localization of pyruvate carboxylase, phosphoenolpyruvate carboxykinase and ' malic enzyme ' and the absence of glyoxylate cycle enzymes in the sea mussel (*Mytilus edulis* L.). *Comparative Biochemistry and Physiology*, **44B**, 1057–66.

De Zwaan, A., Van Marrewijk, W. J. A. & Holwerda, D. A. (1973). Anaerobic carbohydrate metabolism in the sea mussel *Mytilus edulis* L. *Netherlands Journal of Zoology*, **23**, 225–8.

De Zwaan, A. & Zandee, D. I. (1972a). Body distribution and seasonal changes in the glycogen content of the common sea mussel *Mytilus edulis*. *Comparative Biochemistry and Physiology*, **43A**, 53–8.

De Zwaan, A. & Zandee, D. I. (1972*b*). The utilization of glycogen and accumulation of some intermediates during anaerobiosis in *Mytilus edulis* L. *Comparative Biochemistry and Physiology*, **43B**, 47–54.

Dehnel, P. A. (1955). Rates of growth of gastopods as a function of latitude. *Physiological Zoology*, **28**, 115–44.

Dehnel, P. A. (1956). Growth rates in latitudinally and vertically separated populations of *Mytilus californianus*. *Biological Bulletin. Marine Biological Laboratory, Woods Hole, Mass.*, **110**, 43–53.

Dejours, P. (1972). Comparison of gas transport by convection among animals. *Respiration Physiology*, **14**, 96–104.

Dejours, P., Garey, W. F. & Rahn, H. (1970). Comparison of ventilatory and circulatory flow rates between animals in various physiological conditions. *Respiration Physiology*, **9**, 108–17.

Del Vecchio, V., Valori, P., Alasia, A. M. & Gualdi, G. (1962). La determinazione dell arsenico nei molluschi (*Mytilus* Linn.). *Igiene e Sanita Publica*, **18**, 18–30.

Delhaye, W. & Cornet, D. (1975). Contribution to the study of the effect of copper on *Mytilus edulis* during the reproductive period. *Comparative Biochemistry and Physiology*, **50A**, 511–18.

Delsman, H. C. (1910). De voortplanting van der mossel. *Verslagen omtrent den Staat der nederlansche Zeevisscherijen*, extra bijlage, **3**, 75–88.

Dexter, R. W. (1947). The marine communities of a tidal inlet. *Ecological Monographs*, **17**, 262–94.

Dixon, G. H. & Kornberg, H. L. (1959). Assay methods for key enzymes of the glyoxylate cycle. *Biochemical Journal*, **72**, 3P.

Dodd, J. R. (1969). Effect of light on rate of growth of bivalves. *Nature, London*, **224**, 617–18.

Dodge, H. (1952). A historical review of the mollusks of Linnaeus. I. The classes Loricata and Pelecypoda. *Bulletin of the American Museum of Natural History*, **100**, 1–264.

Dodgson, R. W. (1928). Report on mussel purification. *Fishery Investigations. Ministry of Agriculture, Fisheries and Food, London*, Ser. II, **10**, 1–498.

Dral, A. D. G. (1967). The movements of the latero-frontal cilia and the mechanism of particle retention in the mussel (*Mytilus edulis* L.). *Netherlands Journal of Sea Research*, **3**, 391–422.

Dral, A. D. G. (1968). On the feeding of mussels (*Mytilus edulis* L.) in concentrated food suspensions. *Netherlands Journal of Zoology*, **18**, 440–1.

Drinkwaard, A. C. (1972*a*). Het mosselproefstation op Texel operationeel. *Visserij*, **25**, 61–9.

Drinkwaard, A. C. (1972*b*). Het mosselproefstation op Texel operationeel. *Visserij*, **25**, 216–238.

Drinnan, R. E. (1958). The winter feeding of the oystercatcher (*Haematopus ostralegus*) on the edible mussel (*Mytilus edulis*) in the Conway estuary. *Fishery Investigations. Ministry of Agriculture, Fisheries and Food, London*, Ser. II, **22**, 1–15.

References

Drinnan, R. E. (1964). An apparatus for recording the water-pumping behaviour of lamellibranchs. *Netherlands Journal of Sea Research*, **2**, 223–32.

Duchâteau, D. & Florkin, M. (1956). Systèmes intracellulaires d'acides aminés libres et osmorégulation des Crustacés. *Journal de Physiologie, Paris*, **8**, 520.

Duff, M. F. (1967). The uptake of enteroviruses by the New Zealand marine blue mussel, *Mytilus edulis aoteanus*. *American Journal of Epidemiology*, **85**, 486–93.

Dunbar, M. J. (1947). Note on the delimitation of the Arctic and Subarctic zones. *Canadian Field Naturalist*, **61**, 12–14.

Dunthorn, A. A. (1971). The predation of cultivated mussels by eiders. *Bird Study*, **18**, 107–12.

Du Paul, W. D. & Webb, K. L. (1970). The effect of temperature on salinity-induced changes in the free amino acid pool of *Mya arenaria*. *Comparative Biochemistry and Physiology*, **32**, 785–801.

Durfoot, M. (1973). Sur la formation des lamelles annelées dans les ovocytes des *Mytilus edulis* L. *Comptes Rendus, Academie des Sciences, Paris*, **276**, 3175–7.

Ebert, T. A. (1973). Estimating growth and mortality rates from size data. *Oecologia*, **11**, 281–98.

Eble, A. F. (1969). A histochemical demonstration of glycogen, glycogen phosphorylase and branching enzyme in the American oyster. *Proceedings of the National Shellfisheries Association*, **59**, 27–34.

Ebling, F. J., Kitching, J. A., Muntz, L. & Taylor, C. M. (1964). The ecology of Lough Ine. 13. Experimental observations of the destruction of *Mytilus edulis* and *Nucella lapillus* by crabs. *Journal of Animal Ecology*, **33**, 73–82.

Edwards, D. B. (1973). Experiments on the survival and yield of relaid seed mussels (*Mytilus edulis*). International Council for the Exploration of the Sea, CM 1973/ K: 35, pp. 1–6. (Mimeo.)

Edwards, E. (1968). *A review of mussel production by raft culture*. Irish Sea Fisheries Board, Resource Record Paper, 7 pp.

Ellenby, C. (1947). A copepod parasite of the mussel new to the British fauna. *Nature, London*, **159**, 645–46.

Ellis, D. V. (1955). Some observations on the shore fauna of Baffin Island. *Arctic*, **8**, 224–36.

Elmhirst, R. (1923). Notes on the breeding and growth of marine animals in the Clyde Sea area. *Report of the Scottish Marine Biological Station for 1922*, 19–43.

Emerson, D. N. (1969). Influence of salinity on ammonia excretion rates and tissue constituents of euryhaline invertebrates. *Comparative Biochemistry and Physiology*, **29**, 1115–33.

Engel, R. H. & Neat, M. J. (1970). Glycolytic and gluconeogenic enzymes in the quahog, *Mercenaria mercenaria*. *Comparative Biochemistry and Physiology*, **37**, 397–403.

Engel, R. H., Neat, M. J. & Hillman, R. E. (1972). Sublethal, chronic

430

effects of DDT and Lindane on glycolytic and gluconeogenic enzymes of the quahog, *Mercenaria mercenaria*. In: *Marine pollution and sea Life*, pp. 257–9. Fishing News (Books) Ltd, London.

Engle, J. B. & Loosanoff, V. L. (1944). On season of attachment of larvae of *Mytilus edulis* L. *Ecology*, **25**, 433–40.

Erman, P. (1961). Atmungsmessungen an Geweben und Gewebehomogenaten Gewebehomogenaten der Miesmuschel (*Mytilus edulis* L.) aus Brack- und Meerwasser. *Kieler Meeresforschungen*, **17**, 176–89.

FAO (1972). *Yearbook of fishery statistics, 1971*. Food and Agricultural organisation, Rome. 558 pp.

Fagerlund, U. M. M. & Idler, D. R. (1960). Marine sterols. VI. Sterol biosynthesis in molluscs and echinoderms. *Canadian Journal of Biochemistry and Physiology*, **38**, 997–1002.

Fahien, L. A. & Smith, S. E. (1974). The enzyme-enzyme complex of transaminase and glutamate dehydrogenase. *Journal of Biological Chemistry*, **249**, 2696–703.

Farley, C. A. (1969). Sarcomatoid proliferative disease in a wild population of Blue mussels (*Mytilus edulis*). *Journal of the National Cancer Institute*, **4**, 509–16.

Farley, C. A. & Sparks, A. K. (1970). Proliferative diseases of haemocytes, endothelial cells, connective tissue cells in molluscs. *Bibliotheca Haematologica*, **36**, 610–17.

Favretto, L. (1968). Aspetti merceologici della mitilicoltura nel Golfo di Trieste. *Bollettino della Società Adriatica di Scienze naturali in Trieste*, **56**, 243–61.

Feare, C. J. (1966). The winter feeding of the purple sandpiper. *British Birds*, **59**, 165–79.

Feare, C. J. (1971). Predation of limpets and dogwhelks by oystercatchers. *Bird Study*, **18**, 121–9.

Feare, C. J. (1972). The adaptive significance of aggregation behaviour in the dogwhelk *Nucella lapillus* (L.). *Oecologia*, **7**, 117–26.

Feder, H. M. (1970). Growth and predation by the sea star *Pisaster ochraceus* (Brandt) in Monterey Bay California. *Ophelia*, **8**, 161–85.

Fenchel, T. (1972). Aspects of decomposer food chains in marine benthos. In: *Verhandlungsbericht der Deutschen Zoologischen Gesellschaft*, 65. Gustav Fischer Verlag, Berlin.

Feng, S. Y. (1965). Pinocytosis of proteins by oyster leucocytes. *Biological Bulletin. Marine Biological Laboratory, Woods Hole, Mass.*, **129**, 95–105.

Feng, S. Y. (1967). Responses of molluscs to foreign bodies, with special reference to the oyster. *Federation Proceedings. Federation of American Societies for Experimental Biology*, **26**, 1685–92.

Feng, S. Y., Feng, J. S., Burke, C. N. & Khairallah, L. H. (1971). Light and electron microscopy of the leucocytes of *Crassostrea virginica* (Mollusca: Pelecypoda). *Zeitschrift für Zellforschung und mikroskopische Anatomie*, **120**, 222–45.

Feng, S. Y., Khairallah, E. A. & Canzonier, W. J. (1970). Hemolymph-

References

free amino acids and related nitrogenous compounds of *Crassostrea virginica* infected with *Bucephalus* sp. and *Minchinia nelsoni*. *Comparative Biochemistry and Physiology*, **34**, 547–56.

Feng, S. Y. & Van Winkle, W. (1975). The effect of temperature and salinity on the heart beat of *Crassostrea virginica*. *Comparative Biochemistry and Physiology*, **50A**, 473–6.

Field, I. A. (1909). The food value of the sea mussel. *Bulletin of the United States Bureau of Fisheries, Washington*, **29**, 85–128.

Field, I. A. (1916). A community of sea mussels. *The American Museum Journal*, **16**, 357–66.

Field, I. A. (1922). Biology and economic value of the sea mussel *Mytilus edulis*. *Bulletin of the United States Bureau of Fisheries, Washington*, **38**, 127–259.

Fischer, E. (1929). Sur la distribution et les conditions de vie de *Mytilus edulis* L. sur les côtes de la Manche. *Journal de Conchyliologie*, **73**, 109–18.

Fischer-Piette, E. (1935). Histoire d'une moulière. *Bulletin biologique de la France et de la Belgique*, **69**, 153–77.

Fischer-Piette, E. (1939). Sur la croissance et la longévité de *Patella vulgata* L. en fonction du milieu. *Journal de Conchyliologie*, **83**, 303–10.

Fish, C. J. & Johnson, M. W. (1937). The biology of the zooplankton in the Bay of Fundy and Gulf of Maine with special reference to production and distribution. *Journal of the Biology Board of Canada*, **3**, 189–322.

Fisher, L. (1969). An immunological study of pelecypod taxonomy. *Veliger*, **11**, 434–8.

Fleming, C. A. (1959). Notes on New Zealand recent and tertiary mussels (Mytilidae). *Transactions of the Royal Society of New Zealand*, **87**, 165–78.

Florkin, M. (1962). La régulation isosmotique intracellulaire chez les invertébrés marins euryhalins. *Bulletin de l'Académie royale de Belgique, Classe des Sciences*, **48**, 687–94.

Florkin, M. (1966). Nitrogen metabolism. In: *Physiology of Mollusca* (ed. K. M. Wilbur and C. M. Yonge), vol. II, pp. 309–51. Academic Press, New York and London.

Florkin, M. & Bricteux-Grégoire, S. (1972). Nitrogen metabolism in molluscs. In: *Chemical zoology* (ed. M. Florkin and B. T. Scheer), vol. VII, pp. 301–48. Academic Press, New York and London.

Florkin, M. & Duchâteau, G. (1943). Les formes du système enzymatique de l'uricolyse et l'évolution du catabolism purique chez les animaux. *Archives Internationales de Physiologie*, **53**, 267–307.

Florkin, M. & Duchâteau, G. (1948). Sur l'osmoregulation de l'anodonte (*Anodonta cygnea* L.). *Physiologia Comparata et Oecologia*, **1**, 29–45.

Florkin, M. & Schoffeniels, E. (1965). Euryhalinity and the concept of physiological radiation. In: *Studies in comparative biochemistry* (ed. K. A. Munday), pp. 6–40. Pergamon Press, Oxford.

References

Florkin, M. & Schoffeniels, E. (1969). *Molecular approaches to ecology.* Academic Press, New York and London. 203 pp.

Flügel, H. & Schlieper, C. (1962). Der Einfluss physikalischer und chemischer Faktoren auf die cilienaktivität und Pumprate der Miesmuschel *Mytilus edulis* L. *Kieler Meeresforschungen,* **18**, 51–66.

Ford, W. C. L. & Candy, D. J. (1972). The regulation of glycolysis in perfused locust flight muscle. *Biochemical Journal,* **130**, 1101–12.

Foret-Montardo, P. (1970). Etude de l'action des produits de base entrant dans la composition des detergents issus de la petroleochimie vis-a-vis de quelques invertebres benthiques marin. *Tethys,* **2**, 567–614.

Forster, J. R. M. & Gabbott, P. A. (1971). The assimilation of nutrients from compounded diets by the prawns *Palaemon serratus* and *Pandalus platyceros. Journal of the Marine Biological Association of the United Kingdom,* **51**, 943–61.

Fossato, H. & Siviero, J. (1974). Oil pollution monitoring in the lagoon of Venice using the mussel *Mytilus galloprovincialis. Marine Biology,* **25**, 1–6.

Foster-Smith, R. L. (1974). A comparative study of the feeding mechanisms of *Mytilus edulis* L., *Cerastoderma edule* (L.), and *Venerupis pullastra* (Montagu) (Mollusca: Bivalvia). PhD thesis, University of Newcastle-on-Tyne. 153 pp.

Foster-Smith, R. L. (1975a). The effect of concentration of suspension on the filtration rates and pseudofaecal production for *Mytilus edulis* L., *Cerastoderma edule* (L.) and *Venerupis pullastra* (Montagu). *Journal of Experimental Marine Biology and Ecology,* **17**, 1–22.

Foster-Smith, R. L. (1975b). The effect of concentration of suspension and inert material on the assimilation of algae by three bivalves. *Journal of the Marine Biological Association of the United Kingdom,* **55**, 411–18.

Fox, D. L. (1936). The habitat and food of the California sea mussel. *Bulletin of the Scripps Institute of Oceanography,* **4**, 1–64.

Fox, D. L. & Coe, W. R. (1943). Biology of the Californian sea mussel (*Mytilus californianus*). II. Nutrition, metabolism, growth and calcium deposition. *Journal of Experimental Zoology,* **93**, 205–49.

Fox, D. L. & Marks, G. W. (1936). The habitat and food of the California sea mussel; the digestive enzymes. *Bulletin of the Scripps Institute of Oceanography, Technical Series,* **4**, 29–47.

Fox, D. L., Sverdrup, H. U. & Cunningham, J. P. (1937). The rate of water propulsion by the California mussel. *Biological Bulletin. Marine Biological Laboratory, Woods Hole, Mass.,* **72**, 417–38.

Fox, H. M. (1924). Lunar periodicity in reproduction. *Proceedings of the Royal Society of London,* Ser. B, **95**, 523–50.

Fraga, F. (1956). Variación estacional de la composición quimica del mejillón (*Mytilus edulis*). *Investigación Pesquera,* **4**, 109–25.

Fraga, F. & Vives, F. (1960). Retención de partículas orgánicas por el mejillón en los viveros flotantes *Reunión Sobre productividad y Pesquerias Barcelona,* **4**, 71–3.

References

Franzen, A. (1955). Comparative morphological investigations into the spermiogenesis among Mollusca. *Zoologiska bidrag från Uppsala*, **31**, 355–482.

Fraser, J. H. (1938). The fauna of fixed and floating structures in the Mersey estuary and Liverpool Bay. *Proceedings and Transactions of the Liverpool Biological Society*, **51**, 1–21.

Fretter, V. (1967). The prosobranch veliger. *Proceedings of the Malacological Society, London*, **37**, 357–66.

Fretter, V. & Graham, A. (1964). Reproduction. In: *Physiology of Mollusca* (ed. K. M. Wilbur and C. M. Yonge), pp. 127–64. Academic Press, New York and London.

Fry, F. E. J. (1947). *Effects of the environment on animal activity. University of Toronto Studies, Biological Series*, No. 55. Publications of the Ontario Fisheries Research Laboratory, The University of Toronto Press.

Fry, F. E. J. (1957). The aquatic respiration of fish. In: *The physiology of fishes* (ed. M. E. Brown), vol. 1, pp. 1–63. Academic Press, New York and London.

Fry, F. E. J. (1958). Temperature compensation. *Annual Review of Physiology*, **20**, 207–24.

Fullarton, J. H. & Scott, T. (1889). Mussel farming at Montrose. *Scientific Investigations. Fishery Board of Scotland. 7th Annual Report*, pp. 327–41.

Fuller, J. L. (1946). Season of attachment and growth of sedentary marine organisms at Lamoine, Maine. *Ecology*, **27**, 150–8.

Gaarder, T. & Eliassen, E. (1954). The energy metabolism of *Ostrea edulis. Universitetet i Bergen. Arbok*, **3**, 1–7.

Gabbott, P. A. (1975). Storage cycles in marine bivalve molluscs: a hypothesis concerning the relationship between glycogen metabolism and gametogenesis. In: *Proceedings of the ninth European marine biology symposium* (ed. H. Barnes), pp. 191–211. Aberdeen University Press, Aberdeen.

Gabbott, P. A. & Bayne, B. L. (1973). Biochemical effects of temperature and nutritive stress on *Mytilus edulis* L. *Journal of the Marine Biological Association of the United Kingdom*, **53**, 269–86.

Gabbott, P. A. & Holland, D. L. (1973). Growth and metabolism of *Ostrea edulis* larvae. *Nature, London*, **241**, 475–6.

Gabbott, P. A. & Stephenson, R. R. (1974). A note on the relationship between the dry weight condition index and the glycogen content of adult oysters (*Ostrea edulis* L.) kept in the laboratory. *Journal du Conseil. Conseil permanent International pour l'Exploration de la Mer*, **35**, 359–61.

Gabbott, P. A. & Walker, A. J. M. (1971). Changes in the condition index and biochemical content of adult oysters (*Ostrea edulis* L.) maintained under hatchery conditions. *Journal du Conseil. Conseil permanent International pour l'Exploration de la Mer*, **34**, 99–106.

Gabe, M. (1955). Particularités histologiques des cellules neurosécrétrices chez quelques Lamellibranches. *Comptes Rendus des Séances de la Société de Biologie*, **240**, 1810–12.

Gabe, M. (1965). La neurosécrétion chez les mollusques et ses rapports avec la reproduction. *Archives d'Anatomie microscopique et de Morphologie experimentale*, **54**, 371–86.

Gabe, M. (1966). *Neurosecretion*. Pergamon Press, Oxford. 872 pp.

Gäde, G. & Zebe, E. (1973). Über den anaerobiosestoffwechsel von mollusken muskeln. *Journal of Comparative Physiology*, **85**, 291–301.

Galtsoff, P. S. (1938a). Physiology of reproduction of *Ostrea virginica* (I). *Biological Bulletin. Marine Biological Laboratory, Woods Hole, Mass.*, **74**, 461–86.

Galtsoff, P. S. (1938b). Physiology of reproduction of *Ostrea virginica* (II). *Biological Bulletin. Marine Biological Laboratory, Woods Hole, Mass.*, **75**, 286–307.

Galtsoff, P. S. (1940). Physiology of reproduction of *Ostrea virginica* (III). *Biological Bulletin. Marine Biological Laboratory, Woods Hole, Mass.*, **78**, 117–35.

Galtsoff, P. S. (1964). The American oyster. *Bulletin of the United States Fish and Wildlife Service*, **64**, 1–480.

Galtsoff, P. S., Chipman, W. A., Engle, J. B. & Calderwood, H. N. (1947). Ecological and physiological studies of the effects of sulfate pulp mill wastes on oysters in the York river, Virginia. *Fishery Bulletin of the United States Fish and Wildlife Service*, **51**, 59–186.

Gardner, D. & Riley, J. P. (1972). The component fatty acids of the lipids of some species of marine and freshwater molluscs. *Journal of the Marine Biological Association of the United Kingdom*, **52**, 827–38.

Gaston, S. & Campbell, J. W. (1966). Distribution of arginase activity in molluscs. *Comparative Biochemistry and Physiology*, **17**, 259–70.

Genovese, S. (1958). Sulla presenza di *Mytilicola intestinalis* Steuer (Copepoda parasitica) nel lago di Ganzirri. *Atti della Società Peloritana di Scienze Fisiche, Matematiche e Naturali, Messina*, **5**, 47–53.

Genovese, S. (1965). Ulteriore contributo alla sistematica del genere *Mytilus*. Analisi biometrica di due popolazioni proveniente dal Canale di Leme e da Boulogne. *Bolletino di Zoologia*, Fasc. II, **32**, 247–62.

George, W. C. (1952). The digestion and absorption of fat in lamellibranchs. *Biological Bulletin. Marine Biological Laboratory, Woods Hole, Mass.*, **102**, 118–27.

Giese, A. C. (1959). Comparative physiology. Annual reproductive cycles of marine invertebrates. *Annual Review of Physiology*, **21**, 547–76.

Giese, A. C. (1966). Lipids in the economy of marine invertebrates. *Physiological Reviews*, **46**, 244–98.

Giese, A. C. (1967). Some methods for study of the biochemical constituents of marine invertebrates. *Oceanography and Marine Biology Annual Review*, **5**, 159–86.

References

Giese, A. C. (1969). A new approach to the biochemical composition of the mollusc body. *Oceanography and Marine Biology Annual Review*, **7**, 175–229.

Gilles, R. (1970). Intermediary metabolism and energy production in some invertebrates. *Archives Internationales de Physiologie et de Biochimie*, **78**, 313–26.

Gilles, R. (1972a). Biochemical ecology of Mollusca. In: *Chemical zoology* (ed. M. Florkin and B. T. Scheer), vol. VII, pp. 467–99. Academic Press, New York and London.

Gilles, R. (1972b). Osmoregulation in three molluscs: *Acanthochitona discrepans* (Brown), *Glycymeris glycymeris* (L.) and *Mytilus edulis* L. *Biological Bulletin. Marine Biological Laboratory, Woods Hole, Mass.*, **142**, 25–35.

Gilles, R., Hogue, P. & Kearney, E. B. (1971). Effects of various ions on the succinic dehydrogenase activity of *Mytilus californianus. Life Sciences*, Part II, **10**, 1421–7.

Gilmour, T. H. J. (1974). The structure, ciliation and function of the lips of some bivalve molluscs. *Canadian Journal of Zoology*, **52**, 335–43.

Girard, J. C., Huguet, R. & Solère, M. (1969). Les acides aminés soufres et leures dérivés au cours du cycle de *Mytilus galloprovincialis* Lmk. *Travaux de la Société de Pharmacie de Montpellier*, **29**, 193–6.

Gitay, A. (1972). Marine farming prospects on the South African west coast. *South African News and Fishing Industry Review*, May, 50–2.

Giusti, F. (1967). The action of *Mytilicola intestinalis* Steuer on *Mytilus galloprovincialis* Lam. of the Tuscan Coast. *Parasitology*, **28**, 17–26.

Giusti, F. (1970). The fine structure of the style sac and intestine in *Mytilus galloprovincialis* Lam. *Proceedings of the Malacological Society of London*, **39**, 95–104.

Glaister, G. & Kelly, M. (1936). The oxygen consumption and carbohydrate metabolism of the retractor muscle of the foot of *Mytilus edulis. Journal of Physiology*, **87**, 56–68.

Glass, N. R. (1969). Discussion of calculation of power function with special reference to respiratory metabolism in fish. *Journal of the Fisheries Research Board of Canada*, **26**, 2643–50.

Glaus, K. J. (1968). Factors influencing the production of byssus threads in *Mytilus edulis. Biological Bulletin. Marine Biological Laboratory, Woods Hole, Mass.*, **135**, 420.

Goddard, C. K. (1966). Carbohydrate metabolism. II. Pathways of carbohydrate metabolism. In: *Physiology of Mollusca* (ed. K. M. Wilbur and C. M. Yonge), vol. II, pp. 294–308. Academic Press, New York and London.

Gooch, J. L. (1975). Mechanisms of evolution and population genetics. In: *Marine ecology* (ed. O. Kinne), vol. II. Wiley-Interscience, New York. (In press.)

Gooch, J. L. & Schopf, T. J. M. (1970). Population genetics of marine species of the phylum ectoprocta. *Biological Bulletin. Marine Biological Laboratory, Woods Hole, Mass.*, **138**, 138–56.

References

Gooch, J. L. & Schopf, T. J. M. (1971). Genetic variation in the marine ectoproct *Schizoporella errata*. *Biological Bulletin. Marine Biological Laboratory, Woods Hole, Mass.*, **141**, 235–46.

Goodrich, E. S. (1945). The study of nephridia and genital ducts since 1895. *Quarterly Journal of Microscopical Science*, **86**, 115–392.

Gordon, A. H. (1969). *Electrophoresis of proteins in polyacrylamide and starch gels*. North Holland/American Elsevier, Amsterdam. 149 pp.

Goreau, G. F., Goreau, N. I. & Yonge, C. M. (1972). On the mode of boring in *Fungiacava eilatensis* (Bivalvia: Mytilidae). *Journal of Zoology*, **166**, 55–60.

Goreau, T. F., Goreau, N. I., Yonge, C. M. & Neumann, Y. (1970). On feeding and nutrition in *Fungiacava eilatensis* (Bivalvia, Mytilidae), a commensal living in fungiid corals. *Journal of Zoology*, **160**, 159–72.

Gosselin, R. E. (1961). The cilioexcitatory activity of serotonin. *Journal of Cellular and Comparative Physiology*, **58**, 17–25.

Gosselin, R. E. (1966). Physiologic regulators of ciliary motion. *American Review of Respiratory Diseases*, **93**, 41–59.

Gosselin, R. E., Moore, K. E. & Milton, A. S. (1962). Physiological control of molluscan gill cilia by 5-hydroxytryptamine. *Journal of General Physiology*, **46**, 277–96.

Goudsmit, E. M. (1972). Carbohydrates and carbohydrate metabolism in Mollusca. In: *Chemical zoology* (ed. M. Florkin and B. T. Scheer), vol. VII, pp. 219–43. Academic Press, New York and London.

Graham, A. (1949). The molluscan stomach. *Transactions of the Royal Society of Edinburgh*, **61**, 737–78.

Graham, H. W. & Gay, H. (1945). Season of attachment and growth of sedentary marine organisms at Oakland, California. *Ecology*, **26**, 375–86.

Graham, D. L. (1972). Trace metal levels in intertidal mollusks of California. *Veliger*, **14**, 365–72.

Graham, G. M. (1956). *Sea fisheries. Their investigation in the United Kingdom*. Arnold, London.

Grainger, J. N. R. (1951). Notes on the biology of the copepod *Mytilicola intestinalis* Steuer. *Parasitology*, **41**, 135–42.

Gray, J. (1928). *Ciliary movement*. Macmillan, New York.

Greenberg, M. J. (1969). The role of isoreceptors in the neurohormonal regulation of bivalve hearts. In: *Comparative physiology of the heart: current trends* (ed. F. McCann), *Experientia*, Supplement, **15**, pp. 232–48. Birkhauser Verlag, Basel.

Griffith, D. de G. (1972). Toxicity of oil and detergents to two species of edible molluscs under artificial tidal conditions. In: *Marine pollution and sea life* (ed. M. Runo), pp. 224–6. Fishing News (Books) Ltd, London.

Gruffydd, L. D. & Beaumont, A. R. (1972). A method for rearing *Pecten maximus* larvae in the laboratory. *Marine Biology*, **15**, 350–5.

Gruffydd, L. D., Lane, D. J. W. & Beaumont, A. R. (1975). The glands of the larval foot in *Pecten maximus* L. and possible homologues in other

437

References

bivalves. *Journal of the Marine Biological Association of the United Kingdom*, **55**, 463–76.

Guiseppe, C. (1964). Ciclo biologico reproduttivo di *Mytilus galloprovincialis* Lamark del lago di Ganzirri (Messina). *Atti della Società Peloritana di Scienze Fisiche, Matematiche e Naturali, Messina*, **10**, 537–43.

Gunter, G. (1957). Temperature. In: *Treatise on marine ecology and paleoecology* (ed. J. W. Hedgpeth), vol. 1, pp. 159–84. The Geological Society of America, Boulder, Colorado.

Halstead, B. W. (1965). *Poisonous and venomous animals of the world, vol. I. Invertebrates.* US Government Printing Office, Washington. 994 pp.

Hammen, C. S. (1966). Carbon dioxide fixation in marine invertebrates. V. Rate and pathway in the oyster. *Comparative Biochemistry and Physiology*, **17**, 289–96.

Hammen, C. S. (1968). Aminotransferase activities and amino acid excretion of bivalve molluscs and brachiopods. *Comparative Biochemistry and Physiology*, **26**, 697–705.

Hammen, C. S. (1969). Metabolism of the oyster, *Crassostrea virginica*. *American Zoologist*, **9**, 309–18.

Hammen, C. S. (1975). Succinate and lactate oxidoreductases of bivalve molluscs. *Comparative Biochemistry and Physiology*, **50B**, 407–12.

Hammen, C. S., Hanlon, D. P. & Lum, S. C. (1962). Oxidative metabolism of *Lingula*. *Comparative Biochemistry and Physiology*, **5**, 185–91.

Hammen, C. S., Miller, H. F. & Geer, W. H. (1966). Nitrogen excretion of *Crassostrea virginica*. *Comparative Biochemistry and Physiology*, **17**, 1199–1200.

Hancock, D. A. (1965a). Adductor muscle size in Danish and British mussels in relation to starfish predation. *Ophelia*, **2**, 253–67.

Hancock, D. A. (1965b). Graphical estimation of growth parameters. *Journal du Conseil. Conseil permanent International pour l'Exploration de la Mer*, **29**, 340–51.

Hancock, D. A. (1969a). Oyster pests and their control. *Ministry of Agriculture, Fisheries and Food, Laboratory Leaflet*, N.S., **19**, 1–30.

Hancock, D. A. (1969b). The shellfisheries of Chile. *Publicaciones del Instituto de Fomento Pesquero*, No. 45. 94 pp.

Hancock, D. A. (1973). The relationship between stock and recruitment in exploited invertebrates. *Rapport et Procès-verbaux des réunions. Conseil permanent International pour l'Exploration de la Mer*, **164**, 113–31.

Hancock, D. A. & Simpson, A. C. (1962). Parameters of marine invertebrate populations. In: *The exploitation of natural animal populations* (ed. E. D. Le Cren and M. W. Holdgate), pp. 29–50. Blackwell Scientific Publications, Oxford.

Hancock, D. A. & Urquhart, A. E. (1965). The determination of natural mortality and its causes in an exploited population of cockles (*Cardium edule*, L.). *Fishery Investigations. Ministry of Agriculture, Fisheries and Food, London*, Ser. II, **24**, 1–40.

Hanks, J. E. (1957). The rate of feeding of the common oyster drill *Urosalpinx cinerea* (Say) at controlled water temperatures. *Biological Bulletin. Marine Biological Laboratory, Woods Hole, Mass.*, **112**, 330–5.

Harding, C. W. (1884). Molluscs, mussels, whelks etc. used for food or bait. *International Fisheries Exhibition London 1883. Fisheries Exhibition Literature*, **6**, 301–23.

Harding, J. P. (1949). The use of probability paper for the graphical analyses of polymodal distributions. *Journal of the Marine Biological Association of the United Kingdom*, **28**, 141–53.

Hardy, C. D. (1970). Hydrographic data report: Long Island Sound – 1969. *State University of New York: Marine Sciences Research Center Technical Report No. 4.* 129 pp.

Harger, J. R. E. (1968). The role of behavioural traits in influencing the distribution of two species of sea mussel, *Mytilus edulis* and *M. californianus. Veliger*, **11**, 45–9.

Harger, J. R. E. (1970a). The effect of wave impact on some aspects of the biology of sea mussels. *Veliger*, **12**, 401–14.

Harger, J. R. E. (1970b). Comparisons among growth characteristics of two species of sea mussel, *Mytilus edulis* and *M. californianus. Veliger*, **13**, 44–56.

Harger, J. R. E. (1970c). The effects of species composition on the survival of mixed populations of the sea mussels *Mytilus californianus* and *M. edulis. Veliger*, **13**, 147–52.

Harger, J. R. E. (1972a). Competitive coexistence. Maintenance of interacting associations of the sea mussels *Mytilus edulis* and *Mytilus californianus. Veliger*, **14**, 387–410.

Harger, J. R. E. (1972b). Competitive coexistence among intertidal invertebrates. *American Scientist*, **60**, 600–7.

Harger, J. R. E. (1972c). Variation and relative 'niche' size in the sea mussel *Mytilus edulis* in association with *M. californianus. Veliger*, **14**, 275–83.

Harger, J. R. E. & Landenberger, D. E. (1971). The effects of storms as a density dependent mortality factor on populations of sea mussels. *Veliger*, **14**, 195–201.

Harris, D. (1968). A method of separating two superimposed normal distributions using arithmetic probability paper. *Journal of Animal Ecology*, **37**, 315–19.

Haskin, H. H. (1950). The selection of food by the Common oysterdrill *Urosalpinx cinerea* Say. *Proceedings of the National Shellfisheries Association*, **40**, 62–8.

Haskin, H. H. (1964). The distribution of oyster larvae. In: *Symposium on experimental marine ecology* (ed. N. Marshall, H. P. Jeffries, T. A. Napora and J. M. Sieburth), pp. 76–80. Occasional Publication No. 2, Graduate School of Oceanography, University of Rhode Island.

Hatton, H. (1938). Essais de bionomie explicative sur quelques espèces intercotidales d'algues et d'animaux. *Annales de l'Institut Océanographique, Monaco*, **17**, 241–348.

References

Hauschka, S. D. (1963). Purification and characterisation of *Mytilus* egg membrane lysin from sperm. *Biological Bulletin. Marine Biological Laboratory, Woods Hole, Mass.*, **125**, 363.

Havinga, B. (1929). Krebse und weichtiere. *Handbuch der Seefischerei Nordeuropas, Stuttgart*, **3**, 1–147.

Havinga, B. (1956a). Oyster and mussel culture. *Rapports et Procès-verbaux des Réunions. Conseil permanent International pour l'Exploration de la Mer*, **140**, 5–6.

Havinga, B. (1956b). Mussel culture in the Dutch Waddensea. *Rapports et Procès-verbaux des réunions. Conseil permanent international pour l'Exploration de la Mer*, **140**, 49–52.

Havinga, B. (1964). Mussel culture. *Sea Frontiers*, **10**, 155–61.

Hazel, J. R. & Prosser, C. L. (1974). Molecular mechanisms of temperature compensation in poikilotherms. *Physiological Reviews*, **54**, 620–77.

Hazelhoff, E. H. (1938). Uber die Ausnutzung des Sauerstoffs bei verschiedenen Wassertieren. *Zeitschrift für vergleichende Physiologie*, **26**, 306–27.

Hecht, S. (1934). Vision. II. The nature of the photoreceptor process. In: *A handbook of general experimental psychology* (ed. C. Murchison), pp. 704–828. Clark University Press, Worcester, Mass. (Reprinted 1969 by Rusell & Russell, New York.)

Hedrick, P. W. (1971). A new approach to measuring genetic similarity. *Evolution*, **25**, 276–80.

Heinonen, A. (1962). Reproduction of *Mytilus edulis* in the Finnish S.W. archipeligo in summer 1960. *Suomalaisen eläin-ja kasvitieteelisen seuran vanomen julkaisuja*, **16**, 137–43.

Helm, M. M., Holland, D. L. & Stephenson, R. R. (1973). The effect of supplementary algal feeding of a hatchery breeding stock of *Ostrea edulis* L. on larval vigour. *Journal of the Marine Biological Association of the United Kingdom*, **53**, 673–84.

Helm, M. M. & Spencer, B. E. (1972). The importance of the rate of aeration in hatchery cultures of the larvae of *Ostrea edulis* L. *Journal du Conseil. Conseil permanent International pour l'Exploration de la Mer*, **34**, 244–55.

Helm, M. M. & Trueman, E, R. (1967). The effect of exposure on the heart rate of the mussel, *Mytilus edulis* L. *Comparative Biochemistry and Physiology*, **21**, 171–7.

Hemmingsen, A. M. (1960). Energy metabolism as related to body size and respiratory surfaces and its evolution. *Report of the Steno Memorial Hospital and the Nordisk Insulinlaboratorium*, **4**, 7–58.

Hemmingsen, E. A. & Douglas, E. L. (1970). Respiratory characteristics of the haemoglobin-free *Chaenocephalus aceratus*. *Comparative Biochemistry and Physiology*, **33**, 733–44.

Henderson, J. T. (1929). Lethal temperatures of Lamellibranchiata. *Contributions to Canadian Biology and Fisheries*, **4**, 397–412.

Hepper, B. T. (1955). Environmental factors governing the infection of mussels *Mytilus edulis* by *Mytilicola intestinalis. Fishery Investigations. Ministry of Agriculture, Fisheries and Food, London*, Ser. II, **20**, 1–21.

Hepper, B. T. (1957). Notes on *Mytilus galloprovincialis* (Lmk) in Great Britain. *Journal of the Marine Biological Association of the United Kingdom*, **36**, 33–40.

Heppleston, P. B. (1971). The feeding ecology of oyster-catchers (*Haematopus ostralegus* L.) in winter in northern Scotland. *Journal of Animal Ecology*, **40**, 651–72.

Herdman, W. A. (1893). Report upon the methods of oyster and mussel culture on the west coast of France. *Report of the Lancashire Sea Fisheries Laboratory*, 41–80.

Herdman, W. A. & Scott, A. (1895). Report on the investigations carried on in 1894 in connection with the Lancashire Sea-Fisheries Laboratory at University College Liverpool. *Proceedings and Transactions of the Liverpool Biological Society*, **9**, 104–62.

Hers, H. G., De Wulf, H. & Stalmans, W. (1970). The control of glycogen metabolism in the liver. *FEBS Letters*, **12**, 73–82.

Hers, M. J. (1943). Relation entre respiration et circulation chez *Anodonta cygnea* L. *Annales de la Société royale zoologique de Belgique*, **74**, 45–54.

Hewatt, W. G. (1935). Ecological succession in the *Mytilus californianus* habitat as observed in Monterey Bay, California. *Ecology*, **16**, 244–51.

Hickling, C. F. (1968). *The farming of fish.* Pergamon Press, Oxford. 88 pp.

Hickman, R. W. & Gruffydd, L. D. (1971). The histology of the larvae of *Ostrea edulis* during metamorphosis. In: *Fourth European marine biology symposium* (ed. D. J. Crisp), pp. 281–94. Cambridge University Press, London.

Hidu, H. (1965). Effects of synthetic surfactants on the larvae of clams (*Mercenaria mercenaria*) and oysters (*Crassostrea virginica*). *Journal of the Water Pollution Control Federation*, **37**, 262–70.

Hidu, H. & Tubiash, H. S. (1963). A bacterial basis for the growth of anti-biotic-treated bivalve larvae. *Proceedings of the National Shellfisheries Association*, **54**, 25–39.

Hidu, H. & Ukeles, R. (1964). Dried unicellular algae as food for larvae of the hard shell clam, *Mercenaria mercenaria. Proceedings of the National Shellfisheries Association*, **53**, 85–101.

Hill, R. B. & Welsh, J. H. (1966). Heart, circulation and blood cells. In: *Physiology of Mollusca* (ed. K. M. Wilbur and C. M. Yonge), vol. 2, pp. 125–174. Academic Press, New York and London.

Hinchcliffe, P. R. & Riley, J. P. (1972). The effect of diet on the component fatty-acid composition of *Artemia salina. Journal of the Marine Biological Association of the United Kingdom*, **52**, 203–11.

Hirai, E. (1963). On the breeding seasons of invertebrates in the neighbourhood of the marine station of Asamushi. *Science Reports Tohoku University*, Ser. 4, *Biology*, **24**, 369–75.

Hobden, D. J. (1967). Iron metabolism in *Mytilus edulis*. I. Variation in

References

total content and distribution. *Journal of the Marine Biological Association of the United Kingdom*, **47**, 597–606.

Hobden, D. J. (1969). Iron metabolism in *Mytilus edulis*. II. Uptake and distribution of radioactive iron. *Journal of the Marine Biological Association of the United Kingdom*, **49**, 661–8.

Hochachka, P. W. (1973). Comparative intermediary metabolism. In: *Comparative animal physiology*, 3rd edition (ed. C. L. Prosser), vol. I, pp. 212–78. W. B. Saunders, Philadelphia.

Hochachka, P. W., Fields, J. & Mustafa, T. (1973). Animal life without oxygen: basic biochemical mechanisms. *American Zoologist*, **13**, 543–55.

Hochachka, P. W., Freed, J. M., Somero, G. N. & Prosser, C. L. (1971). Control sites in glycolysis of crustacean muscle. *International Journal of Biochemistry*, **2**, 125–30.

Hochachka, P. W. & Mustafa, T. (1972). Invertebrate facultative anaerobiosis. *Science*, **178**, 1056–60.

Hochachka, P. W. & Mustafa, T. (1973). Enzymes in facultative anaerobiosis of molluscs.I. Malic enzyme of oyster adductor muscle. *Comparative Biochemistry and Physiology*, **45B**, 625–37.

Hochachka, P. W. & Somero, G. (1971). Biochemical adaptation to the environment. In: *Fish physiology* (ed. W. S. Hoar and D. J. Randall), vol. VI, pp. 99–156. Academic Press, New York and London.

Hochachka, P. W. & Somero, G. N. (1973). *Strategies of biochemical adaptation*. W. B. Saunders, Philadelphia. 358 pp.

Hockley, A. R. (1951). On the biology of *Mytilicola intestinalis* (Steuer). *Journal of the Marine Biological Association of the United Kingdom*, **30**, 223–32.

Hoggarth, K. R. & Trueman, E. R. (1967). Techniques for recording the activity of aquatic invertebrates. *Nature, London*, **213**, 1050–1.

Holdgate, M. W. (ed.) (1971). *The seabird wreck of 1969 in the Irish Sea; analytical and other data*. Supplement to a report published by the Natural Environment Research Council, London. 18 pp.

Holeton, G. F. (1972). Gas exchange in fish with and without haemoglobin. *Respiratory Physiology*, **14**, 142–50.

Holland, D. L. & Gabbott, P. A. (1971). A micro-analytical scheme for the determination of protein, carbohydrate, lipid and RNA levels in marine invertebrate larvae. *Journal of the Marine Biological Association of the United Kingdom*, **51**, 659–68.

Holland, D. L. & Hannant, P. J. (1973). Addendum to a micro-analytical scheme for the biochemical analysis of marine invertebrate larvae. *Journal of the Marine Biological Association of the United Kingdom*, **53**, 833–8.

Holland, D. L. & Hannant, P. J. (1974). Biochemical changes during growth of the spat of the oyster, *Ostrea edulis* L. *Journal of the Marine Biological Association of the United Kingdom*, **54**, 1004–16.

Holland, D. L. & Spencer, B. E. (1973). Biochemical changes in fed and starved oysters, *Ostrea edulis* L., during larval development, metamorphosis and early spat growth. *Journal of the Marine Biological Association of the United Kingdom*, **53**, 287–98.

Holland, D. L. & Walker, G. (1975). The biochemical composition of the cypris larva of the barnacle *Balanus balanoides* L. *Journal du Conseil. Conseil permanent International pour l'Exploration de la Mer*, in press.

Holme, N. A. (1961). Shell form in *Venerupis rhomboides*. *Journal of the Marine Biological Association of the United Kingdom*, **41**, 705–22.

Holmes, N. (1970). Marine fouling in power stations. *Marine Pollution Bulletin*, **1**, 105–6.

Holwerda, D. A. & De Zwaan, A. (1973). Kinetic and molecular characteristics of allosteric pyruvate kinase from muscle tissue of the sea mussel *Mytilus edulis* L. *Biochimica et Biophysica Acta*, **309**, 296–306.

Holwerda, D. A., De Zwaan, A. & Van Marrewijk, W. J. A. (1973). The regulatory role of pyruvate kinase in the sea mussel *Mytilus edulis* L. *Netherlands Journal of Zoology*, **23**, 232–5.

Hopkins, H. S. (1946). The influence of season, concentration of seawater and environmental temperature upon the oxygen consumption of tissues in *Venus mercenaria*. *Journal of Experimental Zoology*, **102**, 143–58.

Horridge, G. A. (1958). Transmission of excitation through the ganglia of *Mya* (Lamellibranchiata). *Journal of Physiology*, **143**, 553–72.

Hosomi, A. (1968). The ecological study on the adhering, growth and death in the population of *Mytilus edulis* L. at the seashore of Suma in Kobe. *Japanese Journal of Ecology*, **18**, 74–9.

Hosomi, A. (1969). The synecological observation on the biological succession at the surface of vertical wall of the jetty for sand arrestation of intertidal zone at the seashore of Suma in Kobe. *Hyogo Biology*, **6**, 32–4.

Houghton, D. R. (1963). The relationship between tidal level and the occurrence of *Pinnotheres pisum* (Pennant) in *Mytilus edulis* L. *Journal of Animal Ecology*, **32**, 253–7.

Houtteville, P. & Lubet, P. (1974). Analyse expérimentale, en culture organotypique, de l'action des ganglions cérébropleureux et viscéreux sur le manteau de la moule mâle, *Mytilus edulis* L. (Mollusque Pélécypode). *Compte Rendus, Académie de Sciences, Paris*, Ser. D, **278**, 2469–72.

Hrs-Brenko, M. (1967). *Mytilicola intestinalis* Steuer (Copepoda, parasitica) a parasite in mussels in the east Adriatic. *Thalassia Jugoslavica*, **3**, 143–59.

Hrs-Brenko, M. (1971). Observations on the occurrence of planktonic larvae of several bivalves in the northern Adriatic Sea. In: *Fourth European marine biology symposium* (ed. D. J. Crisp), pp. 45–53. Cambridge University Press, London.

References

Hrs-Brenko, M. (1973). The study of mussel larvae and their settlement in Vela Draga Bay (Pula, The Northern Adriatic Sea). *Aquaculture*, **2**, 173–82.

Hrs-Brenko, M. & Calabrese, A. (1969). The combined effects of salinity and temperature on larvae of the mussel *Mytilus edulis*. *Marine Biology*, **4**, 224–6.

Hrs-Brenko, M. & Igić, L. (1968). Effects of fresh water and saturated sea-water brine on the survival of mussels, oysters and some epibionts on them. *General Fisheries Council for the Mediterranean, Studies and Reviews*, **37**, 29–44.

Hughes, G. M. (1973). Respiratory response to hypoxia in fish. *American Zoologist*, **13**, 475–89.

Hughes, G. M. & Shelton, G. (1962). Respiratory mechanisms and their nervous control in fish. *Advances in Comparative Physiology and Biochemistry*, **1**, 275–364.

Hughes, R. N. (1969). A study of feeding in *Scrobicularia plana*. *Journal of the Marine Biological Association of the United Kingdom*, **49**, 802–23.

Hughes, R. N. (1970). An energy budget for a tidal-flat population of the bivalve *Scrobicularia plana* (Da Costa). *Journal of Animal Ecology*, **39**, 357–81.

Human, V. L. (1971). The occurrence of *Urosalpinx cinerea* (Say) in Newport Bay. *Veliger*, **13**, 299.

Humphreys, W. J. (1962). Electron microscope studies on eggs of *Mytilus edulis*. *Journal of Ultrastructure Research*, **7**, 467–87.

Humphreys, W. J. (1964). Electron microscope studies of the fertilised egg and the two-cell stage of *Mytilus edulis*. *Journal of Ultrastructure Research*, **17**, 314–26.

Humphreys, W. J. (1967). The fine structure of cortical granules in eggs and gastrulae of *Mytilus edulis*. *Journal of Ultrastructure Research*, **17**, 314–26.

Hunter, R. L. & Markert, C. L. (1957). Histochemical demonstration of enzymes separated by zone electrophoresis in starch gels. *Science*, **125**, 1294–5.

Hunter, W. R. (1949). The structure and behaviour of ' *Hiatella gallicana*' (Lamarck) and '*H. arctica*' (L.) with special reference to the boring habit. *Proceedings of the Royal Society of Edinburgh*, Ser. B, **63**, 271–89.

Huntsman, A. G. (1921). The effect of light on growth in the mussel. *Transactions of the Royal Society of Canada*, Ser. 3, **15**, 23–8.

Hutchins, L. W. (1927). The bases for temperature zonation in geographical distributions. *Ecological Monographs*, **17**, 325–35.

Idelman, S. (1967). Données récentes sur l'infrastructure du spermatozoide. *Année Biologique, Paris*, **6**, 113–90.

Idler, D. R. & Wiseman, P. (1971). Sterols of molluscs. *International Journal of Biochemistry*, **2**, 516–28.

Idler, D. R. & Wiseman, P. (1972). Molluscan sterols: a review. *Journal of the Fisheries Research Board of Canada*, **29**, 385–98.

Ingham, H. R., Mason, J. & Wood, P. C. (1968). Distribution of toxin in molluscan shellfish following the occurrence of mussel toxicity in north-east England. *Nature, London,* **220**, 25–7.

Isham, L. B. & Tierney, J. Q. (1953). Some aspects of the larval development and metamorphosis of *Teredo (Lyrodus) pedicellata* De Quatrefages. *Bulletin of Marine Science of the Gulf and Caribbean, Miami,* **2**, 574–89.

Ivanov, A. I. (1971). Preliminary results of breeding mussels (*Mytilus galloprovincialis*) in Kerch Bay and some other parts of the Black Sea. *Oceanology, Moscow,* **11**, 733–41.

Iversen, E. S. (1968). *Farming the edge of the sea.* Fishing News (Books) Ltd, London. 301 pp.

Ivlev, V. S. (1934). Eine Mikromethode zur bestimming des Kaloriengehalts von Nahrstoffen. *Biochemische Zeitschrift,* **275**, 49–55.

Iwata, K. S. (1950). Spawning of *Mytilus edulis.* 2. Discharge by electrical stimulation. *Bulletin of the Japanese Society of Scientific Fisheries,* **15**, 443–6.

Iwata, K. S. (1951a). Spawning of *Mytilus edulis.* 4. Discharge by KCl injection. *Bulletin of the Japanese Society of Scientific Fisheries,* **16**, 393–4.

Iwata, K. S. (1951b). Spawning of *Mytilus edulis.* 6. Discharge with alkali treatment. *Bulletin of the Japanese Society of Scientific Fisheries,* **17**, 157–60.

Iwata, K. S. (1952). Mechanism of egg maturation in *Mytilus edulis. Biological Journal of Okayama University,* **1**, 1–11.

Jameson, H. L. & Nicoll, W. (1913). On some parasites of the scoter (*Oedemia nigra*) and their relation to the pearl-inducing trematode in the mussel. *Proceedings of the Zoological Society of London,* 53–63.

Jeffries, H. P. (1972). A stress syndrome in the hard clam, *Mercenaria mercenaria. Journal of Invertebrate Pathology,* **20**, 242–51.

Jensen, A. S. (1912). Lamellibranchiata. *The Danish Ingolf Expedition,* **2**, 1–115.

Jensen, A. S. & Spärck, R. (1934). Bløddyr 2. Saltvandsmuslinger. *Danmarks Fauna, Copenhagen,* **40**, 1–208.

Jensen, A. & Sakshaug, E. (1970). Producer–consumer relationships in the sea. 2. Correlation between *Mytilus* pigmentation and the density and composition of phytoplanktonic populations. *Journal of Experimental Marine Biology and Ecology,* **5**, 246–53.

Jezyk, P. F. & Penicnak, A. J. (1966). Fatty acid relationships in an aquatic food chain. *Lipids,* **1**, 427–9.

Johannes, R. E. & Satomi, M. (1967). Measuring organic matter retained by aquatic invertebrates. *Journal of the Fisheries Research Board of Canada,* **24**, 2467–71.

Johnson, A. G. & Utter, F. M. (1973). Electrophoretic variants of aspartate aminotransferase of the mussel, *Mytilus edulis* (Linneus 1958). *Comparative Biochemistry and Physiology,* **44B**, 317–23.

Johnson, I. (1952). The demonstration of a 'host-factor' in commensal crabs. *Transactions of the Kansas Academy of Science,* **55**, 458–64.

References

Johnstone, J. (1898). The spawning of the mussel (*Mytilus edulis*). *Proceedings and Transactions of the Liverpool Biological Society*, **13**, 104–21.

Jolicoeur, P. & Heusner, A. A. (1971). The allometry equation in the analysis of the standard oxygen consumption and body weight of the white rat. *Bioenergetics*, **27**, 841–55.

Jones, A. M., Jones, Y. & Stewart, W. P. (1972). Mercury in marine organisms of the Tay region. *Nature, London*, **238**, 164–5.

Jones, D. A. & Gabbott, P. A. (1975). Prospects for the use of microcapsules as food particles for marine particulate feeders. In: *Proceedings of the second international symposium on microencapsulation*, Chelsea College, University of London. (In press.)

Jones, D. A., Munford, J. G. & Gabbott, P. A. (1974). Microcapsules as artificial food particles for aquatic filter feeders. *Nature, London*, **247**, 233–5.

Jones, S. (1950). Observations on the bionomics and fishery of the brown mussel (*Mytilus* spp.) of the Cape region of peninsular India. *Journal of the Bombay Natural History Society*, **49**, 519–28.

Jones, S. (1973). On the culture potential of the marine mussel, *Mytilus* spp., in the Indian region. In: *Coastal aquaculture of the Indo-Pacific region* (ed. T. V. R. Pillay), p. 394. Fishing News (Books) Ltd, London.

Jones, S. & Alagarswami, K. (1968). Mussel fishery resources of India. In: *Symposium on the living resources of the seas around India, Indian council of agricultural research, Cochin*, December 1968. Abstract No. 51, p. 28.

Jørgensen, C. B. (1946). In Thorson, G. (1946).

Jørgensen, C. B. (1949). The rate of feeding by *Mytilus* in different kinds of suspension. *Journal of the Marine Biological Association of the United Kingdom*, **28**, 333–44.

Jørgensen, C. B. (1952). Efficiency of growth in *Mytilus edulis* and two gastropod veligers. *Nature, London*, **170**, 714.

Jørgensen, C. B. (1959). On porosity of gill filters and filtration rate in mussels (*Mytilus edulis*). *Bulletin. Mount Desert Island Biological Laboratory*, **4**, 28.

Jørgensen, C. B. (1960). Efficiency of particle retention and rate of water transport in undisturbed lamellibranchs. *Journal du Conseil. Conseil permanent International pour l'Exploration de la Mer*, **26**, 94–116.

Jørgensen, C. B. (1966). *Biology of suspension feeding*. Pergamon Press, Oxford. 357 pp.

Jørgensen, C. B. & Goldberg, E. D. (1953). Particle filtration in some ascidians and lamellibranchs. *Biological Bulletin. Marine Biological Laboratory, Woods Hole, Mass.*, **105**, 477–89.

Joyner, T. & Spinelli, J. (1969). Mussels – a potential source of high-quality protein. *Commercial Fisheries Review*, Aug.–Sept., 31–5.

Kändler, R. (1926). Muschellarven aus dem Helgoländer Plankton. *Wissenschaftliche Meeresuntersuchungen der Kommission sur*

Wissenschaftlichen Untersuchung der Deutschen Meere, Abt. Helgoland, **16**, 1–9.

Kanwisher, J. W. (1955). Freezing in intertidal animals. *Biological Bulletin. Marine Biological Laboratory, Woods Hole, Mass.,* **109**, 56–63.

Kanwisher, J. W. (1966). Freezing in intertidal animals. In: *Cryobiology* (ed. H. T. Merryman), pp. 487–94. Academic Press, New York and London.

Karandeeva, O. G. (1959). Certain aspects of the metabolism of *Modiola phaseolina* and *Mytilus galloprovincialis* in anaerobic and arranged aerobic conditions. *Trudy Sevastopolskoi Biologicheskoi Stantsii. Akademiya Nauk SSSR,* **11**, 238–53. (In Russian.)

Kasuga, N. & Ishida, S. (1957). Metabolic patterns in bivalves. XII. Histological distribution of uric acid in the mussel, *Mytilus edulis. Journal of the College of Arts and Sciences, Chiba University. Natural Sciences Series,* **2**, 243–6.

Kawai, K. (1959). The cytochrome system in marine lamellibranch tissues. *Biological Bulletin. Marine Biological Laboratory, Woods Hole, Mass.,* **117**, 125–32.

Keckes, S., Pucar, Z. & Marazovic, L. (1967). Accumulation of electrodialytically separated physicochemical forms of [106]Ru by mussels. *International Journal of Oceanology and Limnology,* **1**, 246–53.

Kennedy, D. (1960). Neural photoreception in a lamellibranch mollusc. *Journal of General Physiology,* **44**, 277–99.

Kennedy, V. S. & Mihursky, J. A. (1971). Upper temperature tolerances of some estuarine bivalves. *Chesapeake Science,* **12**, 193–204.

Kennedy, V. S. & Mihursky, J. A. (1972). Effects of temperature on the respiratory metabolism of three Chesapeake Bay bivalves. *Chesapeake Science,* **13**, 1–22.

Kennedy, V. S., Roosenburg, W. H., Castagna, M. & Mihursky, J. A. (1974). *Mercenaria mercenaria* (Mollusca: Bivalvia): temperature–time relationships for survival of embryos and larvae. *Fishery Bulletin. Fish and Wildlife Service. United States Department of Interior,* **72**, 1160–6.

Kerr, S. R. (1971a). Analysis of laboratory experiments on growth efficiency of fishes. *Journal of the Fisheries Research Board of Canada,* **28**, 801–8.

Kerr, S. R. (1971b). Prediction of growth efficiency in nature. *Journal of the Fisheries Research Board of Canada,* **28**, 809–14.

Kerswill, C. J. (1949). Effect of water circulation on the growth of quahaugs and oysters. *Journal of the Fisheries Research Board of Canada,* **7**, 545–51.

Kinne, O. (1963). The effects of temperature and salinity on marine and brackish water animals. I. Temperature. *Oceanography and Marine Biology Annual Review,* **1**, 301–40.

Kinne, O. (1964a). The effects of temperature and salinity on marine and brackish-water animals. II. Salinity and temperature–salinity combinations. *Oceanography and Marine Biology Annual Review,* **2**, 281–339.

References

Kinne, O. (1964b). Non-genetic adaptation to temperature and salinity. *Helgoländer wissenschaftliche Meeresuntersuchungen*, **9**, 433–58.

Kinne, O. (1970a). Opening address of international symposium on cultivation of marine organisms and its importance for marine biology. *Helgoländer wissenschaftliche Meeresuntersuchungen*, **20**, 1–5.

Kinne, O. (1970b). Closing address of international symposium on cultivation of marine organisms and its importance for marine biology. *Helgoländer wissenschaftliche Meeresuntersuchungen*, **20**, 707–10.

Kinne, O. (1970c). Temperature – invertebrates. In: *Marine ecology* (ed. O. Kinne), vol. 1, part 1, pp. 407–514. Wiley-Interscience, New York.

Kirby-Smith, W. W. (1972). Growth of the bay scallop: the influence of experimental water-currents. *Journal of Experimental Marine Biology and Ecology*, **8**, 7–18.

Kirschner, L. B. (1967). Comparative physiology: invertebrate excretory organs. *Annual Review of Physiology*, **29**, 169–96.

Kiseleva, G. A. (1966a). Some problems of the ecology of the larvae of Black Sea mussels. In: *Distribution of the benthos and biology of benthic organisms in the southern seas. Akademiya Nauk Ukrainskoi, SSR. Seriya Biologiya Marya, Kiev,* pp. 16–20. (In Russian.)

Kiseleva, G. A. (1966b). Factors stimulating larval metamorphosis of the lamellibranch *Brachyodontes lineatus* (Gmelin). *Zoologicheskii Zhurnal*, **45**, 1571–3.

Kitching, J. A. & Ebling, F. J. (1967). Ecological studies at Lough Ine. *Advances in Ecological Research*, **4**, 197–291.

Kitching, J. A. Sloane, J. F. & Ebling, F. J. (1959). The ecology of Lough Ine. 8. Mussels and their predators. *Journal of Animal Ecology*, **28**, 331–41.

Klappenbach, M. A. (1965). Lista preliminar de los Mytilidae brasilienos conclaves para su determinacion y notas sobre su distribucion. *Anales de la Academia Brasileira de Ciencias*. Suppl., **37**, 327–52.

Kleiber, M. (1961). *The fire of life; an introduction to animal energetics.* Wiley, New York. 454 pp.

Knight, W. (1968). Asymptotic growth: an example of nonsense disguised as mathematics. *Journal of the Fisheries Research Board of Canada*, **25**, 1303–7.

Knight-Jones, E. W. & Moyse, J. (1961). Intraspecific competition in sedentary marine animals. In: *Mechanisms in biological competition* (ed. F. L. Milthorpe), pp. 72–95. *Symposia of the Society for Experimental Biology*, **15**. Cambridge University Press, London.

Kobayashi, I. (1969). Internal microstructure of the shell of bivalve molluscs. *American Zoologist*, **9**, 663–72.

Koehman, J. H. & van Genderen, H. (1972). Tissue levels in animals and effects caused by chlorinated hydrocarbon insecticides, chlorinated biphenyls and mercury in the marine environment along the Netherlands coast. In: *Marine pollution and sea life* (ed. M. Ruiva), pp. 428–35. Fishing News (Books) Ltd, London.

Koehman, J. H., Veen, J., Brouwer, E., Brouwer, L. H. & Koolen, J. L. (1968). Residues of chlorinated hydrocarbon insecticides in the North Sea environment. *Helgoländer wissenschaftliche Meeresuntersuchungen*, **17**, 375–80.

Koehn, R. K. (1975). Migration and population structure in the pelagically dispersing marine invertebrate, *Mytilus edulis*. In: *Proceedings of the third international congress on isozymes* (ed. C. Markert). Academic Press, New York and London. (In press.)

Koehn, R. K. & Mitton, J. B. (1972). Population genetics of marine pelecypods. 1. Ecological heterogeneity and evolutionary strategy at an enzyme locus. *American Naturalist*, **106**, 47–56.

Koehn, R. K., Milkman, R. & Mitton, J. B. (1975). Population genetics of marine pelecypods. IV. Selection, migration and genetic differentiation in the Blue Mussel, *Mytilus edulis*. *Evolution*, in press.

Koehn, R. K., Turano, F. J. & Mitton, J. B. (1973). Population genetics of marine pelecypods. II. Genetic differences in microhabitats of *Modiolus demissus*. *Evolution*, **27**, 100–5.

Konstantinova, M. I. (1966). Characteristics of the motion of pelagic larvae of marine invertebrates. *Doklady (Proceedings) of the Academy of Sciences of the USSR*, **170**, 753–6.

Kornberg, H. L. & Elsden, S. R. (1961). The metabolism of 2-carbon compounds by micro-organisms. *Advances in Enzymology*, **23**, 401–70.

Korringa, P. (1947). Relations between the moon and periodicity in the breeding of marine animals. *Ecological Monographs*, **17**, 347–81.

Korringa, P. (1951). Le *Mytilicola intestinalis* Steuer (copepoda parasitica) menace l'industrie moulière en Zelande. *Revue des Travaux de l'Office (scientifique et technologique) des Pêches maritimes*, **17**, 9–13.

Korringa, P. (1957). Water temperature and breeding throughout the geographical range of *Ostrea edulis*. *L'Annee Biologique*, 3rd ser. **33**, 1–17.

Korringa, P. (1968). Biological consequences of marine pollution with special reference to the North Sea fisheries. *Helgoländer wissenschaftliche Meeresuntersuchungen*, **17**, 126–40.

Korringa, P. (1970). Shellfish farming on the continental coast of Europe. *Proceedings of the Symposium on Mollusca, Marine Biological Association of India*, pp. 818–23.

Korringa, P. (1971). Mariculture. In: *McGraw-Hill Yearbook of Science and Technology 1971*, pp. 13–23. McGraw-Hill, New York.

Korringa, P. & Postma, H. (1957). Investigations into the fertility of the Gulf of Naples and adjacent salt water lakes, with special reference to shellfish cultivation. *Publicazioni della Stazione zoologica di Napoli*, **29**, 229–84.

Korringa, P. & Roskam, R. J. (1961). An unusual case of shell-fish poisoning. *ICES Shellfish Committee Document No. 49*. 2 pp. (Mimeo.)

References

Krebs, H. A. (1972). Some aspects of the regulation of fuel supply in omnivorous animals. In: *Advances in enzyme regulation* (ed. G. Weber), vol. 8, pp. 397–420. Pergamon Press, Oxford.

Kriaris, N. (1967). La vie larvaire et la croissance de la moule en Bretagne. *Penn ar bed*, **6**, 25–30.

Krijgsman, B. J. & Divaris, G. A. (1955). Contractile and pacemaker mechanisms of the heart of molluscs. *Biological Reviews*, **30**, 1–39.

Kristensen, J. H. (1972). Structure and function of crystalline styles of bivalves. *Ophelia*, **10**, 91–108.

Krogh, A. (1914). The quantitative relation between temperature and standard metabolism in animals. *Internationale Zeitschrift für Physikalisch-chemische Biologie*, **1**, 491–508.

Krogh, A. (1916). *The respiratory exchange of animals and man*. Longmans, Green & Co., London. 173 pp.

Krüger, F. (1960). Zur Frage der Grössenabhängigkeit des Sauerstoffverbrauchs von *Mytilus edulis* L. *Helgölander wissenschaftliche Meeresuntersuchungen*, **7**, 125–48.

Kuenen, D. J. (1942). On the distribution of mussels on the intertidal flats near den Helder. *Archives Néerlandaises Zoologie*, **6**, 117–60.

Kuenzler, E. J. (1961a). Structure and energy flow of a mussel population in a Georgia salt marsh. *Limnology and Oceanography*, **6**, 191–204.

Kuenzler, E. J. (1961b). Phosphorus budget of a mussel population. *Limnology and Oceanography*, **6**, 400–15.

Kühl, H. (1972). Hydrography and biology of the Elbe estuary. *Oceanography and Marine Biology Annual Reviews*, **10**, 225–309.

Kuznetzov, V. V. & Mateeva, T. A. (1948). Details of the bioecological features of marine invertebrates of eastern Murman. *Trudy Murmanskoi Biologicheskoi Stanstii*, **1**, 241–60. (In Russian.)

Lagerspetz, K. & Sirkka, A. (1959). Versuche über den Sauerstoffverbrauch von *Mytilus edulis* aus dem Brackwasser der finnischen Kuste. *Kieler Meeresforschungen*, **15**, 89–96.

Lamana, M. A. (1973). Algunos aspectos del metabolismo del glucogeno en el mejillon. DSc thesis, University of Barcelona.

Lambert, L. (1935). La culture de la moule en Hollande. *Revue des Travaux de l'Office (scientifique et technologique) des Pêches maritimes*, **8**, 431–80.

Lambert, L. (1939). *La Moule et la mytiliculture*. A. Guillot, Versailles. 55 pp.

Lambert, L. (1950). Les coquillages comestibles, huîtres, moules, coquillages variés. '*Que sais-je?*' 416. Presses Universitaires de France.

Lambert, L. (1951). L'ostreiculture et la mytiliculture en Zelande (Pays Bas). *Revue des Travaux de l'Office (scientifique et technologique) des Pêches maritimes*, **16**, 111–28.

Lamy, E. (1936). Révision des Mytilidae vivants du Muséum National d'Histoire Naturelle de Paris. *Journal de Conchyliologie*, **80**, 66–102.

Landenberger, D. E. (1968). Studies on selective feeding in the Pacific Starfish *Pisaster* in south California. *Ecology*, **49**, 1062–75.

Lane, D. J. W. & Nott, J. A. (1975). A study of the morphology, fine structure and histochemistry of the foot of the pediveliger of *Mytilus edulis* L. *Journal of the Marine Biological Association of the United Kingdom*, **55**, 477–96.

Lange, R. (1963). The osmotic function of amino acids and taurine in the mussel, *Mytilus edulis*. *Comparative Biochemistry and Physiology*, **10**, 173–9.

Lange, R. (1964). The osmotic adjustment in the echinoderm *Strongylocentrotus droebachiensis*. *Comparative Biochemistry and Physiology*, **13**, 205–16.

Lange, R. (1968). The relation between the oxygen consumption of isolated gill tissue of the common mussel, *Mytilus edulis* L., and salinity. *Journal of Experimental Marine Biology and Ecology*, **2**, 37–45.

Lange, R. (1970). Isosmotic intracellular regulation and euryhalinity in marine bivalves. *Journal of Experimental Marine Biology and Ecology*, **5**, 170–9.

Lange, R. (1972). Some recent work on osmotic, ionic and volume regulations in marine animals. *Oceanography and Marine Biology Annual Review*, **10**, 97–136.

Lange, R. & Mostad, A. (1967). Cell volume regulation in osmotically adjusting marine animals. *Journal of Experimental Marine Biology and Ecology*, **1**, 209–19.

Lange, R., Staaland, H. & Mostad, A. (1972). The effect of salinity and temperature on solubililty of oxygen and respiratory rate in oxygen-dependent marine invertebrates. *Journal of Experimental Marine Biology and Ecology*, **9**, 217–29.

Langton, R. W. (1975). Synchrony in the digestive diverticula of *Mytilus edulis* L. *Journal of the Marine Biological Association of the United Kingdom*, **55**, 221–30.

Langton, R. W. & Gabbott, P. A. (1974). The tidal rhythm of extracellular digestion and the response to feeding in *Ostrea edulis* L. *Marine Biology*, **24**, 181–7.

Largen, M. J. (1967). The diet of the dogwhelk *Nucella lapillus* (Gastropoda, Prosobranchia) *Journal of Zoology*, **151**, 123–7.

Laseron, C. F. (1956). New South Wales mussels. A taxonomic review of the family Mytilidae from the Peronian zoogeographical province. *Australian Zoology*, **12**, 263–83.

Lavallard, R., Balas, G. & Schlenz, R. (1969). Contribution a l'étude de la croissance relative chez *Mytilus perna* L. *Boletin da Faculdade de Filosofia, Ciências e Letras, Universidade de Sao Paulo. Zoologia e Biologia Marhina*, N.S., **26**, 19–31.

Laverack, M. S. (1968). On the receptors of marine invertebrates. *Oceanography and Marine Biology Annual Review*, **6**, 249–324.

Lebour, M. V. (1906). The mussel beds of Northumberland. *Northumberland Sea Fisheries Commission. Report No. 28.*

Lebour, M. V. (1912). A review of British marine cercaria. *Parasitology*, **4**, 416–56.

References

Lebour, M. V. (1933). The importance of larval mollusca in the plankton. *Journal du Conseil. Conseil permanent International pour l'Exploration de la Mer*, **8**, 335–43.

Lebour, M. V. (1938). Notes on the breeding of some lamellibranchs from Plymouth, and their larvae. *Journal of the Marine Biological Association of the United Kingdom*, **23**, 119–44.

Lee, R. F., Sauerheber, R. & Benson, A. A. (1972). Petroleum hydrocarbons: uptake and discharge by marine mussel, *Mytilus edulis*. *Science*, **177**, 344–6.

Le Gall, P. (1970). Etude des moulières Normandes; renouvellement, croissance. *Vie et Milieu*, **21B**, 545–90.

Legaré, J. E. H. & Maclellan, D. C. (1960). A qualitative and quantitative study of the plankton of the Quoddy region in 1957 and 1958 with special reference to the food of herring. *Journal of the Fisheries Research Board of Canada*, **17**, 409–48.

Leloup, E. (1960). Recherches sur la répartition de *Mytilicola intestinalis* Steuer 1905 le long de la cote Belge. *Bulletin de l'Institut royal des sciences naturelles de Belgique*, **36**, 1–12.

Lent, C. M. (1968). Air gaping by the ribbed mussel, *Modiolus demissus* (Dillwyn). Effects and adaptive significance. *Biological Bulletin. Marine Biological Laboratory, Woods Hole, Mass.*, **134**, 60–73.

Lent, C. M. (1969). Adaptations of the ribbed mussel, *Modiolus demissus* (Dillwyn) to the intertidal habitat. *American Zoologist*, **9**, 283–92.

Leonard, C. (1975). Isocitrate lyase activity in marine bivalves. MSc thesis, University of Wales.

Levene, H. (1953). Genetic equilibrium when more than one niche is available. *American Naturalist*, **87**, 331–3.

Levinton, J. S. (1972). Stability and trophic structure in deposit feeding and suspension feeding communities. *American Naturalist*, **106**, 472–86.

Levinton, J. S. (1973). Genetic variation in a gradient of environmental variability: marine Bivalvia (Mollusca). *Science*, **130**, 75–6.

Levinton, J. S. & Fundiller, D. (1975). An ecological and physiological approach to the study of biochemical polymorphisms. In: *Proceedings of the ninth European marine biology symposium* (ed. H. Barnes), pp. 165–76. Aberdeen University Press, Aberdeen.

Lewis, J. R. (1964). *The ecology of rocky shores*. English Universities Press, London.

Lewis, J. R. (1972). Problems and approaches to baseline studies in coastal communities. In: *Marine pollution and sea life* (ed. M. Ruiva), pp. 401–3. Fishing News (Books) Ltd, London.

L-Fando, J. J., García-Fernández, M. C. & R-Candela, J. L. (1972). Glycogen metabolism in *Ostrea edulis* (L.). Factors affecting glycogen synthesis. *Comparative Biochemistry and Physiology*, **43B**, 807–14.

Ling, S. W. (1973). A review of the status and problems of coastal aquaculture in the Indo-Pacific region. In: *Coastal aquaculture in the Indo-Pacific region* (ed. T. V. R. Pillay), pp. 2–25. Fishing News (Books) Ltd, London.

List, T. (1902). Fauna und Flora des Golfes von Neapel und der angrenzenden Meeres-Abschnitte. I. Die Mytiliden des Golfes von Neapel und der agrenzenden Meeres-Abschnitte. *Mitteilungen aus der Zoologischen Station zu Neapel*, **27**, 1–312.

Livingstone, D. R. (1975). A comparison of the kinetic properties of pyruvate kinase in two populations of Mytilus edulis L. In: *Proceedings of the ninth European marine biology symposium* (ed. H. Barnes). pp. 151–64. Aberdeen University Press. Aberdeen.

Livingstone, D. R. & Bayne, B. L. (1974). Pyruvate kinase from the mantle tissue of Mytilus edulis L. *Comparative Biochemistry and Physiology*, **48 B**, 481–97.

Lloyd, R. F. & Lloyd, K. O. (1963). Sulphatases and sulphated polysaccharides in the viscera of marine molluscs. *Nature, London*, **199**, 287.

Lo Bianco, S. (1899). Notizie biologische reguardanti specialmente il periodo di maturità sessuale degli animali del Golfo di Napoli. *Mitteilungen aus der Zoologischen Station zu Neapel*, **13**, 448–573.

Longcamp, D., Lubet, P. & Drosdowsky, M. (1974). The *in vitro* biosynthesis of steroids by the gonad of the mussel (*Mytilus edulis*). *General and Comparative Endocrinology*, **22**, 116–27.

Longo, F. J. & Anderson, E. (1969a). Cytological aspects of fertilisation in the lamellibranch, Mytilus edulis. 1. Polar body formation and development of the female pronucleus. *Journal of Experimental Zoology*, **172**, 69–96.

Longo, F. J. & Anderson, E. (1969b). Cytological aspects of fertilisation in the lamellibranch, Mytilus edulis. 2. Development of the male pronucleus and the association of the maternally and paternally derived chromosomes. *Journal of Experimental Zoology*, **172**, 97–120.

Longo, F. J. & Dornfeld, E. (1967). The fine structure of the spermatid differentiation in the mussel Mytilus edulis. *Journal of Ultrastructure Research*, **20**, 462–80.

Loosanoff, V. L. (1942). Shell movements of the edible mussel Mytilus edulis in relation to temperature. *Ecology*, **23**, 231–4.

Loosanoff, V. L. (1954). New advances in the study of bivalve larvae. *American Scientist*, **42**, 607–24.

Loosanoff, V. L. (1962). Effects of turbidity on some larval and adult bivalves. *Proceedings of the Gulf and Caribbean Fisheries Institute, 14th Annual Session*, pp. 80–94.

Loosanoff, V. L. & Davies, H. C. (1950). Conditioning Venus mercenaria for spawning in winter and breeding its larvae in the laboratory. *Biological Bulletin. Marine Biological Laboratory, Woods Hole, Mass.*, **98**, 60–5.

Loosanoff, V. L. & Davis, H. C. (1951). Delayed spawning of lamellibranchs by low temperature. *Journal of Marine Research*, **10**, 197–202.

Loosanoff, V. L. & Davis, H. C. (1963). Rearing of bivalve molluscs. *Advances in Marine Biology*, **1**, 1–136.

Loosanoff, V. L., Davis, H. C. & Chanley, P. E. (1966). Dimensions and shapes of larvae of some marine bivalve molluscs. *Malacologia*, **4**, 351–435.

453

References

Loosanoff, V. L., Miller, W. S. & Smith, P. B. (1951). Growth and setting of larvae of *Venus mercenaria* in relation to temperature. *Journal of Marine Research*, **10**, 59–81.

Loosanoff, V. L. & Nomejko, C. A. (1951). Existence of physiologically different races of oysters, *Crassostrea virginica*. *Biological Bulletin. Marine Biological Laboratory, Woods Hole, Mass.*, **101**, 151–6.

Lopes, C. R. & Lorenzo, R. (1957). Colorimetric determination of lead in the mussel *Mytilus edulis* and the seawater of the estuary of Vigo. *Boletin del Instituto espanel de Oceanografia*, **84**, 1–13.

Lough, R. G. (1974). A re-evaluation of the combined effects of temperature and salinity on survival and growth of *Mytilus edulis* larvae using response surface techniques. *Proceedings of the National Shellfisheries Association*, **64**, 73–6.

Lough, R. G. & Gonor, J. J. (1971). Early embryonic stages of *Adula californiensis* (Pelecypoda: Mytilidae) and the effect of temperature and salinity on development rate. *Marine Biology*, **8**, 118–25.

Lough, R. G. & Gonor, J. J. (1973a). A response-surface approach to the combined effects of temperature and salinity on the larval development of *Adula californiensis* (Pelecypoda: Mytilidae). I. Survival and growth of three- and fifteen-day old larvae. *Marine Biology*, **22**, 241–50.

Lough, R. G. & Gonor, J. J. (1973b). A response-surface approach to the combined effects of temperature and salinity on the larval development of *Adula californiensis* (Pelecypoda: Mytilidae). II. Long-term larval survival and growth in relation to respiration. *Marine Biology*, **22**, 295–305.

Lowe, G. A. (1974). Effect of temperature change on the heart rate of *Crassostrea gigas* and *Mya arenaria* (Bivalvia). *Proceedings of the Malacological Society of London*, **41**, 29–36.

Lowe, G. A. & Trueman, E. R. (1972). The heart and water flow rates of *Mya arenaria* (Bivalvia: Mollusca) at different metabolic levels. *Comparative Biochemistry and Physiology*, **41A**, 487–94.

Lowe, J. I., Wilson, P. D., Rick, A. J. & Wilson, A. J. (1971). Chronic exposure of oysters to DDT, toxaphene and parathion. *Proceedings of the National Shellfisheries Association*, **61**, 71–9.

Loxton, J. & Chaplin, A. E. (1974). The metabolism of *Mytilus edulis* L. during facultative anaerobiosis. *Biochemical Society Transactions*, **2**, 419–21.

Lozada, L. E., Rolleri, C. J. & Yañez, N. R. (1971). Consideraciones biologicas de *Choromytilus chorus* en dos sustratos differentes. *Biologia Pesquera*, **5**, 61–108.

Lubet, P. (1955a). Cycle neurosécrétoire chez *Chlamys varia* et *Mytilus edulis*. *Compte Rendus, Academie des Sciences, Paris*, **241**, 119–21.

Lubet, P. (1955b). La déterminisme de la ponte chez les lamellibranches (*Mytilus edulis* L.) *Compte Rendus, Academie des Sciences, Paris*, **241**, 254–6.

Lubet, P. (1956). Effets de l'ablation des centres nerveux sur l'émission des gametes chez *Mytilus edulis* L. et *Chlamys varia* L. (Mollusques, Lamellibranches). *Annales des sciences naturelles*, **18**, 175–83.

Lubet, P. (1957). Cycle sexuel de *Mytilus edulis* L. et de *Mytilus galloprovincialis* Lmk. dans le Bassin d'Arcachon (Gironde). *Année biologique*, **33**, 19–29.

Lubet, P. (1959). Recherches sur le cycle sexuel et l'émission des gamètes chez les Mytilidae et les Pectinidae (Moll. Bivalves). *Revue des travaux de l'Office (scientifique et technologique) des Pêches maritimes*, **23**, 387–548.

Lubet, P. E. (1961). *Rapport au Gouvernement de la Yougoslavie sur l'Ostreiculture et la Mytiliculture, Programme elargi d'Assistance Technique.* Report 1334. FAO, Rome, 55 pp.

Lubet, P. (1965). Incidences de l'ablation bilatérale des ganglion cérébroides sur la gamétogénèse et la développement du tissue conjonctif chez la Moule *Mytilus galloprovincialis* Lmk. (Mollusca: Lamillibranches). *Comptes rendus des séances de la Société de Biologie*, **159**, 397–399.

Lubet, P. (1966). Essai d'analyse experimentale des perturbations produites par les ablations de ganglions nerveux chez *Mytilus edulis* L. et *Mytilus galloprovincialis* Lmk. (Mollusques: Lamellibranches). *Annales d'Endocrinologie*, **27**, 353–65.

Lubet, P. (1973). Exposé synoptique des données biologique sur la moule *Mytilus galloprovincialis* (Lamarck 1819). *Synopsis FAO sur les pêches No. 88.* (SAST-Moule, 3, 16(10), 028, 08, pag. var.) FAO, Rome.

Lubet, P. & Chappuis, J. (1966). Etude du débit palléal et de la filtration par une méthode directe chez *Mytilus edulis* et *M. galloprovincialis*. *Bulletin de la Société linnéenne de Normandie*, **10**, 210–16.

Lubet, P. & Chappuis, J. G. (1967). Action de la témperature sur le rythme cardiaque de *Mytilus edulis* L. (Mollusque, Lamellibranche). *Comptes Rendus des Séances de la Société de biologie*, **161**, 1544–7.

Lubet, P. & Choquet, C. (1971). Cycles et rhythmes sexuels chez les mollusques bivalves et gastropodes. Influence du milieu et étude experimentale. *Haliotis*, **1**, 129–49.

Lubet, P. & Le Feron de Longcamp, D. (1969). Etude des variations annuelles des constituants lipidiques chez *Mytilus edulis* L. de la Baie de Seine (Calvados). *Compte Rendus des Séances de la Société de Biologie*, **163**, 1110–12.

Lubet, P. & Le Gall, P. (1967). Observations sur le cycle sexuel de *Mytilus edulis* L. à Luc-s-Mer *Bulletin de la Société Linnéene de Normandie*, **10**, 303–7.

Lubet, P. & Pujol, J. P. (1963). Sur l'evolution du système neurosécréteur de *Mytilus galloprovincialis* Lmk. (Mollusque: Lamellibranche) lors de variations de la salinité. *Compte Rendus, Académie des Sciences*, **257**, 4032–4034.

Lubet, P. & Pujol, J. P. (1965). Incidence de la neurosécrétion sur l'euryhalinité de *Mytilus galloprovincialis* Lmk. variation de la teneur

References

en eau. *Rapport et procès-verbaux des réunions. Commission International pour l'Exploration scientifique de la Mer Mediterraneé*, **18**, 149–154.

Lubinsky, I. (1958). Studies on *Mytilus edulis* of the 'Calanus' expedition to Hudson Bay and Ungava Bay. *Canadian Journal of Zoology*, **36**, 869–81.

Lucas, A. M. (1931*a*). An investigation of the nervous system as a possible factor in the regulation of ciliary activity of the lamellibranch gill. *Journal of Morphology and Physiology*, **51**, 147–93.

Lucas, A. M. (1931*b*). The distribution of the branchial nerve in *Mytilus edulis* and its relation to the problem of nervous control of ciliary activity. *Journal of Morphology and Physiology*, **51**, 195–205.

Lucas, A. (1971). Les gametes des mollusques. *Haliotis*, **1**, 185–214.

Lucu, C. & Jelisavcic, O. (1970). Uptake of ^{137}Cs in some marine animals in relation to temperature, salinity, weight and moulting. *Internationale Revue der Gesamten Hydrobiologie, Berlin*, **55**, 783–96.

Lucu, C., Jelisavcic, O., Lulic, S. & Strohal, P. (1969). Interactions of ^{239}Pa with tissues of *Mytilus galloprovincialis* and *Carcinus mediterraneus*. *Marine Biology*, **2**, 103–4.

Lum, S. C. & Hammen, C. S. (1964). Ammonia excretion of *Lingula*. *Comparative Biochemistry and Physiology*, **12**, 185–90.

Lunetta, J. E. (1969). Reproductive physiology of the mussel *Mytilus perna*. *Boletin da Faculdade de Filosofia, Ciências e Letras, Universidade de São Paulo. Zoologia e Biologia Marhina*, **26**, 33–111.

Lynch, M. P. & Wood, L. (1966). Effects of environmental salinity on free amino acids of *Crassostrea virginica* Gmelin. *Comparative Biochemistry and Physiology*, **19**, 783–90.

Maas Geesteranus, R. A. (1942). On the formation of banks by *Mytilus edulis*. *Archives Néerlandaises Zoologie*, **6**, 283–325.

MacArthur, R. H. (1972). *Geographical ecology. Patterns in the distribution of species.* Harper & Row, New York.

McCann, F. (1969). Comparative Physiology of the Heart: current Trends. *Experientia*, Supplement, **15**,

McDermott, J. (1966). The incidence and host–parasite relations of pinnotherid crabs (Decapoda, Pinnotheridae). *Proceedings of the National Coastal Shallow Water Research Conference*, **1**, 162–4.

McDougal, K. D. (1943). Sessile marine invertebrates of Beaufort, North Carolina. *Ecological Monographs*, **13**, 321–74.

MacGinitie, G. E. (1941). On the method of feeding of four pelecypods. *Biological Bulletin. Marine Biological Laboratory, Woods Hole, Mass.*, **80**, 18–25.

McIntosh, W. C. (1885). On the British species of *Cyanea* and the reproduction of *Mytilus edulis* L. *Annals and Magazine of Natural History, London*, **5**, 148–52.

McIntosh, W. C. (1891). Report on the mussel and cockle beds in the estuaries of the Tees, the Esk and the Humber, pp. 1–87.

Maclean, J. L. (1972). Mussel culture: methods and prospects. *Australian Fisheries Paper*, No. 20. 12 pp.

Madsen, H. (1940). A study of the littoral fauna of north west Greenland. *Meddelelser om Grønland*, **124**, 1–24.

Majori, L. & Petroni, F. (1973a). Su di un modello semplificato per l'equilibrio di ripartizione dei metalli dra mitilo (*Mytilus galloprovincialis* Lmk) e aqua marina. Il fattore di accumulo. *Igiene Moderna*, **66**, 19–38.

Majori, L. & Petroni, F. (1973b). Fenomeno di accumulo nel mitilo (*Mytilus galloprovincialis*) stabulato semplificato per l'equilibrio dinamico di ripartizione dei metalli fra mitilo e aqua marina. I. Inquinamento da cadmio. *Igiene Moderna*, **66**, 39–63.

Majori, L. & Petroni, F. (1973c). Fenomeno di accumulo nel mitilo (*Mytilus galloprovincialis*) satbulato in ambiente artificialmente inquinato. Verifica di un modello semplificato per l'equilibrio dinamico di ripartizione dei metalli fra mitilo e aqua marina. II. Inquinamento da rame. *Igiene Moderna*, **66**, 64–78.

Majori, L. & Petroni, F. (1973d). Fenomeno di accumulo nel mitilo (*Mytilus galloprovincialis*) stabulato in ambiente artificialmente inquinato. Verifica di un modello semplificato per l'equilibrio dinamico di ripartizione dei metalli fra mitilo e aqua marina. III. Inquinamento da piombo. *Igiene Moderna*, **66**, 79–98.

Majori, L. & Petroni, F. (1973e). Fenomeno di accumulo nel mitilo (*Mytilus galloprovincialis*) stabulato in ambiente artificialmente inquinato. Verifica di un modello semplificato per l'equilibrio dinamico di ripartizione dei metalli fra mitilo e aqua marina. IV. Inquinamento da mercurio. *Igiene Moderna*, **66**, 99–122.

Malanga, C. J. & Aiello, E. (1971). Anaerobic cilio-excitation and metabolic stimulation by 5–hydroxytryptamine in bivalve gill. *Comparative and General Pharmacology*, **2**, 456–68.

Malanga, C. J. & Aiello, E. L. (1972). Succinate metabolism in the gills of the mussels *Modiolus demissus* and *Mytilus edulis*. *Comparative Biochemistry and Physiology*, **43B**, 795–806.

Maloeuf, N. S. R. (1937). Studies on the respiration (and osmoregulation) of animals. I. Animals without an oxygen transporter in their internal medium. *Zeitschrift für vergleichende Physiologie*, **25**, 1–28.

Mangum, C. P. & Burnett, L. E. (1975). The extraction of oxygen by estuarine invertebrates. In: *Physiological ecology of estuarine organisms* (ed. F. J. Vernberg), pp. 147–63. University of South Carolina Press, Columbia, South Carolina.

Mangum, C. & Van Winkle, W. (1973). Responses of aquatic invertebrates to declining oxygen conditions. *American Zoologist*, **13**, 529–41.

Manikowski, S. (1968). Observations on the occurrence and distribution of birds in the Baltic near the Hel Peninsula (Poland). *Acta ornithologica*, *Warsaw*, **11**, 45–59.

Mann, H. (1956). The influence of *Mytilicola intestinalis* on the development of the gonads of *Mytilus edulis*. *Rapport et procès-verbaux des réunions. Conseil permanent International pour l'Exploration de la Mer*, **140**, 57–8.

Manwell, C., Baker, C. M. A., Ashton, P. A. & Corner, E. D. S. (1967).

References

Biochemical differences between *Calanus finmarchicus* and *C. helgolandicus*. *Journal of the Marine Biological Association of the United Kingdom*, **47**, 145–69.

Marks, G. W. (1938). The copper tolerance of some species of molluscs of the southern California coast. *Biological Bulletin. Marine Biological Laboratory, Woods Hole, Mass.*, **75**, 224–37.

Mars, P. (1950). Euryhalinité de quelques Mollusques méditerranéens. *Vie et Milieu*, **1**, 441–8.

Marteil, L. (1961). Un essai de culture de moules sur cordes en Loire-Atlantique de 1959 a 1961. *Science et Pêche*, **94**, 5 pp.

Martin, A. W. (1966). Carbohydrate metabolism. I. Sugar and polysaccharides. In: *Physiology of Mollusca* (ed. K. M. Wilbur and C. M. Yonge), vol. II, pp. 275–93. Academic Press, New York and London.

Martin, A. W. & Harrison, F. M. (1966). Excretion. In: *Physiology of Mollusca* (ed. K. M. Wilbur and C. M. Yonge), vol. I, pp. 353–86. Academic Press, New York and London.

Martin, A. W., Harrison, F. M., Huston, M. J. & Stewart, D. M. (1958). The blood volumes of some representative molluscs. *Journal of Experimental Biology*, **35**, 260–79.

Martoja, M. (1972). Endocrinology of Mollusca. In: *Chemical zoology* (ed. M. Florkin and B. T. Scheer), vol. 7, pp. 349–92. Academic Press, New York and London.

Mason, J. (1968). Cultivation of mussels, *Mytilus edulis* L., in Scotland. *International Council for the Exploration of the Sea*, C.M. 1968/E.4, pp. 1–5. (Mimeograph.)

Mason, J. (1969). Mussel raft trials succeed in Scotland. *World Fishing*, **18**(4), 22–4.

Mason, J. (1971). Mussel cultivation. *Underwater Journal*, **3**, 52–9.

Mason, J. (1972). The cultivation of the European mussel, *Mytilus edulis* Linnaeus. *Oceanography and Marine Biology Annual Review*, **10**, 437–60.

Mason, J. O. & McLean, W. R. (1962). Infectious hepatitis traced to the consumption of raw oysters. *American Journal of Hygiene*, **75**, 90–111.

Mateeva, T. A. (1948). The biology of *Mytilus edulis* L. in eastern Murman. *Trudȳ Murmanskogo biologicheskogo Instituta*, **1**, 215–41. (In Russian.)

Matthews, A. (1913). Notes on the development of *Mytilus edulis* and *Alcyonium digitatum* in the Plymouth Laboratory. *Journal of the Marine Biological Association of the United Kingdom*, **9**, 557–60.

Mattisson, A. G. M. & Beechey, R. B. (1966). Some studies on cellular fractions of the adductor muscle of *Pecten maximus*. *Experimental Cell Research*, **41**, 227–43.

Mayzaud, P. (1973). Respiration and nitrogen excretion of zooplankton. II. Studies of the metabolic characteristics of starved animals. *Marine Biology*, **21**, 19–28.

Meaney, R. A. (1970*a*). Investigations on the mussel (*Mytilus edulis* L.) resources in the River Boyne estuary Ireland during 1969. *Fisheries Development Division, Resource Record Paper, Dublin*, Jan., 1–15.

Meaney, R. A. (1970*b*). Investigations on a seed mussel bed off the coast of Wexford during 1970. *Fisheries Development Division, Resource Record Paper, Dublin*, Dec., 1–5.

Meaney, R. A. (1970*c*). *Mussels in Ireland*. Irish Sea Fisheries Board, Resource Development Note. 12 pp.

Meisenheimer, J. (1901). Entwicklungsgeschichte von *Dreissena polymorpha* Pall. *Zeitschrift für wissenschaftliche Zoologie,* **69**, 1–137.

Meixner, R. (1971). Wachstum und Ertrag von *Mytilus edulis* bei Flosskultur in der Flensburger Förde. *Archiv für Fischereiwissenschaft*, **23**, 41–50.

Mellon, D. (1965). Complex electrical responses from ganglion cells in the surf clam (*Spisula*). *Biological Bulletin. Marine Biological Laboratory, Woods Hole, Mass.*, **129**, 415–16.

Mellon, D. (1972). Electrophysiology of touch sensitive neurons in a mollusc. *Journal of Comparative Physiology*, **79**, 63–78.

Menge, B. A. (1972). Foraging strategy of a starfish in relation to actual prey availability and environmental predictability. *Ecological Monographs*, **42**, 25–50.

Menzel, R. W. (1968). Chromosome number in nine families of pelecypod mollusks. *Nautilus*, **82**, 53–8.

Merrill, A. S. & Turner, R. D. (1967). Nest building in the bivalve genera *Musculus* and *Lima*. *Veliger*, **6**, 55–9.

Meyer-Waarden, P. F. & Mann, H. (1951). Recherches allemandes relatives au ' *Mytilicola*', copepode parasite de la moule existant dans les watten allemandes 1950–1951. *Revue des travaux de l'Office (scientifique et technologique) des Pêches maritimes*, **17**, 63–74.

Meyer-Waarden, P. F. & Mann, H. (1954*a*). Ein weiterer Beitrag zur Epidemiologie von *Mytilicola intestinalis*. *Archiv für Fischereiwissenschaft*, **5**, 26–34.

Meyer-Waarden, P. F. & Mann, H. (1954*b*). Untersuchungen über die Bestände von *Mytilus galloprovincialis* an der italienischen Küste auf ihren befall mit *Mytilicola intestinalis* (Copepoda parasitica). *Bolletino di Pesca, Piscicoltura e Idrobiologia, Rome*, **8**, 201–20.

Mikhailova, I. G. & Prazduikar, E. V. (1961). On the problem of reactions of the tissues of *Mytilus edulis*. *Trudȳ Murmanskogo biologicheskogo Instituta*, **3**, 125–30. (In Russian.)

Mikhailova, I. G. & Prazduikar, E. V. (1962). Inflammatory reactions in the Barent's Sea mussel (*Mytilus edulis*). *Trudȳ Murmanskogo biologicheskogo Instituta*, **4**, 208–20. (In Russian.)

Mileikovski, S. A. (1959). Interrelations between the pelagic larvae of *Nephthys ciliata* (O. F. Muller), *Macoma baltica* and *Mya arenaria* of the White Sea. *Zoologicheskii Zhurnal*, **38**, 1889–91. (In Russian.)

Mileikovsky, S. A. (1968). Distribution of pelagic larvae of bottom invertebrates of the Norwegian and Barents Seas. *Marine Biology*, **1**, 161–7.

References

Mileikovsky, S. A. (1970). Seasonal and daily dynamics in pelagic larvae of marine shelf bottom invertebrates in nearshore waters of Kandalaksha Bay (White Sea). *Marine Biology*, **5**, 180–94.

Mileikovsky, S. A. (1971). Types of larval development in marine bottom invertebrates, their distribution and ecological significance: a re-evaluation. *Marine Biology*, **10**, 193–213.

Mileikovsky, S. A. (1972). The ' pelagic larvaton ' and its role in the biology of the world ocean, with special reference to pelagic larvae of marine bottom invertebrates. *Marine Biology*, **16**, 13–21.

Mileikovsky, S. A. (1973). Speed of active movement of pelagic larvae of marine bottom invertebrates and their ability to regulate their vertical position. *Marine Biology*, **23**, 11–17.

Mileikovsky, S. A. (1974). On predation of pelagic larvae and early juveniles of marine bottom invertebrates by adult benthic invertebrates and their passing alive through their predators. *Marine Biology*, **26**, 303–11.

Milkman, R. & Beaty, L. D. (1970). Large-scale electrophoretic studies of allelic variation in *Mytilus edulis*. *Biological Bulletin. Marine Biological Laboratory, Woods Hole, Mass.*, **139**, 450.

Millar, R. H. (1968). Growth lines in the larvae and adults of bivalve molluscs. *Nature, London*, **217**, 683.

Millar, R. H. & Scott, J. M. (1967*a*). Bacteria-free culture of oyster larvae. *Nature, London*, **216**, 1139–40.

Millar, R. H. & Scott, J. M. (1967*b*). The larvae of the oyster *Ostrea edulis* during starvation. *Journal of the Marine Biological Association of the United Kingdom*, **47**, 475–84.

Millar, R. H. & Scott, J. M. (1968). An effect of water quality on the growth of cultured larvae of the oyster *Ostrea edulis* L. *Journal du Conseil. Conseil international pour l'Exploration de la Mer*, **32**, 123–30.

Millard, N. (1952). Observations and experiments on fouling organisms in Table Bay Harbour, South Africa. *Transactions of the Royal Society of South Africa*, **33**, 415–46.

Miller, A. T. (1966). The role of oxygen in metabolic regulation. *Helgoländer wissenschaftliche Meeresuntersuchungen*, **14**, 392–406.

Millott, N. (1968). The dermal light sense. *Symposia of the Zoological Society of London*, **23**, 1–36.

Milne, H. A. (1940). Some ecological aspects of the intertidal area of the estuary of the Aberdeenshire Dee. *Transactions of the Royal Society of Edinburgh*, **50**, 107–40.

Milne, H. A. & Dunnet, G. M. (1972). Standing crop, productivity and trophic relations of the fauna of the Ythan estuary. In: *The estuarine environment* (ed. R. S. K. Barnes and J. Green), pp. 86–106. Associated Scientific Publishers, Amsterdam.

Milne, P. H. (1972). *Fish and shellfish farming in coastal waters*. Fishing News (Books) Ltd, London. 208 pp.

Milton, A. S. & Gosselin, R. E. (1960). Metabolism and cilio-accelerator action of 5-hydroxytryptamine (5-HTP) in gill plates of *Mytilus* and

Modiolus. Federation Proceedings. Federation of American Societies for Experimental Biology, **19**, 126.

Mitton, J. B., Koehn, R. K. & Prout, T. (1973). Population genetics of marine pelecypods. III. Epistasis between functionally related isoenzymes of *Mytilus edulis. Genetics*, **73**, 487–96.

Miyazaki, I. (1935). On the development of some marine bivalves, with special reference to the shelled larvae. *Journal of the Imperial Fisheries Institute, Tokyo*, **31**, 1–10.

Miyazaki, I. (1938*a*). On fouling organisms in the oyster farm. *Bulletin of the Japanese Society of Scientific Fisheries*, **6**, 223–32.

Miyazaki, I. (1938*b*). On a substance which is contained in green algae and induces spawning action of the male oyster. *Bulletin of the Japanese Society of Scientific Fisheries*, **7**, 137–8.

Molinier, R. & Picard, J. (1957). Aperçu bionomique sur les peuplements marins littoraux des côtes rocheuses méditerranéens de l'Espagne. *Extrait du Bulletin des Travaux publiés par la Station d'Aquiculture et de Pêche de Castiglione*, N.S., **8**, 1–18.

Molins, L. R. (1957). Colorometric determination of copper in seawater and in the mussel *Mytilus edulis. Boletin del Instituto espanol de Oceanografia*, **86**, 1–11.

Moon, T. W. & Pritchard, A. W. (1970). Metabolic adaptations in vertically separated populations of *Mytilus californianus* Conrad. *Journal of Experimental Marine Biology and Ecology*, **5**, 35–46.

Moore, D. R. & Reish, D. J. (1969). Studies on the *Mytilus edulis* community in Alimitos Bay, California. 4. Seasonal variation in gametes from different regions in the Bay. *Veliger*, **11**, 250–5.

Moore, H. J. (1971). The structure of the latero-frontal cirri on the gills of certain lamellibranch molluscs and their role in suspension feeding. *Marine Biology*, **11**, 23–7.

Moore, K. E. & Gosselin, R. E. (1962). Effects of 5-hydroxytryptamine on the anaerobic metabolism and phosphorylase activity of lamellibranch gill. *Journal of Pharmacology and Experimental Therapeutics*, **138**, 145–53.

Moore, K. E., Milton, A. S. & Gosselin, R. E. (1961). Effect of 5-hydroxytryptamine on the respiration of excised lamellibranch gill. *British Journal of Pharmacology*, **17**, 278–85.

Moore, M. & Lowe, D. (1975). The cytology and cytochemistry of the hemocytes of *Mytilus edulis* and their responses to injected carbon particles. *Journal of Invertebrate Pathology*, in press.

Moreno, de Aizpun, Moreno, V. J. & Malaspina, A. M. (1971). Estudios sobre el mejillón (*Mytilus platensis* d'Orb) en explotación comercial del sector bonaerense Mar Argentino. 2. Ciclo anual en los principales componentes bioquimicos. *Contributions. Institute of Marine Biology, Mar de Plata*, **156**, 1–15.

Morgan, P. R. (1972). The influence of prey availability on the distribution and predatory behaviour of *Nucella lapillus* (L.). *Journal of Animal Ecology*, **41**, 257–74.

Mori, K., Tamate, H. & Imai, T. (1966). Histochemical study on the change of 17β-hydroxysteroid dehydrogenase activity in the oyster during the stages of sexual maturation and spawning. *Tohoku Journal of Agricultural Research*, **17**, 179–87.

Morris, R. J. & Sargent, J. R. (1973). Studies on the lipid metabolism of some oceanic crustaceans. *Marine Biology*, **22**, 77–83.

Morton, B. (1969*a*). Studies on the biology of *Dreissena polymorpha* Pall. I. General anatomy and morphology. *Proceedings of the Malacological Society of London*, **38**, 301–21.

Morton, B. (1969*b*). Studies on the biology of *Dreissena polymorpha* Pall. II. Population dynamics. *Proceedings of the Malacological Society of London*, **38**, 471–82.

Morton, B. (1970). The tidal rhythm and rhythm of feeding and digestion in *Cardium edule*. *Journal of the Marine Biological Association of the United Kingdom*, **50**, 499–512.

Morton, B. (1971). The diurnal rhythm and tidal rhythm of feeding and digestion in *Ostrea edulis*. *Biological Journal of the Linnean Society, London*, **3**, 329–42.

Morton, B. (1973). Some aspects of the biology and functional morphology of the organs of feeding and digestion of *Limnoperna fortunei* (Dunker) (Bivalvia: Mytilacea). *Malacologia*, **12**, 265–81.

Morton, J. E. & Miller, M. (1968). *The New Zealand shore*. Collins, London and Auckland. 638 pp.

Mossop, B. K. E. (1921). A study of the sea mussel (*Mytilus edulis* Linn.). *Contributions to Canadian Biology and Fisheries*, **2**, 17–48.

Mossop, B. K. E. (1922). The rate of growth of the sea mussel (*Mytilus edulis* L.) at St. Andrews New Brunswick, Digby Nova Scotia and Hudson Bay. *Transactions of the Royal Canadian Institute*, **4**, 3–21.

Mostert, S. A. (1972). Preliminary report on black mussel culture in the Langebaan lagoon. *South African Shipping News and Fishing Industry Review*, **27**(9), 59–63.

Motwani, M. P. (1955). Experimental and ecological studies on the adaptation of *Mytilus edulis* L to salinity fluctuations. *Proceedings of the National Institute of Sciences of India*, **21**, 227–46.

Mpitsos, G. J. (1973). Physiology of vision in the mollusk *Lima scabra*. *Journal of Neurophysiology*, **36**, 371–83.

Murakami, A. (1962). On the mechanism of ciliary junctions in gill of *Mytilus*. *Journal of the Faculty of Science, Tokyo University, Sect. 4 (Zoology)*, **9**, 319–32.

Murdoch, W. W. (1969). Switching in general predators. Experiments on predator specificity and stability of prey populations. *Ecological Monographs*, **39**, 335–54.

Murray, P. J. (1971). Mussel culture. *FAO Aquaculture Bulletin*, **3**(4), 8.

Mustafa, T. & Hochachka, P. W. (1971). Catalytic and regulatory properties of pyruvate kinases in tissues of a marine bivalve. *Journal of Biological Chemistry*, **246**, 3196–203.

Mustafa, T. & Hochachka, P. W. (1973*a*). Enzymes in facultative anaerobiosis of molluscs. II. Basic catalytic properties of phosphoenolpyruvate carboxykinase in oyster adductor muscle. *Comparative Biochemistry and Physiology*, **45 B**, 639–55.

Mustafa, T. & Hochachka, P. W. (1973*b*). Enzymes in facultative anaerobiosis of molluscs. III. Phosphoenolpyruvate carboxykinase and its role in aerobic–anaerobic transition. *Comparative Biochemistry and Physiology*, **46 B**, 657–67.

Muus, B. J. (1967). The fauna of Danish estuaries and lagoons. *Meddelelser fra Danmarks Fiskeri-og Havundersøgelser*, **5**, 1–316.

Muus, K. (1973). Setting, growth and mortality of young bivalves in the Oresund. *Ophelia*, **12**, 79–116.

Nagabhushanam, R. (1963). Neurosecretory cycle and reproduction in the bivalve, *Crassostrea virginica*. *Indian Journal of Experimental Biology*, **1**, 161–2.

Nagabhushanam, R. (1964). Effect of removal of neurosecretory cells on spawning in the mussel *Modiolus demissus*. *Current Science*, **7**, 215–6.

Nair, N. B. (1962). Ecology of marine fouling and wood-boring organisms of Western Norway. *Sarsia*, **8**, 1–88.

Nakahara, H. & Bevelander, G. (1969). An electron microscope study of ingestion of thorotrast by amoebocytes of *Pinctada radiata*. *Texas Reports on Biology and Medicine*, **27**, 101–9.

Naylor, E. (1962). Seasonal changes in the population of *Carcinus maenas* in the littoral zone. *Journal of Animal Ecology*, **31**, 601–9.

Naylor, E. (1965). The effects of heated effluents upon marine and estuarine organisms. *Advances in Marine Biology*, **3**, 63–103.

Nelson, T. C. (1912). *Report of the Biological Department of the New Jersey Agricultural Experiment Station for the year 1911.*

Nelson, T. C. (1928*a*). On the distribution of critical temperatures for spawning and for ciliary activity in bivalve mollusks. *Science*, **67**, 220–1.

Nelson, T. C. (1928*b*). Pelagic dissoconchs of the common mussel *Mytilus edulis* with observations on the behaviour of the larvae of allied genera. *Biological Bulletin. Marine Biological Laboratory, Woods Hole, Mass.*, **55**, 180–92.

Nelson, T. C. & Allison, J. B. (1940). On the nature and action of diantlin; a new hormone-like substance carried by the spermatozoa of the oyster. *Journal of Experimental Zoology*, **85**, 299–338.

Newcombe, C. L. (1935). A study of the community relationships of the sea mussel *Mytilus edulis* L. *Ecology*, **16**, 234–43.

Newell, B. S. (1953). Cellulolytic activity in the lamellibranch crystalline style. *Journal of the Marine Biological Association of the United Kingdom*, **32**, 491–5.

Newell, N. D. (1942). Late Palaeozoic pelecypods: Mytilacea. *Geological Survey of Kansas*, **10**, part 2.

Newell, R. C. (1967). Oxidative activity of poikilotherm mitochondria as a function of temperature. *Journal of Zoology*, **151**, 299–311.

References

Newell, R. C. (1969). The effect of temperature fluctuation on the metabolism of intertidal invertebrates. *American Zoologist*, **9**, 293–307.

Newell, R. C. (1970). *Biology of intertidal animals*. Logos Press, London. 555 pp.

Newell, R. C. (1973). Environmental factors affecting the acclimatory responses of ectotherms. In: *Effects of temperature on ectothermic organisms* (ed. W. Wieser), pp. 151–64. Springer-Verlag, Berlin.

Newell, R. C. & Bayne, B. L. (1973). A review on temperature and metabolic acclimation in intertidal marine invertebrates. *Netherlands Journal of Sea Research*, **7**, 421–33.

Newell, R. C. & Brown, B. (1971). Effects of chemical wastes on marine organisms. *Marine Technology*, **2**, 217–22.

Newell, R. C. & Pye, V. I. (1970a). Seasonal changes in the effect of temperature on the oxygen consumption of the winkle *Littorina littorea* (L.) and the mussel *Mytilus edulis* L. *Comparative Biochemistry and Physiology*, **34**, 367–83.

Newell, R. C. & Pye, V. I. (1970b). The influence of thermal acclimation on the relation between oxygen consumption and temperature in *Littorina littorea* (L.) and *Mytilus edulis* L. *Comparative Biochemistry and Physiology*, **34**, 385–97.

Newell, R. C. & Pye, V. I. (1971a). Quantitative aspects of the relationship between metabolism and temperature in the winkle *Littorina littorea* (L.). *Comparative Biochemistry and Physiology*, **38B**, 635–50.

Newell, R. C. & Pye, V. I. (1971b). Temperature-induced variations in the respiration of mitochondria from the winkle, *Littorina littorea* (L.). *Comparative Biochemistry and Physiology*, **40B**, 249–61.

Newell, R. C. & Roy, A. (1973). A statistical model relating the oxygen consumption of a mollusk (*Littorina littorea*) to activity, body size, and environmental condition. *Physiological Zoology*, **46**, 253–75.

Newsholme, E. A. & Start, C. (1973). *Regulation in metabolism*. Wiley, New York. 349 pp.

New Zealand Fishing Industry Board (1971). *Report on mussel cultivation seminar, Wellington, October 1971*. 46 pp.

Nicol, W. (1906). Notes on trematode parasites of the cockle and mussel. *Annals and Magazine of Natural History, London*, **7**, 148–55.

Niijima, L. (1963). Acrosome reaction and sperm entry in *Mytilus edulis*. *Bulletin of the Marine Biological Station of Asamushi*, **11**, 217–21.

Niijima, L. & Dan, J. (1965). The acrosome reaction in *Mytilus edulis* L. 1. Fine structure of the intact acrosome. 2. Stages in the reaction, observed in supernumerary and calcium-treated spermatozoa. *Journal of Cell Biology*, **25**, 243–59.

Nikolić, M. & Stojnić, I. (1963). A system of mussel culture. *Proceedings and Technical Papers of the General Fisheries Council for the Mediterranean*, **7**, 251–5.

Nikolskii, G. V. (1966). On the structure of the population and on the mortality character of mussels *Mytilus edulis* in the littoral of the White Sea. *Zoologicheskii Zhurnal*, **45**, 1878–80. (In Russian.)

Nilsson, L. (1969). Food consumption of diving ducks wintering at the coast of southern Sweden in relation to food resources. *Oikos*, **20**, 128–35.

Nixon, S. W., Oviatt, C. A., Rogers, C. & Taylor, K. (1971). Mass and metabolism of a mussel bed. *Oecologia*, **8**, 21–30.

Norton-Griffiths, M. (1967). Some ecological aspects of the feeding behaviour of the oystercatcher *Haematopus ostralegus* on the edible mussel *Mytilus edulis*. *Ibis*, **109**, 412–24.

Obusan, R. A. & Urbano, E. E. (1968). Tahong – food for the millions. *Technical Paper of the Indo-Pacific Fisheries Council, No. 29.* 25 pp. (Mimeo.)

Ockelmann, K. W. (1965). Developmental types in marine bivalves and their distribution along the Atlantic coast of Europe. In: *Proceedings of the First European Malacological Congress (1962)*, pp. 25–35.

O'Connor, R. J. (1975). A re-examination of curve-fitting in growth studies. *Condor*, in press.

O'Doherty, P. J. A. & Feltham, L. A. W. (1971). Glycolysis and gluconeogenesis in the giant scallop, *Plactopen magellanicus* (Gmelin). *Comparative Biochemistry and Physiology*, **38B**, 543–51.

O'Gower, A. K. & Nicol, P. I. (1968). A latitudinal cline of haemoglobins in a bivalve mollusc. *Heredity*, **23**, 485–92.

Okubo, K. & Okubo, T. (1962). Study on the bioassay method for the evaluation of water pollution. II. Use of fertilized eggs of sea urchins and bivalves. *Bulletin of the Tokai Fishery Research Laboratory*, **32**, 131–40.

Oldham, C. (1930). The shell smashing habit of gulls. *Ibis*, **12**, 239–43.

Orton, J. H. (1920). Sea temperatures, breeding and distribution in marine animals. *Journal of the Marine Biological Association of the United Kingdom*, **12**, 339–66.

Orton, J. H. (1933). Strange spatfall of the common mussel on the common cockle. *Nature, London*, **131**, 513–14.

Orton, J. H., Southward, A. J. & Dodd, J. M. (1956). Studies on the biology of limpets. 2. The breeding of *Patella vulgata* L. in Britain. *Journal of the Marine Biological Association of the United Kingdom*, **35**, 149–76.

O'Sullivan, A. J. (1971). Ecological effects of sewage discharge in the marine environment. *Proceedings of the Royal Society of London*, Ser. B, **177**, 331–51.

Owen, G. (1955). Observations on the stomach and digestive diverticula of the Lamellibranchia. I. The Anisomyaria and Eumellibranchia. *Quarterly Journal of Microscopical Science*, **96**, 517–37.

Owen, G. (1966a). Feeding. In: *Physiology of Mollusca* (ed. K. M. Wilbur and C. M. Yonge), vol. II, pp. 1–51. Academic Press, New York and London.

References

Owen, G. (1966*b*). Digestion. In: *Physiology of Mollusca* (ed. K. M. Wilbur and C. M. Yonge), vol. II, pp. 53–96. Academic Press, New York and London.

Owen, G. (1972). Lysosomes, peroxisomes and bivalves. *Science Progress*, **60**, 299–318.

Owen, G. (1974*a*). Studies on the gill of *Mytilus edulis*: the eu-latero-frontal cirri. *Proceedings of the Royal Society of London*, Ser. B, **187**, 83–91.

Owen, G. (1974*b*). Feeding and digestion in the Bivalvia. *Advances in Comparative Physiology and Biochemistry*, **5**, 1–35.

Paine, R. T. (1966). Food web complexity and species diversity. *American Naturalist*, **100**, 65–75.

Paine, R. T. (1969). The *Pisaster–Tegula* interaction: prey patches, predator food preference and intertidal community structure. *Ecology*, **50**, 635–44.

Paine, R. T. (1971). A short term experimental investigation of resource partitioning in a New Zealand rocky intertidal habitat. *Ecology*, **52**, 1096–1106.

Paine, R. T. (1974). Intertidal community structure. Experimental studies on the relationship between a dominant competitor and its principal predator. *Oecologia*, **15**, 93–120.

Palichenko, Z. G. (1948). On the biology of *Mytilus edulis* in the White Sea. *Zoologicheskii Zhurnal*, **27**, 411–20. (In Russian.)

Paloheimo, J. E. & Dickie, L. M. (1965). Food and growth of fishes. I. A growth curve derived from experimental data. *Journal of the Fisheries Research Board of Canada*, **22**, 521–42.

Paloheimo, J. E. & Dickie, L. M. (1966*a*). Food and growth of fishes. II. Effects of food and temperature on the relation between metabolism and body weight. *Journal of the Fisheries Research Board of Canada*, **23**, 869–908.

Paloheimo, J. E. & Dickie, L. M. (1966*b*). Food and growth of fishes. III. Relations among food, body size and growth efficiency. *Journal of the Fisheries Research Board of Canada*, **23**, 1209–48.

Pamatmat, M. M. (1969). Seasonal respiration of *Transennella tantilla* Gould. *American Zoologist*, **9**, 418–26.

Paparo, A. (1972). Innervation of the lateral cilia in the mussel, *Mytilus edulis* L. *Biological Bulletin. Marine Biological Laboratory, Woods Hole, Mass.*, **143**, 592–604.

Paparo, A. & Aiello, E. (1970). Cilio-inhibitory effects of branchial nerve stimulation in the mussel, *Mytilus edulis*. *Comparative and General Pharmacology*, **1**, 241–50.

Paparo, A. & Finch, C. E. (1972). Catecholamine localization, content, and metabolism in the gill of two lamellibranch molluscs. *Comparative and General Pharmacology*, **3**, 303–9.

Parsons, T. R. & Takahashi, M. (1973). *Biological oceanographic processes*. Pergamon Press, Oxford. 186 pp.

Pasteels, J-J. (1967). Absorption et athrocytose par l'épithélium branchial de *Mytilus edulis*. *Comptes Rendus, Académie des Sciences*, **264**D, 2505–7.

Pasteels, J-J. (1968). Pinocytose et athrocytose par l'épithélium branchial de *Mytilus edulis* L. *Zeitschrift für Zellforschung und mikroskopische Anatomie*, **92**, 339–59.

Pasteels, J-J. (1969). Excrétion de phosphatase acide par les cellules mucipares de la branchie de *Mytilus edulis* L. *Zeitschrift für Zellforschung und mikroskopische Anatomie*, **102**, 594–600.

Paul, M. D. (1942). Studies on the growth and breeding of certain sedentary organisms in Madras Harbour. *Proceedings of the Indian Academy of Sciences*, Sect. B, **15**, 1–42.

Pauley, G. B. & Heaton, L. H. (1969). Experimental wound repair in the freshwater mussel *Anodonta oregonensis*. *Journal of Invertebrate Pathology*, **13**, 241–9.

Pavlović, V., Kekic, H. & Mladenović, O. (1970). Glycogen in hepatopancreas and in muscles in *Ostrea edulis* L. and in *Mytilus galloprovincialis* Lmk. in the season summer–winter. *Bulletin scientifique. Conseil des Académies de la RPF Yougoslavie*, **15**, 76–7.

Paz-Andrade, A. & Waugh, G. D. (1968). Raft cultivation of mussels is big business in Spain. *World Fishing*, **17**(3), 50–2.

Pearce, J. B. (1966a). The biology of the mussel crab *Fabia subquadrata* from the waters of the San Juan Archipeligo, Washington. *Pacific Science*, **20**, 3–35.

Pearce, J. B. (1966b). On *Pinnixia faba* and *P. littoralis* (Decapoda: Pinnotheridae) symbiotic with the clam *Tresus capax* (Pelecypoda: Mactridae). In: *Some contemporary studies in marine science* (ed. H. Barnes), pp. 565–89. Allen & Unwin, London.

Pearce, J. B. (1969). Thermal addition and the benthos; Cape Cod canal. *Chesapeake Science*, **10**, 227–33.

Pearl, R. & Miner, J. R. (1935). Experimental studies on the duration of life. 14. The comparative mortality of lower organisms. *Quarterly Review of Biology*, **10**, 60–79.

Penchaszadeh, P. E. (1971). Estudios sobre el mejillón (*Mytilus platensis* d'Orb) en exploitacion commercial del sector Bona erense Mar Argentino. 1. Reproducción, crecimiento y estructura de la población. *Contributions. Institute of Marine Biology, Mar del Plata*, **153**, 1–15.

Pentraeth, R. J. (1973). The accumulation from water of ^{65}Zn, ^{54}Mn, ^{58}Co and ^{59}Fe by the mussel, *Mytilus edulis*. *Journal of the Marine Biological Association of the United Kingdom*, **53**, 127–43.

Péquignat, E. (1973). A kinetic and autoradiographic study of the direct assimilation of amino-acids and glucose by organs of the mussel *Mytilus edulis*. *Marine Biology*, **19**, 227–44.

Pequegnat, W. E. (1963). Population dynamics in a sub-littoral fauna. *Pacific Science*, **17**, 424–30.

Percy, J. A. & Aldrich, F. A. (1971). Metabolic rate independent of temperature in mollusc tissue. *Nature, London*, **231**, 393–4.

Percy, J. A., Aldrich, F. A. & Marcus, T. R. (1971). Influence of environmental factors on respiration of excised tissues of American oysters, *Crassostrea virginica* (Gmelin). *Canadian Journal of Zoology*, **49**, 353–60.

References

Perkins, E. J. (1967). Some aspects of the biology of *Carcinus maenas* (L.). *Transactions of the Dumfriesshire and Galloway Natural History and Antiquarian Society*, Ser. 3, **44**, 47–56.

Perkins, E. J. (1968). The toxicity of oil emulsifiers to some inshore fauna. *Field Studies*, **2**, suppl., 81–90.

Pickens, L. E. R. (1937). The mechanism of urine formation in invertebrates. II. The excretory mechanism in certain molluscs. *Journal of Experimental Biology*, **14**, 20–34.

Pickens, P. E. (1965). Heart rate of mussels as a function of latitude, intertidal height and acclimation temperature. *Physiological Zoology*, **38**, 390–405.

Pierce, S. K. (1970). Water balance in the genus *Modiolus* (Mollusca: Bivalvia: Mytilidae): osmotic concentrations in changing salinities. *Comparative Biochemistry and Physiology*, **36**, 521–33.

Pierce, S. K. (1971a). A source of solute for volume regulation in marine mussels. *Comparative Biochemistry and Physiology*, **38A**, 619–35.

Pierce, S. K. (1971b). Volume regulation and valve movements by marine mussels. *Comparative Biochemistry and Physiology*, **39A**, 103–17.

Pierce, S. K. & Greenberg, M. J. (1972). The nature of cellular volume regulation in marine bivalves. *Journal of Experimental Biology*, **57**, 681–92.

Pierce, S. K. & Greenberg, M. J. (1973). The initiation and control of free amino acid regulation of cell volume in salinity-stressed marine bivalves. *Journal of Experimental Biology*, **59**, 435–40.

Piiper, J. & Schumann, D. (1967). Efficiency of oxygen exchange in the gills of the dogfish, *Scyliorhinus stellaris*. *Respiration Physiology*, **2**, 135–48.

Pike, R. B. (1971). Report on mussel farming and mussel biology for the fishing industry board. *Technical Report*. *New Zealand Fisheries Industry Board*, **71**, 1–7.

Pilgrim, R. L. C. (1953a). Osmotic relations in Molluscan contractile tissues. I. Isolated ventricle strip preparations from Lamellibranchs (*Mytilus edulis* L., *Anodonta cygnea* L., *Ostrea edulis* L.). *Journal of Experimental Biology*, **30**, 297–317.

Pilgrim, R. L. C. (1953b). Osmotic relations in Molluscan contractile tissues. II. Isolated gill preparations from Lamellibranchs (*Mytilus edulis* L., *Anodonta cygnea* L.). *Journal of Experimental Biology*, **30**, 318–30.

Pleissis, Y. (1958). Note preliminaire sur l'étude statistique des coquilles vides de bivalves en particulier de *Mytilus edulis*. *Bulletin du Musée national d'Histoire naturelle*, **30**, 454–7.

Polikarpov, G. G. (1966). *Radioecology of aquatic organisms*. North Holland, Amsterdam. 314 pp.

Portmann, J. E. (1968). Progress report on a programme of insecticide analysis and toxicity testing in relation to the marine environment. *Helgoländer wissenschaftliche Meeresuntersuchungen*, **17**, 247–56.

Portmann, J. E. (1970). The toxicity of 120 substances to marine organisms. *Shellfish information leaflet No. 19*. Ministry of Agriculture, Fisheries and Food, Burnham-on-Crouch. 10 pp.

Potts, W. T. W. (1954). The inorganic composition of the blood of *Mytilus edulis* and *Anodonta cygnea. Journal of Experimental Biology*, **31**, 376–85.

Potts, W. T. W. (1958). The inorganic and amino acid composition of some lamellibranch muscles. *Journal of Experimental Biology*, **35**, 749–64.

Potts, W. T. W. (1967). Excretion in the molluscs. *Biological Reviews*, **42**, 1–41.

Potts, W. T. W. (1968). Aspects of excretion in the molluscs. In: *Studies in the structure, physiology and ecology of molluscs* (ed. V. Fretter), pp. 187–92. Academic Press, New York and London.

Poulet, S. A. (1974). Seasonal grazing of *Pseudocalanus minutus* on particles. *Marine Biology*, **25**, 109–23.

Prakash, A., Medcof, J. C. & Tennant, A. D. (1971). Paralytic shellfish poisoning in Eastern Canada. *Bulletin of the Fisheries Research Board of Canada*, **177**, 87 pp.

Prater, A. J. (1972). The ecology of Morecambe Bay. 3. The food and feeding habits of Knot (*Calidris canutus* L.) in Morecambe Bay. *Journal of Applied Ecology*, **9**, 179–94.

Precht, H. (1958). Concepts of the temperature adaptations of unchanging reaction systems of cold-blooded animals. In: *Physiological adaptation* (ed. C. L. Prosser), pp. 50–78. Ronald Press Co., New York.

Price, T. J. (1962). *Accumulation of radionuclides and the effect of radiation on molluscs*. United States Department of Health, Education and Welfare, Public Health Service Publication No. 999-WP-25. 8 pp.

Prichard, R. K. & Schofield, P. J. (1968). The glyoxylate cycle, fructose 1,6-diphosphatase and gluconeogenesis in *Fasciola hepatica. Comparative Biochemistry and Physiology*, **29**, 581–90.

Pringle, B. H., Hissong, D. E., Katz, E. L. & Mulawka, S. T. (1968). Trace metal accumulation by estuarine mollusks. *Journal of the Sanitary Engineering Division, Proceedings of the American Society of Civil Engineers*, **94**, 455–75.

Prior, D. J. (1972*a*). Electrophysiological analysis of peripheral neurones and their possible role in the local reflexes of a mollusc. *Journal of Experimental Biology*, **57**, 133–45.

Prior, D. J. (1972*b*). A neural correlate of behavioural stimulus intensity discrimination in a mollusc. *Journal of Experimental Biology*, **57**, 147–60.

Prosser, C. L. (1955). Physiological variation in animals. *Biological Reviews*, **30**, 229–62.

Prosser, C. L. (1957). Proposal for study of physiological variation in marine animals. *L'Année Biologique*, 3rd ser., **33**, 191–7.

Prosser, C. L. (1958). The nature of physiological adaptation. In: *Physiological adaptation* (ed. C. L. Prosser), pp. 167–80. American Physiological Society, Washington.

References

Prosser, C. L. & Brown, F. A. (1961). *Comparative animal physiology*, 2nd edition. W. B. Saunders, Philadelphia. 688 pp.

Pucci, I. (1961). Cytochemical investigations on the egg of *Mytilus. Acta Embryologiae et Morphologiae Experimentalis*, **4**, 96–101.

Pujol, J. P. (1968). La physiologie cardiaque des mollusques bivalves. *Bulletin de la Société linnéenne de Normandie*, **9**, 158–99.

Purchon, R. D. (1957). The stomach in the Filibranchia and Pseudolamellibranchia. *Proceedings of the Zoological Society of London*, **129**, 27–60.

Purchon, R. D. (1960). Phylogeny in the Lamellibranchia. *Proceedings of the Centennial and Bicentennial Congress of Biology, Singapore, 1958*, pp. 69–82. University of Malaya Press, Singapore.

Purchon, R. D. (1963). Phylogenetic classification of the Bivalvia with special reference to the Septibranchia. *Proceedings of the Malacological Society of London*, **35**, 71–80.

Purchon, R. D. (1968). *The biology of the Mollusca*. Pergamon Press, Oxford.

Purchon, R. D. (1971). Digestion in filter feeding bivalves – a new concept. *Proceedings of the Malacological Society of London*, **39**, 253–62.

Purdie, A. (1887). *Studies in biology for New Zealand students. 3. The anatomy of the common mussels* (Mytilus latus, edulis *and* magellanicus). New Zealand Colonial Museum and Geological Survey Dept. 45 pp.

Pye, V. (1973). Acute temperature change and the oxidative rates of ectotherm mitochondria. In: *Effects of temperature on ectothermic organisms* (ed. W. Wieser), pp. 83–95. Springer-Verlag, Berlin.

Pye, V. I. & Newell, R. C. (1973). Factors affecting thermal compensation in the oxidative metabolism of the winkle *Littorina littorea. Netherlands Journal of Sea Research*, **7**, 411–20.

Pyefinch, K. A. (1950). Notes on the ecology of ship-fouling organisms. *Journal of Animal Ecology*, **19**, 29–35.

Pyefinch, K. A. & Downing, F. S. (1949). Notes on the general biology of *Tubularia larynx*, Ellis and Solander. *Journal of the Marine Biological Association of the United Kingdom*, **28**, 21–43.

Quayle, D. B. (1969). Paralytic shellfish poisoning in British Columbia. *Bulletin of the Fisheries Research Board of Canada*, **168**, 68 pp.

Quick, J. A. (1971). *A preliminary investigation: the effect of elevated temperature on the American oyster* Crassostrea virginica (*Gmelin*). Professional Papers, Florida Department of Natural Resources, Ser. 15. 190 pp.

Quraishi, F. O. (1964). The effects of temperature on the feeding behaviour of mussels. PhD thesis, University of Wales Marine Science Laboratories, Menai Bridge, Wales.

Radford, P. J. (1972). The simulation language as an aid to ecological modeling. In: *Mathematical models in ecology* (ed. J. N. R. Jeffers), pp. 277–95. Blackwell Scientific Publications, Oxford.

Rafail, S. Z. (1968). A statistical analysis of ration and growth of plaice

(*Pleuronectes platessa*). *Journal of the Fisheries Research Board of Canada*, **25**, 717–32.

Rahn, H. (1966). Aquatic gas exchange: theory. *Respiration Physiology*, **1**, 1–12.

Rahn, H., Wangensteen, O. D. & Farhi, L. E. (1971). Convection and diffusion gas exchange in air or water. *Respiration Physiology*, **12**, 1–6.

Raimbault, R. & Tournier, M. (1973). Les cultures marines sur le littoral français de la Mediterranée. *Science et Pêche*, **223**, 20 pp.

Ramsay, J. A. (1952). *A physiological approach to the lower animals*. Cambridge: University Press, London. 149 pp.

Randall, D. J. (1970). Gas exchange in fish. In: *Fish physiology* (ed. W. S. Hoar & D. J. Randall), vol. IV, pp. 252–92. Academic Press, New York and London.

Rao, K. P. (1953). Rate of water propulsion in *Mytilus californianus* as a function of latitude. *Biological Bulletin. Marine Biological Laboratory, Woods Hole, Mass.*, **104**, 171–81.

Rao, K. P. (1954). Tidal rhythmicity of rate of water propulsion in *Mytilus* and its modificibility by transplantation. *Biological Bulletin. Marine Biological Laboratory, Woods Hole, Mass.*, **106**, 353–9.

Rao, K. P. & Bullock, T. H. (1954). Q_{10} as a function of size and habitat temperature in poikilotherms. *American Naturalist*, **88**, 33–43.

Rattenbury, J. C. & Berg, W. E. (1954). Embryonic segregation during early development of *Mytilus edulis. Journal of Morphology*, **95**, 393–414.

Raven, C. P. (1966). *Morphogenesis: the analysis of molluscan development*, 2nd edition. Pergamon Press, Oxford. 365 pp.

Raven, C. P. (1972). Chemical embryology of Mollusca. In: *Chemical zoology* (ed. M. Florkin and B. T. Scheer), vol. VII, pp. 155–85. Academic Press, New York and London.

Rawitz, B. (1887). Das centrale Nervensystem der Acephalen. *Jenaische Zeitschrift für Medizin und Naturwissenschaft*, **20**, 384–460.

Rawitz, B. (1890). Der Mantelrand der Acephalen. 2. Theil Arcacea, Mytilacea, Unionacea. *Jenaische Zeitschrift für Medizin und Naturwissenschaft*, **24**, 549–631.

Raymont, J. E. G. (1955). The fauna of an intertidal mud flat. In: *Papers in marine biology and oceanography*, pp. 178–203. *Deep Sea Research*, **3** (Suppl.).

Raymont, J. E. G. (1963). *Plankton and productivity of the oceans*. Pergamon Press, Oxford.

Raymont, J. E. G. & Carrie, B. G. A. (1964). The production of zooplankton in Southampton Waters. *Internationale Revue der gesamten Hydrobiologie und Hydrographie*, **49**, 185–232.

Read, K. R. H. (1962*a*). Respiration of the bivalve molluscs *Mytilus edulis* L. and *Brachidontes demissus plicatulus* Lamarck as a function of size and temperature. *Comparative Biochemistry and Physiology*, **7**, 89–101.

Read, K. R. H. (1962*b*). Transamination in certain tissue homogenates

References

of the bivalve molluscs *Mytilus edulis* L. and *Modiolus modiolus* L. *Comparative Biochemistry and Physiology*, **7**, 15–22.

Read, K. R. H. (1964). Ecology and environmental physiology of some Puerto Rican bivalve molluscs and a comparison with boreal forms. *Caribbean Journal of Science*, **4**, 459–65.

Read, K. R. H. & Cumming, K. B. (1967). Thermal tolerance of the bivalve molluscs, *Modiolus modiolus* L. and *Brachidontes demissus* Dillwyn. *Comparative Biochemistry and Physiology*, **22**, 149–55.

Reade, P. & Reade, E. (1972). Phagocytosis in invertebrates. II. The clearance of carbon particles by the clam *Tridacna maxima*. *Research Journal of the Reticuloendothelial Society*, **12**, 349–60.

Rees, C. B. (1951). The interpretation and classification of lamellibranch larvae. *Bulletin of Marine Ecology*, **3**, 73–104.

Rees, C. B. (1954). Continuous plankton records: the distribution of lamellibranch larvae in the North Sea, 1950–51. *Bulletin of Marine Ecology*, **4**, 21–46.

Reid, R. G. B. (1965). The structure and function of the stomach in bivalve molluscs. *Journal of Zoology*, **147**, 156–84.

Reid, R. G. B. (1968). The distribution of digestive tract enzymes in lamellibranchiate bivalves. *Comparative Biochemistry and Physiology*, **24**, 727–44.

Reish, D. J. (1964*a*). Studies on the *Mytilus edulis* community in Alimitos Bay, California. 1. Development and destruction of the community. *Veliger*, **6**, 124–31.

Reish, D. J. (1964*b*). Studies on the *Mytilus edulis* community in Alimitos Bay, California. 2. Population variations and discussion of the associated organisms. *Veliger*, **6**, 202–7.

Reish, D. J. (1972). The use of invertebrates as indicators of varying degrees of marine pollution. In: *Marine pollution and sea life* (ed. M. Ruivo), pp. 203–7. Fishing News (Books) Ltd, London.

Reish, D. J. & Ayers, J. L. (1968). Studies on the *Mytilus edulis* community of Alimitos Bay, California. 3. The effects of reduced dissolved oxygen and chlorinity concentrations on survival and byssus thread formation. *Veliger*, **10**, 384–8.

Remane, A. & Schlieper, C. (1971). *Biology of brackish water*. Wiley-Interscience, New York. 372 pp.

Remmert, H. (1969). Uber Poikilosmotie und Isoosmotie. *Zeitschrift für vergleichende Physiologie*, **65**, 424–7.

Renzoni, A. (1961). Variazione istologische stagionali dell gonadi di *Mytilus galloprovincialis* Lam. in rapporto al ciclo riproduttivo. *Rivista di Biologia (Perugia) Rome*, **54**, 45–59.

Renzoni, A. (1962). Ulteriori dati sul ciclo biologico riproduttivo di *Mytilus galloprovincialis* Lam. *Rivista di Biologia (Perugia) Rome*, **55**, 37–47.

Renzoni, A. (1963). Ricerche ecologiche e idrobiologiche su *Mytilus galloprovincialis* Lam. nel Golfo di Napoli. *Bolletino di Pesca, Piscicoltura e Idrobiologia, Rome*, **18**, 187–238.

Renzoni, A. & Sacchi, C. F. (1961). Notes sur l'écologie de la moule

(*Mytilus galloprovincialis* Lam.) dans le lac Fusaro (Naples). *Rapports et Procès-verbaux des Réunions. Conseil permanent International pour l'Exploration de la Mer*, **16**, 811–14.

Reshöft, K. (1961). Untersuchungen zur zellulären osmotischen und thermischen Resistenz verschiedener Lamellibranchier der deutschen Küstengewässer. *Kieler Meeresforschungen*, **17**, 65–84.

Reverberi, G. (1971). Mytilus. In: *Experimental embryology of marine and fresh-water invertebrates* (ed. G. Reverberi), pp. 175–87. North-Holland Publishing Co., Amsterdam.

Reynolds, N. (1956). A simplified system of mussel purification. *Fishery Investigations. Ministry of Agriculture, Fisheries and Food, London*, Ser. II, **20**, 1–18.

Reynolds, N. (1969). The settlement and survival of young mussels in the Conway fishery. *Fishery Investigations. Ministry of Agriculture, Fisheries and Food, London*, Ser. II, **26**, 1–24.

Ricci, E. (1957). Contribution à la biometrie, à la biologie et la physicochimie de la moule commune (*Mytilus galloprovincialis* Lmk). *Annales Station océanographique de Salammbô*, **9**, 166 pp.

Rice, T. R. & Wolfe, D. A. (1971). Radioactivity – chemical and biological aspects. In: *Impingement of man on the oceans* (ed. D. W. Hood), pp. 325–80. Wiley, New York.

Richards, O. W. (1928). The growth of the mussel *Mytilus californianus*. *Nautilus*, **41**, 99–101.

Richards, O. W. (1946). Comparative growth of *Mytilus californianus* at La Jolla, California and *M. edulis* at Woods Hole, Massachusetts. *Ecology*, **27**, 370–2.

Richardson, I. D. (1967). Which fish to farm? *Hydrospace*, **1**, 72–6.

Ricketts, E. F. & Calvin, J. (1965). *Between pacific tides*. Stanford University Press, Stanford, California.

Rigg, G. R. & Miller, R. C. (1949). Intertidal plant and animal zonation in the vicinity of Neah Bay, Washington. *Proceedings of the California Academy of Sciences*, **26**, 323–57.

Rigopoulos, C. (1972). Observations on the culture of *Mytilus galloprovincialis*. *Report of the General Fisheries Council for the Mediterranean*, **11**, 71 pp.

Ritchie, J. (1927). Reports on the prevention of the growth of mussels in submarine shafts and tunnels at Westbank Electric Station, Portobello. *Transactions of the Royal Scottish Society of Arts*, **19**, 1–20.

Roberts, D. (1972). The assimilation and chronic effects of sub-lethal concentrations of endosulfan on condition and spawning in the common mussel, *Mytilus edulis. Marine Biology*, **16**, 119–25.

Roberts, D. (1973). Some sub-lethal effects of pesticides on the behaviour and physiology of bivalved molluscs. PhD thesis, University of Liverpool.

Robertson, J. D. (1953). Further studies on ionic regulation in marine invertebrates. *Journal of Experimental Biology*, **30**, 277–96.

Robertson, J. D. (1964). Osmotic and ionic regulation. In: *Physiology of*

References

Mollusca (ed. K. M. Wilbur and C. M. Yonge), vol. 1, pp. 283–311. Academic Press, New York and London.

Robin, Y. & Thoai, N. V. (1961). Métabolisme des dérivés guanidylés. X. Métabolisme de l'octopine: son rôle biologique. *Biochimica et Biophysica Acta*, **52**, 233–40.

Robinson, G. A. (1968). Distribution of *Gonyaulax tamarensis* Lebour in the western North Sea in April, May and June 1968. *Nature, London*, **220**, 22–3.

Robinson, J., Richardson, A., Crabtree, A. N., Coulson, J. C. & Potts, G. R. (1967). Organochlorine residues in marine organisms. *Nature, London*, **214**, 1307–11.

Rodegker, W. & Nevenzel, J. C. (1964). The fatty acid composition of three marine invertebrates. *Comparative Biochemistry and Physiology*, **11**, 53–60.

Rooth, J. (1957). Over het voedsel van de zilvermeeuwen. (On the food of herring gulls.) *Levende natuur, Amsterdam*, **60**, 209–13.

Ropes, J. W. (1968). The feeding habits of the green crab *Carcinus maenas* (L.). *Fishery Bulletin. Fish and Wildlife Service, US Department of the Interior, Washington*, **67**, 183–203.

Ross, J. P. & Goodman, D. (1974). Vertical intertidal distribution of *Mytilus edulis*. *Veliger*, **16**, 388–95.

Rothschild, M. (1936). Gigantism and variation in *Peringia ulvae* Pennant caused by larval trematodes. *Journal of the Marine Biological Association of the United Kingdom*, **20**, 537–46.

Rotthauwe, H. W. (1958). Untersuchungen zur Atmungsphysiologie und Osmoregulation bei *Mytilus edulis* mit einem kurzen Anhang über die Blutkonzentration von *Dreissena polymorpha* in Abhängigkeit vom Elektrolytgehalt des Aussenmediums. *Veröffentlichungen des Instituts für Meeresforschung in Bremerhaven*, **5**, 143–59.

Ruddell, C. L. (1971). The fine structure of oyster agranular amoebocytes from regenerating mantle wounds in the Pacific oyster, *Crassostrea gigas*. *Journal of Invertebrate Pathology*, **18**, 260–8.

Runnström, S. (1929). Weitere studien über die Temperaturanpassung der Fortplanzung und Einwicklung mariner Tiere. *Bergens museums Årbog*, **10**, 1–46.

Ryan, C. A. & King, T. E. (1962). The succinic oxidase and DPNH oxidase systems of the bay mussel, *Mytilus edulis*. *Archives of Biochemistry and Biophysics*, **85**, 450–6.

Ryther, J. H. (1968). *The status and potential of aquaculture, particularly invertebrate and algae culture, part II*. US Department of Commerce, National Bureau of Standards, Washington. 261 pp.

Ryther, J. H. (1969). The potential of the estuary for shellfish production. *Proceedings of the National Shellfisheries Association*, **59**, 18–22.

Ryther, J. H. & Bardach, J. E. (1968). *The status and potential of aquaculture, particularly invertebrate and algae culture, part I*. US Department of Commerce, National Bureau of Standards, Washington. 45 pp.

Sacchi, C. F. & Renzoni, A. (1962). L'écologie de *Mytilus galloprovincialis* (Lam.) dans l'étang littoral du Fusaro et les rythmes annuels et nycthéméraux des facteurs environnants. *Pubblicazioni della Stazione Zoologica di Napoli*, **32** (Supplement), 255–93.

Sadykhova, I. A. (1967). Some data on the biology and growth of *Mytilus grayanus* Dunker in experimental cages in Peter the Great Bay. *Proceedings of the Symposium on Mollusca*, **2**, 431–5.

Sadykhova, I. A. (1970*a*). On the determination of the duration of life in *Crenomytilus grayanus*. In: *Osnovi biologicheskoy productivnosti okeana e eē ispolsovanie, Nauka, Moscva*, pp. 263–76. (In Russian.)

Sadykhova, I. A. (1970*b*). A contribution to the biology of *Crenomytilus grayanus*. (Dysodonta, Mytilidae). *Zoologicheskii Zhurnal*, **49**, 1408–10. (In Russian.)

Sadykhova, I. A. (1970*c*). On the allometry of growth of *Crenomytilus grayanus* (Dunker) from the Gulf of Peter the Great. *Trudȳ molodykh uchēnykh VNIRO*, **3**, 108–15. (In Russian.)

Sagara, J. (1958). Artificial discharge of reproductive elements of certain bivalves caused by treatment of seawater and by injection with NH_4OH. *Bulletin of the Japanese Society of Scientific Fisheries*, **23**, 505–10.

Sagiura, Y. (1959). Seasonal change in sexual maturity and sexuality of *Mytilus edulis* L. *Bulletin of the Japanese Society of Scientific Fisheries*, **25**, 1–6.

Sagiura, Y. (1962). Electrical induction of spawning in two marine invertebrates *Urechis unicinctus* and hermaphroditic *Mytilus edulis*. *Biological Bulletin. Marine Biological Laboratory, Woods Hole, Mass.*, **123**, 203–6.

Salanki, J. (1966). Daily activity rhythm of two Mediterranean Lamellibranchia. *Annales de l'Institut Biologie, Tihany*, **33**, 135–42.

Salaque, A., Barbier, M. & Lederer, E. (1966). Sur la biosynthèse des sterols de l'huître (*Ostrea gryphea*) et de l'oursin (*Paracentrotus lividus*). *Comparative Biochemistry and Physiology*, **19**, 45–51.

Sanders, H. L. (1968). Marine benthic diversity: a comparative study. *American Naturalist*, **102**, 245–82.

Sassaman, C. & Mangum, C. P. (1972). Adaptations to environmental oxygen levels in infaunal and epifaunal anemones. *Biological Bulletin. Marine Biological Laboratory, Woods Hole, Mass.*, **143**, 657–78.

Sastry, A. N. (1966). Temperature effects in reproduction of the bay scallop *Aequipecten irradians* Lamarck. *Biological Bulletin. Marine Biological Laboratory, Woods Hole, Mass.*, **130**, 118–34.

Sastry, A. N. & Blake, N. J. (1971). Regulation of gonad development in the bay scallop, *Aequipecten irradians* Lamarck. *Biological Bulletin. Marine Biological Laboratory, Woods Hole, Mass.*, **140**, 274–83.

Sautet, J., Ollivier, H. & Quicke, J. (1964). Contribution à l'étude de la fixation et de l'élimination biologiques de l'arsenic par *Mytilus edulis*. *Annales de Médecine légale*, **44**, 466–71.

Savage, R. E. (1956). The great spatfall of mussels in the River Conway

References

Estuary in Spring 1940. *Fishery Investigations. Ministry of Agriculture, Fisheries and Food, London,* Ser. II, **20**, 1–21.

Savilov, I. A. (1953). The growth and variations in growth of the white Sea invertebrates *Mytilus edulis, Mya arenaria* and *Balanus balanoides. Trudȳ Instituta Okeanologii. Akademiya nauk SSSR, Moscow,* **7**, 198–259. (In Russian.)

Sawaya, P. (1965). Physiological aspects of the ecology of the mussel *Mytilus perna* L. *Anales de la Academia Brasileira de Ciencias,* **37** (Supplement), 176–8.

Schafer, R. D. (1963). Effects of pollution on the amino acid content of *Mytilus edulis. Pacific Science,* **17**, 246–50.

Scheer, B. T. (1945). The development of marine fouling communities. *Biological Bulletin. Marine Biological Laboratory, Woods Hole, Mass.,* **89**, 103–21.

Scheltema, R. S. (1961). Metamorphosis of the veliger larvae of *Nassarius obsoletus* (Gastropoda) in response to bottom sediment. *Biological Bulletin. Marine Biological Laboratory, Woods Hole, Mass.,* **120**, 92–109.

Scheltema, R. S. (1965). The relationship of salinity to larval survival and development in *Nassarius obsoletus* (Gastropoda). *Biological Bulletin. Marine Biological Laboratory, Woods Hole, Mass.,* **129**, 340–54.

Schlieper, C. (1955*a*). Uber die physiologischen Wirkungen des Brackwassers. (Nach Versuchen an der Meismuschel *Mytilus edulis.*) *Kieler Meeresforschungen,* **11**, 22–33.

Schlieper, C. (1955*b*). Die regulation des Herzschlages der Miesmuschel *Mytilus edulis* L. bei geöfneten und bei geschlossenen Schalen. *Kieler Meeresforschungen,* **11**, 139–48.

Schlieper, C. (1966). Genetic and nongenetic cellular resistance adaptation in marine invertebrates. *Helgoländer wissenschaftliche Meeresuntersuchungen,* **14**, 482–502.

Schlieper, C. (1967). Cellular ecological adaptations and reactions, demonstrated by surviving isolated gill tissues of bivalves. In: *The cell and environmental temperature* (ed. A. S. Troshin; English edition ed. C. L. Prosser), pp. 191–9. Pergamon Press, Oxford.

Schlieper, C. (1971). Physiology of brackish water. In: *Biology of brackish water* (A. Remane and C. Schlieper), pp. 211–350. Wiley-Interscience, New York.

Schlieper, C. & Kowalski, R. (1956). Uber den Einflus des Mediums auf die thermische und osmotische Resistenz des Kiemengewebes der Miesmuschel *Mytilus edulis* L. *Kieler Meeresforschungen,* **12**, 37–45.

Schlieper, C. & Kowalski, R. (1958). Ein zellularer Regulationsmechanismus für erhöhte Kiemenventilation nach Anoxybiose bei *Mytilus edulis* L. *Kieler Meeresforschungen,* **14**, 42–7.

Schlieper, C., Flugel, H. & Rudolf, J. (1960). Temperature and salinity relationship in marine bottom invertebrates. *Experientia,* **16**, 470–4.

Schlieper, C., Flugel, H. & Theede, H. (1967). Experimental investigations of the cellular resistance ranges of marine temperate and tropical bivalves. Results of the Indian Ocean Expedition of the German Research Association. *Physiological Zoology*, **40**, 345–61.

Schlieper, C., Kowalski, R. & Erman, P. (1958). Beitrag zur ökologisch-zellphysiologischen Charakterisierung des borealen Lamellibranchier *Modiolus modiolus* L. *Kieler Meeresforschungen*, **14**, 3–10.

Schoffeniels, E. (1964). Effect of inorganic ions on the activity of L-glutamic acid dehydrogenase. *Life Sciences*, **3**, 845–50.

Schoffeniels, E. & Gilles, R. (1972). Ionoregulation and osmoregulation in Mollusca. In: *Chemical zoology* (ed. M. Florkin and B. T. Scheer), vol. VII, pp. 393–420. Academic Press, New York and London.

Schram, J. A. (1970). Studies on the meroplankton in the inner Oslofjord. II. Seasonal differences and seasonal changes in the specific distribution of larvae. *Nytt Magasin Zoologi*, **18**, 1–21.

Schultz-Baldes, M. (1973). Die Miesmuschel *Mytilus edulis* als Indikator für die Bleikonzentration im Weserestuar und in der Deutschen Bucht. *Marine Biology*, **21**, 37–44.

Schultz-Baldes, M. (1974). Lead uptake from seawater and food, and lead loss in the common mussel, *Mytilus edulis*. *Marine Biology*, **25**, 177–93.

Sclufer, E. (1955). The respiration of *Spisula* eggs. *Biological Bulletin. Marine Biological Laboratory, Woods Hole, Mass.*, **109**, 113–22.

Scott, A. (1901). Note on the spawning of the mussel *Mytilus edulis*. *Proceedings and Transactions of the Liverpool Biological Society*, **15**, 161–4.

Scott, D. M. & Major, C. W. (1972). The effect of copper (II) on survival, respiration and heart rate in the common blue mussel *Mytilus edulis*. *Biological Bulletin. Marine Biological Laboratory, Woods Hole, Mass.*, **143**, 679–88.

Scrutton, C. M. & Utter, M. F. (1968). The regulation of glycolysis and gluconeogenesis in animal tissues. *Annual Review of Biochemistry*, **37**, 249–302.

Seck, C. (1958). Untersuchungen zur Frage der Ionenregulation bei in Brackwasser lebenden Evertebraten. *Kieler Meeresforschungen*, **13**, 220–43.

Seed, R. (1968). Factors influencing shell shape in *Mytilus edulis* L. *Journal of the Marine Biological Association of the United Kingdom*, **48**, 561–84.

Seed, R. (1969*a*). The ecology of *Mytilus edulis* L. (Lamellibranchiata) on exposed rocky shores. 1. Breeding and settlement. *Oecologia*, **3**, 277–316.

Seed, R. (1969*b*). The ecology of *Mytilus edulis* L. (Lamellibranchiata) on exposed rocky shores. 2. Growth and mortality. *Oecologia*, **3**, 317–50.

Seed, R. (1969*c*). The incidence of the pea crab *Pinnotheres pisum* in the two types of *Mytilus* (Mollusca: Bivalvia) from Padstow, south-west England. *Journal of Zoology*, **158**, 413–20.

References

Seed, R. (1971). A physiological and biochemical approach to the taxonomy of *Mytilus edulis* L. and *M. galloprovincialis* Lmk. from south-west England. *Cahier de Biologie Marine, Roscoff*, **12**, 291–322.

Seed, R. (1973). Absolute and allometric growth in the mussel *Mytilus edulis* L. (Mollusca: Bivalvia). *Proceedings of the Malacological Society of London*, **40**, 343–57.

Seed, R. (1974). Morphological variations in *Mytilus* from the Irish coasts in relation to the occurrence and distribution of *M. galloprovincialis* Lmk. *Cahier de Biologie Marine, Roscoff*, **15**, 1–25.

Seed, R. (1975). Reproduction in *Mytilus* (Mollusca: Bivalvia) in European waters. *Pubblicazioni della Stazione Zoologica di Napoli, Milan*, in press.

Segal, E. (1961). Acclimation in molluscs. *American Zoologist*, **1**, 235–44.

Segal, E. (1962). Initial response of the heart rate of a gastropod, *Acmaea limatula*, to abrupt changes in temperature. *Nature, London*, **195**, 674–5.

Segal, E., Rao, K. K. & James, T. W. (1953). Rate of activity as a function of intertidal height within populations of some littoral molluscs. *Nature, London*, **172**, 1108–9.

Segar, D. A., Collins, J. D. & Riley, J. P. (1971). The distribution of the major and some minor elements in marine animals. II. Molluscs. *Journal of the Marine Biological Association of the United Kingdom*, **51**, 131–6.

Seidman, I. & Entner, N. (1961). Oxidative enzymes and their role in phosphorylation in sarcosomes of adult *Ascaris lumbricoides*. *Journal of Biological Chemistry*, **236**, 915–19.

Selander, R. K. (1970). Behavior and genetic variation in natural populations. *American Zoologist*, **10**, 53–66.

Selander, R. K., Yang, S. Y., Lewontin, R. C. & Johnson, W. E. (1970). Genetic variation in the Horseshoe crab (*Limulus polyphemus*) a phylogenetic 'relic'. *Evolution*, **24**, 402–14.

Seraydarian, M. W. & Kalvaitis, Z. (1964). Phosphagen in *Mytilus californianus*. *Comparative Biochemistry and Physiology*, **12**, 1–12.

Shapiro, A. Z. (1964). The effect of certain inorganic poisons on the respiration of *Mytilus galloprovincialis* L. *Trudȳ Sevastopolskoi Biologicheskoi Stantsii*, **17**, 334–41. (In Russian.)

Shaw, J. (1958). Further studies on ionic regulation in the muscle fibres of *Carcinus maenas*. *Journal of Experimental Biology*, **35**, 902–19.

Sheldon, R. W. & Parsons, T. R. (1967). *A practical manual on the use of the Coulter Counter in marine research*. Coulter Electronics Sales Company, Toronto, Canada.

Shelford, V. E. (1930). Geographical extent and succession in Pacific North American intertidal (*Balanus*) communities. *Publications of the Puget Sound Marine Biological Station of the University of Washington, Seattle*, **7**, 217–24.

Shelford, V. E. (1935). Some marine biotic communities of the Pacific coast of North America. *Ecological Monographs*, **5**, 251–354.

Sherwood, H. P. (1957). The sterilization of cockles and mussels by boiling. *Monthly Bulletin of the Ministry of Health*, **16**, 80–6.

Shimizu, M., Kajihara, T., Suyama, I. & Hiyama, Y. (1971). Uptake of ^{58}Co by mussel *Mytilus edulis*. *Journal of Radiation Research*, **12**, 17–28.

Shuster, C. N. (1956). On the shell of bivalved molluscs. *Proceedings of the National Shellfisheries Association*, **47**, 34–42.

Sievers, A. (1969). Comparative toxicity of *Gonyaulax monilata* and *Gymnodinium breve* to annelids, crustaceans, molluscs and a fish. *Journal of Protozoology*, **16**, 401–4.

Simpson, A. C. (1968). Oil, emulsifiers and commercial shellfish. *Field Studies*, **2** (Supplement), 91–8.

Simpson, A. C. (1960). Fisheries Laboratory, Burnham-on-Crouch. Report following a visit in October 1960.

Sindermann, C. J. (1970). *Principle diseases of marine fish and shellfish*. Academic Press, New York and London.

Singer, T. P. (1971). Evolution of the respiratory chain and of its flavo-proteins. In: *Biochemical evolution and the origin of life* (ed. E. Schoffeniels), pp. 203–23. North-Holland, Amsterdam.

Sleigh, M. A. (1969). Coordination of the rhythm of beat in some ciliary systems. *International Review of Cytology*, **25**, 31–54.

Sleigh, M. A. (ed.) (1974). *Cilia and flagella*. Academic Press, New York and London. 500 pp.

Sleigh, M. A. & Aiello, E. (1972). The movement of water by cilia. *Acta Protozoologica*, **11**, 265–77.

Smith, J. E. (ed.) (1968). *'Torrey Canyon' pollution and marine life*. Cambridge University Press, London. 196 pp.

Smith, L. S. & Davis, J. C. (1965). Haemodynamics in *Tresus nuttallii* and certain other bivalves. *Journal of Experimental Biology*, **43**, 171–80.

Snedecor, G. W. & Cochran, W. G. (1972). *Statistical methods*. The Iowa State University Press, Iowa. 593 pp.

Soot-Ryen, T. (1955). A report on the family Mytilidae (Pelecypoda). *Allan Hancock Pacific Expeditions*, **20**, 1–174.

Soot-Ryen, T. (1969). Family Mytilidae Rafinesque 1815. In: *Treatise on invertebrate palaeontology, part N. Mollusca 6* (ed. R. C. Moore), pp. 271–80. Geological Society of America/University of Kansas Press.

Sova, V. V., Elyakova, L. A. & Vaskovsky, V. E. (1970). The distribution of laminarinases in marine invertebrates. *Comparative Biochemistry and Physiology*, **32**, 459–64.

Spärck, R. (1920). Nogle bemaerkninger angående yngleforhold hos *Mytilus edulis* L. *Videnskabelige Meddelelser fra Dansk naturhistorisk Forening i Kjøbenhavn*, **71**, 161–4.

Spärck, R. (1936). On the relation between metabolism and temperature in some marine lamellibranchs and its ecological and zoogeographical importance. *Biologiske Meddelelser*, **13**, 1–27.

References

Sparks, A. K. (1972). *Invertebrate pathology. Noncommunicable diseases.* Academic Press, New York and London. 387 pp.

Speakman, J. N. & Krunkel, P. A. (1972). Quantification of the effects of rate of temperature change on aquatic biota. *Water Research,* 6, 1283–90.

Speeg, K. V. & Campbell, J. W. (1969). Arginase and urea metabolism in terrestrial snails. *American Journal of Physiology,* 216, 1003–12.

Spraque, J. B. (1969). Measurement of pollutant toxicity to fish. I. Bioassay methods for acute toxicity. *Water Research,* 3, 793–821.

Sprague, J. B. (1970). Measurement of pollutant toxicity to fish. II. Utilising and applying bioassay results. *Water Research,* 4, 3–32.

Sprague, J. B. (1971). Measurement of pollutant toxicity to fish. III. Sublethal effects and 'safe' concentrations. *Water Research,* 5, 245–66.

Sprague, J. B. & Duffy, J. R. (1971). DDT residues in Canadian Atlantic fishes and shellfishes in 1967. *Journal of the Fisheries Research Board of Canada,* 28, 59–64.

Sribhibhadh, A. (1973). Status and problems of coastal aquaculture in Thailand. In: *Coastal aquaculture in the Indo-Pacific region* (ed. T. V. R. Pillay), pp. 74–83. Fishing News (Books) Ltd, London.

Stafford, J. (1912). On the recognition of bivalve larvae in plankton collections. *Contributions to Canadian Biology and Fisheries,* 1906–1910, 221–42.

Stanley, S. M. (1970). Relation of shell form to life habits of the Bivalvia (Mollusca). *Memoirs of the Geological Society of America,* 125, 496 pp.

Stanley, S. M. (1972). Functional morphology and evolution of byssally attached bivalve molluscs. *Journal of Palaeontology,* 46, 165–212.

Stegeman, J. J. & Teal, J. M. (1973). Accumulation, release and retention of petroleum hydrocarbons by the oyster *Crassostrea virginica. Marine Biology,* 22, 37–44.

Steidinger, K. A., Burklew, M. A. & Ingle, R. M. (1973). The effects of *Gymnodinium breve* toxin on estuarine animals. In: *Marine pharmacognosy* (ed. D. F. Martin and G. M. Padilla), pp. 179–202. Academic Press, New York and London.

Stephens, G. C. (1967). Dissolved organic material as a nutritional source for marine and estuarine invertebrates. In: *Estuaries* ed. G. H. Lauff), pp. 367–73. American Association for the Advancement of Science, Washington.

Stephens, G. C. (1968). Dissolved organic matter as a potential source of nutrition for marine organisms. *American Zoologist,* 8, 95–106.

Stephenson, T. A. & Stephenson, A. (1972). *Life between tidemarks on rocky shores.* W. H. Freeman, San Francisco.

Steuer, A. (1902). *Mytilicola intestinalis,* n.gen. n.sp. aus dem Darme von *Mytilus galloprovincialis* Lam. *Zoologischer Anzeiger,* 25, 635–7.

Stinnakre, J. & Tauc, L. (1969). Central neuronal response to the activation of osmoreceptors in the osphradium of *Aplysia. Journal of Experimental Biology,* 51, 347–61.

Stohler, R. (1930). Beitrag zur Kenntnis des Geschlechtszyklus von *Mytilus californianus* Conrad. *Zoologischer Anzeiger*, **90**, 263–8.

Stokes, T. M. & Awapara, J. (1968). Alanine and succinate as end-products of glucose degradation in the clam *Rangia cuneata*. *Comparative Biochemistry and Physiology*, **25**, 883–92.

Strathmann, R. S. (1971). The feeding behaviour of planktotrophic echinoderm larvae: mechanisms, regulation and rates of suspension feeding. *Journal of Experimental Marine Biology and Ecology*, **6**, 109–60.

Strathmann, R. (1974). The spread of sibling larvae of sedentary marine invertebrates. *American Naturalist*, **108**, 29–44.

Strathmann, R. S., Jahn, T. L. & Fonesca, J. R. C. (1972). Suspension feeding by marine invertebrate larvae: clearance of particles by ciliated bands of a rotifer, pluteus and trochophore. *Biological Bulletin. Marine Biological Laboratory, Woods Hole, Mass.*, **142**, 505–29.

Stubbings, H. G. (1954). The biology of the common mussel in relation to fouling problems. *Research*, **7**, 222–9.

Suburo, S. & Sakamoto, E. (1951). On the reproduction of *Mytilus grayanus* Dunker. *Bulletin of the Japanese Society of Scientific Fisheries*, **17**, 63–6.

Sugiura, Y. (1962). Electrical induction of spawning in two marine invertebrates (*Urechis unicinctus*, hermaphroditic *Mytilus edulis*). *Biological Bulletin. Marine Biological Laboratory, Woods Hole, Mass.*, **123**, 203–6.

Sullivan, C. M. (1948). Bivalve larvae of Malpeque Bay, P.E.I. *Bulletin of the Fisheries Research Board of Canada*, No. 77.

Sumner, A. T. (1966). The cytology and histochemistry of the digestive gland cells of some freshwater lamellibranchs. *Journal of the Royal Microscopical Society*, **85**, 201–11.

Sumner, A. T. (1969). The distribution of some hydrolytic enzymes in the cells of the digestive gland of certain lamellibranchs and gastropods. *Journal of Zoology*, **158**, 277–91.

Suryanarayanan, H. & Alexander, K. M. (1971). Fuel reserves of molluscan muscle. *Comparative Biochemistry and Physiology*, **40A**, 55–60.

Swedmark, M., Braaten, B., Emanuelsson, E. & Granmo, A. (1971). Biological effects of surface active agents on marine animals. *Marine Biology*, **9**, 183–201.

Sweeney, D. (1963). Dopamine: its occurrence in molluscan ganglia. *Science*, **139**, 1051.

Tabata, K. (1970). Studies on the toxicity of heavy metals to aquatic animals and the factors to decrease the toxicity. I. On the formation and toxicity of precipitate of heavy metals. *Bulletin of the Tokai Regional Fishery Research Laboratory*, **58**, 203–14.

Takahashi, K. (1971). Abrupt stoppage of *Mytilus* cilia caused by chemical stimulation. *Journal of the Faculty of Science, Tokyo University*, Sect. 4, *Zoology*, **12**, 219–28.

Takahashi, K. & Murakami, A. (1968). Nervous inhibition of ciliary motion in the gill of the mussel, *Mytilus edulis. Journal of the Faculty of Science, Tokyo University*, Sect. 4, *Zoology*, **11**, 359–72.

Takatsuki, S. (1934). On the nature and functions of the amoebocytes of *Ostrea edulis. Quarterly Journal of Microscopical Science*, **76**, 379–431.

Tamarin, A. & Keller, P. J. (1972). An ultrastructural study of the byssal thread forming system in *Mytilus. Journal of Ultrastructure Research*, **40**, 401–16.

Tamarin, A., Lewis, P. & Askey, J. (1974). Specialised cilia of the byssus attachment plaque forming region in *Mytilus californianus. Journal of Morphology*, **142**, 321–7.

Tammes, P. M. L. & Dral, A. D. G. (1955). Observations on the straining of suspensions by mussels. *Archives Néerlandaises de Zoologie*, **11**, 87–112.

Tan, E. L. (1971). Nutritive value of *Mytilus viridus* as a potential protein source for animal feeds. *Journal of the Singapore National Academy of Science*, **1**, 82–4.

Tanaka, Y. (1958). Studies on molluscan larvae. *Venus*, **20**, 207–19.

Tang, P. S. (1933). On the rate of oxygen consumption by tissues and lower organisms as a function of oxygen tension. *Quarterly Review of Biology*, **8**, 260–74.

Tappel, A. L. (1960). Cytochromes of muscles of marine invertebrates. *Journal of Cellular and Comparative Physiology*, **55**, 111–26.

Taylor, A. (1974). Aspects of the respiratory physiology of the bivalve *Arctica islandica* (L.). PhD thesis, Liverpool University.

Tenore, K. R. & Dunstan, W. M. (1973). Comparison of feeding and biodeposition of three bivalves at different food levels. *Marine Biology*, **21**, 190–5.

Tenore, K. R., Goldman, J, C. & Clarner, D. P. (1973). The food chain dynamics of the oyster, clam and mussel in an aquaculture food chain. *Journal of Experimental Marine Biology and Ecology*, **12**, 157–65.

Teshima, S. & Kanazawa, A. (1974). Biosynthesis of sterols in abalone, *Haliotus gurneri*, and mussel, *Mytilus edulis. Comparative Biochemistry and Physiology*, **47B**, 555–61.

Tham, A. K., Yang, S. W. & Tan, W. H. (1973). Experiments in coastal aquaculture in Singapore. In: *Coastal aquaculture in the Indo-Pacific region* (ed. T. V. R. Pillay), pp. 375–83. Fishing News (Books) Ltd, London.

Thamdrup, H. M. (1935). Beiträge zur Okologie de Wattenfauna auf experimenteller Grundlage. *Meddelelser fra Danmarks Fiskeri- og Havundersøgelser*, **10**, 1–125.

Theede, H. (1963). Experimentelle Untersuchungen über die Filtrations leistung der Miesmuschel *Mytilus edulis* L. *Kieler Meeresforschungen*, **19**, 20–41.

Theede, H. (1965). Vergleichende experimentelle Untersuchungen über die zellulare Gefrierresistenz mariner Muscheln. *Kieler Meeresforschungen*, **21**, 153–66.

References

Theede, H. (1973). Resistance adaptations of marine invertebrates and fish to cold. In: *Effects of temperature on ectothermic organisms* (ed. W. Wieser), pp. 249–69. Springer-Verlag, Berlin.

Theede, H. & Lassig, J. (1967). Comparative studies on cellular resistance of bivalves from marine and brackish waters. *Helgoländer wissenschaftliche Meeresunteruchungen*, **16**, 119–29.

Theede, H., Ponat, A., Hiroki, K. & Schlieper, C. (1969). Studies on the resistance of marine bottom invertebrates to oxygen-deficiency and hydrogen sulphide. *Marine Biology*, **2**, 325–37.

Thiesen, B. F. (1968). Growth and mortality of culture mussels in the Danish Wadden Sea. *Meddelelser fra Danmarks Fiskeri-og Havundersøgelser*, N.S., **6**, 47–78.

Thiesen, B. F. (1972). Shell cleaning and deposit feeding in *Mytilus edulis* L. *Ophelia*, **10**, 49–55.

Thiesen, B. F. (1973). The growth of *Mytilus edulis* L. (Bivalvia) from Disko and Thule district, Greenland. *Ophelia*, **12**, 59–77.

Thoai, N. V. & Robin, Y. (1959). Métabolisme des dérivés guanidylés. VIII. Biosynthèse de l'octopine et répartition de l'enzyme l'opérant chez les invertebrés. *Biochimica et Biophysica Acta*, **35**, 446–53.

Thoai, N. V. & Robin, Y. (1961). Métabolisme des dérivés guanidylés. IX. Biosynthèse de l'octopine: Étude du mécanismes de la réaction et de quelques propriétés de l'octopine synthétase. *Biochimica et Biophysica Acta*, **52**, 221–33.

Thomas, H. T. (1953). Mussels. *Rapport et Procès-verbaux dex Réunions. Conseil permanent International pour l'exploration de la Mer*, **134**, 34.

Thompson, R. J. (1972). Feeding and metabolism in the mussel *Mytilus edulis*, L. PhD thesis, University of Leicester.

Thompson, R. J. & Bayne, B. L. (1972). Active metabolism associated with feeding in the mussel *Mytilus edulis* L. *Journal of Experimental Marine Biology and Ecology*, **8**, 191–212.

Thompson, R. J. & Bayne, B. L. (1974). Some relationships between growth, metabolism and food in the mussel, *Mytilus edulis*. *Marine Biology*, **27**, 317–26.

Thompson, R. J., Ratcliffe, N. A. & Bayne, B. L. (1974). Effects of starvation on structure and function in the digestive gland of the mussel (*Mytilus edulis* L.). *Journal of the Marine Biological Association of the United Kingdom*, **54**, 699–712.

Thornberger, E. J., Oliver, I. T. & Scutt, P. B. (1968). Comparative electrophoretic patterns of dehydrogenases in different species. *Comparative Biochemistry and Physiology*, **25**, 973–87.

Thorson, G. (1936). The larval development, growth, and metabolism of Arctic marine bottom invertebrates, compared with those of other seas. *Meddelelser om Grønland*, **100b**, 1–155.

Thorson, G. (1946). Reproduction and larval development of Danish marine bottom invertebrates. *Meddelelser fra Danmarks Fiskeri- og Havundersøgelser, Serie Plankton*, **4**, 1–523.

Thorson, G. (1950). Reproductive and larval ecology of marine bottom invertebrates. *Biological Reviews*, **25**, 1–45.

References

Thorson, G. (1957). Bottom communities (sublittoral or shallow shelf). *Memoirs of the Geological Society of America*, **67**, 461–534.

Thorson, G. (1966). Some factors influencing the recruitment and establishment of marine benthic communities. *Netherlands Journal of Sea Research*, **3**, 267–93.

Tiffany, W. J. (1972). Aspects of excretory ultrafiltration in the bivalved molluscs. *Comparative Biochemistry and Physiology*, **43A**, 527–36.

Tinbergen, N. & Kruuk, H. (1962). Van scholesksters en mosselen. (About oystercatchers and mussels.) *Levende natuur, Amsterdam*, **65**, 1–8.

Tinbergen, N. & Norton-Griffiths, M. (1964). Oystercatchers and mussels. *British Birds*, **57**, 64–70.

Tomita, G. (1955a). A modification of the Nomurs and Tomita method for measuring the mechanical activity of cilia. *Scientific Reports of Tôhuku University*, Ser. IV, **21**, 1–7.

Tomita, G. (1955b). On the Nomura and Tomita method for measuring the mechanical activity of cilia. *Bulletin of the Marine Biology Station, Asamushi, Tôhuku University*, **7**, 159–63.

Torres, J. J. & Mangum, C. P. (1974). Effects of hyperoxia on survival of benthic marine invertebrates. *Comparative Biochemistry and Physiology*, **47A**, 17–22.

Tripp, M. R. (1963). Cellular responses of molluscs. *Annals of the New York Academy of Sciences*, **113**, 467–74.

Trueman, E. R. (1966). The fluid dynamics of the bivalve mollusca *Mya* and *Margaritifera*. *Journal of Experimental Biology*, **45**, 369–82.

Trueman, E. R. (1967). Activity and heart rate of bivalve molluscs in their natural habitat. *Nature, London*, **214**, 832–3.

Trueman, E. R. (1968). The burrowing activities of bivalves. *Symposia of the Zoological Society of London*, **22**, 167–86.

Trueman, E. R. & Lowe, G. A. (1971). The effect of temperature and littoral exposure on the heart rate of a bivalve mollusc, *Isognomon alatus*, in tropical conditions. *Comparative Biochemistry and Physiology*, **38A**, 555–64.

Tsuzuki, K. (1957). Metabolic patterns in bivalves. XI. Xanthine dehydrogenases in *Venerupis philippinarium, Mactra sulcataria, Anadara inflata, Ostrea gigas, Meretrix meretrix lusoria* and *Mytilus edulis*. *Journal of the College of Arts and Sciences, Chiba University. Natural Sciences Series*, **2**, 239–42.

Tubiash, H. S., Chanley, P. E. & Leifson, E. (1965). Bacterial necrosis, a desease of larval and juvenile bivalve molluscs. *Journal of Bacteriology*, **90**, 1036–44.

Ukeles, R. (1962). The effect of several toxicants on five genera of marine phytoplankton. *Applied Microbiology*, **10**, 532–7.

Ukeles, R. & Sweeney, B. M. (1969). Influence of dinoflagellate trichocysts and other factors on the feeding of *Crassostrea virginica* larvae on *Monochrysis lutheri*. *Limnology and Oceanography*, **14**, 403–10.

Umiji, S. (1969). Neurosecretion in the mussel *Mytilus perna*. *Boletin da*

References

Faculdade de Filosofia, Ciências e Letras, Universidade de São Paula. Zoologia e Biologia Marhina, N.S., **26**, 181–254.

Ursin, E. (1963). On the incorporation of temperature in the von Bertalanffy growth equation. *Meddelelser fra Danmarks Fiskeri- og Havundersøgelser*, N.S., **4**, 1–16.

Ushakov, B. P. (1965). Cellular resistance adaptation to temperature and thermostability of somatic cells with special reference to marine animals. *Marine Biology*, **1**, 153–60.

Usuki, I. (1962a). Energy source for the ciliary movement and the respiration and its metabolism in oyster gill. *Scientific Reports of Tôhoku University*, Ser. 4, **28**, 53–7.

Usuki, I. (1962b). Distribution of glycogen, fat and phospholipid in the gill tissues of certain marine bivalves. *Scientific Reports of Tôhoku University*, Ser. 4 **28**, 53–7.

Uysal, H. (1970). Biological and ecological investigations on the mussel (*Mytilus galloprovincialis* Lamarck) living on the coastline of Turkey. *Scientific Reports of the Faculty of Science, Ege University*, **79**, 1–79.

Vahl, O. (1972). Efficiency of particle retention in *Mytilus edulis* L. *Ophelia*, **10**, 17–25.

Vahl, O. (1973). Pumping and oxygen consumption rates of *Mytilus edulis* L. of different sizes. *Ophelia*, **12**, 45–52.

Valli, G. (1971). Ciclo di maturita sessuale in *Mytilus galloprovincialis* Lmk. di Duino (Trieste). *Bolletino di Pesca, Piscicoltura e Idrobiologia, Rome*, **26**, 259–65.

Vance, R. R. (1973a). On reproductive strategies in marine benthic invertebrates. *American Naturalist*, **107**, 339–52.

Vance, R. R. (1973b). More on reproductive strategies in marine benthic invertebrates. *American Naturalist*, **107**, 353–61.

Van Dam, K. (1938). *On the utilisation of oxygen and regulation of breathing in some aquatic animals*. Dissertation: Drukkerij 'Volharding', Groningen.

Van Dam, L. (1954). On the respiration in scallops (Lamellibranchiata). *Biological Bulletin. Marine Biological Laboratory, Woods Hole, Mass.*, **107**, 192–202.

Van Marrewijk, W. J. A., Holwerda, D. A. & De Zwaan, A. (1973). A comparative study of the enzyme activity of CO_2-fixing enzymes, pyruvate kinase and lactic dehydrogenase in the mussel *Mytilus edulis* L., the crayfish *Astacus leptodactylus* Esch. and the rat *Rattus* sp. *Netherlands Journal of Zoology*, **3**, 229–31.

Van Weel, P. B. (1961). The comparative physiology of digestion in molluscs. *American Zoologist*, **1**, 245–52.

Van Weel, P. B. (1974). 'Hepatopancreas'? *Comparative Biochemistry and Physiology*, **47**A, 1–9.

Van Weers, A. W. (1973). Uptake and loss of ^{65}Zn and ^{60}Co by the mussel *Mytilus edulis* L. In: *Radioactive contamination of the marine environment*, pp. 385–401. International Atomic Energy Agency, Vienna.

485

References

Van Winkle, W. (1968). The effects of season, temperature and salinity on the oxygen consumption of bivalve gill tissue. *Comparative Biochemistry and Physiology*, **26**, 69–80.

Van Winkle, W. (1970). Effect of environmental factors on byssal thread formation. *Marine Biology*, **7**, 143–8.

Van Winkle, W. (1972). Ciliary activity and oxygen consumption of excised bivalve gill tssue. *Comparative Biochemistry and Physiology*, **42A**, 473–85.

Vassallo, M. T. (1973). Lipid storage and transfer in the scallop *Chlamys hericia* Gould. *Comparative Biochemistry and Physiology*, **44A**, 1169–75.

Verduin, J. (1969). Hard clam pumping rates: energy requirements. *Science*, **166**, 1309–10.

Vernberg, F. J. (1962). Comparative physiology: latitudinal effects on physiological properties of animal populations. *Annual Review of Physiology*, **24**, 517–46.

Vernberg, W. B. & Vernberg, F. J. (1972). *Environmental physiology of marine animals*. Springer-Verlag, Berlin. 346 pp.

Vernberg, F. J., Schlieper, C. & Schneider, D. E. (1963). The influence of temperature and salinity on ciliary activity of excised gill tissue of molluscs from North Carolina. *Comparative Biochemistry and Physiology*, **8**, 271–85.

Verwey, J. (1952). On the ecology and distribution of cockle and mussel in the Dutch Wadden Sea, their role in sedimentation and the source of their food supply. *Archives Néerlandaises Zoologie*, **10**, 171–239.

Verwey, J. (1954). De mossel en Zijn Eisen *Faraday*, **24**, 13 pp. (With English summary.)

Verwey, J. (1966). The role of some external factors in the vertical migration of marine animals. *Netherlands Journal of Sea Research*, **3**, 245–66.

Vibe, C. (1951). The marine mammals and the marine fauna of the Thule district. *Meddelelser om Grønland*, **150**, 1–115.

Vilela, H. & Monteiro, M. C. (1958). Infestation of the mussel *Mytilus edulis* in the Tejo River by the copepod *Mytilicola intestinalis* Steuer. *Notas e Estudos do Instituto de Biologia Maritima, Lisbon*, **20**, 1–20.

Vilela, H. & Monteiro, M. C. (1960). Infestacao de mexilhoes *Mytilus edulis* L. do Rio Tejo pelo copepode *Mytilocola* intestinalis Steuer. *Boletín de la Sociedad Espãnola de Historia Natural, Madrid*, **58**, 375–88.

Vinogradov, A. P. (1953). The elementary chemical composition of marine organisms. (Translated from the Russian.) Sears Foundation Marine Research Memoir II, Yale University. 647 pp.

Virkar, R. A. & Webb, K. L. (1970). Free amino acid composition of the soft-shell clam *Mya arenaria* in relation to the salinity of the medium. *Comparative Biochemistry and Physiology*, **32**, 775–83.

Vives, F. & Fraga, F. (1961). Producción básica en la Rîa de Vigo (NW de España). *Investigación Pesquera*, **19**, 129–37.

Von Bertalanffy, L. (1957). Quantitative laws in metabolism and growth. *Quarterly Review of Biology*, **32**, 217–31.

Voogt, P. A. (1972). Lipid and sterol components and metabolism in Mollusca. In: *Chemical zoology* (ed. M. Florkin and B. T. Scheer), vol. VII, pp. 245–300. Academic Press, New York and London.

Wada, S. K. (1955). Fertilisation of *Mytilus edulis* with special reference to the acrosome reaction of the spermatozoa. *Memoirs of the Faculty of Fisheries, Kagoshima*, **4**, 105–12.

Wagge, L. E. (1955). Amoebocytes. In: *International review of cytology* (ed. G. Bourne and J. F. Danielli), vol. 4, pp. 31–78. Academic Press, New York and London.

Wallengren, H. (1905). Zur Biologie der Muscheln. I. Die Wasserströmungen. *Lunds Universitets Arsskrift*, **2**, 1–64.

Walne, P. R. (1956). Experimental rearing of the larvae of *Ostrea edulis* L. in the laboratory. *Fishery Investigations. Ministry of Agriculture, Food and Fisheries, London*, Ser. II, **20**, 1–23.

Walne, P. R. (1958). Growth of oysters (*Ostrea edulis* L.). *Journal of the Marine Biological Association of the United Kingdom*, **37**, 591–602.

Walne, P. R. (1961). Observations on the mortality of *Ostrea edulis* L. *Journal of the Marine Biological Association of the United Kingdom*, **41**, 113–22.

Walne, P. R. (1963). The culture of marine molluscs and crustacea. In: *The better use of the world's fauna for food* (ed. J. D. Ovington), pp. 147–75. Institute of Biology, London.

Walne, P. R. (1964). The culture of marine bivalve larvae. In: *Physiology of the Mollusca* (ed. K. M. Wilbur and C. M. Yonge), vol. I, pp. 197–210. Academic Press, New York and London.

Walne, P. R. (1965). Observations on the influence of food supply and temperature on the feeding and growth of the larvae of *Ostrea edulis* L. *Fishery Investigations. Ministry of Agriculture, Fisheries and Food, London*, Ser. II, **24**, 1–45.

Walne, P. R. (1966). Experiments in the large-scale culture of the larva of *Ostrea edulis* L. *Fishery Investigations. Ministry of Agriculture, Fisheries and Food, London*, Ser. II, **25**, 1–53.

Walne, P. R. (1970*a*). Present problems in the culture of the larvae of *Ostrea edulis*. *Helgoländer wissenschaftliche Meeresuntersuchungen*, **20**, 514–25.

Walne, P. R. (1970*b*). The seasonal variation of meat and glycogen content of seven populations of oysters *Ostrea edulis* L. and a review of the literature. *Fishery Investigations. Ministry of Agriculture, Fisheries and Food, London*, Ser. II, **26**, 35 pp.

Walne, P. R. (1972). The influence of current speed, body size and water temperature on the filtration rate of five species of bivalves. *Journal of the Marine Biological Association of the United Kingdom*, **52**, 345–74.

Walne, P. R. & Dean, G. D. (1972). Experiments on predation by the shore crab, *Carcinus maenas* L. on *Mytilus* and *Mercenaria*. *Journal du Conseil. Conseil permanent International pour l'Exploration de la Mer*, **34**, 190–9.

References

Wang, D-H. & Scheer, B. T. (1963). UDPG-glycogen transglucosylase and a natural inhibitor in crustacean tissues. *Comparative Biochemistry and Physiology*, **9**, 263–74.

Wangersky, P. J. (1975). Production of dissolved organic matter. In: *Marine Ecology* vol. 2 (ed. O. Kinne). Wiley-Interscience, New York. (In press.)

Ward, M. E. & Aiello, E. (1973). Water pumping, particle filtration, and neutral red absorption in the bivalve mollusc *Mytilus edulis*. *Physiological Zoology*, **46**, 157–67.

Warren, A. E. (1936). An ecological study of the sea mussel (*Mytilus edulis* Linn). *Journal of the Biological Board of Canada, Ottawa*, **2**, 89–94.

Warren, C. E. & Davis, G. E. (1967). Laboratory studies on the feeding, bioenergetics and growth of fish. In: *The biological basis of freshwater fish production* (ed. S. D. Gerking), pp. 175–214. Blackwell Scientific Publications, Oxford.

Watabe, N. & Kobayashi, S. (1961). Quoted by Wilbur, K. M. (1964).

Watanabe, T. & Ackman, R. G. (1974). Lipids and fatty acids of the American (*Crassostrea virginica*) and European flat (*Ostrea edulis*) oysters from a common habitat, and after one feeding with *Dicrateria inornata* or *Isochrysis galbana*. *Journal of the Fisheries Research Board of Canada*, **31**, 403–9.

Watkinson, J. G. & Smith, R. (1972). *New Zealand fisheries*. Marine Department of New Zealand, Wellington. 91 pp.

Waugh, D. L. (1972). Upper lethal temperatures of the pelecypod *Modiolus demissus* in relation to declining environmental temperatures. *Canadian Journal of Zoology*, **50**, 523–7.

Waugh, D. L. & Garside, E. T. (1971). Upper lethal temperatures in relation to osmotic stress in the ribbed mussel, *Modiolus demissus*. *Journal of the Fisheries Research Board of Canada*, **28**, 527–32.

Waugh, G. D. (1954). The occurrence of *Mytilicola intestinalis* (Steuer) on the east coast of England. *Journal of Animal Ecology*, **23**, 364–7.

Waugh, G. D. (1966). A crop from the sea. *Geographical Magazine, London*, **39**, 263–7.

Webber, H. (1968). Mariculture. *Bioscience*, **18**, 940–5.

Webster, J. D. (1941). Feeding habits of the black oystercatcher. *Condor*, **43**, 175–80.

Wells, H. W. & Gray, I. E. (1960). The seasonal occurrence of *Mytilus edulis* on the Carolina coast as a result of transport around Cape Hatteras. *Biological Bulletin. Marine Biological Laboratory Woods Hole, Mass.*, **119**, 550–9.

Welsh, J. H. (1961). Neurohormones in Mollusca. *American Zoologist*, **1**, 267–72.

Welsch, V. & Storch, V. (1969). Zum Aufbau und zur Innervation des wimperepithelis der Bivalvia–Palpen. (Fine structure and innervation of the ciliated epithelium of the lamellibranch palps.) *Zeitschrift für Zellforschung und Mikroskopische Anatomie*, **97**, 383–91.

Werner, B. (1939). Uber die Entwicklung und Artuntersscheidung von

References

Muschellarven des Norseeplanktons, unter Gesonderer Beruchsichtigung der Scholenentwicklung. *Zoologische Jahrbücher (Anatomie)*, **66**, 237–70.

Weymouth, F. W. & McMillin, H. C. (1931). The relative growth and mortality of the Pacific Razor clam (*Siliqua patula* Dixon) and their bearing on the commercial fishery. *Bulletin of the Bureau of Fisheries, Washington*, **46**, 543–67.

Whedon, W. F. (1936). Spawning habits of the mussel *Mytilus californianus* with notes on the possible relation to mussel poison. *University of California Publications in Zoology*, **41**, 35–44.

Whedon, W. F. & Sommer, H. (1937). Respiratory exchange of *Mytilus californianus*. *Zeitschrift für vergleichende Physiologie*, **25**, 523–8.

White, K. M. (1937). *Mytilus. Liverpool Marine Biology Committee Memoirs, No. 31.* University of Liverpool Press, Liverpool. 177 pp.

White, K. M. (1942). The pericardial cavity and the pericardial gland of the Lamellibranchia. *Proceedings of the Malacological Society of London*, **25**, 37–88.

White, W. R. (1966). *Effect of low-level chlorination on mussels at Poole Power Station.* Central Electricity Generating Board Research and Development Department, Laboratory Note RD/L/N17/66.

White, W. R. (1968). *A method for measuring the pumping rate of mussels* (*Mytilus edulis* L.). Central Electricity Generating Board Research and Development Department, Laboratory Note RD/L/N 116/68.

Whiteley, H. R. (1960). The distribution of the formate activating enzyme and other enzymes involving tetrahydrofolic acid in animal tissues. *Comparative Biochemistry and Physiology*, **1**, 227–47.

Wiborg, K. F. & Bøhle, B. (1968). Den spanske blåskjellindustri: dyrking og foredling, somt notater om østersdyrking og skjell-graving i Vigoomradet i Nord-vest Spania. *Fiskets Gang*, **54**, 91–5.

Widdows, J. (1972). Thermal acclimation by *Mytilus edulis* L. PhD thesis, University of Leicester, England.

Widdows, J. (1973*a*). Effect of temperature and food on the heart beat, ventilation rate and oxygen uptake of *Mytilus edulis*. *Marine Biology*, **20**, 269–76.

Widdows, J. (1973*b*). The effects of temperature on the metabolism and activity of *Mytilus edulis* L. *Netherlands Journal of Sea Research*, **7**, 387–98.

Widdows, J. (1976). Physiological adaptation of *Mytilus edulis* L. to fluctuating temperatures. *Journal of Comparative Physiology*, in press.

Widdows, J. & Bayne, B. L. (1971). Temperature acclimation of *Mytilus edulis* with reference to its energy budget. *Journal of the Marine Biological Association of the United Kingdom*, **51**, 827–43.

Wiederhold, M. L., MacNichol, E. F. & Bell, A. L. (1973). Photoreceptor spike responses in the hardshell clam, *Mercenaria mercenaria*. *Journal of General Physiology*, **61**, 24–55.

Wieser, W. (1952). Investigations on the microfauna inhabiting seaweeds

489

References

on rocky coasts. *Journal of the Marine Biological Association of the United Kingdom*, **31**, 145–74.

Wieser, W. (1973). Temperature relations of ectotherms: a speculative review. In: *Effects of temperature on ectothermic organisms* (ed. W. Wieser), pp. 1–23. Springer-Verlag, Berlin.

Wilber, C. G. (1969). *The biological aspects of water pollution.* C. C. Thomas, Springfield, Illinois. 296 pp.

Wilbur, K. M. (1964). Shell formation and regeneration. In: *Physiology of Mollusca* (ed. K. M. Wilbur and C. M. Yonge), vol. I, pp. 243–82. Academic Press, New York and London.

Wilbur, K. M. & Owen, G. (1964). Growth. In: *Physiology of Mollusca* (ed. K. M. Wilbur and C. M. Yonge), vol. I, pp. 211–42. Academic Press, New York and London.

Wilkins, N. P. & Mathers, N. F. (1973). Enzyme polymorphisms in the European oyster, *Ostrea edulis* L. *Animal Blood Groups: Biochemistry and Genetics*, **4**, 41–7.

Willemsen, J. (1952). Quantities of water pumped by mussels (*Mytilus edulis*) and cockles (*Cardium edule*). *Archives Néerlandaises de Zoologie*, **10**, 153–60.

Williams, C. S. (1967). The parasitism of young mussels by *Mytilicola intestinalis. Journal of Natural History*, **1**, 299–301.

Williams, C. S. (1968). The influence of *Polydora ciliata* (Johnst) on the degree of parasitism of *Mytilus edulis* L. by *Mytilicola intestinalis* Steuer. *Journal of Animal Ecology*, **37**, 709–12.

Williams, C. S. (1969*a*). The life history of *Mytilicola intestinalis* Steuer. *Journal du Conseil. Conseil permanent International pour l'Exploration de la Mer*, **32**, 419–29.

Williams, C. S. (1969*b*). The effect of *Mytilicola intestinalis* on the biochemical composition of mussels. *Journal of the Marine Biological Association of the United Kingdom*, **49**, 161–73.

Williams, R. J. (1970). Freezing tolerance in *Mytilus edulis. Comparative Biochemistry and Physiology*, **35**, 145–61.

Williamson, H. C. (1907). The spawning, growth and movement of the mussel (*Mytilus edulis*), horse mussel (*Modiolus modiolus*) and the spoutfish (*Solen siliqua*). *Scientific Investigations. Fishery Board, Scotland, 25th Annual Report*, 221–55.

Wilson, B. R. (1968). Survival and reproduction of the mussel *Xenostrobus securis* (Lam.) in a Western Australian estuary. 1. Salinity tolerance. *Journal of Natural History*, **2**, 307–28.

Wilson, B. R. (1969). Survival and reproduction of the mussel *Xenostrobus securis* (Lamarck) (Mollusca; Bivalvia; Mytilidae) in a Western Australian estuary. *Journal of Natural History*, **3**, 93–120.

Wilson, B. R. & Hodgkin, E. P. (1967). A comparative account of the reproductive cycles of five species of marine mussels (Bivalvia; Mytilidae) in the vicinity of Freemantle, W. Australia. *Australian Journal of Marine and Freshwater Research*, **18**, 175–203.

Wilson, D. P. (1952). The influence of the nature of the substratum on the metamorphosis of the larvae of marine animals, especially the larvae of *Ophelia bicornis* Savigny. *Annales de l'Institut Océanographique*, **27**, 49–156.

Wilson, J. (1886). On the development of the common mussel (*Mytilus edulis*). *Scientific Investigations. Fishery Board, Scotland, 5th annual report*, 247–56.

Wilson, J. H. (1971). A study of the feeding, growth and reproduction of *Mytilus edulis* L. in Carlingford Lough. PhD thesis, The Queen's University, Belfast, Ireland.

Wilson, J. H. & Seed, R. (1975). Reproduction in *Mytilus edulis* L. (Mollusca; Bivalvia) in Carlingford Lough, Northern Ireland. *Irish Fisheries Investigations, Dublin*, Ser. B, *Marine*, in press.

Winberg, G. G. (1956). Rate of metabolism and food requirements of fishes. *Trudȳ Belorusskogo gosudarstvennogo universiteta Minske*, 253 pp. (Translated from Russian by *Fisheries Research Board of Canada, Translation Series*, **194**, 1960.)

Winter, J. E. (1969). Uber den Einfluss der Nahrungskonzentration und anderer Faktoren auf Filtrierleistung und Nahrungsausnutzung der Muscheln *Arctica islandica* und *Modiolus modiolus*. *Marine Biology*, **4**, 7–135.

Winter, J. E. (1970). Filter feeding and food utilisation in *Arctic islandica* L. and *Modiolus modiolus* L. at different food concentrations. In: *Marine food chains* (ed. J. H. Steele), pp. 196–206. University of California Press. Berkeley, California.

Winter, J. E. (1972). Long-term laboratory experiments on the influence of ferric hydroxide flakes on the filtering behaviour, growth, ionic content and mortality of *Mytilus edulis* L. In: *Marine pollution and sea life* (ed. M. Ruivo), pp. 392–5. Fishing News (Books) Ltd, London.

Winter, J. E. (1973). The filtration rate of *Mytilus edulis* and its dependence on algal concentration, measured by a continuous automatic recording apparatus. *Marine Biology*, **22**, 317–28.

Winter, J. E. (1974). Growth in *Mytilus edulis* using different types of food. *Berichte der Deutschen Wissenschaftlichen Kommission für Meeresforschung*, **23**, 360–75.

Wisely, B. (1963). Detection and avoidance of a cuprous oxide anti-fouling paint by bivalve and gastropod larvae (Mollusca). *Australian Journal of Marine and Freshwater Research*, **18**, 63–72.

Wisely, B. (1964). Aspects of reproduction, settling and growth in the mussel *Mytilus edulis planulatus*. *Journal of the Malacological Society of Australia*, **8**, 25–30.

Wisely, B. & Blick, R. A. P. (1967). Mortality of marine invertebrate larvae in Hg, Cu, and Zn solutions. *Australian Journal of Marine and Freshwater research*, **18**, 63–72.

Wojtowicz, M. B. (1972). Carbohydrases of the digestive gland and the crystalline style of the Atlantic deep-sea scallop (*Placopecten magellanicus*, Gmelin). *Comparative Biochemistry and Physiology*, **43A**, 131–42.

References

Wolfe, D. A. & Schelske, C. L. (1969). Accumulation of fallout from the lower Trent and Neuse rivers. In: *Proceedings of the second national symposium on radioecology* (ed. D. J. Nelson and F. C. Evans), pp. 493–504.

Wood, L. & Hargis, W. J. (1971). Transport of bivalve larvae in a tidal estuary. In: *Fourth European marine biology symposium* (ed. D. J. Crisp), pp. 29–44. Cambridge University Press, London.

Wood, P. C. (1969). *The production of clean shellfish*. Laboratory Leaflet, No. 20 (N.S.). Fisheries Laboratory, Lowestoft. 15 pp.

Wood, P. C. (1972). The principles and methods employed for the sanitary control of molluscan shellfish. In: *Marine pollution and sea life* (ed. M. Ruivo). pp. 560–4. Fishing News (Books) Ltd, London.

Woodhead, D. S. (1971). The biological effects of radioactive waste. *Proceedings of the Royal Society of London*, Ser. B, **177**, 423–37.

Woortmann, K-D. (1926). Beiträge zur Nervenphysiologie von *Mytilus edulis*. *Journal of Comparative Physiology*, **4**, 488–527.

Worley, L. G. (1944). Studies of the vitally stained Golgi apparatus. II. Yolk formation and pigment concentration in the mussel *Mytilus californianus* Conrad. *Journal of Morphology*, **75**, 77–101.

Wright, F. (1917). Mussel beds, their production and maintenance. *Annals of Applied Biology*, **4**, 123–5.

Yoneda, T. (1967). Cysteamine as a metabolite in marine mollusc, *Mytilus edulis*. *Bulletin of the Faculty of Fisheries, Hokkaido University*, **18**, 88.

Yoneda, T. (1968). Enzymatic oxidation of cysteamine to taurine in the marine mollusc *Mytilus edulis*. *Bulletin of the Faculty of Fisheries, Hokkaido University*, **19**, 140–6.

Yonge, C. M. (1926a). The digestive diverticula in the lamellibranchs. *Transactions of the Royal Society of Edinburgh*, **54**, 703–18.

Yonge, C. M. (1926b). Structure and physiology of the organs of feeding and digestion in *Ostrea edulis*. *Journal of the Marine Biological Association of the United Kingdom*, **14**, 295–386.

Yonge, C. M. (1946). On the habits and adaptations of *Aloidis* (*Corbula*) *gibba*. *Journal of the Marine Biological Association of the United Kingdom*, **26**, 358–76.

Yonge, C. M. (1947). The pallial organs in the aspidobranch Gastropoda and their evolution throughout the Mollusca. *Philosophical transactions of the Royal Society of London*, Ser. B, **232**, 443–518.

Yonge, C. M. (1951). Observations on *Spenia binhami* Turton. *Journal of the Marine Biological Association of the United Kingdom*, **30**, 387–92.

Yonge, C. M. (1952). Studies on Pacific coast molluscs. V. Structure and adaptation in *Entodesma saxicola* (Baird) and *Mytilimeria nuttallii* Conrad. *University of California Publications in Zoology*, **55**, 439–50.

Yonge, C. M. (1953a). The monomyarian condition in the Lamellibranchia. *Transactions of the Royal Society of Edinburgh*, **62**, 443–78.

Yonge, C. M. (1953b). Form and habit in *Pinna carnea* Gmelin. *Philosophical Transactions of the Royal Society of London*, Ser. B, **237**, 335–74.

Yonge, C. M. (1955). Adaptation to rock boring in *Botula* and *Lithophaga* (Lamellibranchia, Mytilidae). *Quarterly Journal of Microscopical Science*, **96**, 383–410.

Yonge, C. M. (1958). Observations on *Petricola carditoides* (Conrad). *Proceedings of the Malacological Society of London*, **33**, 25–31.

Yonge, C. M. (1962a). On *Etheria elliptica* Lam. and the course of evolution, including assumption of monomyarianism, in the family Etheriidae (Bivalvia: Unionacea). *Philosophical Transactions of the Royal Society of London* Ser. B, **244**, 423–58.

Yonge, C. M. (1962b). On the primitive significance of the byssus in the Bivalvia and its effects in evolution. *Journal of the Marine Biological Association of the United Kingdom*, **42**, 113–25.

Yonge, C. M. (1968). Form and habit in species of *Malleus* (including the 'hammer oysters') with comparative observations on *Isognomon isognomon*. *Biological Bulletin. Marine Biological Laboratory, Woods Hole, Mass.*, **135**, 378–405.

Yonge, C. M. (1969). Functional morphology and evolution within the Carditacea (Bivalvia). *Proceedings of the Malacological Society of London*, **38**, 493–527.

Yonge, C. M. (1975). The status of the Plicatulidae and the Dimyidae in relation to the superfamily Pectinacea (Mollusca: Bivalvia). *Journal of Zoology*, in press.

Yonge, C. M. & Campbell, J. I. (1968). On the heteromyarian condition in the Bivalvia with special reference to *Dreissena polymorpha* and certain Mytilacea. *Transactions of the Royal Society of Edinburgh*, **68**, 21–43.

Yoo, S. K. (1969). Food and growth of the larvae of certain important bivalves. *Bulletin of the Pusan Fisheries College (Natural Sciences), Korea*, **9**, 65–87.

Yoshida, H. (1953). Studies on larvae and young shells of industrial bivalves in Japan. *Journal of the Shimonoseki College of Fisheries*, **3**, 1–106.

Young, D. R. & Folsom, T. R. (1967). Loss of ^{65}Zn from the California sea mussel *Mytilus californianus*. *Biological Bulletin. Marine Biological Laboratory, Woods Hole, Mass.*, **153**, 438–47.

Young, R. T. (1941). The distribution of the mussel *Mytilus californianus* in relation to the salinity of its environment. *Ecology*, **22**, 379–86.

Young, R. T. (1942). Spawning season of the Californian mussel *Mytilus californianus*. *Ecology*, **23**, 490–2.

Young, R. T. (1945). Stimulation of spawning in the mussel (*Mytilus californianus*). *Ecology*, **26**, 58–69.

Young, R. T. (1946). Spawning and setting season of the mussel *Mytilus californianus*. *Ecology*, **27**, 354–63.

Zachs, S. I. & Welsh, J. H. (1953). Cholinesterase and lipase in the amoebocytes, intestinal epithelium and heart muscle of the quahog, *Venus mercenaria*. *Biological Bulletin. Marine Biological Laboratory, Woods Hole, Mass.*, **105**, 200–11.

References

Zakhvatkina, K. A. (1959). Larvae of bivalve molluscs of the Sevastopol region of the Black Sea. *Trudȳ Sevastopolskoi Biologicheskoi Stantsii*, **11**, 108–52. (Translated to English; Virginia Institute of Marine Science Translation Series No. 15, April, 1966.)

Zenkevitch, L. (1963). *Biology of the seas of the USSR*. Allen & Unwin, London.

Zeuthen, E. (1947). Body size and metabolic rate in the animal kingdom, with special regard to the marine microfauna. *Comptes Rendus des Travaux du Laboratoire Carlsberg, Série Chimique*, **26**, 17–161.

Zeuthen, E. (1953). Oxygen uptake as related to body size in organisms. *Quarterly Review of Biology*, **28**, 1–12.

Zitko, V. (1971). Determination of residual fuel oil contamination of aquatic animals. *Bulletin of Environmental Contamination and Toxicology*, **5**, 559–64.

Zobell, C. E. & Feltham, C. B. (1938). Bacteria as food for certain marine invertebrates. *Journal of Marine Research*, **1**, 312–27.

Zs-Nagy, I. (1974). Some quantitive aspects of oxygen consumption and anaerobic respiration of molluscan tissues – a review. *Comparative Biochemistry and Physiology*, **49A**, 399–405.

Zs-Nagy, I. & Ermini, M. (1973a). Oxidation of $NADH_2$ by the lipochrome pigment of the tissues of the bivalve *Mytilus galloprovincialis* (Mollusca, Pelecypoda). *Comparative Biochemistry and Physiology*, **43B**, 39–46.

Zs-Nagy, I. & Ermini, M. (1973b). ATP production in the tissues of the bivalve *Mytilus galloprovincialis* (Pelecypoda) under normal and anoxic conditions. *Comparative Biochemistry and Physiology*, **43B**, 593–600.

Index

495

Index

Index

Index

pigments, photosensitive, 258–9
Pinctada radiata, haemocytes of, 210
Pinna, large heteromyarian, 9–10
Pinna nobilis, oxygen utilisation efficiency of, 265
Pinnidae, 9
Pinnotheres (pea crab), parasite of mussels, 56–7
pinocytosis, by digestive cells, 151
Pisaster ochraceus, predator on *M. californianus*, 54, 64, 65
Placopecten magellanicus: enzymes of carbohydrate metabolism in, 154, 322, 323, 335, 336, 345; lacks octopine dehydrogenase? 334
plankton, mussel larvae in, 81–2, 357
planktotrophy, 118, 119
Planorbis, osphradia of, 259
plantigrades (post-larval mussels), 85, 96; crabs as predators on, 55; movement of, from filamentous substrates to established mussel beds, 32, 33, 36–8, 358; number of, required per breeding pair for static population of *Mya*, 116; oxygen uptake of, 100, 101
Plicatulacea, 7
pollutants, mussels and, 67–80
Polydora (polychaete), infection of *Mercenaria* by, 304
polysaccharides: in larvae, 110; sulphonated, 237; *see also* glycogen
potassium: maintained at higher level in blood than in medium, 211, 212; in muscle, 240; in osmotic regulation, 238
predators on mussels, 51, 52–6; on cultivated mussels, 390, 393, 401; on larvae, 116; percentage of production taken by, 59; stabilisation of competing prey species by, 64
pressure, hydrostatic: responses of swimming larvae to changes of, 113–14, 115
prodissoconch I, shell of veliger larva, 82, 85, 90; in identification, 83
prodissoconch II, shell of veliconcha larva, 82, 85, 95
production by mussel populations, 58–9, 406–7; simulated ratio of, to respiration, 288, 289
proline: among major free amino acids, 315; formation of, from glutamate, 317; in osmoregulation, 243
Prorocentrum, paralytic poison from, 405
proteases, in haemocytes, 210
proteins: in blood, 213; catabolism of, and ratio of oxygen uptake to nitrogen excretion, 268; constant ratio of oxygen uptake to content of, 191; in eggs, 297; in larvae, 110, 111, 297; in osmotic regulation, 237,

238; in starvation, 233, 270, 298, 301, 302, 330, 333
protonephridia, 224
provinculum, of larval shell, 85, 86
pseudofaeces (excess particles rejected by palps), 10, 141, 144, 145, 146; components of, 148; production of, in CSMP calculations, 283, 289
Pteriacea, 7
purinolysis, enzymes of, 228–9, 230
Pyrodinium pheneus, paralytic poison from, 68
pyruvate carboxylase, 320, 322, 323, 334; as metabolic regulator, 346
pyruvate kinase (PK), 320, 321, 322, 327; kinetic properties of, 346–7; in mantle, 345; as metabolic regulator, 338, 346, 348–51, 355; pH and activity of, 350, 352; reciprocal operation of phosphoenolpyruvate carboxykinase and? 354–5; regulation of, 353

radiation, lethal dose of, 75
radioactive isotopes, uptake of, 74–5
Rangia cuneata: amino acids in, 244, 246, 247; growth of larvae of, 105; osmotic regulation in, 237, 238, 242; phosphoenolpyruvate carboxykinase in, 322; preparation of mitochondria from, 335
redox balance, 330–1, 331–2; in anaerobiosis, 333–6
rejection tracts, on palps, 145, 146
reno-pericardial canals, ciliated, 224
reproduction: efficiency of (*Modiolus*), 58; strategy of, 118–20
reproductive cycle, 19; factors controlling, 29, 30–1; gonads in, 22–4; methods of assessing, 19–20; neurosecretion and, 248–9; *see also* gametogenesis, spawning
reproductive ducts, 20, 89
respiration: energy cost of, 159; total annual, of a population of mussels, 58; *see also* oxygen uptake
respiratory quotient, of developing embryo, 91
ruthenium (radioactive), uptake of, 75

salinity: acclimation to change in, 201, 202, 203–5, 239; and amino acid metabolism, 241, 243, 244, 246; in CSMP calculations, 286; and excretion of ammonia, 234–5; and filtration rate, 143; and growth, 47–8, (of larvae) 104, 105, 117; and heart beat, 223; and metamorphosis, 112; and neurosecretion, 250; and osmotic regulation, 212, 237; and oxygen uptake, 200–5; range of, for embryo development, 92–3, and for fertilisation, 89; tolerances of different species

504

9

Marine mussels,
their ecology and
physiology
